Photochemical Pioneers: Feature Papers in Photochemistry Volumes I & II

Photochemical Pioneers: Feature Papers in Photochemistry Volumes I & II

Selected Articles Published by MDPI

Basel • Beijing • Wuhan • Barcelona • Belgrade • Novi Sad • Cluj • Manchester

This is a reprint of articles published open access by MDPI, freely accessible at: http://www.mdpi.com.

For citation purposes, cite each article independently as indicated on the article page online and as indicated below:

Lastname, A.A.; Lastname, B.B. Article Title. *Journal Name* **Year**, *Volume Number*, Page Range.

ISBN 978-3-7258-3335-1 (Hbk)
ISBN 978-3-7258-3336-8 (PDF)
https://doi.org/10.3390/books978-3-7258-3336-8

© 2025 by the authors. Articles in this book are Open Access and distributed under the Creative Commons Attribution (CC BY) license.

Contents

Preface vii

Marcelo I. Guzman
Feature Papers in Photochemistry
Reprinted from: *Photochem* **2024**, *12*, 511–517, https://doi.org/10.3390/photochem4040032 . . . 1

Rohini Vallavoju, Ranjith Kore, Radhika Parikirala, Mahesh Subburu, Ramesh Gade, Vipin Kumar, et al.
Synthesis and Characterization of New Tetradentate N_2O_2-Based Schiff's Base Cu (II) Complexes for Dye Photodegradation
Reprinted from: *Photochem* **2023**, *3*, 274–287, https://doi.org/10.3390/photochem3020016 8

Alexandre Mau, Guillaume Noirbent, Céline Dietlin, Bernadette Graff, Didier Gigmes, Frédéric Dumur and Jacques Lalevée
Panchromatic Copper Complexes for Visible Light Photopolymerization
Reprinted from: *Photochem* **2021**, *1*, 167–189, https://doi.org/10.3390/photochem1020010 22

Miroslava Edelmannová, Martin Reli, Kamila Kočí, Ilias Papailias, Nadia Todorova, Tatiana Giannakopoulou, et al.
Photocatalytic Reduction of CO_2 over Iron-Modified g-C_3N_4 Photocatalysts
Reprinted from: *Photochem* **2021**, *1*, 462–476, https://doi.org/10.3390/photochem1030030 45

Daniele Malpicci, Clelia Giannini, Elena Lucenti, Alessandra Forni, Daniele Marinotto and Elena Cariati
Mono-, Di-, Tri-Pyrene Substituted Cyclic Triimidazole: A Family of Highly Emissive and RTP Chromophores
Reprinted from: *Photochem* **2021**, *1*, 477–487, https://doi.org/10.3390/photochem1030031 60

Felipe S. Stefanello, Jean C. B. Vieira, Juliane N. Araújo, Vitória B. Souza, Clarissa P. Frizzo, Marcos A. P. Martins, et al.
Solution and Solid-State Optical Properties of Trifluoromethylated 5-(Alkyl/aryl/heteroaryl)-2-methyl- pyrazolo[1,5-*a*]pyrimidine System
Reprinted from: *Photochem* **2022**, *2*, 345–357, https://doi.org/10.3390/photochem2020024 71

Min Hee Joo, So Jeong Park, Hye Ji Jang, Sung-Min Hong, Choong Kyun Rhee and Youngku Sohn
Enhanced Photoluminescence of Electrodeposited Europium Complex on Bare and Terpyridine-Functionalized Porous Si Surfaces
Reprinted from: *Photochem* **2021**, *1*, 38–52, https://doi.org/10.3390/photochem1010004 84

Alberto Gimenez-Gomez, Lucien Magson, Beatriz Peñin, Nil Sanosa, Jacobo Soilán, Raúl Losantos and Diego Sampedro
A Photochemical Overview of Molecular Solar Thermal Energy Storage
Reprinted from: *Photochem* **2022**, *2*, 694–716, https://doi.org/10.3390/photochem2030045 99

Paula M. Crespo, Oscar F. Odio and Edilso Reguera
Photochemistry of Metal Nitroprussides: State-of-the-Art and Perspectives
Reprinted from: *Photochem* **2022**, *2*, 390–404, https://doi.org/10.3390/photochem2020027 122

İsa Sıdır, Sándor Góbi, Yadigar Gülseven Sıdır and Rui Fausto
Infrared Spectrum and UV-Induced Photochemistry of Matrix-Isolated Phenyl 1-Hydroxy-2-Naphthoate
Reprinted from: *Photochem* **2021**, *1*, 10–25, https://doi.org/10.3390/photochem1010002 137

Alison G. Barnes, Nicolas Richy, Anissa Amar, Mireille Blanchard-Desce, Abdou Boucekkine, Olivier Mongin and Frédéric Paul
Electronic Absorption, Emission, and Two-Photon Absorption Properties of Some Extended 2,4,6-Triphenyl-1,3,5-Triazines
Reprinted from: *Photochem* **2022**, *2*, 326–344, https://doi.org/10.3390/photochem2020023 153

Aleksey A. Vasilev, Stanislav Baluschev, Sonia Ilieva and Diana Cheshmedzhieva
E–Z Photoisomerization in Proton-Modulated Photoswitchable Merocyanine Based on Benzothiazolium and o-Hydroxynaphthalene Platform
Reprinted from: *Photochem* **2023**, *3*, 301–312, https://doi.org/10.3390/photochem3020018 172

Konstantin Moritz Knötig, Domenic Gust, Thomas Lenzer and Kawon Oum
Excited-State Dynamics of Carbazole and *tert*-Butyl-Carbazole in Organic Solvents
Reprinted from: *Photochem* **2024**, *4*, 163–178, https://doi.org/10.3390/photochem4020010 184

Konstantin Moritz Knötig, Domenic Gust, Kawon Oum and Thomas Lenzer
Excited-State Dynamics of Carbazole and *tert*-Butyl-Carbazole in Thin Films
Reprinted from: *Photochem* **2024**, *4*, 179–197, https://doi.org/10.3390/photochem4020011 200

Denis Tikhonov, Diksha Garg and Melanie Schnell
Inverse Problems in Pump–Probe Spectroscopy
Reprinted from: *Photochem* **2024**, *4*, 57–110, https://doi.org/10.3390/photochem4010005 219

Tânia M. Gonçalves, Inês S. Martins, Hugo F. Silva, Valery V. Tuchin and Luís M. Oliveira
Spectral Optical Properties of Rabbit Brain Cortex between 200 and 1000 nm
Reprinted from: *Photochem* **2021**, *1*, 190–208, https://doi.org/10.3390/photochem1020011 273

Victoria C. Frederick, Thomas A. Ashy, Barbara Marchetti, Michael N. R. Ashfold and Tolga N. V. Karsili
Photoprotective Properties of Eumelanin: Computational Insights into the Photophysics of a Catechol:Quinone Heterodimer Model System
Reprinted from: *Photochem* **2021**, *1*, 26–37, https://doi.org/10.3390/photochem1010003 292

Javier Segarra-Martí, Sara M. Nouri and Michael J. Bearpark
Modelling Photoionisations in Tautomeric DNA Nucleobase Derivatives 7H-Adenine and 7H-Guanine: Ultrafast Decay and Photostability
Reprinted from: *Photochem* **2021**, *1*, 287–301, https://doi.org/10.3390/photochem1020018 304

Kohei Kawabata, Ayano Miyoshi and Hiroyuki Nishi
Photoprotective Effects of Selected Polyphenols and Antioxidants on Naproxen Photodegradability in the Solid-State
Reprinted from: *Photochem* **2022**, *2*, 880–890, https://doi.org/10.3390/photochem2040056 319

Dilini Kodikara, Zhongyu Guo and Chihiro Yoshimura
Effect of Benzophenone Type UV Filters on Photodegradation of Co-existing Sulfamethoxazole in Water
Reprinted from: *Photochem* **2023**, *3*, 288–300, https://doi.org/10.3390/photochem3020017 330

Jacobo Soilán, Leonardo López-Cóndor, Beatriz Peñín, José Aguilera, María Victoria de Gálvez, Diego Sampedro and Raúl Losantos
Evaluation of MAA Analogues as Potential Candidates to Increase Photostability in Sunscreen Formulations
Reprinted from: *Photochem* **2024**, *4*, 128–137, https://doi.org/10.3390/photochem4010007 343

Preface

This volume, *Photochemical Pioneers: Feature Papers in Photochemistry Volumes I & II*, highlights the recent groundbreaking research in the field of photochemistry. By bringing together a diverse collection of seminal works, we aim to take readers on a path that sheds light on the interactions between light and matter, showcasing the direct impact of photochemical processes on various scientific disciplines.

The papers included in this compilation delve into the fundamental mechanisms of light-induced molecular transformations, exploring their applications in organic synthesis, environmental science, materials science, and medical therapies. From the development of efficient photocatalysts to the design of novel photo-responsive materials, these works highlight the versatility and potential of photochemistry to address global challenges related to energy, health, and environmental sustainability.

This collection is intended for researchers and students interested in the fields of chemistry, physics, materials science, and environmental science. By showcasing the latest advancements and emerging trends in photochemistry, we hope to inspire future generations of scientists to explore the limitless possibilities of light-induced processes. We extend our sincere gratitude to the authors and reviewers who contributed to the success of these Special Issues. Their dedication and expertise have been instrumental in shaping the content of these two volumes. We also acknowledge the support of the *Photochem* Editorial team for their guidance and assistance throughout the publication process.

As we continue to unravel the mysteries of photochemistry, we believe that this collection will serve as a valuable resource for researchers and students alike. By fostering collaboration and inspiring innovation, we can collectively advance the field of photochemistry and its applications to help improve society.

Through the exploration of topics such as photocatalysis, photodynamic therapy, solar energy conversion, and organic synthesis, this selection of papers offers a comprehensive overview of the current state of the art in photochemistry. It highlights the importance of interdisciplinary research and the potential for groundbreaking discoveries in this field. We invite you to embark on this journey through the world of photochemistry, where light and matter intertwine to create a future filled with innovation and sustainability.

Marcelo Guzman

Editorial

Feature Papers in Photochemistry

Marcelo I. Guzman [1,2]

[1] Department of Chemistry, College of Arts and Sciences, University of Kentucky, Lexington, KY 40506, USA; marcelo.guzman@uky.edu; Tel.: +1-(859)-323-2892
[2] Lewis Honors College, University of Kentucky, Lexington, KY 40506, USA

1. Introduction

As the Special Issues "Feature Papers in Photochemistry" and "Feature Papers in Photochemistry II" conclude, it is crucial to acknowledge the remarkable progress and persistent gaps that continue to shape the journey of photochemistry research. The field of photochemistry has seen unprecedented advancements, driven by novel techniques and interdisciplinary approaches. The study of light-induced molecular transformations [1,2], the application of photochemical processes in organic synthesis [1,3,4], the development of high-efficiency photocatalysts [5], and the discovery of novel environmental photochemical mechanisms [2,6] have all marked significant progress in recent years. Moreover, photochemical reactions in astrochemical mimics have been investigated to understand the chemical evolution of interstellar environments [7].

Groundbreaking synthetic applications of light-induced photodecarboxylation reactions have been made through the use of ligand-to-iron charge transfer for hydrodifluoromethylation and hydromethylation of alkenes [1]. The mechanistic contributions of ligand-to-metal charge-transfer (LMCT) complexes in the photocatalysis of adsorbates at the air-solid interface of TiO_2 and the role of water vapor have been recently revealed [6]. Similarly, molecular electron donor–acceptor systems have been examined to shed light on charge transfer processes capable of advancing electronic and photonic applications, including solar energy and organic electronics [4]. Novel reaction pathways of diazoalkanes excited with visible light have expanded the synthetic toolkit for constructing complex molecules [8].

Innovative medical applications such as targeted drug delivery using light have been proposed to enable the controlled release of therapeutic agents with unmatched spatial and temporal resolution [9]. In the realm of skincare, the depth penetration of light into skin has been investigated across various wavelengths, providing critical data for medical and cosmetic applications [10]. Moreover, combined photodynamic and photothermal therapy using a bacteria-responsive porphyrin might facilitate more effective treatments for bacterial infections [11].

Environmental sciences have also seen significant advancements in photochemistry. The catalyst-free photochemical activation of peroxymonosulfate in xanthene-rich systems demonstrated the efficacy of proton transfer processes in Fenton-like synergistic decontamination for environmental cleanup [12]. The photochemistry of 2-oxocarboxylic acids in aqueous atmospheric aerosols resulting in the formation of secondary organic aerosols has significant implications for atmospheric chemistry and climate models [2]. The photodegradation of organic micropollutants in aquatic environments has garnered interest for its potential in managing water quality and reducing the environmental impact of emerging pollutants [13]. The design of sustainable covalent organic frameworks (COFs) as heterogeneous photocatalysts and their reaction mechanisms for organic synthesis have been reviewed [14]. Innovative solutions for sustainable energy storage technologies have been proposed, such as the application of photoelectrochemistry in oxygen evolution for rechargeable Li-O_2 batteries [15]. The use of periodic illumination advanced the understanding of photoelectrocatalytic systems for CO_2 reduction, offering a sustainable strategy for reducing greenhouse gasses and generating fuel feedstock [16].

Innovative photochemical techniques for controlled polymerizations might enable precise polymer architectures and advanced materials with tailored properties [17]. The field of asymmetric synthesis has seen progress through enantioselective photochemical reactions, which were enabled by triplet energy transfer [18]. The advanced understanding of the $E \rightarrow Z$ isomerization of alkenes using small-molecule photocatalysts might facilitate precise control in organic synthesis and materials science [19].

The challenges of scaling up photochemical reactions from lab scale to industrial production has been initially tackled from a technical and practical viewpoint [20]. Technological innovations in photochemistry for organic synthesis have been reviewed, such as flow chemistry, high-throughput experimentation, scale-up, and photoelectrochemistry implementation, driving significant progress in the field [21].

Despite the multiple advancements in the field of photochemistry, many knowledge gaps existed. Among these gaps were the creation of novel metal complexes for efficient and photostable organic dye degradation and for visible light photopolymerization reactions. Moreover, metal-doped photocatalysts for CO_2 reduction with enhanced methane and hydrogen production were desired. Synthetic molecules with polycyclic aromatic hydrocarbon moieties, if created, could provide materials with a high quantum yield for bioimaging and anti-counterfeiting applications. Progress in the research of molecular solar fuels, fundamental absorption and fluorescence, the light-induced transitions of metal nitroprussides materials, and the photoluminescence of lanthanide complexes for advanced displays was also needed. The gaps included problems in spectroscopy and the study of photophysical properties, applications to bioimaging, the $E \rightarrow Z$ photoisomerization of dyes, relaxation processes, and inverse problems in pump–probe spectroscopy. Photobiology and photoprotection gaps included targeted analysis of the spectral properties of biological tissues for disease diagnosis, photostability mechanisms, DNA nucleobase photoprotection, and the photostabilization and degradation of pharmaceuticals and sunscreens.

The Special Issues, "Feature Papers in Photochemistry" and "Feature Papers in Photochemistry II", have played a pivotal role in addressing these gaps. By bringing together a diverse collection of research papers, they highlighted groundbreaking work in photochemistry, showcased novel methodologies, and provided insights into emerging trends. They created a platform for exchanging ideas and fostering collaborations, thereby accelerating the pace of discovery in the field. Twenty high-quality contributions in these Special Issues highlight advances in the field of photochemistry by providing both research articles and reviews, which offer a comprehensive overview of innovative results and methodologies in the field of photochemistry.

Each of the twenty contributions advances the understanding of how photochemical processes can be harnessed for practical applications, such as pollution control, renewable energy, and the synthesis of complex molecules with functional properties. The Special Issues have been specially curated to ensure that they represent the cutting edge of today's photochemical research. The inclusion of diverse perspectives and interdisciplinary approaches underscores the importance of photochemistry in addressing global challenges. Researchers, educators, and students can explore these photochemistry advancements in the provided open-access platform of these Special Issues. By bringing together high-quality research and review articles, these Special Issues serve as valuable resources for anyone interested in the latest developments in photochemistry. Highlights of the twenty important contributions are presented in the section below, categorized into three themes.

2. An Overview of the Published Articles

2.1. Contributions on Applications in Environmental Science and Materials Science

Vallavoju et al. (Contribution 1) presented the synthesis and characterization of novel tetradentate Schiff's base Cu (II) complexes, which may find potential application as photocatalysts for environmental remediation, e.g., for organic dye degradation. The produced complexes with superior photocatalytic performance were of high photostability, low band-gap energy, and efficient visible-light activity. A different study by Mau et al. (Contribution 2) using copper complexes suggested their possible use as photoinitiators for visible light photopolymerization. The work of Edelmannová et al. (Contribution 3) explored the use of iron-modified graphitic carbon nitride (g-C_3N_4) photocatalysts for CO_2 reduction. The systematic study in Contribution 3 determined that the photocatalyst with the lowest iron content showed superior gas evolution of methane and hydrogen as compared to higher iron content samples. Optimizing iron content in g-C_3N_4 was demonstrated by Edelmannová et al. as an important task for efficient CO_2 photoreduction with advanced materials for sustainable energy solutions.

Malpicci et al. (Contribution 4) studied the synthesis and photophysical properties of cyclic triimidazole derivatives substituted with pyrene. The study revealed that these chromophoric compounds exhibit impressive quantum yields and room-temperature phosphorescence (RTP) properties, with phosphorescence lifetimes increasing with the number of pyrene moieties. Promising applications of these materials for bioimaging, anti-counterfeiting, and display technologies could be recognized based on the work of Malpicci et al. The research of Stefanello et al. (Contribution 5) examined the photophysical properties of a series of 5-(alkyl/aryl/heteroaryl)-2-methyl-7-(trifluoromethyl)pyrazolo[1,5-a]pyrimidines, reporting their UV-visible absorption and fluorescence properties in solution and solid state and assessing their thermal stability using thermogravimetric analysis.

An attractive study conducted by Joo et al. (Contribution 6) demonstrated that a europium complex electrodeposited on bare and terpyridine-functionalized porous silicon surfaces exhibited enhanced photoluminescence properties. The characterization of the surfaces created in Contribution 6 by amperometry electrodeposition was performed using scanning electron microscopy (SEM), X-ray diffraction (XRD), Fourier transform infrared spectroscopy (FTIR), and X-ray photoelectron spectroscopy (XPS). These characterizations supported the materials' composition and properties, which are valuable for developing advanced materials for display technologies and photoelectrochemical applications. Gimenez-Gomez et al. (Contribution 7) provided a comprehensive review of molecular solar fuels, focusing on their design challenges and recent advances. The review highlighted the optical and photochemical properties of compounds such as norbornadiene, azobenzene, and dihydroazulene, which are promising candidates for practical applications in solar energy storage. Crespo et al. (Contribution 8) provided a valuable perspective delving into the photochemistry of metal nitroprussides and emphasized the pivotal role of the nitrosyl group's electronic structure. The work in Contribution 8 explained light-induced electronic transitions of metal nitroprussides and their impact on the stability and functional properties of these compounds, which may find applications in tuning the magnetic, electrical, and optical properties of advanced materials.

2.2. Contributions on Spectroscopy and Photophysical Properties

Sider et al. (Contribution 9) investigated the infrared spectrum and UV-induced photochemistry of matrix-isolated phenyl 1-hydroxy-2-naphthoate using matrix isolation infrared spectroscopy and density functional theory (DFT) computations. The insights offered about the fundamental photochemical processes and spectral properties of 1-hydroxy-2-naphthoate contribute to the understanding of photochemical reactions in isolated environments. The work in Contribution 9 identified the most stable conformer and studied its UV photodecarbonylation yielding 2-phenoxynaphthalen-1-ol and carbon monoxide, a reaction that showed dependence on the excitation wavelength. Barnes et al. (Contribution 10) examined the electronic absorption, emission, and two-photon absorption properties of extended 2,4,6-triphenyl-1,3,5-triazines. DFT calculations were used by Barnes et al. to examine their linear optical properties and two-photon absorption cross-sections, which may be valuable for fluorescence bioimaging applications. Vasilev et al. (Contribution 11) studied the reversible $E \rightarrow Z$ photoisomerization of a photoswitchable merocyanine dye present in a polar acidic medium, which was induced with visible light at λ = 450 nm. A combination of UV-visible spectroscopy measurements and DFT calculations demonstrated in Contribution 11 the potential of this dye for harvesting visible light and for developing light-responsive materials.

Moritz Knötig et al. (Contribution 12) shed light on the excited state dynamics of carbazole and 3,6-di-tert-butylcarbazole in organic solvents, an intricate photophysical problem. The use of advanced ultrafast spectroscopy techniques in Contribution 12 revealed the mechanisms behind the compounds' relaxation processes, providing valuable insights for more stable organic optoelectronic materials. An additional study by Moritz Knötig et al. (Contribution 13) provided a detailed photophysical understanding of the intermolecular pathways and energy transfer mechanisms of carbazole derivatives in thin films used in organic electronics, such as organic light-emitting diodes (OLEDs). The work of Tikhonov et al. (Contribution 14) addressed the inverse problems in pump–probe spectroscopy, an ultrafast kinetic technique that provides deep insights into photophysical and photochemical processes. The research in Contribution 14 presented a consistent approach for solving these inverse problems, avoiding the pitfalls of simply using least-squares fitting. By employing regularized Markov Chain Monte Carlo sampling, Tikhonov et al. offered a robust solution to the nonlinear inverse problem. Python-based software for implementing the fitting routine was made available by Contribution 14, together with a discussion of critical experimental parameters, such as pulse overlap and minimal time resolution.

2.3. Contributions on Photobiology and Photoprotection

In the photobiology context, Gonçalves et al. (Contribution 15) investigated the spectral optical properties of brain cortex from rabbit tissues using transmittance and reflectance spectroscopy. The study identified melanin and lipofuscin pigments and evaluated their impact on the absorption properties of the cortex. Insights into the aging of brain tissues were provided in Contribution 15 to enable the potential development of optical methods to diagnose and monitor brain diseases. The article by Frederick et al. (Contribution 16) offered new computational insights into the photophysics of a heterodimer model system made of eumelanin, which contains the catechol and benzoquinone functionalities. The study investigated the mechanisms behind the photostability of eumelanin, a skin pigment that protects against UV damage, using multi-reference computational methods. The results of Contribution 16 revealed a photoinduced intermolecular hydrogen transfer that is key to the photoprotective properties of eumelanin. The work of Frederick et al. improved the understanding of eumelanin's photophysics and its role in protecting chromosomal deoxyribonucleic acid (DNA) from UV-induced damage. Moreover, Segarra-Martí et al. (Contribution 17) investigated the photoionization processes of DNA nucleobase derivatives using advanced computational methods. The study by Segarra-Martí et al. revealed ultrafast decay and photostability mechanisms, providing insights into the photoprotection of DNA and potential implications for understanding UV-induced DNA damage and related health concerns.

Kawabata et al. (Contribution 18) developed a photostabilization strategy to ensure the quality and quantity of photodegradable pharmaceuticals. In their study, polyphenols with antioxidant and photostabilizing properties were evaluated during the UV photodegradation of naproxen (NPX) in the solid state. The molecules of quercetin, curcumin, and resveratrol efficiently suppressed NPX photodegradation by behaving as antioxidants with strong UV filtering activity. The findings in Contribution 18 might be applied to photostabilize crushed or decapsulated medicines in the future. In a related context, Kodikara et al. (Contribution 19) investigated the impact of benzophenone-type UV filters on the photodegradation of sulfamethoxazole in water. The research revealed that benzophenone and its derivative oxybenzone significantly enhance the degradation rate of sulfamethoxazole by sensitizing reactive intermediates. Indeed, the work in Contribution 19 showed that coexisting benzophenone and its derivative oxybenzone possess promising potential to be used as effective photosensitizers, which may facilitate the photodegradation processes of organic micropollutants in aquatic environments. Soilán et al. (Contribution 20) studied the photostability of avobenzone, a popular sunscreen agent in the skin care industry, which under UV light loses effectiveness as it is degraded. Soilán et al. revealed that mycosporine-like amino acid (MAA) analogs were demonstrated as potential alternative stabilizers to octocrylene, resulting in enhanced avobenzone photostability and a promising next generation of sunscreens.

3. Conclusions

Overall, the "Feature Papers in Photochemistry" and "Feature Papers in Photochemistry II" Special Issues are a testament to the dynamic and impactful nature of photochemical research, offering insights and innovations that have the potential to shape the future of science and technology in the field of photochemistry. Multiple advancements in the field of photochemistry have been provided in the Special Issues. However, several areas for future exploration remain, including the intricacies of light–matter interactions at the quantum level, the stability and efficiency of photocatalytic systems under practical conditions, and the environmental impact of photochemical processes. Continued research and innovation in photochemistry is further needed to address these matters. Looking ahead, it is imperative that future research in photochemistry focuses on several key areas.

First, there is a need for more in-depth studies on the mechanistic aspects of photochemical reactions. Understanding the fundamental processes that govern these reactions will enable the development of more efficient and selective photochemical systems. Second, the integration of photochemistry with other disciplines, such as materials science, biology, food science, and environmental science, holds great promise. Interdisciplinary approaches can lead to the discovery of new photochemical applications and the optimization of existing ones. Third, the development of sustainable photochemical processes remains a critical goal. Researchers should aim to design photocatalysts and photoreactors that are not only highly efficient but also environmentally benign and economically viable. Finally, advancing computational and theoretical methods to predict and design photochemical reactions will be crucial. These methods can provide valuable insights into the behavior of photochemical systems and guide experimental efforts.

In conclusion, the Special Issues have laid a strong foundation by addressing current gaps and highlighting innovative research in photochemistry. Moving forward, a concerted focus on fundamental understanding, interdisciplinary integration, sustainability, and advanced computational methods will be essential to unlocking the full potential of photochemistry. This concerted strategy should drive the field toward a future filled with new discoveries and transformative technologies.

Gratitude is extended to all authors and reviewers for their contributions, which have significantly enriched these Special Issues.

Acknowledgments: M.I.G. acknowledges support under NSF award 2403875.

Conflicts of Interest: The author declares no conflicts of interest.

List of Contributions

1. Vallavoju, R.; Kore, R.; Parikirala, R.; Subburu, M.; Gade, R.; Kumar, V.; Raghavender, M.; Chetti, P.; Pola, S. Synthesis and Characterization of New Tetradentate N_2O_2-Based Schiff's Base Cu (II) Complexes for Dye Photodegradation. *Photochem* **2023**, *3*, 274–287.
2. Mau, A.; Noirbent, G.; Dietlin, C.; Graff, B.; Gigmes, D.; Dumur, F.; Lalevée, J. Panchromatic Copper Complexes for Visible Light Photopolymerization. *Photochem* **2021**, *1*, 167–189.
3. Edelmannová, M.; Reli, M.; Kočí, K.; Papailias, I.; Todorova, N.; Giannakopoulou, T.; Dallas, P.; Devlin, E.; Ioannidis, N.; Trapalis, C. Photocatalytic Reduction of CO_2 over Iron-Modified g-C3N4 Photocatalysts. *Photochem* **2021**, *1*, 462–476.
4. Malpicci, D.; Giannini, C.; Lucenti, E.; Forni, A.; Marinotto, D.; Cariati, E. Mono-, Di-, Tri-Pyrene Substituted Cyclic Triimidazole: A Family of Highly Emissive and RTP Chromophores. *Photochem* **2021**, *1*, 477–487.

5. Stefanello, F.S.; Vieira, J.C.B.; Araújo, J.N.; Souza, V.B.; Frizzo, C.P.; Martins, M.A.P.; Zanatta, N.; Iglesias, B.A.; Bonacorso, H.G. Solution and Solid-State Optical Properties of Trifluoromethylated 5-(Alkyl/aryl/heteroaryl)-2-methyl-pyrazolo[1,5-a]pyrimidine System. *Photochem* **2022**, *2*, 345–357.
6. Joo, M.H.; Park, S.J.; Jang, H.J.; Hong, S.-M.; Rhee, C.K.; Sohn, Y. Enhanced Photoluminescence of Electrodeposited Europium Complex on Bare and Terpyridine-Functionalized Porous Si Surfaces. *Photochem* **2021**, *1*, 38–52.
7. Gimenez-Gomez, A.; Magson, L.; Peñin, B.; Sanosa, N.; Soilán, J.; Losantos, R.; Sampedro, D. A Photochemical Overview of Molecular Solar Thermal Energy Storage. *Photochem* **2022**, *2*, 694–716.
8. Crespo, P.M.; Odio, O.F.; Reguera, E. Photochemistry of Metal Nitroprussides: State-of-the-Art and Perspectives. *Photochem* **2022**, *2*, 390–404.
9. Sidir, İ.; Góbi, S.; Gülseven Sıdır, Y.; Fausto, R. Infrared Spectrum and UV-Induced Photochemistry of Matrix-Isolated Phenyl 1-Hydroxy-2-Naphthoate. *Photochem* **2021**, *1*, 10–25.
10. Barnes, A.G.; Richy, N.; Amar, A.; Blanchard-Desce, M.; Boucekkine, A.; Mongin, O.; Paul, F. Electronic Absorption, Emission, and Two-Photon Absorption Properties of Some Extended 2,4,6-Triphenyl-1,3,5-Triazines. *Photochem* **2022**, *2*, 326–344.
11. Vasilev, A.A.; Baluschev, S.; Ilieva, S.; Cheshmedzhieva, D. E–Z Photoisomerization in Proton-Modulated Photoswitchable Merocyanine Based on Benzothiazolium and o-Hydroxynaphthalene Platform. *Photochem* **2023**, *3*, 301–312.
12. Knötig, K.M.; Gust, D.; Lenzer, T.; Oum, K. Excited-State Dynamics of Carbazole and tert-Butyl-Carbazole in Organic Solvents. *Photochem* **2024**, *4*, 163–178.
13. Knötig, K.M.; Gust, D.; Oum, K.; Lenzer, T. Excited-State Dynamics of Carbazole and tert-Butyl-Carbazole in Thin Films. *Photochem* **2024**, *4*, 179–197.
14. Tikhonov, D.S.; Garg, D.; Schnell, M. Inverse Problems in Pump–Probe Spectroscopy. *Photochem* **2024**, *4*, 57–110.
15. Gonçalves, T.M.; Martins, I.S.; Silva, H.F.; Tuchin, V.V.; Oliveira, L.M. Spectral Optical Properties of Rabbit Brain Cortex between 200 and 1000 nm. *Photochem* **2021**, *1*, 190–208.
16. Frederick, V.C.; Ashy, T.A.; Marchetti, B.; Ashfold, M.N.R.; Karsili, T.N.V. Photoprotective Properties of Eumelanin: Computational Insights into the Photophysics of a Catechol:Quinone Heterodimer Model System. *Photochem* **2021**, *1*, 26–37.
17. Segarra-Martí, J.; Nouri, S.M.; Bearpark, M.J. Modelling Photoionisations in Tautomeric DNA Nucleobase Derivatives 7H-Adenine and 7H-Guanine: Ultrafast Decay and Photostability. *Photochem* **2021**, *1*, 287–301.
18. Kawabata, K.; Miyoshi, A.; Nishi, H. Photoprotective Effects of Selected Polyphenols and Antioxidants on Naproxen Photodegradability in the Solid-State. *Photochem* **2022**, *2*, 880–890.
19. Kodikara, D.; Guo, Z.; Yoshimura, C. Effect of Benzophenone Type UV Filters on Photodegradation of Co-existing Sulfamethoxazole in Water. *Photochem* **2023**, *3*, 288–300.
20. Soilán, J.; López-Cóndor, L.; Peñín, B.; Aguilera, J.; de Gálvez, M.V.; Sampedro, D.; Losantos, R. Evaluation of MAA Analogues as Potential Candidates to Increase Photostability in Sunscreen Formulations. *Photochem* **2024**, *4*, 128–137.

References

1. Qi, X.K.; Yao, L.J.; Zheng, M.J.; Zhao, L.L.; Yang, C.; Guo, L.; Xia, W.J. Photoinduced Hydrodifluoromethylation and Hydromethylation of Alkenes Enabled by Ligand-to-Iron Charge Transfer Mediated Decarboxylation. *ACS Catal.* **2024**, *14*, 1300–1310. [CrossRef]
2. Guzman, M.I.; Eugene, A.J. Aqueous Photochemistry of 2-Oxocarboxylic Acids: Evidence, Mechanisms, and Atmospheric Impact. *Molecules* **2021**, *26*, 5278. [CrossRef]
3. Piedra, H.F.; Plaza, M. Advancements in visible-light-induced reactions via alkenyl radical intermediates. *Photochem. Photobiol. Sci.* **2024**, *23*, 1217–1228. [CrossRef] [PubMed]
4. Chen, X.; Zhang, X.; Xiao, X.; Wang, Z.J.; Zhao, J.Z. Recent Developments on Understanding Charge Transfer in Molecular Electron Donor-Acceptor Systems. *Angew. Chem. Int. Ed.* **2023**, *62*, 27. [CrossRef]
5. Kumari, H.; Sonia; Suman; Ranga, R.; Chahal, S.; Devi, S.; Sharma, S.; Kumar, S.; Kumar, P.; Kumar, S.; et al. A Review on Photocatalysis Used For Wastewater Treatment: Dye Degradation. *Water Air Soil Pollut.* **2023**, *234*, 46. [CrossRef]
6. Hoque, M.A.; Barrios Cossio, J.; Guzman, M.I. Photocatalysis of Adsorbed Catechol on Degussa P25 TiO_2 at the Air–Solid Interface. *J. Phys. Chem C* **2024**, *128*, 17470–17482. [CrossRef] [PubMed]
7. Garrod, R.T.; Jin, M.; Matis, K.A.; Jones, D.; Willis, E.R.; Herbst, E. Formation of Complex Organic Molecules in Hot Molecular Cores through Nondiffusive Grain-surface and Ice-mantle Chemistry. *Astrophys. J. Suppl. Ser.* **2022**, *259*, 71. [CrossRef]

8. Empel, C.; Pei, C.; Koenigs, R.M. Unlocking novel reaction pathways of diazoalkanes with visible light. *Chem. Commun.* **2022**, *58*, 2788–2798. [CrossRef]
9. Rapp, T.L.; DeForest, C.A. Targeting drug delivery with light: A highly focused approach. *Adv. Drug Deliv. Rev.* **2021**, *171*, 94–107. [CrossRef]
10. Finlayson, L.; Barnard, I.R.M.; McMillan, L.; Ibbotson, S.H.; Brown, C.T.A.; Eadie, E.; Wood, K. Depth Penetration of Light into Skin as a Function of Wavelength from 200 to 1000 nm. *Photochem. Photobiol.* **2022**, *98*, 974–981. [CrossRef]
11. Hu, H.; Wang, H.; Yang, Y.C.; Xu, J.F.; Zhang, X. A Bacteria-Responsive Porphyrin for Adaptable Photodynamic/Photothermal Therapy. *Angew. Chem. Int. Ed.* **2022**, *61*, 7. [CrossRef]
12. Qin, F.Z.; Almatrafi, E.; Zhang, C.; Huang, D.L.; Tang, L.; Duan, A.B.; Qin, D.Y.; Luo, H.Z.; Zhou, C.Y.; Zeng, G.M. Catalyst-Free Photochemical Activation of Peroxymonosulfate in Xanthene-Rich Systems for Fenton-Like Synergistic Decontamination: Efficacy of Proton Transfer Process. *Angew. Chem. Int. Ed.* **2023**, *62*, 9. [CrossRef]
13. Guo, Z.Y.; Kodikara, D.; Albi, L.S.; Hatano, Y.; Chen, G.; Yoshimura, C.; Wang, J.Q. Photodegradation of organic micropollutants in aquatic environment: Importance, factors and processes. *Water Res.* **2023**, *231*, 16. [CrossRef]
14. López-Magano, A.; Daliran, S.; Oveisi, A.R.; Mas-Ballesté, R.; Dhakshinamoorthy, A.; Alemán, J.; Garcia, H.; Luque, R. Recent Advances in the Use of Covalent Organic Frameworks as Heterogenous Photocatalysts in Organic Synthesis. *Adv. Mater.* **2023**, *35*, 66. [CrossRef]
15. Du, D.F.; Zhu, Z.; Chan, K.Y.; Li, F.J.; Chen, J. Photoelectrochemistry of oxygen in rechargeable Li-O_2 batteries. *Chem. Soc. Rev.* **2022**, *51*, 1846–1860. [CrossRef]
16. Zhou, R.; Guzman, M.I. CO_2 Reduction under Periodic Illumination of ZnS. *J. Phys. Chem C* **2014**, *118*, 11649–11656. [CrossRef]
17. Aydogan, C.; Yilmaz, G.; Shegiwal, A.; Haddleton, D.M.; Yagci, Y. Photoinduced Controlled/Living Polymerizations. *Angew. Chem., Int. Ed.* **2022**, *61*, 21. [CrossRef] [PubMed]
18. Grosskopf, J.; Kratz, T.; Rigotti, T.; Bach, T. Enantioselective Photochemical Reactions Enabled by Triplet Energy Transfer. *Chem. Rev.* **2022**, *122*, 1626–1653. [CrossRef] [PubMed]
19. Nevesely, T.; Wienhold, M.; Molloy, J.J.; Gilmour, R. Advances in the E → Z Isomerization of Alkenes Using Small Molecule Photocatalysts. *Chem. Rev.* **2022**, *122*, 2650–2694. [CrossRef]
20. Zondag, S.D.A.; Mazzarella, D.; Nöel, T. Scale-Up of Photochemical Reactions: Transitioning from Lab Scale to Industrial Production. *Annu. Rev. Chem. Biomol. Eng.* **2023**, *14*, 283–300. [CrossRef]
21. Buglioni, L.; Raymenants, F.; Slattery, A.; Zondag, S.D.A.; Nöel, T. Technological Innovations in Photochemistry for Organic Synthesis: Flow Chemistry, High-Throughput Experimentation, Scale-up, and Photoelectrochemistry. *Chem. Rev.* **2022**, *122*, 2752–2906. [CrossRef] [PubMed]

Disclaimer/Publisher's Note: The statements, opinions and data contained in all publications are solely those of the individual author(s) and contributor(s) and not of MDPI and/or the editor(s). MDPI and/or the editor(s) disclaim responsibility for any injury to people or property resulting from any ideas, methods, instructions or products referred to in the content.

Article

Synthesis and Characterization of New Tetradentate N₂O₂-Based Schiff's Base Cu (II) Complexes for Dye Photodegradation

Rohini Vallavoju [1], Ranjith Kore [1], Radhika Parikirala [1], Mahesh Subburu [2], Ramesh Gade [2], Vipin Kumar [3], Matta Raghavender [1], Prabhakar Chetti [3,*] and Someshwar Pola [1,*]

[1] Department of Chemistry, Osmania University, Hyderabad 500007, India
[2] Department of Humanities & Sciences, Vardhaman College of Engineering, Hyderabad 500007, India
[3] Department of Chemistry, National Institute of Technology, Kurukshetra 136119, India
* Correspondence: chetti@nitkkr.ac.in (P.C.); somesh.pola@gmail.com (S.P.)

Abstract: We have reported tetradentate ligands (salophen) coordinated with N and O atoms that led to the Cu (II) complexes. These Cu (II) complexes (C-1 and C-2) were firstly established by using elemental analysis and confirmed by using mass spectra. At the same time, the characterization of C-1 and C-2 complexes is performed by using several spectroscopic methods and morphological analysis. The bandgap values of the C-1 and C-2 complexes are evaluated with UV-vis DRS spectra. The PL spectral data and photocurrent curves clearly indicated the small recombination rate of the hole–electron pair. The synthesized C-1 and C-2 complexes' photocatalytic properties were examined for the degradation of cationic dyes such as methylene blue (MB λ_{max} = 654 nm) and methyl violet (MV λ_{max} = 590 nm) below visible-light action. The C-2 complex is more active than the C-1 complex because of its high photostability, small band-gap energy, and low recombination rate for hole–electron pair separation, and improved visible-light character, which encourages the generation of hydroxyl radical species throughout the photodegradation process. Scavenger probes were used to identify the dynamic species for the photodegradation of dyes, and a mechanism investigation was established.

Keywords: Cu (II) complexes; photocatalysis; surface area; rate of recombination; methyl violet dye

1. Introduction

The use of Schiff base ligands in coordination chemistry has been extensively studied, particularly those with tetradentate ligands containing nitrogen and oxygen donor atoms [1–5]. Metal (II) complexes of these ligands have shown promising catalytic activity in oxidative catalysis [6] and deoxygenation [7] reactions. Cu (II) Schiff base complexes, in particular, have received interest for their potential catalytic activity (conventional or photocatalytic) in the oxidation of organic substrates [8,9]. The use of Schiff base ligands derived from 2-hydroxy-1-naphthaldehyde and aryl-1,2-diamines in the synthesis of M(II) complexes has been extensively studied [10–12]. Among these, Cu (II) Schiff base complexes have been reported to exhibit various catalytic activities [13,14]. However, despite the large number of metal–Schiff base complexes available, Cu (II) Schiff base complexes are relatively less studied for their photocatalytic applications in organic transformations. The recent study on the piezo-photomineralization of MR and RhB cationic dyes using Cu (II) Schiff base complexes obtained from 4-chlorobenzene-1,2-diamine or 4-fluorobenzene-1,2-diamine with 2-hydroxy-l-naphthaldehyde is of great significance. The study can provide insights into the potential of Cu (II) Schiff base complexes as photocatalyst in the degradation of organic pollutants [15–17].

Under visible-light irradiation, mono- and bimetallic complexes have been used to photodegrade organic pollutants [18]. Photooxidation of aromatic hydrocarbons and methylstyrenes have also been accomplished using Ru (II) and Zn (II) complexes with

Schiff base ligands [19]. Other studies have investigated the potential of Pd (II) Schiff base complexes for the synthesis of fused heterocyclic aromatic hydrocarbons and the activation of allylic C-H bonds and Ag-doped Pd (II) complexes under visible-light irradiation for the degradation of dyes [20,21]. Zn (II) complexes have also been investigated for their possible use in the photooxidation of 2,2'-(Ethyne-1,2-diyl) dianilines [22], the photodecomposition of organic dye pollutants [23], and the piezo-photomineralization of dyes and industrial waste [24]. More recently, Ti (IV) complexes obtained using salophen-based ligands have been explored for their potential piezo-photocatalytic degradation of methyl red and rhodamine B dyes [25].

In this study, the reactions of tetradentate Schiff base ligands obtained from 4-chlorobenzene-1,2-diamine or 4-fluorobenzene-1,2-diamine with 2-hydroxy-1-naphthaldehyde and Cu^{2+} ions were investigated for their potential use in the photomineralization of MR and MV cationic dyes under visible-light treatment. The structures of ligands L-1 and L-2, and respective complexes C-1 and C-2 are given in Figure 1.

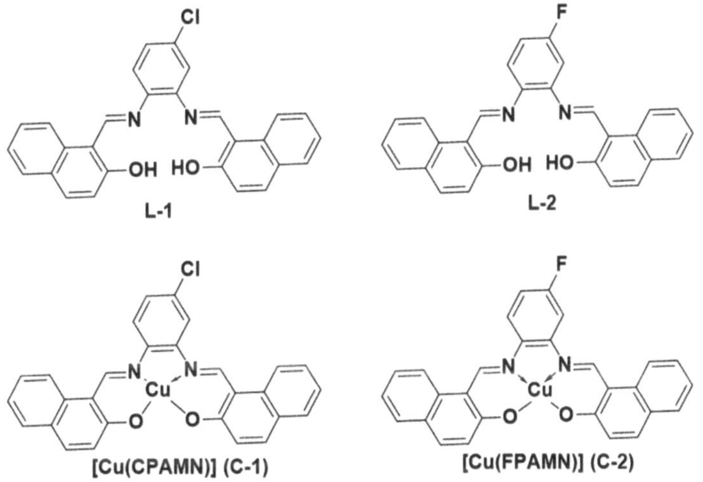

Figure 1. Structures of Schiff's base ligands and new Cu(II) complexes.

2. Materials and Methods

All the experimental processes for preparation of ligands were performed by following earlier methods [23]. The detailed experimental methods and characterization techniques for the analysis of Cu (II) complexes are shown in the Supplemental Materials. The carefully purified and dried Cu (II) complexes C-1 and C-2 were analyzed using elemental mass spectra, XPS, TGA, FTIR, FESEM, UV-visible emission, ESR, and photocurrent measurements.

Experimental Procedure for Photomineralization of Dye Pollutants

The photocatalyst (30 mg) was placed into 80 mL dye pollutant solution (5×10^{-4} M) in a 100 mL quartz glass cylindrical photoreactor. The mineralization of MR and MV cationic dyes was kept in the presence of visible light (300 Watts tungsten light with photon flux 8.02×10^{13} Einstein/s and wavelength 380–780 nm), as found by utilizing the chemical actinometric method with Reinecke salt [18] for 30 min. The adsorption–desorption equilibrium technique for the specific dye was attained in the absence of light for 10 min; the light was then permitted to fall on the reaction dye solution. Once the light was turned on, then a sample was taken every 5 min for UV-visible spectroscopic analysis. To verify the reusability of the catalyst, C-1/C-2 was separated by the centrifugal method, used again for the degradation of the dye, and its concentration studied using a UV-visible spectrophotometer at dye λ_{max}.

3. Results and Discussion

Molar conductivity experiments were used to determine if the new Cu (II) complexes were ionic or covalent. The conductivity of both C-1 and C-2 complexes was measured with the 10^{-3} M concentrations and showed the values 12.07 and 9.17 ohm^{-1} mol^{-1} cm^2, respectively. These values indicated that both the complexes are covalent and nonelectrolytic in nature (Table S1) [25]. Further, after 72 hours, the complexes were reexamined for molar conductance and exhibited the same conductance. Hence, the Cu (II) complexes are highly stable as Cu^{2+} ions are strongly complexed with the salophen ligands. The elemental and physicochemical data for the C-1 and C-2 complexes are depicted in Table S1. The analytical data show that various elements such as Cu, Cl/F, N, C, and H agree with the theoretical data, which are empirically formulated as C-1 and C-2 complexes. The MALDI mass spectral data for complexes C-1 and C-2 coincide with the exact molecular weight and are outlined in Figures S1 and S2, respectively. The Cu (II) complexes spectra show that the m/z values of 511.1475, and 495.1571 are in agreement with the molecular ions of C-1 and C-2, respectively (Figures S1 and S2) [24].

X-ray photoelectron spectroscopy (XPS) can be a useful tool for analyzing Cu (II) complexes with nitrogen and oxygen donors. XPS can provide information about the chemical composition and electronic state of the elements in the sample. When performing XPS analysis, the sample is bombarded with X-rays, which cause the emission of electrons from the surface of the sample. The energies of the emitted electrons are then measured to determine the binding energy of each element in the sample. By analyzing the binding energies of the Cu, nitrogen, and oxygen atoms in the complex, information can be obtained about the chemical bonding in the sample. In Cu (II) complexes with nitrogen and oxygen donors, the Cu atom is typically coordinated to one or more nitrogen or oxygen atoms, forming a complex with specific coordination geometry. The XPS spectra of these complexes show peaks corresponding to the Cu 2p, nitrogen 1s, and oxygen 1s electron orbitals. The positions and intensities of these peaks can be used to determine the electronic state of the Cu atom and the chemical environment of the nitrogen and oxygen atoms.

The Cu 2p XPS peak for a Cu (II) complex with Cl-substituted Schiff's base ligand shifted to a slightly higher binding energy compared to the peak for a complex with a nonsubstituted Schiff's base ligand. The nitrogen 1s and oxygen 1s peaks can also provide information about the bonding between the ligand and Cu atom, and any changes in the electron density around the ligand due to the presence of the Cl atom, as shown in Figure 2. Similarly, the Cu 2p XPS peak for a Cu (II) complex with an F-substituted Schiff's base ligand shifted to a slightly lower binding energy compared to the peak for a complex with a nonsubstituted Schiff's base ligand. The nitrogen 1s and oxygen 1s peaks can also provide information about the bonding between the ligand and Cu atom, and any changes in the electron density around the ligand due to the presence of the F atom, which is shown in Figure 3; details of analysis are mentioned below.

The XPS spectrum of the C-1 complex displays peaks at 965.09, 955.76, 945.9, and 935.99 eV for the $2p_{1/2}$ and $2p_{3/2}$ states of Cu of Cu-O and Cu-N bonds, respectively [26,27]. The remaining peaks, such as 203.24 eV for Cl (2p); 534.09 and 531.91 eV for O (1s); and 399.13 and 400.33 eV for Cu-N and Ar-C=N-Ar N (1s) states [21,28], are also displayed. In the case of the C-2 complex, 965.13, 955.55, 945.09 and 935.71 eV are indicated for the Cu-N and Cu-O ($2p_{1/2}$ and $2p_{3/2}$) bonds. Additionally, 688.01 for F (1s), 533.52 and 531.81 eV for O (1s), and 399.24 and 400.48 eV N (1s) peaks are observed in C-2; three different peaks are present in both the C-1 and C-2 complexes for C (1s), such as 285.27, 285.55, and 285.99 eV for C-Cl, C-OAr, and -C=N-, respectively, for C-1; and for the C-2 complex, 284.80, 285.35, and 286.31 eV for C-F, C-OAr and -C=N-, respectively, as revealed in Figures 2 and 3.

To know the thermal stability and thermal decomposition path for both Cu (II) complexes, thermogravimetric (TG) analysis of both complexes was investigated under a nitrogen atmosphere from 50 to 750 °C with an evaporating rate of 10 °C min^{-1}. The TGA of the C-1 complex decomposed at 385 °C, whereas the C-2 complex started at 395 °C. Both the Cu (II) complexes were decomposed in a single step, which designated that the ratio of ligand to metal is equal (1:1) [25] and revealed the conforming disintegration thermogram in Figure 4. The experimental TG plots display that C-1 and C-2 complexes decompose ligands in the solo stage with a weightiness loss of 84.38% (calcd. 84.56%) between 385 and 500 °C [25]. The TG curvature displays a mesa between 350–500 °C, and then there is no further disintegration up to 750 °C. The values obtained agree, indicating that the end material is pure CuO, which is confirmed by XPS and p-XRD data, as shown in Figure 5 and Figure S3 for the respective complex's residual substance.

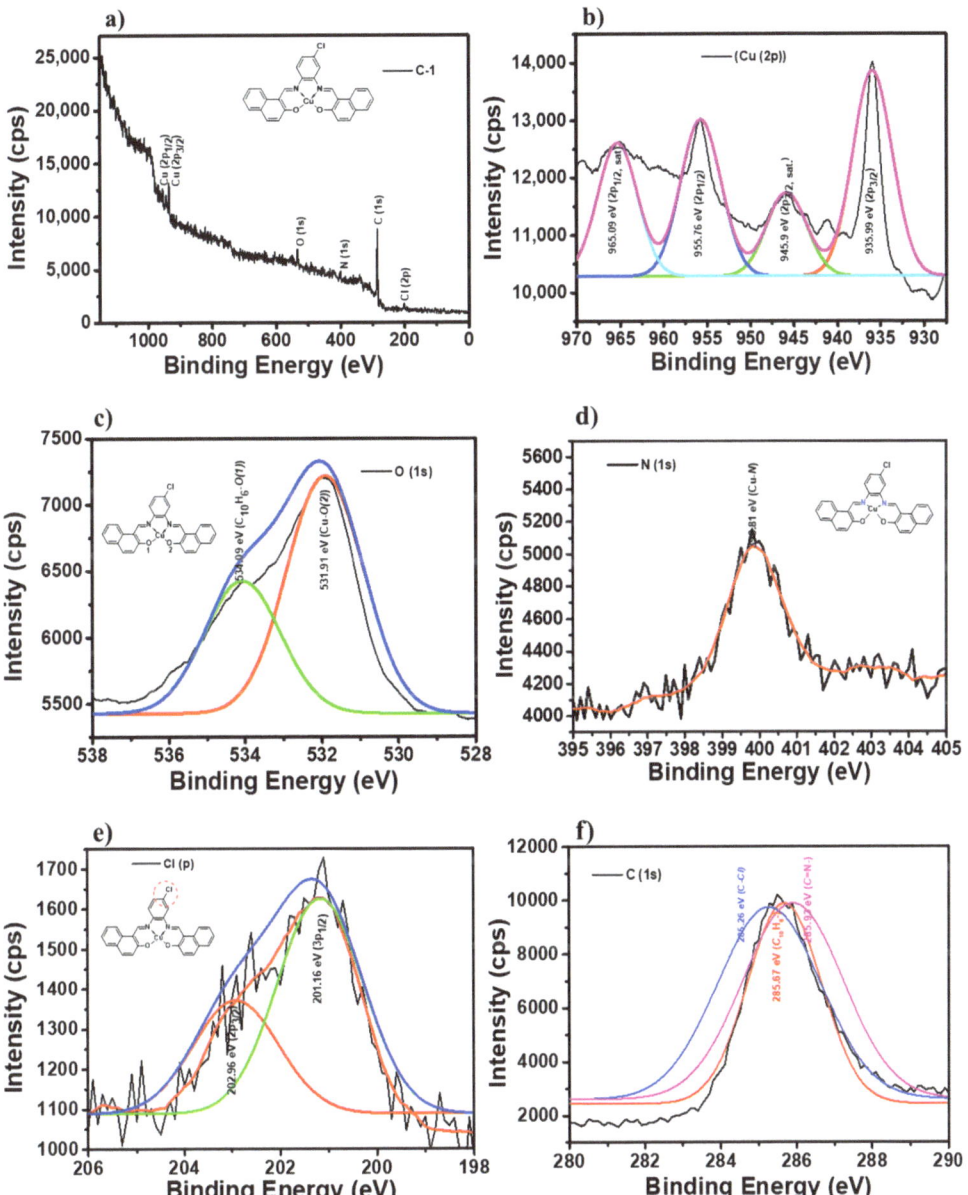

Figure 2. XPS spectra: (**a**) overall survey spectra; (**b**) Cu (2p) spectra; (**c**) O (1s) spectra; (**d**) N (1s) spectra; (**e**) Cl (2p) spectra; and (**f**) C (1s) spectra of the C-1 complex.

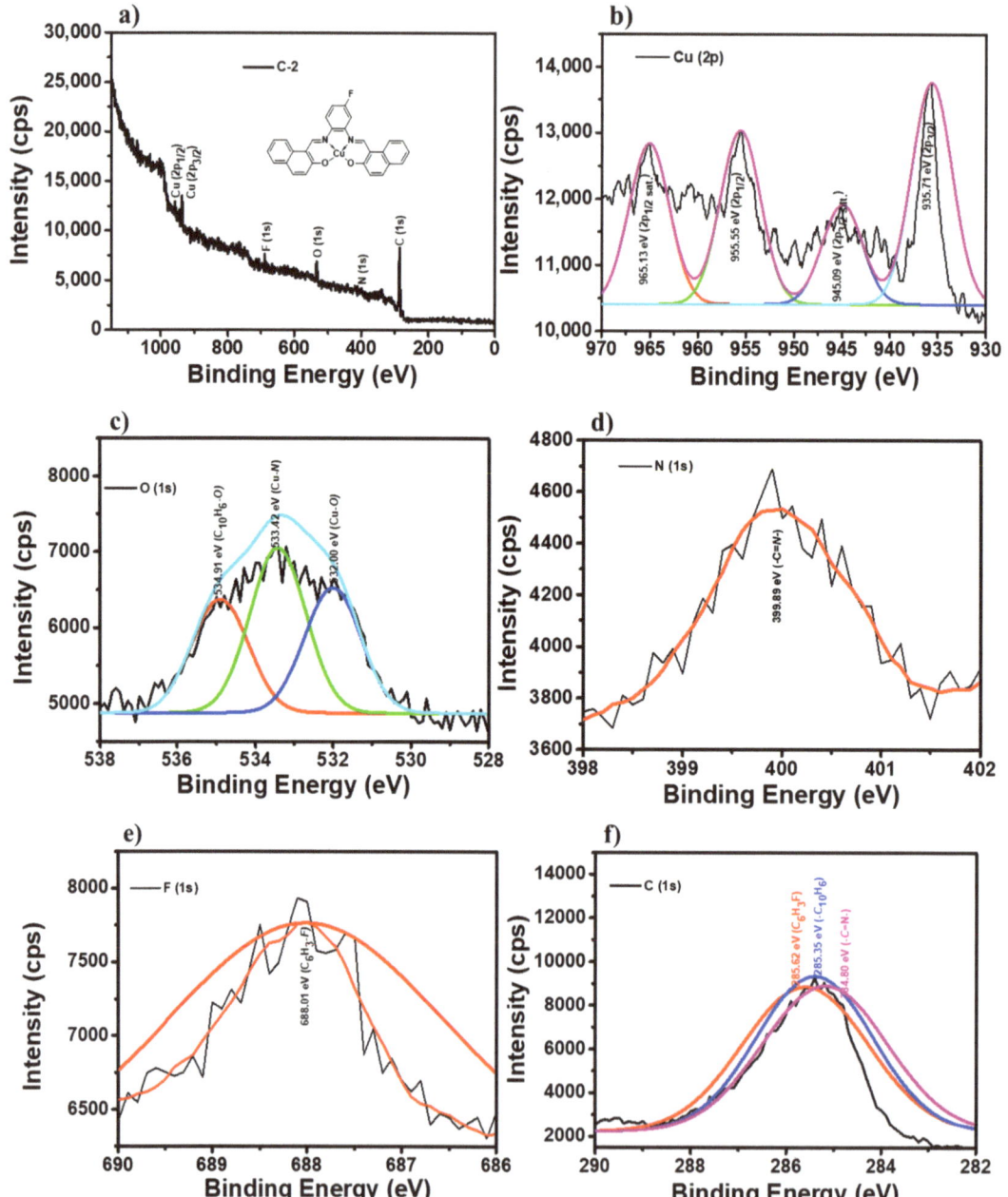

Figure 3. XPS spectra: (**a**) overall survey spectra; (**b**) Cu (2p) spectra; (**c**) O (1s) spectra; (**d**) N (1s) spectra; (**e**) F (1s) spectra; and (**f**) C (1s) spectra of the C-2 complex.

FTIR was used to confirm the bonding vibrational modes of all the ligands and Cu (II) complexes. The weak and strong peaks noticed at 1578 and 1452 cm^{-1} in the spectrum of the ligand L-1 is known to be the vibrational stretching modes of the imine (-C=N-) group [24]. In both the Cu (II) complexes, vanishing of the phenolic groups as correlated to ligand spectra and azomethine group peaks are shifted to the higher wavenumber side, which designates the complexation of the azomethine group. The two novel stretching modes in the FTIR spectra of C-1 and C-2 complexes, one around 403 and 406 cm^{-1} and the other around 504 and 502 cm^{-1}, have been identified as n(Cu-N) and n(Cu-O), respectively. The vibrational modes shown in the FTIR spectra of Cu (II) complexes signify that the Cu^{2+} ions are complexed with L-1/L-2 via two -C=N-functional and two *Ar-OH* sets. The illustrative FTIR spectra of Cu-complexes, C-1 and C-2, are revealed in Figures S4 and S5, respectively.

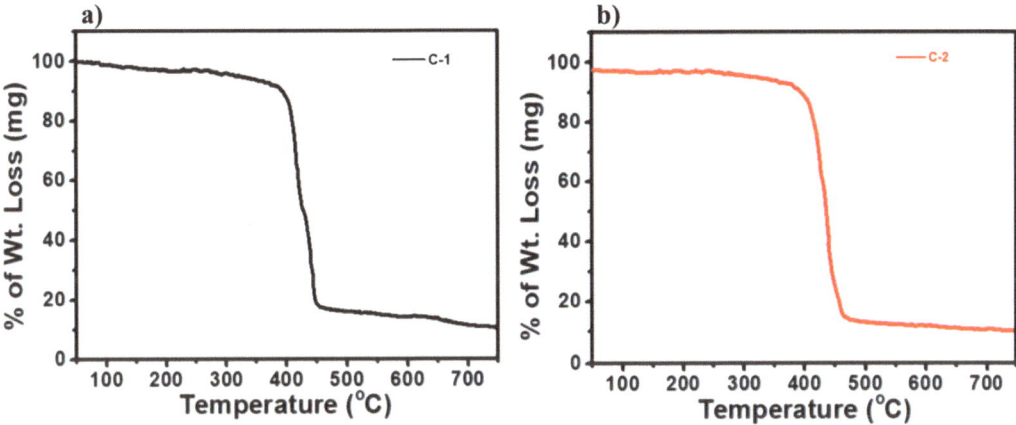

Figure 4. Thermograms of (**a**) C-1 and (**b**) C-2 complexes.

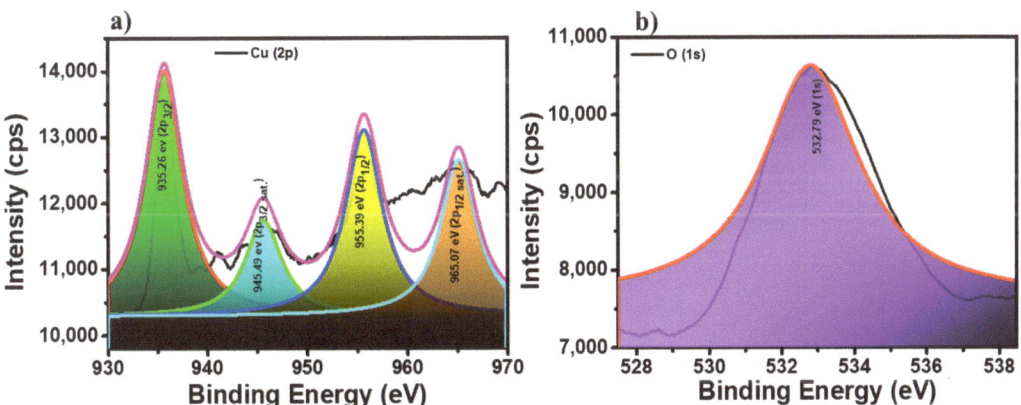

Figure 5. XPS spectra of CuO residue (**a**) Cu (2p) and (**b**) O (1s) obtained from TG analysis.

The electron spin resonance (ESR) spectra of the Cu (II) complexes are presented in Figure 6a,b. The spectra of the two complexes are anisotropic in nature, and each one exhibits three peaks. The $g_{||}$ and $g_\perp$ were calculated from the spectra (Table 1). ESR spectra of the complexes provide an excellent basis for distinguishing the unpaired electron as being present in either the ground state $dx^2 - y^2$ or the ground state dz^2. Thus, using the equation $(g2 - g1)/(g3 - y2)$, if the value of R is greater than one, the electron is present in $dz2$, and if the value is less than one, the electron is present in $dx^2 - y^2$. In the case of C-1, the R value is 0.263 and was found that $g_{||} > g_\perp > 2$ for both the complexes, indicating that the unpaired electron is present predominantly in the $dx^2 - y^2$ orbital of the Cu (II) ion. This indicates that these complexes are monomeric in nature, and that there are no metal–metal interactions and exchange couplings [29].

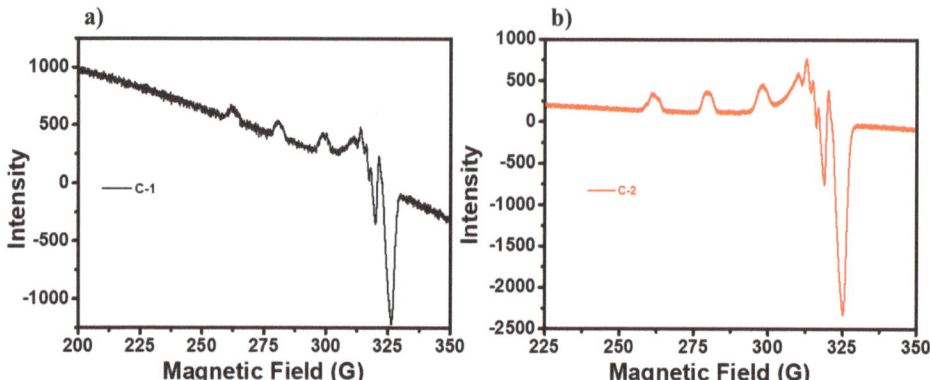

Figure 6. ESR spectra of (**a**) C-1 and (**b**) C-2 complexes.

Table 1. ESR parameters of Cu (II) complexes.

Complex	$g_{\parallel}$	$g_{\perp}$	g_{ave}
C-1	2.36	2.088	2.224
C-2	2.38	2.099	2.239

3.1. Absorption and Emission Studies

The UV-visible spectra of the Schiff base ligands L-1 and L-2 were analyzed in DMSO and showed absorption bands between 330 and 425 nm [24,30]. The onset bands of the Schiff bases are at 504 and 512 nm for L-1 and L-2, respectively, indicating that the bandgap of the Schiff bases falls between 2.43 and 2.42 eV. Upon complexation with Cu(II) ions, the absorption bands of the Cu(II)-complex shifted to the higher wavelength side (Figure 7) [28–30], indicating the occurrence of ligand–metal charge transfer and the strong interaction of lone-pair electrons of donor N-atoms with Cu^{2+} ions. The bandgap energies of the Cu(II) complexes were found to be lower than those of the pure Schiff base compounds, as shown in Table 2. This suggests that complexation with Cu(II) ions could enhance the absorption and utilization of visible light by the Schiff base ligands for photocatalytic applications.

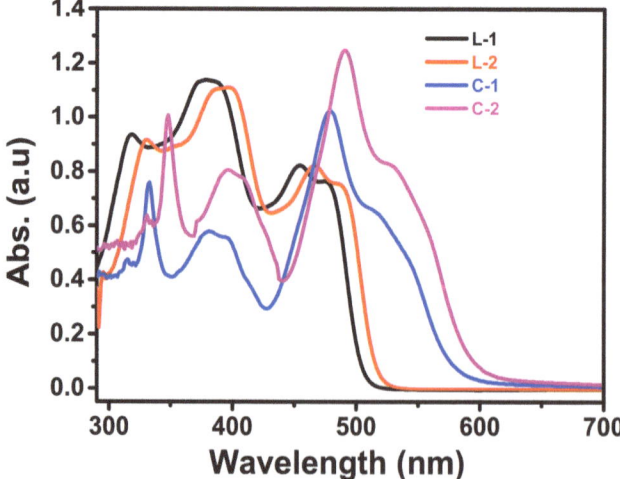

Figure 7. UV-vis spectra of the L-1 and L-2 ligands and the C-1 and C-2 complexes.

Table 2. Experimental λ_{onset} (nm), bandgap energies (in eV), and surface area of Cu (II) complexes.

Name	λ_{onset} (nm)	Bandgap Energy (eV)	Surface Area (m^2/g)
L-1	504	2.46	5.8
L-2	512	2.42	6.9
C-1	584	2.12	45.6
C-2	599	2.07	62.3

Surface areas of the ligands L-1 and L-2 were far smaller than Cu (II) complexes C-1 and C-2, as given in Table 2 [31].

DFT and TD-DFT calculations were also performed to gain a deeper understanding of the electronic excitations observed in the experimental UV-vis spectra of C-1 and C-2 complexes. The B3LYP functional with a mixed basis set was used for the Cu(II) complexes, with the 6-31G (d, p) basis set applied to the H, C, N, F, Cl, and O atoms, and the LANL2DZ basis set used for the copper metal [30,32,33].

It is noteworthy that the electronic excitations for both C-1 and C-2 complexes exhibit a narrow range from 445 to 448 nm, as shown in Table S2. C-1 has an intense absorption at 445 nm. When chlorine atom is replaced with fluorine atom, the absorption energy red-shifts slightly, and C-2 has an intense absorption at 448 nm. In both complexes, the main transition is from HOMO to LUMO.

The decrease in HOMO and LUMO energies was observed after replacing chlorine with fluorine (Table S3). The HOMO and LUMO orbitals of the C-1 and C-2 complexes are spread out over the ligand and the copper ion. This suggests that charge is transferred from the ligand to the copper ion during the electronic transitions. The HOMO is primarily located on the ligand, while the LUMO is localized on the copper ion (Figure S6), indicating that the ligand donates electrons to the copper ion upon excitation.

The photoluminescence spectra of the Cu (II) complexes (C-1 and C-2) were measured in solid phase using a 2 nm slit. The complexes were excited at their respective maximum wavelengths, 550 nm for C-1 and 555 nm for C-2. The spectra showed that the rate of recombination of electron–hole pairs was low, [30] indicating that there was a potential for charge separation or charge trapping in the Cu (II) complexes [24]. The spectral emission was caused by the excited h+ and e- recombination. A lower emission indicates a lower rate of charge carriers' recombination. The emission strength of C-2 was found to be six-fold lower than that of C-1, indicating that C-1 had a lower rate of recombination compared to C-2. The smaller emission energy of the C-1 complex suggests that it has a lower rate of recombination of charge carriers, as shown in Figure 8.

As shown in Figure 9, FESEM images were said to be the most important evidence after coordination for determining the shape of Cu (II) complexes. The consequence showed that there were ample Cu (II) complexes in nanostrips and nanoribbons with a smaller facet ratio, similar to dm (diameter) and flat surfaces. This was an imaginable spectacle in the solvo-thermal preparation of Cu (II) complex nanostrips and nanoribbons [25]. Figure 9 exhibits the morphology of reported complexes stimulated in ethanol solution. The outcome designated that Cu^{2+} ions were successfully coordinated with ligands, and the morphology compared with ligands was entirely changed.

3.2. Photocatalysis

The photocatalytic activity of Cu (II) complexes is investigated by the photodegradation of methylene blue (MB) and methyl violet (MV) dyes below the visible-light treatment. The time-based change in MB [20] and MV [24] concentrations with treatment time in the blank (no catalyst) and existence of C-1/C-2 are revealed in Figure 10. As the irradiation time rises, MB and MV photodegrade more quickly. In the absence of a photocatalyst, MB and MV are observed to experience 2% photodegradation, which may be the result of photolysis. Under the same experimental circumstances, the photodegradation of MB dye is 5% and 2%, respectively, in the presence of simple ligands and blank [24]. However, after 30 min of exposure to visible light, the photodegradation of MB and MV dyes with

C-1 was approximately 56% and 72%, respectively. Similarly, with the C-2 complex, the photodegradation was approximately 97% (MB) and 99% (MV) after 30 min of visible-light irradiation. As a result, C-2 exhibits greater photocatalytic activity in the current study against MB and MV photodegradation compared to the C-1 complex.

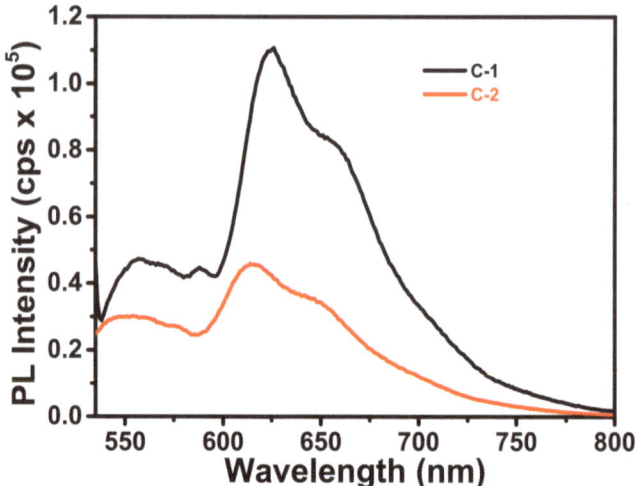

Figure 8. Photoluminescence spectra of C-1 and C-2 complexes.

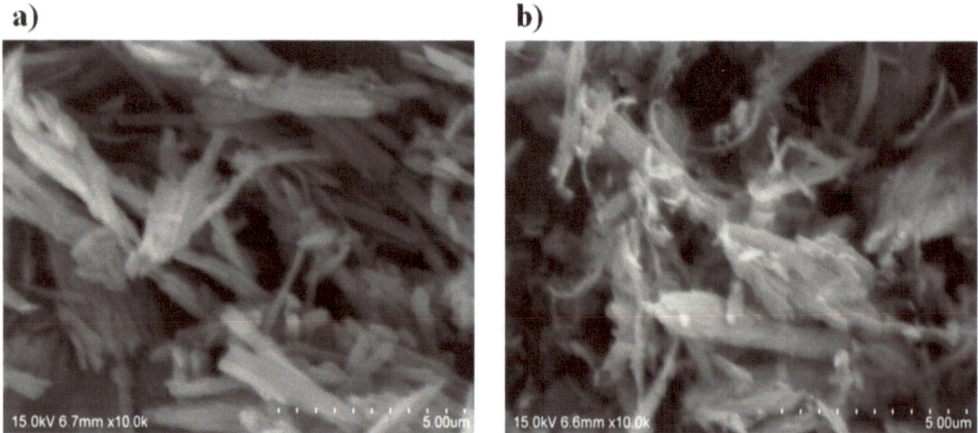

Figure 9. FESEM images of (a) C-1 and (b) C-2 complexes.

The photocatalytic decay of organic dye contaminants in the existence of Cu (II) complexes was initiated by the photogenerated electrons and holes, which ultimately produced free radicals such as $^{\bullet}OH$ and $O_2^{\bullet}$. The photodegradation rate is straightly proportional to the possibility for the generation of these radicals on the photocatalyst surface and their reaction with the dye molecules [24]. The formation of hydroxyl radicals throughout the MB and MV photodegradation processes in the occurrence of Cu (II) complexes has been determined experimentally using benzoquinone (BQ) as a superoxide radical quencher [23,24,34,35]. Under equal conditions, photodegradation of MB or MV is carried out by adding 0.5 g of BQ in the presence of C-2. The photocatalytic activity of C-2 was inhibited for the first 5 min of irradiation with the addition of BQ. Even after 30 min of irradiation in presence of BQ, photodegradation of the MB dye was about 20%

(Figure 11a). As a result, in addition to O$_2^{\bullet}$, other active species such as holes were also involved in dye degradation. The possible MB and MV photodegradation processes using Cu (II) complexes as photocatalysts are shown in Figure 11b. The photocatalytic performance of the Cu (II) photocatalyst to photodegrade the organic dyes is directly related to the degree of generation of active O$_2^{\bullet}$ radicals throughout the visible-light irradiation method [25]. Therefore, the greater photocatalytic activity of C-2 than C-1 can be attributed to observations in this study.

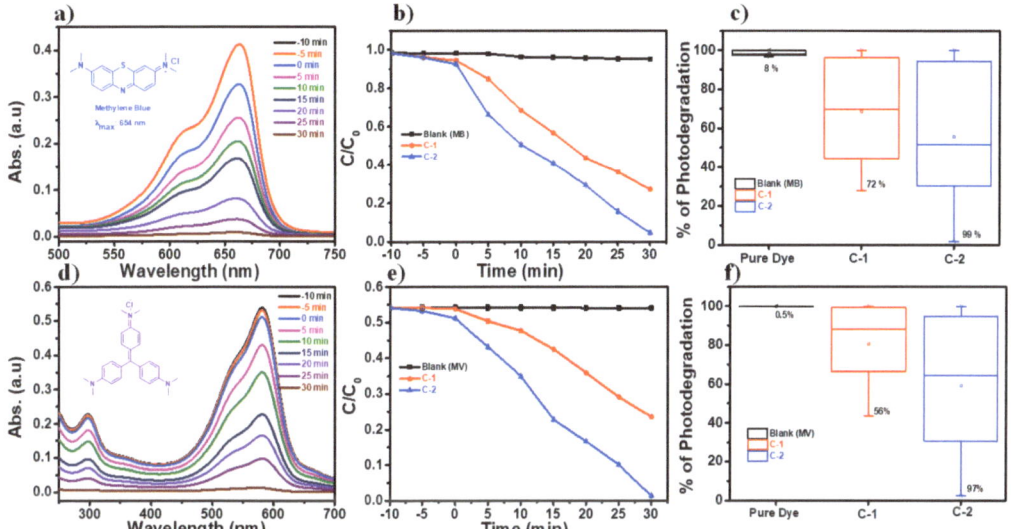

Figure 10. (a,d) Chronological absorbance curves; (b,e) photocatalytic plots, and (c,f) % photodegradation of MB and MV dyes in the presence of blank, C-1, and C-2 under visible-light irradiation.

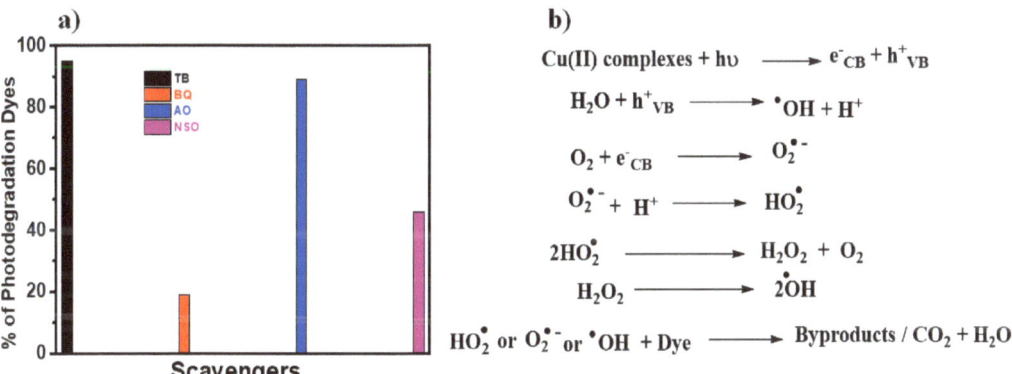

Figure 11. (a) Scavengers under visible-light irradiation and (b) Mechanistic studies in the presence of C-2 complex.

The kinetic curves of the C-2 photocatalytic reaction indicated that it had enhanced catalytic efficacy over C-1 and blank photocatalysts (Figure 12). The photoreaction kinetic graph was plotted against ln(C/Co) vs. time because of the photodegradation efficiency of MB and MV dye solutions, as shown in Figure 12a,b, and followed the pseudo-first order rate constant (Equation (1)). The rate constant (k) of photodegradation of MB dye was 4.4 and 158.4 min^{-1}, which is more than 35 times higher for C-2 as compared with

C-1 complex [24]. Similarly, the k value of photodegradation for MV dye was 2.3 and 201.2 min^{-1}, which is more than 87 times higher for C-2 as compared with C-1 complex [24].

$$\ln(C_0/C) = kt \quad (1)$$

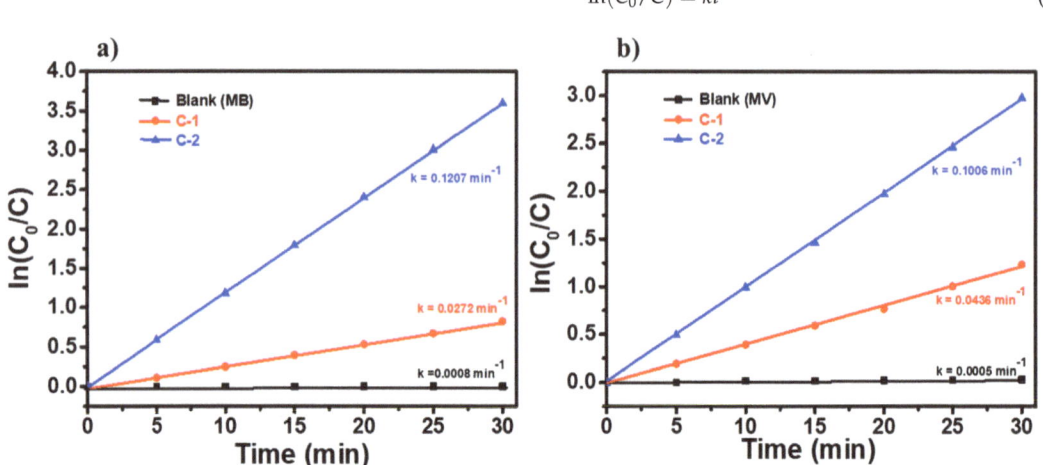

Figure 12. Kinetic plots for (**a**) MB and (**b**) MV dye solution in the presence of C-1 and C-2 complexes under visible-light irradiation.

The results from the transient photocurrent measurement suggest that the C-2 complex has a higher charge-species separation and relocation efficiency than the C-1 complex under visible-light irradiation (Figure 13a). The high transient photocurrent response of C-2 indicates that the complex can generate and separate charge carriers effectively. In contrast, C-1 showed a lower transient photocurrent response and a quicker decay, which suggests a greater recombination rate for the charge carriers. These results suggest that C-2 can utilize visible light more efficiently than C-1 to generate and separate charge carriers, leading to a higher photocurrent response. Overall, the transient photocurrent measurement provides valuable insights into the charge separation and recombination dynamics of the Cu-complexes under visible-light irradiation [24].

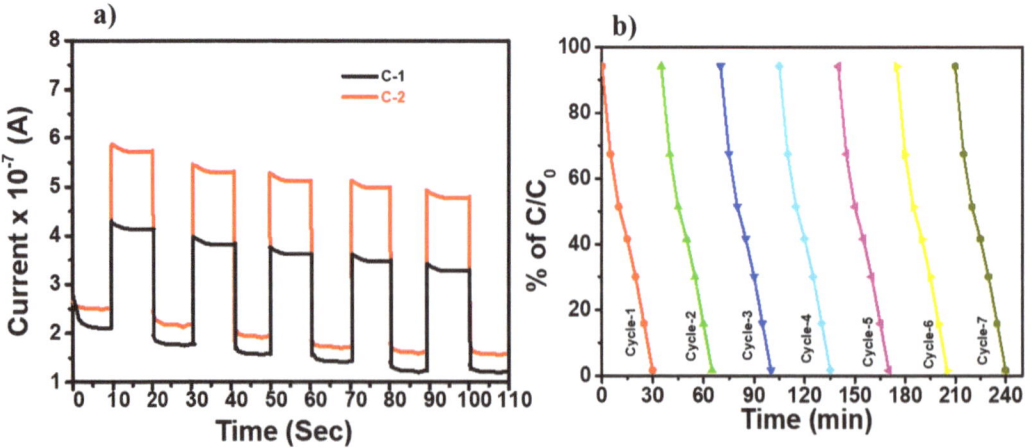

Figure 13. (**a**) The photocurrent response of C-1 and C-2 complexes and (**b**) recycling tests of C-2 complex for MB dye photodegradation.

Seven cycles of photocatalytic degradation were utilized to study the reusable nature of the C-2 catalyst. The catalyst C-2 utilized in every cycle of the photodegradation reaction was separated and cleaned with double-distilled water, vacuum desiccated, and reprocessed in the succeeding sequence of the degradation reaction. As revealed in Figure 13b, the photocatalytic degradation amount of the MB dye solutions still accomplished 97% after seven sequences of photoreaction. It has been established that the C-2 catalyst has an outstanding reuse characteristic, to a certain point. After the seventh cycle of photocatalysis, the photocatalyst C-2 was collected, verified by FTIR, and compared with pure C-2 catalyst, as revealed in Figure 14 [23,24]. As a result, the C-2 complex is highly stable under photocatalysis.

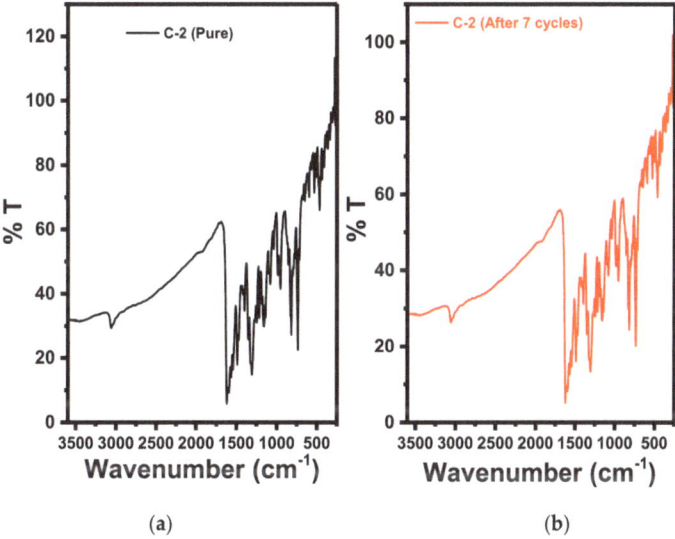

Figure 14. FTIR spectra of the C-2 complex: (**a**) pure; (**b**) photocatalysis following seven cycles of the photodegradation process.

4. Conclusions

In conclusion, the copper (II) complexes were effectively synthesized and characterized using a variety of different methodologies. Cationic dyes were used in the experiment to test the photocatalytic potential of the complexes by seeing how they were degraded in the light. According to the findings, the photocatalytic performance of the C-2 complex was superior to that of the C-1 complex. This superiority may be attributed to the C-2 complex's narrower bandgap energy, bigger surface area, lower rate of recombination of charge carriers, lower emission strength, and stronger photocurrent sensing. Based on these observations, it would seem that the C-2 complex has the potential to function as an efficient photocatalyst in the process of degrading organic dye pollutants.

Supplementary Materials: The following supporting information can be downloaded at: https://www.mdpi.com/article/10.3390/photochem3020016/s1, Table S1: Physicochemical and elemental data of C-1 and C-2 complexes; Table S2: Electronic excitations (λ_{CAL} in nm), oscillator strength (f), major transitions (MT), and % weight (%Ci) of C-1 and C-2 complexes by using TD-B3LYP/6-31G (d, p) method; Table S3: Calculated HOMO and LUMO energies, and HLG (in eV) for the Cu (II) complexes; Figure S1: HR MALDI mass spectrum of the C-1 complex; Figure S2: HR MALDI mass spectrum of the C-2 complex; Figure S3: Powder XRD image of CuO residue obtained from TG analysis; Figure S4: FTIR spectra of the L-1 ligand (a,c) and C-1 complex (b,d); Figure S5: FTIR spectra of L-2 ligand (a,c) and C-2 complex (b,d); Figure S6: Frontier molecular orbitals of the C-1 and C-2 complexes.

Author Contributions: Conceptualization, S.P. and P.C.; methodology, R.V.; software, P.C.; validation, S.P. and P.C.; formal analysis, R.K., R.P., M.S., R.G., M.R. and V.K.; investigation, M.S.; resources, S.P.; data curation, S.P.; writing—original draft preparation, M.S.; writing—review and editing, S.P. and P.C.; visualization, P.C.; supervision, S.P. and P.C.; project administration, S.P. and P.C.; funding acquisition, S.P. and P.C. All authors have read and agreed to the published version of the manuscript.

Funding: This research was funded by SERB (EMR//2014/000452), UGC-UPE-FAR and DST-PURSE, New Delhi, India, and CSIR (02(0339)/18/EMR-II), New Delhi, India.

Data Availability Statement: All the data related to this article is provided in the Supplemental Materials.

Acknowledgments: Authors especially thank DST-FIST schemes and UGC, New Delhi. Rohini Vallaboju thanks the DST-PURSE-II program, New Delhi, for its support.

Conflicts of Interest: The authors declare no conflict of interest.

References

1. Mehmet, T. Polydentate Schiff-base ligands and their Cd(II) and Cu(II) metal complexes: Synthesis, characterization, biological activity and electrochemical properties. *J. Coord. Chem.* **2007**, *60*, 2051–2065.
2. Tunde, L.Y.; Segun, D.O.; Sizwe, Z.; Hezekiel, M.K.; Isiaka, A.L.; Monsurat, M.L.; Nonhlangabezo, M. Design of New Schiff-Base Copper(II) Complexes: Synthesis, Crystal Structures, DFT Study, and Binding Potency toward Cytochrome P450 3A4. *ACS Omega* **2021**, *6*, 13704–13718.
3. Liu, X.; Manzur, C.; Novoa, N.; Celedón, S.; Carrillo, D.; Hamon, J.-R. Multidentate unsymmetrically-substituted Schiff bases and their metal complexes: Synthesis, functional materials properties, and applications to catalysis. *Coord. Chem. Rev.* **2018**, *357*, 144–172. [CrossRef]
4. Wesley, J.A.; Kalidasa, M.K.; Neelakantan, M.A. Review on Schiff bases and their metal complexes as organic photovoltaic materials. *Renew. Sustain. Energy Rev.* **2014**, *36*, 220–227. [CrossRef]
5. Alberto, A.-M.; Viviana, R.-M.; Farrah, C.-B.; Jesús, R.P.-U.; Fernando, C.-C.; Dorian, P.-C.; Raúl, C.-P.; Galdina, V.S.-M.; Bethsy, A.A.-C.; David, M.-M. Pincer Complexes Derived from Tridentate Schiff Bases for Their Use as Antimicrobial Metallopharmaceuticals. *Inorganics* **2022**, *10*, 134.
6. Andreas, W.; Ulrich, S.S. Metal-Terpyridine Complexes in Catalytic Application—A Spotlight on the Last Decade. *ChemCatChem* **2020**, *12*, 2890–2941.
7. Takuya, S.; Ken-ichi, F. Recent Advances in Homogeneous Catalysis via Metal–Ligand Cooperation Involving Aromatization and Dearomatization. *Catalysts* **2020**, *10*, 635.
8. Kazimer, L.S.; Travis, R.B.; Tehshik, P.Y. Dual Catalysis Strategies in Photochemical Synthesis. *Chem. Rev.* **2016**, *116*, 10035–10074.
9. Manas, S.; Tannistha, R.B.; Armando, J.L.P.; Luísa, M.D.R.S.M. Aroylhydrazone Schiff Base Derived Cu(II) and V(V) Complexes: Efficient Catalysts towards Neat Microwave-Assisted Oxidation of Alcohols. *Int. J. Mol. Sci.* **2020**, *21*, 2832.
10. Ebrahimipour, S.Y.; Maryam, M.; Masoud, T.M.; Jim, S.; Joel, T.M.; Iran, S. Synthesis and structure elucidation of novel salophen-based dioxo-uranium(VI) complexes: In-vitro and in-silico studies of their DNA/BSA-binding properties and anticancer activity. *Eur. J. Med. Chem.* **2017**, *140*, 172–186. [CrossRef]
11. Atkins, R.; Brewer, G.; Kokot, E.; Mockler, G.M.; Sinn, E. Copper (II) and nickel (II) complexes of unsymmetrical tetradentate Schiff base ligands. *Inorg. Chem.* **1985**, *24*, 127–134. [CrossRef]
12. Kushwah, N.P.; Pal, M.K.; Wadawale, A.P.; Jain, V.K. Diorgano-gallium and -indium complexes with salophen ligands: Synthesis, characterization, crystal structure and C–C coupling reactions. *J. Organomet. Chem.* **2009**, *694*, 2375–2379. [CrossRef]
13. Santarupa, T.; Partha, R.; Ray, J.B.; Fallah, M.S.E.; Javier, T.; Eugenio, G.; Samiran, M. Ferromagnetic Coupling in a New Copper(II) Schiff Base Complex with Cubane Core: Structure, Magnetic Properties, DFT Study and Catalytic Activity. *Eur. J. Inorg. Chem.* **2009**, *2009*, 4385–4395.
14. Rong, M.; Wang, J.; Shen, Y.; Han, J. Catalytic oxidation of alcohols by a novel manganese Schiff base ligand derived from salicylaldehyd and l-Phenylalanine in ionic liquids. *Catal. Commun.* **2012**, *20*, 51–53. [CrossRef]
15. Teerawat, K.; Duangdao, C.; Bussaba, P.; Ratanon, C. Degradation of Methylene Blue with a Cu(II)–Quinoline Complex Immobilized on a Silica Support as a Photo-Fenton-Like Catalyst. *ACS Omega* **2022**, *7*, 33258–33265.
16. Soroceanu, A.; Cazacu, M.; Shova, S.; Turta, C.; Kožíšek, J.; Gall, M.; Breza, M.; Rapta, P.; Mac Leod, T.C.; Pombeiro, A.J. Copper (II) complexes with Schiff bases containing a disiloxane unit: Synthesis, structure, bonding features and catalytic activity for aerobic oxidation of benzyl alcohol. *Eur. J. Inorg. Chem.* **2013**, *2013*, 1458–1474. [CrossRef]
17. Ran, J.; Li, X.; Zhao, Q.; Qu, Z.; Li, H.; Shi, Y.; Chen, G. Synthesis, structures and photocatalytic properties of a mononuclear copper complex with pyridine-carboxylato ligands. *Inorg. Chem. Commun.* **2010**, *13*, 526–528. [CrossRef]
18. Mahesh, S.; Ramesh, G.; Venkanna, G.; Prabhakar, C.; Koteshwar, R.R.; Someshwar, P. Effective photodegradation of organic pollutantsin the presence of mono and bi-metallic complexes under visible-light irradiation. *J. Photochem. Photobiol. A Chem.* **2021**, *406*, 112996.
19. Venkanna, G.; Jakeer, A.; Mahesh, S.; Bhongiri, Y.; Ritu, M.; Chetti, P.; Someshwar, P. Evolution of physical and photocatalytic properties of new Zn(II) and Ru(II) complexes. *Polyhedron* **2019**, *170*, 412–423.

20. Someshwar, P.; Mahesh, S.; Ravinder, G.; Vithal, M.; Yu, T.T. New photocatalyst for allylic aliphatic C–H bond activation and degradation of organic pollutants: Schiff base Ti(IV) complexes. *RSC Adv.* **2015**, *5*, 58504–58513.
21. Guguloth, V.; Ahemed, J.; Subburu, M.; Guguloth, V.C.; Chetti, P.; Pola, S. A very fast photodegradation of dyes in the presence of new Schiff's base N_4-macrocyclic Ag-doped Pd (II) complexes under visible-light irradiation. *J. Photochem. Photobiol. A Chem.* **2019**, *382*, 111975. [CrossRef]
22. Mahesh, S.; Ramesh, G.; Prabhakar, C.; Someshwar, P. Photooxidation of 2,2′-(Ethyne-1,2-diyl)dianilines: An Enhanced Photocatalytic Properties of New Salophen-Based Zn(II) Complexes. *Photochem* **2022**, *2*, 358–375.
23. Jakeer, A.; Jakeer, P.; Venkateshwar, R.D.; Ranjith, K.; Ramesh, G.; Yadagiri, B.; Prabhakar, C.; Someshwar, P. Synthesis of new Zn (II) complexes for photo decomposition of organic dye pollutants, industrial wastewater and photo-oxidation of methyl arenes under visible-light. *J. Photochem. Photobiol. A Chem.* **2021**, *419*, 113455.
24. Venkateshwar, R.D.; Mahesh, S.; Ramesh, G.; Manohar, B.; Prabhakar, C.; Babu, N.S.; Penumaka, N.; Yadagiri, B.; Someshwar, P. A new Zn(II) complex-composite material: Piezoenhanced photomineralization of organic pollutants and wastewater from the lubricant industry. *Environ. Sci. Water Res. Technol.* **2021**, *7*, 1737–1747.
25. Rohini, V.; Ranjith, K.; Radhika, P.; Mahesh, S.; Ramesh, G.; Manohar, B.; Someshwar, P.; Prabhakar, C. Enhanced piezophotocatalytic properties of new salophen based Ti (IV) complexes. *Inorg. Chem. Commun.* **2023**, *148*, 110272.
26. Reddy, G.R.; Balasubramanian, S.; Chennakesavulu, K. Zeolite encapsulated Ni(ii) and Cu(ii) complexes with tetradentate N_2O_2 Schiff base ligand: Catalytic activity towards oxidation of benzhydrol and degradation of rhodamine-B. *J. Mater. Chem. A* **2014**, *2*, 15598–15610. [CrossRef]
27. Hafsa, S.; Qureshi Fozia, M.S.; Haque, Z. Biosynthesis of Flower-Shaped CuO Nanostructures and Their Photocatalytic and Antibacterial Activities. *Nano-Micro Lett.* **2020**, *12*, 29.
28. Jiang, D.; Jianbin, X.; Liqiong, W.; Wei, Z.; Yuegang, Z.; Xinheng, L. Photocatalytic performance enhancement of CuO/Cu_2O heterostructures for photodegradation of organic dyes: Effects of CuO morphology. *Appl. Catal. B Environ.* **2017**, *211*, 199–204. [CrossRef]
29. Janusz, G.; Katarzyna, Ś.; Julia, K.; Maciej, W. Multinuclear Ni(ii) and Cu(ii) complexes of a meso 6 + 6 macrocyclic amine derived from trans-1,2-diaminocyclopentane and 2,6-diformylpyridine. *Dalton Trans.* **2022**, *51*, 9735–9747.
30. Gülnur, K.K. Synthesis of new Schiff base and its Ni(II), Cu(II), Zn(II) and Co(II) complexes; photophysical, fluorescence quenching and thermal studies. *J. Mol. Struct.* **2022**, *1256*, 132534.
31. Nooshin, K.; Alison, Z.; Sheida, E. Bioactive Ni(II), Cu(II) and Zn(II) complexes with an N3 functionalized Schiff base ligand: Synthesis, structural elucidation, thermodynamic and DFT calculation studies. *Inorg. Chim. Acta* **2022**, *541*, 121083.
32. Rasha, M.K.; Ahmed, S.; Aly, H.A.; Mohamed, M.A.F.-A. Development of a novel and potential chemical sensor for colorimetric detection of Pd(II) or Cu(II) in E-wastes. *Microchem. J. Part A* **2022**, *172*, 106951.
33. Frisch, M.J.; Trucks, G.W.; Schlegel, H.B.; Scuseria, G.E.; Robb, M.A.; Cheeseman, J.R.; Scalmani, G.; Barone, V.; Petersson, G.A.; Nakatsuji, H.; et al. *Gaussian 16W, Revision B.01*; Gaussian, Inc.: Wallingford, CT, USA, 2016.
34. Minji, Y.; Youngtak, O.; Sugyeong, H.; June, S.L.; Ramireddy, B.; Sun, H.K.; Filipe, M.M.; Sang, O.K.; Dong, H.K. Synergistically enhanced photocatalytic activity of graphitic carbonnitride and WO_3 nanohybrids mediated by photo-Fenton reaction and H_2O_2. *Appl. Catal. B Environ.* **2017**, *206*, 263–270.
35. Jiahui, L.; Xinghui, L.; Yan, C.; Huijun, L.; Ruochang, L.; Xiaoli, D.; Hyoyoung, L.; Hongchao, M. Highly Enhanced Photoelectrocatalytic Oxidation via Cooperative Effect of Neighboring Two Different Metal Oxides for Water Purification. *J. Phys. Chem. C* **2020**, *124*, 11525–11535.

Disclaimer/Publisher's Note: The statements, opinions and data contained in all publications are solely those of the individual author(s) and contributor(s) and not of MDPI and/or the editor(s). MDPI and/or the editor(s) disclaim responsibility for any injury to people or property resulting from any ideas, methods, instructions or products referred to in the content.

Article

Panchromatic Copper Complexes for Visible Light Photopolymerization

Alexandre Mau [1,2], Guillaume Noirbent [3], Céline Dietlin [1,2], Bernadette Graff [1,2], Didier Gigmes [3], Frédéric Dumur [3,*] and Jacques Lalevée [1,2,*]

[1] Université de Haute-Alsace, CNRS, IS2M UMR 7361, F-68100 Mulhouse, France; alexandre.mau@uha.fr (A.M.); celine.dietlin@uha.fr (C.D.); bernadette.graff@uha.fr (B.G.)
[2] Université de Strasbourg, F-67081 Strasbourg, France
[3] Aix Marseille Univ, CNRS, ICR, UMR 7273, F-13397 Marseille, France; guillaume.noirbent@outlook.fr (G.N.); didier.gigmes@univ-amu.fr (D.G.)
* Correspondence: frederic.dumur@univ-amu.fr (F.D.); jacques.lalevee@uha.fr (J.L.)

Citation: Mau, A.; Noirbent, G.; Dietlin, C.; Graff, B.; Gigmes, D.; Dumur, F.; Lalevée, J. Panchromatic Copper Complexes for Visible Light Photopolymerization. *Photochem* **2021**, *1*, 167–189. https://doi.org/10.3390/photochem1020010

Academic Editors: Marcelo I. Guzman and Elena Cariati

Received: 17 June 2021
Accepted: 1 August 2021
Published: 4 August 2021

Publisher's Note: MDPI stays neutral with regard to jurisdictional claims in published maps and institutional affiliations.

Copyright: © 2021 by the authors. Licensee MDPI, Basel, Switzerland. This article is an open access article distributed under the terms and conditions of the Creative Commons Attribution (CC BY) license (https://creativecommons.org/licenses/by/4.0/).

Abstract: In this work, eleven heteroleptic copper complexes were designed and studied as photoinitiators of polymerization in three-component photoinitiating systems in combination with an iodonium salt and an amine. Notably, ten of them exhibited panchromatic behavior and could be used for long wavelengths. Ferrocene-free copper complexes were capable of efficiently initiating both the radical and cationic polymerizations and exhibited similar performances to that of the benchmark G1 system. Formation of acrylate/epoxy IPNs was also successfully performed even upon irradiation at 455 nm or at 530 nm. Interestingly, all copper complexes containing the 1,1′-bis(diphenylphosphino)ferrocene ligand were not photoluminescent, evidencing that ferrocene could efficiently quench the photoluminescence properties of copper complexes. Besides, these ferrocene-based complexes were capable of efficiently initiating free radical polymerization processes. The ferrocene moiety introduced in the different copper complexes affected neither their panchromatic behaviors nor their abilities to initiate free radical polymerizations.

Keywords: copper complex; ferrocene; photopolymerization; visible light; photoinitiators; free radical polymerization; heteroleptic complexes

1. Introduction

Photopolymerization or polymerization initiated through light exposure is a process widely used industrially. Various applications such as adhesives, dentistry [1], medicine, coatings [2], composites [3], laser writing and 3D printing [4] are already based on this process. However, a major drawback of industrial applications of photopolymerization is the use of UV light which is a source of safety concerns [5]. Moreover, due to the easy availability of affordable, compact, lightweight and safe visible irradiation light sources, such as light-emitting diodes (LEDs), the development of photoinitiating systems activable under visible light and low light intensities has become a highly active research field.

Over the past decades, few photoinitiating systems capable of initiating both the free radical polymerization of acrylates and the cationic polymerization of epoxides have been developed [3,6–10]. Among them, a copper complex named G1 (Scheme 1) was developed as a highly efficient photoinitiator for free radical and cationic polymerization under visible LED irradiation [3,11–13]. Photoinitiating systems based on G1 can produce reactive species (radicals and cations) following a catalytic cycle through successive reactions. Considering the efficiency of the photocatalyst G1, other copper complexes, with similar behavior, are still developed with the aim of concomitantly initiating both radical and cationic polymerizations.

However, as exemplified with G1, most of the copper complexes that have been designed to date as photoinitiators of polymerization were mainly designed for being activated with irradiation wavelengths below 450 nm. These wavelengths are still very energetic as they range close to the ultraviolet spectrum. Working with higher irradiation wavelengths could be highly beneficial since this involves milder experimental conditions, safer systems and less energetic irradiation sources. Especially, by working at longer wavelengths, a better light penetration can also be expected. To this end, commercial irradiation sources already exist for wavelengths around 532 nm [11,14].

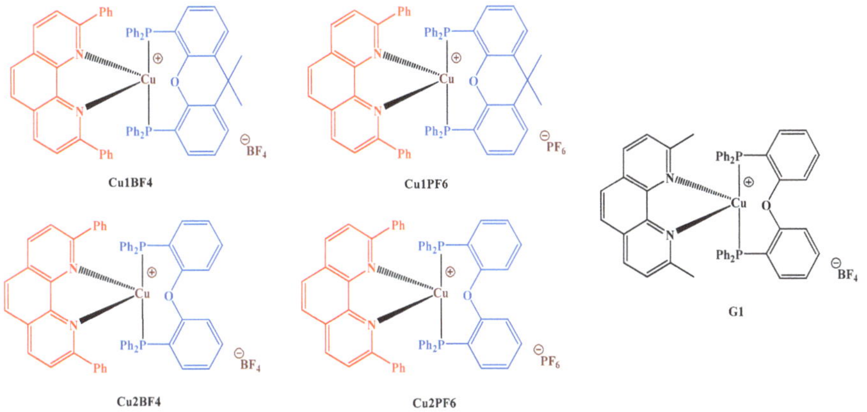

Scheme 1. Investigated copper complexes Cu1BF4, Cu1PF6, Cu2BF4, Cu2PF6 and the benchmark copper (I) complex G1.

In this work, two strategies were investigated to design panchromatic copper complexes which absorbed light above 450 nm: (i) modification of the electron donating substituent attached to the phenanthroline ligand and used as proligand for the design of heteroleptic copper (I) complexes inspired by complex G1; (ii) introduction of a ferrocenyl group in the bulky phosphorylated ligand, namely 1,1′-bis(diphenylphosphino)ferrocene ligand, also used for the design of heteroleptic copper (I) complexes. Eleven copper complexes were designed, synthesized and investigated as visible light photoinitiators of polymerization. Among this series of copper complexes, only one complex, i.e., Cu3BF4 was previously reported in the literature. However, if the structure is known, this complex has never been used in photopolymerization prior to this work [15–17]. The different copper complexes were investigated as photoinitiators of polymerization upon irradiation with visible light-emitting diodes (LEDs) for the free radical polymerization of acrylates, the cationic polymerization of epoxides and the formation of interpenetrated polymer networks (IPN). The efficiency of the three-component photoinitiating systems based on these panchromatic copper complexes, an iodonium salt (Iod) and ethyl 4-(dimethylamino)benzoate (EDB) was investigated and compared to the highly efficient system G1/Iod/EDB, which was used as a reference in this work [18,19]. Photoinitiating abilities of the ferrocene-containing copper complexes were compared to that of their ferrocene-free analogues.

2. Materials and Methods

2.1. Chemical Compounds

All reagents and solvents were purchased from Aldrich (St. Louis, MO, USA) or Alfa Aesar (Tewksbury, MA, USA) and used as received without further purification. Mass spectroscopy was performed by the Spectropole of Aix-Marseille University. ESI mass spectral analyses were recorded with a 3200 QTRAP (Applied Biosystems SCIEX, Waltham, MA, USA) mass spectrometer. The HRMS mass spectral analysis was performed with a QStar Elite (Applied Biosystems SCIEX) mass spectrometer. Elemental analyses were recorded with a Thermo Finnigan EA 1112 elemental analysis apparatus driven by the Eager 300 software. ^{1}H- and ^{13}C-NMR spectra were determined at room temperature in 5 mm o.d. tubes on a Bruker Avance 400 spectrometer of the Spectropole: ^{1}H (400 MHz) and ^{13}C

(100 MHz). The ^{1}H chemical shifts were referenced to the solvent peaks DMSO (2.49 ppm), CDCl$_3$ (7.26 ppm) and the ^{13}C chemical shifts were referenced to the solvent peak DMSO (49.5 ppm), CDCl$_3$ (77.0 ppm). All photoinitiators were prepared with analytical purity up to accepted standards for new organic compounds (>98%); this was checked by high field NMR analysis. 2,9-Diphenyl-1,10-phenanthroline [20], 2,9-dibutyl-1,10-phenanthroline [21], 2-tert-butyl-1,10-phenanthroline [22] and complex G1 [3] were synthesized as previously reported in the literature, without modification and in similar yields.

2.1.1. Compounds Used as Photoinitiators

The copper complex G1 was synthesized according to the procedure reported in [9]. The new copper complexes Cu1BF4, Cu1PF6, Cu2BF4 and Cu2PF6, inspired by G1 and presented in Scheme 1, were synthesized according to the procedure depicted below. All complexes were obtained in a one-step synthesis, by first complexing the phosphine-derived ligand with copper (I), followed by the introduction of the ancillary ligand. All complexes were isolated as solids (See Scheme 2).

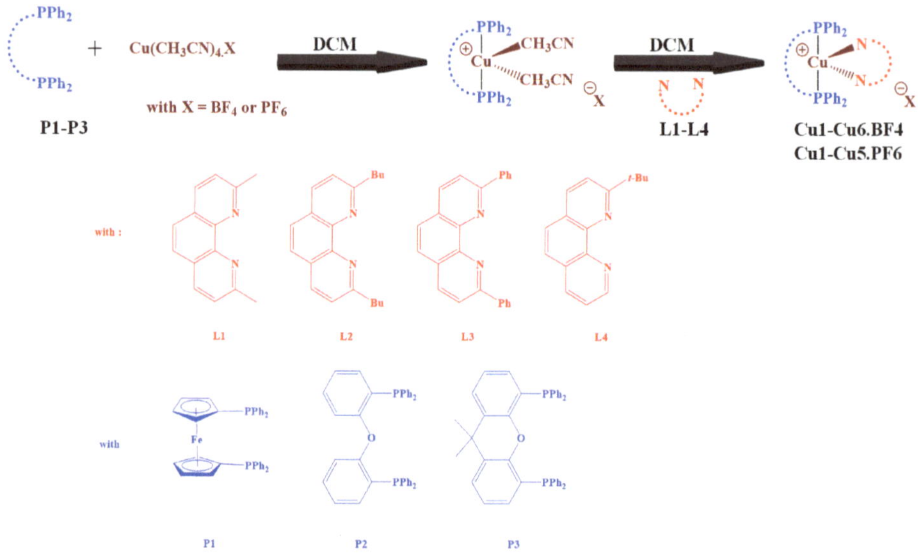

Scheme 2. Synthetic routes to Cu1-Cu6.BF4, Cu1-Cu5.PF6.

Synthesis of Cu1BF4 (Figure 1). A mixture of [Cu(CH$_3$CN)$_4$]BF$_4$ (310 mg, 1 mmol) and Xantphos (578 mg, 1 mmol, M = 578.62 g/mol) was dissolved in dichloromethane (200 mL). The solution was allowed to stir at room temperature for 1h and a solution of 2,9-diphenyl-1,10-phenanthroline (332 mg, 1 mmol, M = 332.41 g/mol) in DCM (20 mL) was added in one portion. The resulting solution was stirred overnight. The solution was concentrated under reduced pressure so that its volume could be reduced to ca. 5 mL. Diethyl ether was added into the resulting solution, affording red crystals of the complex (934 mg, 88% yield). ^{1}H-NMR (400 MHz, CDCl$_3$) δ(ppm): 1.67 (s, 3H), 2.19 (s, 3H), 6.54 (s, 6H), 6.51–6.60 (m, 6H), 6.70–6.96 (m, 10H), 7.02–7.12 (m, 6H), 7.38 (dd, 4H, J = 17.6, 7.2 Hz), 7.59 (d, 2H, J = 7.5 Hz), 8.10–8.30 (m, 4H), 8.55 (d, 4H, J = 7.5 Hz); ^{13}C-NMR (101 MHz, CDCl$_3$) δ(ppm): 26.23, 32.86, 122.28, 122.83, 124.45, 124.83, 125.10, 125.35, 126.39, 126.60, 126.67, 126.74, 126.81, 127.36, 127.61, 127.86, 128.22, 129.31, 130.88, 130.98, 131.09, 135.34, 150.99, 151.92; ^{19}F-NMR (CDCl$_3$) δ: −153.9; ^{31}P-NMR (CDCl$_3$) δ: −14.43; HRMS (ESI MS) m/z: theor: 1060.2567 found: 1060.2565 (M$^+$ detected).

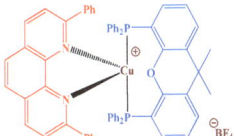

Figure 1. Chemical structure of Cu1BF4.

Synthesis of Cu1PF6 (Figure 2). A mixture of [Cu(CH$_3$CN)$_4$]PF$_6$ (372 mg, 1 mmol, M = 372.72 g/mol) and Xantphos (578 mg, 1 mmol, M = 578.62 g/mol) was dissolved in dichloromethane (200 mL). The solution was allowed to stir at room temperature for 1 h and a solution of 2,9-diphenyl-1,10-phenanthroline (332 mg, 1 mmol, M = 332.41 g/mol) in dichloromethane (50 mL) was added in one portion. The resulting solution was stirred overnight. The solution was concentrated under reduced pressure so that its volume could be reduced to ca. 5 mL. Diethyl ether was added into the resulting solution, affording red crystals of the complex (929 mg, 83% yield). ^{1}H-NMR (400 MHz, CDCl$_3$) δ(ppm): 1.67 (s, 3H), 2.19 (s, 3H), 6.54 (s, 6H), 6.51–6.60 (m, 6H), 6.70–6.96 (m, 10H), 7.02–7.12 (m, 6H), 7.38 (dd, 4H, J = 17.6, 7.2 Hz), 7.59 (d, 2H, J = 7.5 Hz), 8.10–8.30 (m, 4H), 8.55 (d, J = 7.5 Hz, 4H); ^{13}C-NMR (101 MHz, CDCl$_3$) δ(ppm): 26.3, 35.2, 124.6, 126.7, 127.2, 127.4, 128.1, 129.0, 129.8, 130.4, 131.6, 133.4, 137.6, 153.3; ^{19}F-NMR (CDCl$_3$) δ: −72.3, −74.8; ^{31}P-NMR (CDCl$_3$) δ: −15.3, −144.2 (qt, J = 287 Hz); HRMS (ESI MS) m/z: theor: 1060.2567 found: 1060.2562 (M$^+$·detected).

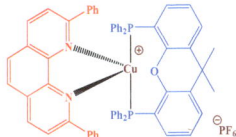

Figure 2. Chemical structure of Cu1PF6.

Synthesis of Cu2BF4 (Figure 3). A mixture of [Cu(CH$_3$CN)$_4$]BF$_4$ (310 mg, 1 mmol) and bis[2-(diphenylphosphino)phenyl]ether (540 mg, 1 mmol) in 200 mL of dichloromethane was stirred at 25 °C for 2 h and then treated with a solution of 2,9-diphenyl-1,10-phenanthroline (332 mg, 1 mmol, M = 332.41 g/mol) in dichloromethane (50 mL). The resulting solution was stirred overnight. The solution was concentrated under reduced pressure so that its volume could be reduced to ca. 5 mL. Diethyl ether was added into the resulting solution, affording red crystals of the complex (908 mg, 89% yield). ^{1}H-NMR (400 MHz, CDCl$_3$) δ(ppm): 6.47–6.65 (m, 6H), 6.78–6.80 (m, 10H), 6.95–7.10 (m, 10H), 7.12–7.18 (m, 2H), 7.20–7.50 (m, 12H), 8.13 (s, 2H), 8.58 (d, 2H, J = 8.1 Hz); ^{13}C-NMR (101 MHz, CDCl$_3$) δ(ppm): 119.2, 119.7, 124.7, 126.6, 127.1, 127.5, 128.0, 128.7, 128.8, 129.0, 129.9, 130.3, 130.7, 132.0, 133.5, 134.2, 137.5, 138.7, 156.5, 158.1; ^{19}F-NMR (CDCl$_3$) δ: −154.0; ^{31}-NMR (CDCl$_3$) δ: −13.7; HRMS (ESI MS) m/z: theor: 933.2219 found: 933.2221 (M$^+$·detected).

Figure 3. Chemical structure of Cu2BF4.

Synthesis of Cu2PF6 (Figure 4). A mixture of [Cu(CH$_3$CN)$_4$]PF$_6$ (372 mg, 1 mmol, M = 372.72 g/mol) and bis[2-(diphenylphosphino)phenyl]ether (540 mg, 1 mmol) in dichloromethane (200 mL) was stirred at 25 °C for 2 h and then treated with a solution of

2,9-diphenyl-1,10-phenanthroline (332 mg, 1 mmol, M = 332.41 g/mol) in dichloromethane (50 mL). The resulting solution was stirred overnight. The solution was concentrated under reduced pressure so that its volume could be reduced to ca. 5 mL. Diethyl ether was added into the resulting solution, affording red crystals of the complex (1 g, 93% yield). ^{1}H-NMR (400 MHz, CDCl$_3$) δ(ppm): 6.47–6.65 (m, 6H), 6.78–6.80 (m, 10H), 6.95–7.10 (m, 10H), 7.12–7.18 (m, 2H), 7.20–7.50 (m, 12H), 8.13 (s, 2H), 8.58 (d, 2H, J = 8.1 Hz); ^{13}C-NMR (101 MHz, CDCl$_3$) δ(ppm): 119.2, 119.7, 124.7, 126.6, 127.1, 127.5, 128.0, 128.7, 128.8, 129.0, 129.9, 130.3, 130.7, 132.0, 133.5, 134.2, 137.5, 138.7, 156.5, 158.1; ^{19}F-NMR (CDCl$_3$) δ: −72.3, −74.9; ^{31}P-NMR (CDCl$_3$) δ: −13.7, −144.2 (qt, J = 287 Hz); HRMS (ESI MS) m/z: theor: 933.2219 found: 933.2220 (M$^{+ \cdot}$ detected).

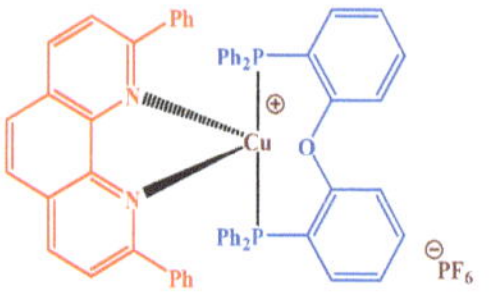

Figure 4. Chemical structure of Cu2PF6.

The investigated copper complexes Cu3BF4, Cu3PF6, Cu4BF4, Cu4PF6, Cu5BF4, Cu5PF6, and Cu6, presented in Scheme 3 were synthesized according to the procedure depicted below.

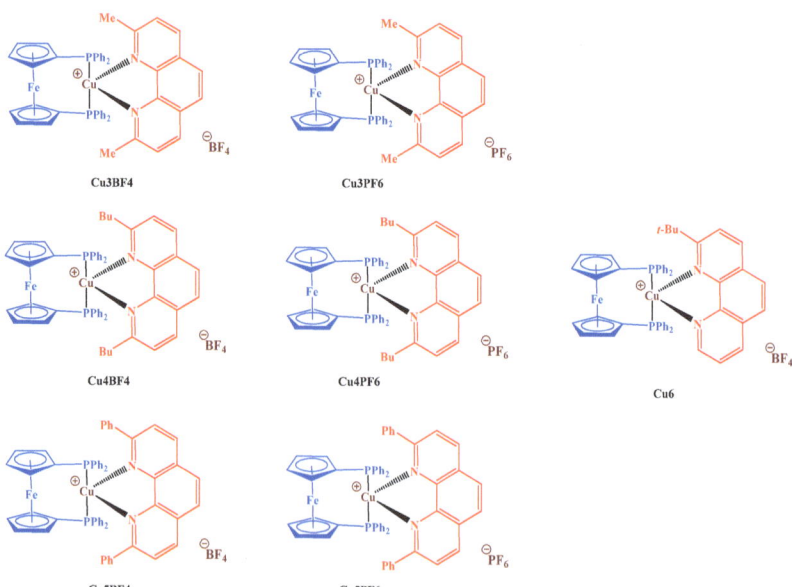

Scheme 3. Investigated copper complexes with ferrocene derivative ligand: Cu3BF4, Cu3PF6, Cu4BF4, Cu4PF6, Cu5BF4, Cu5PF6 and Cu6BF4.

Synthesis of Cu3BF4 (Figure 5). A mixture of tetrakis(acetonitrile)copper(I) tetrafluoroborate (314 mg, 1.0 mmol, M = 314.56 g/mol) and 1,1'-bis(diphenylphosphino)ferrocene (555 mg, 1.0 mmol, M = 554.39 g/mol) in dichloromethane (200 mL) was stirred at 25 °C for 2 h and then treated with a solution of neocuproine (208 mg, 1.0 mmol, M = 208.26 g/mol) in dichloromethane (50 mL). The resulting solution was stirred overnight. The solution was concentrated under reduced pressure so that its volume could be reduced to ca. 5 mL. Diethyl ether was added into the resulting solution, affording red crystals of the complex (895 mg, 98% yield). ^{1}H-NMR (400 MHz, CDCl$_3$) δ(ppm): 2.46 (s, 6H), 4.77 (s, 4H),

4.88 (s, 4H), 7.18–28 (m, 8H), 7.35–7.47 (m, 12H), 7.69 (d, 2H, J = 7.1 Hz), 8.19 (s, 2H), 8.62 (d, 2H, J = 7.3 Hz); ^{13}C-NMR (101 MHz, CDCl$_3$) δ(ppm): 27.6, 72.8 (t, J = 2.4 Hz), 74.5 (t, J = 19.5 Hz), 74.8 (t, J = 5.4 Hz), 125.7, 126.6, 128.2, 128.7 (t, J = 4.5 Hz), 130.1, 132.2 (t, J = 7.5 Hz), 133.8 (t, J = 14.9 Hz), 138.4, 142.9, 159.4; HRMS (ESI MS) m/z: theor: 825.1307 found: 825.1309 (M$^+$·detected). Analyses are consistent with those reported in the literature [15,17].

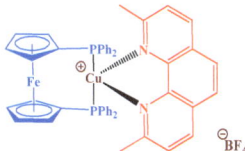

Figure 5. Chemical structure of Cu3BF4.

Synthesis of Cu3PF6 (Figure 6). A mixture of tetrakis(acetonitrile)copper(I) hexafluorophosphate (372 mg, 1.0 mmol, M = 372.72 g/mol) and 1,1'-*bis*(diphenylphosphino)ferrocene (555 mg, 1.0 mmol, M = 554.39 g/mol) in 200 mL of dichloromethane was stirred at 25 °C for 2 h and then treated with a solution of neocuproine (208 mg, 1.0 mmol, M = 208.26 g/mol) in 50 mL of dichloromethane. The resulting solution was stirred overnight. The solution was concentrated under reduced pressure so that its volume could be reduced to ca. 5 mL. Diethyl ether was added into the resulting solution, affording red crystals of the complex (932 mg, 96% yield). ^{1}H-NMR (400 MHz, CDCl$_3$) δ(ppm): 2.45 (s, 6H), 4.77 (s, 4H), 4.88 (s, 4H), 7.18–28 (m, 8H), 7.35–7.47 (m, 12H), 7.69 (d, 2H, J = 8.2 Hz), 8.22 (s, 2H), 8.65 (d, 2H, J = 8.2 Hz); ^{13}C-NMR (101 MHz, CDCl$_3$) δ(ppm): 27.6, 72.8 (t, J = 2.4 Hz), 74.5 (t, J = 19.5 Hz), 74.8 (t, J = 5.4 Hz), 125.7, 126.6, 128.2, 128.7 (t, J = 4.5 Hz), 130.1, 132.2 (t, J = 7.5 Hz), 133.8 (t, J = 14.9 Hz), 138.4, 142.9, 159.4; HRMS (ESI MS) m/z: theor: 825.1307 found: 825.1310 (M$^+$·detected).

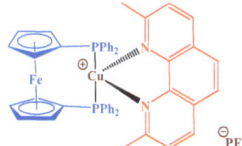

Figure 6. Chemical structure of Cu3PF6.

Synthesis of Cu4BF4 (Figure 7). A mixture of tetrakis(acetonitrile)copper(I) tetrafluoroborate (314 mg, 1.0 mmol, M = 314.56 g/mol) and 1,1'-*bis*(diphenylphosphino)ferrocene (555 mg, 1.0 mmol, M = 554.39 g/mol) in dichloromethane (200 mL) was stirred at 25 °C for 2 h and then treated with a solution of 2,9-dibutyl-1,10-phenanthroline (292 mg, 1.0 mmol, M = 292.42 g/mol) in dichloromethane (50 mL). The resulting solution was stirred overnight. The solution was concentrated under reduced pressure so that its volume could be reduced to ca. 5 mL. Diethyl ether was added into the resulting solution, affording red crystals of the complex (907 mg, 91% yield). ^{1}H-NMR (400 MHz, CDCl$_3$) δ(ppm): 0.71 (t, 6H, J = 7.1 Hz), 0.89–0.99 (m, 4H), 1.01–1.11 (m, 4H), 2.60 (t, 4H, J = 7.8 Hz), 4.72 (s, 4H), 4.81 (s, 4H), 7.10–7.15 (m, 16H), 7.28–7.31 (m, 4H), 7.59 (d, 2H, J = 8.4 Hz), 8.14 (s, 2H), 8.60 (d, 2H, J = 8.4 Hz); ^{13}C-NMR (101 MHz, CDCl$_3$) δ(ppm): 13.9, 22.7, 30.0, 40.9, 72.9 (t, J = 2.5 Hz), 74.0 (t, J = 19.4 Hz), 75.0 (t, J = 5.4 Hz), 123.4, 126.8, 128.7 (t, J = 4.4 Hz), 130.0, 132.1 (t, J = 7.3 Hz), 133.9 (t, J = 14.6 Hz), 138.7, 142.8, 162.0; HRMS (ESI MS) m/z: theor: 909.2246 found: 909.2243 (M$^+$ detected).

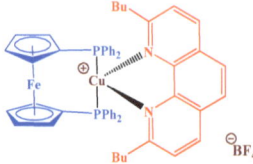

Figure 7. Chemical structure of Cu4BF4.

Synthesis of Cu4PF6 (Figure 8). A mixture of tetrakis(acetonitrile)copper(I) hexafluorophosphate (372 mg, 1.0 mmol, M = 372.72 g/mol) and 1,1′-bis(diphenylphosphino)ferrocene (555 mg, 1.0 mmol, M = 554.39 g/mol) in dichloromethane (200 mL) was stirred at 25 °C for 2 h and then treated with a solution of 2,9-dibutyl-1,10-phenanthroline (292 mg, 1.0 mmol, M = 292.42 g/mol) in dichloromethane (50 mL). The resulting solution was stirred overnight. The solution was concentrated under reduced pressure so that its volume could be reduced to ca. 5 mL. Diethyl ether was added into the resulting solution, affording red crystals of the complex (929 mg, 88% yield). ^{1}H-NMR (400 MHz, CDCl$_3$) δ(ppm): 0.70 (t, 6H, J = 7.1 Hz), 0.89–0.99 (m, 4H), 1.01–1.11 (m, 4H), 2.58 (t, 4H, J = 7.8 Hz), 4.70 (s, 4H), 4.82 (s, 4H), 7.10–7.15 (m, 16H), 7.28–7.31 (m, 4H), 7.58 (d, 2H, J = 8.4 Hz), 8.08 (s, 2H), 8.54 (d, 2H, J = 8.4 Hz); ^{13}C-NMR (101 MHz, CDCl$_3$) δ(ppm): 13.9, 22.7, 30.0, 40.9, 72.9 (t, J = 2.5 Hz), 74.0 (t, J = 19.4 Hz), 75.0 (t, J = 5.4 Hz), 123.4, 126.8, 128.7 (t, J = 4.4 Hz), 130.0, 132.1 (t, J = 7.3 Hz), 133.9 (t, J = 14.6 Hz), 138.7, 142.8, 162.0; HRMS (ESI MS) m/z: theor: 909.2246 found: 909.2244 (M$^+$ detected).

Figure 8. Chemical structure of Cu4PF6.

Synthesis of Cu5BF4 (Figure 9). A mixture of [Cu(CH$_3$CN)$_4$]BF$_4$ (310 mg, 1 mmol) and 1,1′-bis(diphenylphosphino)ferrocene (555 mg, 1 mmol, M = 554.39 g/mol) in dichloromethane (200 mL) was stirred at 25 °C for 2 h and then treated with a solution of 2,9-diphenyl-1,10-phenanthroline (332 mg, 1 mmol, M = 332.41 g/mol) in dichloromethane (50 mL). The resulting solution was stirred overnight. The solution was concentrated under reduced pressure so that its volume could be reduced to ca. 5 mL. Diethyl ether was added into the resulting solution, affording red crystals of the complex (964 mg, 93% yield). ^{1}H-NMR (400 MHz, CDCl$_3$) δ(ppm): 4.19 (s, 4H), 4.40 (s, 4H), 6.53 (t, 4H, J = 7.4 Hz), 6.80 (t, 2H, J = 7.2 Hz), 7.23–7.32 (m, 20H), 7.39 (d, 4H, J = 7.3 Hz), 7.87 (d, 2H, J = 9.2 Hz), 8.04 (s, 2H), 8.54 (d, 2H, J = 8.1 Hz); ^{13}C-NMR (101 MHz, CDCl$_3$) δ(ppm): 73.0 (brs), 74.2 (brs), 124.7, 126.6, 127.2, 128.3, 128.8 (brs), 130.9 (brs), 133.4 (brs), 137.6, 138.8, 143.4, 156.6; ^{19}F-NMR (CDCl$_3$) δ: −153.3; ^{31}P-NMR (CDCl$_3$) δ: −8.88; HRMS (ESI MS) m/z: theor: 979.2089 found: 979.2088 (M$^+$ detected).

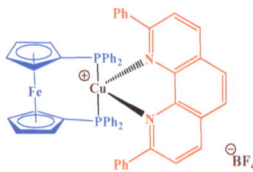

Figure 9. Chemical structure of Cu5BF4.

Synthesis of Cu5PF6 (Figure 10). A mixture of [Cu(CH$_3$CN)$_4$]PF$_6$ (372 mg, 1 mmol, M = 372.72 g/mol) and 1,1'-*bis*(diphenylphosphino)ferrocene (555 mg, 1 mmol, M = 554.39 g/mol) in dichloromethane (200 mL) was stirred at 25 °C for 2 h and then treated with a solution of 2,9-diphenyl-1,10-phenanthroline (332 mg, 1 mmol, M = 332.41 g/mol) in dichloromethane (50 mL). This reaction mixture was stirred for 48 h at room temperature. The resulting solution was stirred overnight. The solution was concentrated under reduced pressure so that its volume could be reduced to ca. 5 mL. Diethyl ether was added into the resulting solution, affording red crystals of the complex. ^{1}H-NMR (400 MHz, CDCl$_3$) δ(ppm): 4.01 (s, 4H), 4.25 (s, 4H), 6.53 (t, 4H, J = 7.4 Hz), 6.80 (t, 2H, J = 7.2 Hz), 7.23–7.32 (m, 20H), 7.39 (d, 4H, J = 7.3 Hz), 7.87 (d, 2H, J = 9.2 Hz), 8.02 (s, 2H), 8.51 (d, 2H, J = 8.1 Hz); ^{13}C-NMR (101 MHz, CDCl$_3$) δ(ppm): 73.0 (brs), 74.1 (brs), 124.7, 126.6, 127.2, 127.6, 128.3, 128.8, 129.0 (brs), 130.8 (brs), 133.3 (brs), 137.5, 138.8, 143.4, 156.7; ^{19}F-NMR (CDCl$_3$) δ: −72.0, −74.5; ^{31}P-NMR (CDCl$_3$) δ: −9.39, −144.1 (qt, J = 287.4 Hz); HRMS (ESI MS) m/z: theor: 979.2089 found: 979.2083 (M$^+$ detected).

Figure 10. Chemical structure of Cu5PF6.

Synthesis of Cu6BF4 (Figure 11). A mixture of [Cu(CH$_3$CN)$_4$]BF$_4$ (310 mg, 1 mmol) and 1,1'-*bis*(diphenylphosphino)ferrocene (555 mg, 1 mmol, M = 554.39 g/mol) in dichloromethane (200 mL) was stirred at 25 °C for 2 h and then treated with a solution of 2-(tert-butyl)-1,10-phenanthroline (236 mg, 1 mmol, M = 236.32 g/mol) in dichloromethane (50 mL). This reaction mixture was stirred for 48 h at room temperature. The solvent was removed under reduced pressure. The residue was dissolved in a minimum of dichloromethane and addition of pentane precipitated the expected complex, that was filtered off, washed several times with pentane and dried under vacuum. ^{1}H-NMR (400 MHz, CDCl$_3$) δ(ppm): 1.22 (s, 9H), 4.34 (s, 4H), 4.55 (s, 4H), 6.90 (brs, 8H), 7.07 (t, 8H, J = 7.5 Hz), 7.24 (t, 4H, J = 7.3 Hz), 7.65 (dd, J = 8.1, 4.8 Hz, 1H), 7.89 (d, 1H, J = 8.6 Hz), 7.94 (d, 1H, J = 8.8 Hz), 8.04 (dd, 1H, J = 8.7, 5.5 Hz), 8.11 (d, 1H, J = 3.7 Hz), 8.45–8.50 (m, 1H), 8.58 (d, 1H, J = 8.6 Hz); ^{13}C-NMR (101 MHz, CDCl$_3$) δ(ppm): 30.61, 38.09, 72.40, 74.54, 123.1, 124.56, 126.73, 127.73, 128.28, 128.68, 129.81, 130.21, 132.67, 138.68 (d, J = 30.8 Hz), 143.57 (d, J = 38.7 Hz), 148.98, 170.51; ^{19}F-NMR (CDCl$_3$) δ: −153.9; ^{31}P-NMR (CDCl$_3$) δ: −15.3; HRMS (ESI MS) m/z: theor: 853.1620 found: 853.1615 (M$^+$ detected).

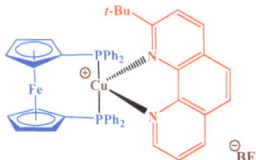

Figure 11. Chemical structure of Cu6BF4.

2.1.2. Other Chemical Compounds

Bis-(4-tert-butylphenyl)iodonium hexafluorophosphate (Iod; SpeedCure 938), ethyl 4-(dimethylamino)benzoate (EDB; SpeedCure EDB) were obtained from Lambson Ltd. (UK). Trimethylolpropane triacrylate (TMPTA) and (3,4-epoxycyclohexane)methyl-3,4-epoxycyclohexylcarboxylate (EPOX; Uvacure 1500) were obtained from Allnex and used as benchmark monomers for radical and cationic photopolymerization, respectively (Scheme 4). Dichloromethane (DCM, purity ≥ 99%) and acetonitrile were used as solvents.

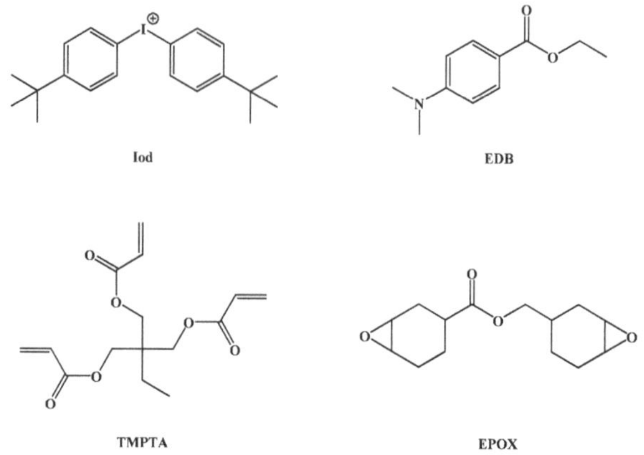

Scheme 4. Chemical structures of additives and monomer.

2.2. UV-Visible Absorption Spectroscopy

UV-visible absorption spectra were acquired in either DCM or acetonitrile in a quartz cell at room temperature using a Jasco V-750 spectrophotometer. The molar extinction coefficients were determined using the Beer–Lambert law with experimental data obtained on solutions of known concentrations.

2.3. Steady-State Fluorescence

Fluorescence spectra were acquired in a quartz cell at room temperature using a JASCO® FP-750 spectrofluorometer. Excitation and emission spectra were recorded in DCM in a quartz cell.

2.4. Photopolymerization Kinetics (FTIR)

Experimental conditions for each photosensitive formulation are given in the captions of the figures. The weight percent of the photoinitiating system is calculated from the monomer content. Photoinitiator concentrations in each photosensitive formulation were chosen to ensure the same light absorption at 405 nm. For the investigated concentrations (see Figure captions), no solubility issues were noticed.

All polymerizations were performed at ambient temperature (21–25 °C) and irradiation was started at t = 10 s. Two LEDs, with an intensity around 50 mW·cm^{-2} at the sample position, were used for the photopolymerization experiments: a LED@405 nm (M405L3—Thorlabs, Newton, NJ, USA) centered at 405 nm and a LED@455 nm (M455L3—Thorlabs) centered at 452 nm. A LED, with an intensity around 11 mW·cm^{-2} at the sample position, was used for the photopolymerization experiments: a LED@530 nm (M530L3—Thorlabs) centered at 530 nm. The emission spectrum is already available in the literature [23].

A Jasco 4100 real-time Fourier transform infrared spectrometer (RT-FTIR) was used to follow the conversion of the acrylate functions of the TMPTA and of the epoxide group of EPOX. The photocurable formulations were deposited on a polypropylene film inside a 1.4 mm-thick mold under air. Evolutions of the C=C double bond band and the epoxide group band were continuously followed from 6117 to 6221 cm^{-1} and from 3710 to 3799 cm^{-1}, respectively.

3. Results and Discussion

3.1. Copper Complexes Structures Inspired by Copper Complex G1

3.1.1. Light Absorption Properties of the Studied Photoinitiators

The ground state absorption spectra of Cu1BF4, Cu1PF6, Cu2BF4, Cu2PF6 and G1 in dichloromethane are presented in Figure 12. These compounds are characterized by a broad and strong absorption band in the near UV spectral region (350–400 nm) which extends up to 450 nm for G1 and up to 650 nm for the others. Absorption maxima (λ_{max}) and the molar extinction coefficients (ε) for λ_{max} and at the nominal wavelength of the used LEDs (405 nm, 455 nm and 530 nm) are gathered in Table 1. For the five photoinitiators, the absorption maxima ensure a good overlap with the emission spectrum of the violet LED (centered at 405 nm) used in this work. The new copper complexes Cu1BF4, Cu1PF6, Cu2BF4 and Cu2PF6 are characterized by rather similar absorption properties to that determined for the copper complex G1 which is used as a reference compound ($\lambda_{405nm} = 1.0 \times 10^3$, 1.1×10^3, 1.1×10^3, 1.2×10^3 and 1.9×10^3 L mol^{-1} cm^{-1}, respectively). Interestingly, an increase in the conjugation length of the phenanthroline ligand could red-shift the absorptions of the four new copper complexes which seem compatible for use at higher wavelengths ($\lambda_{455nm} = 1.2 \times 10^3$, 1.3×10^3, 1.2×10^3 and 1.3×10^3 L mol^{-1} cm^{-1}, respectively; $\lambda_{530nm} = 6.9 \times 10^2$, 7.4×10^2, 6.6×10^2 and 7.3×10^2 L mol^{-1} cm^{-1}, respectively).

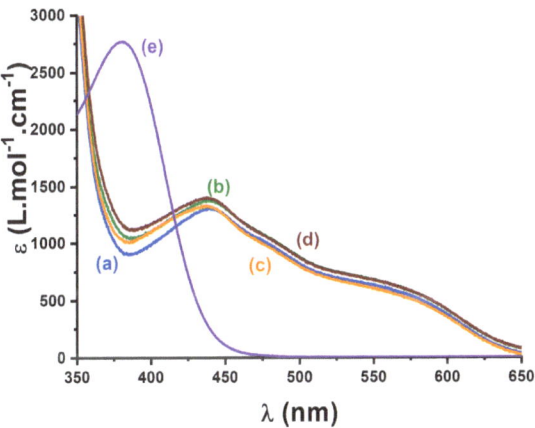

Figure 12. UV-visible absorption spectra of (a) Cu1BF4, (b) Cu1PF6, (c) Cu2BF4, (d) Cu2PF6 and (e) G1 in dichloromethane.

Table 1. Maximum absorption wavelengths λ_{max}, extinction coefficients at λ_{max} and at the nominal emission wavelength of the LEDs (405nm, 455nm and 530nm) for Cu1BF4, Cu1PF6, Cu2BF4, Cu2PF6 and G1.

Compound	λ_{max} (nm)	$\varepsilon_{\lambda max}$ (L mol^{-1} cm^{-1})	ε_{405nm} (L mol^{-1} cm^{-1})	ε_{455nm} (L mol^{-1} cm^{-1})	ε_{530nm} (L mol^{-1} cm^{-1})
Cu1BF4	438	1.3×10^3	1.0×10^3	1.2×10^3	6.9×10^2
Cu1PF6	438	1.4×10^3	1.1×10^3	1.3×10^3	7.4×10^2
Cu2BF4	438	1.3×10^3	1.1×10^3	1.2×10^3	6.6×10^2
Cu2PF6	438	1.4×10^3	1.2×10^3	1.3×10^3	7.3×10^2
G1	380	2.8×10^3	1.9×10^3	7.4×10^1	7.0×10^0

3.1.2. Luminescence Experiments and Reaction Pathway

To assess the properties of the excited states of the investigated copper complexes, steady state fluorescence analyses were carried out. Unfortunately, as presented in Figure 13 with the emission spectra of Cu1PF6 and Cu2PF6 in dichloromethane, a degradation of the copper complexes was observed after several minutes of analysis under irradiation. This possible photolysis of the copper complexes induces a permanent modification of the excited states and, thus, the photoluminescence properties. These facts are in line with the results reported by Korn et al. [24], which demonstrated the presence of an equilibrium in solution between the heteroleptic and the homoleptic copper complexes through ligand exchange.

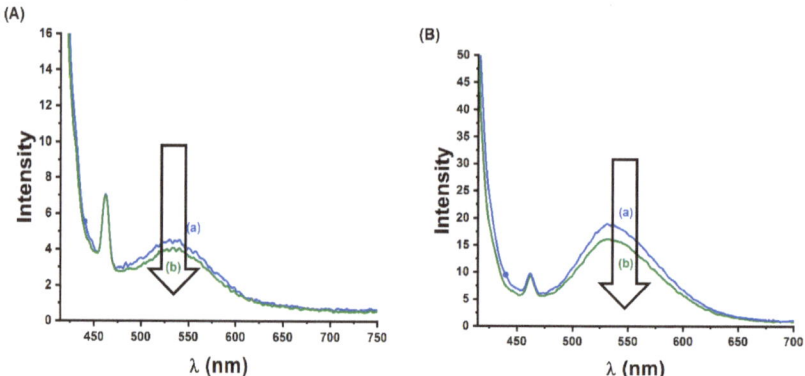

Figure 13. Photoluminescence spectra in dichloromethane under air (curve a) before photolysis, (curve b) after photolysis λ_{exc} (**A**) [Cu1PF6] = 4.8 × 10^{-5} mol L^{-1} (**B**) [Cu2PF6] = 5.3 × 10^{-5} mol L^{-1}.

Due to the similarity of structures of the four new copper complexes with that of the benchmark photoinitiator G1, a similar photoredox catalytic cycle can be considered. As depicted in Scheme 5, this catalytic cycle is based on three species: the photoinitiator G1, the iodonium salt Ar$_2$I$^+$ and the amine EDB. Upon irradiation, the copper complex G1 is excited and reacts with the iodonium salt to generate, via electron transfer reaction, radicals and Cu (II) complex noted G1$^+$. A reaction of the generated radicals with EDB induces the formation of EDB$^·_{(-H)}$ which after reaction with G1$^+$ leads simultaneously to the regeneration of copper complex G1 and the generation of cation EDB$^+_{(-H)}$ capable of initiating cationic polymerization.

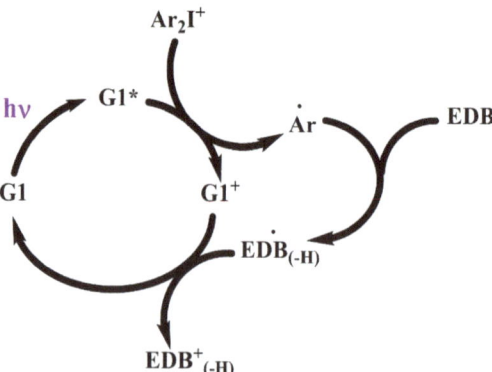

Scheme 5. Photoredox catalytic cycle for the three-component system G1/Iod/EDB—adapted from [3,25].

3.1.3. Experimental Approach for the Concomitant Initiation of the Free Radical and Cationic Polymerization

In this polymerization part and for the sake of direct comparison, the photoinitiator concentrations in each photosensitive formulation were chosen to ensure the same light absorption at 405 nm.

Free Radical Polymerization

The free radical polymerization of TMPTA in the presence of Cu1BF4/Iod/EDB, Cu1PF6/Iod/EDB, Cu2BF4/Iod/EDB or Cu2PF6/Iod/EDB was performed under air using a LED emitting at 405 nm. To fully characterize the performance of the new copper complexes, two photoinitiating systems were used as standards: (i) the three-component system G1/Iod/EDB was used as a reference system to illustrate the performance of one of the best copper complexes ever reported to date as a photoinitiator of polymerization; and (ii) the two-component system Iod/EDB was also used as a standard since the N-aromatic amine EDB can form a charge transfer complex with the iodonium salt (Iod) and produce EDB$^{\cdot+}$ and the radical Ar$^{\cdot}$ capable of initiating polymerizations [26,27]. The concentration of the copper complex used as photoinitiator is chosen to ensure the same light absorption at 405 nm. Photopolymerization profiles of TMTPA are presented in Figure 14. Among the tested systems, Cu1BF4/Iod/EDB, Cu1PF6/Iod/EDB, Cu2BF4/Iod/EDB, Cu2PF6/Iod/EDB and G1/Iod/EDB exhibited similar polymerization rates and final C=C double bond conversions which are higher than those obtained with the reference Iod/EDB system. Indeed, the reaction with Cu1BF4, Cu1PF6, Cu2BF4, Cu1PF6 or G1 is really fast, and it ended in less than 50 s with final conversion around 85%, while the charge transfer complex leads to a final conversion around 65% after 100 s of reaction. Moreover, the counter ion, either BF_4^- or PF_6^-, seems not to have an impact on the polymerization processes. Despite the inhibition of radicals by the oxygen in the medium, the investigated photoinitiating systems were all able to lead to fully polymerized samples. Therefore, the four new copper complexes are equivalent to G1 and highly efficient at initiating free radical polymerizations. For the selected conditions, the developed systems are able to overcome the oxygen inhibition.

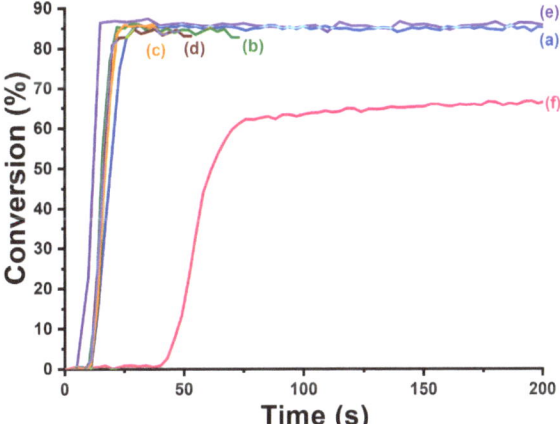

Figure 14. Polymerization profiles (acrylate function conversion vs. irradiation time) of TMPTA upon irradiation with a LED at 405 nm, under air, sample thickness = 1.4 mm; the irradiation starts at 10 s, 50 mW cm^{-2}. Photoinitiating systems: (curve a) Cu1BF4/Iod/EDB (0.72/2.0/2.0 $w/w/w$%), (curve b) Cu1PF6/Iod/EDB (0.69/2.0/2.0 $w/w/w$%), (curve c) Cu2BF4/Iod/EDB (0.62/2.0/2.0 $w/w/w$%), (curve d) Cu2PF6/Iod/EDB (0.63/2.0/2.0 $w/w/w$%), (curve e) G1/Iod/EDB (0.33/2.0/2.0 $w/w/w$%) and (curve f) Iod/EDB (2.0/2.0 w/w%).

Cationic Polymerization

The cationic polymerization of EPOX in the presence of Cu1BF4/Iod/EDB, Cu1PF6/Iod/EDB, Cu2BF4/Iod/EDB or Cu2PF6/Iod/EDB was performed under air using a LED centered at 405 nm. Similar to that carried out for the free radical polymerization experiments, two photoinitiating systems were used as standards: the three-component G1/Iod/EDB system and the two-component Iod/EDB system. Photopolymerization profiles of EPOX are presented in Figure 15. Among the tested systems, Cu1BF4/Iod/EDB, Cu1PF6/Iod/EDB, Cu2BF4/Iod/EDB and Cu2PF6/Iod/EDB are slightly less efficient than the three-component G1/Iod/EDB system but better than the standard two-component Iod/EDB system. Indeed, their polymerization rates and final epoxy group conversions are lower than those obtained for the efficient reference G1/Iod/EDB system but higher than those for the Iod/EDB system. However, the four new copper complexes are still capable of initiating the cationic polymerization.

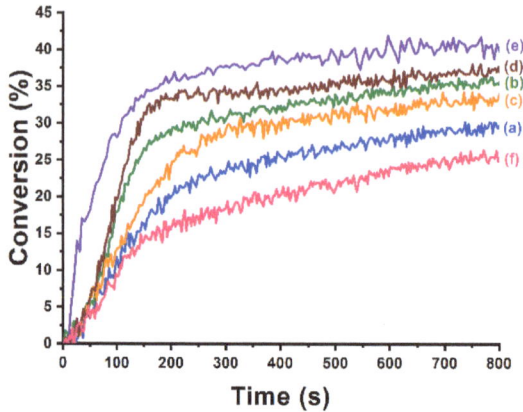

Figure 15. Polymerization profiles (epoxide function conversion vs. irradiation time) of EPOX upon irradiation with a LED at 405 nm, under air, sample thickness = 1.4 mm; the irradiation starts at 10 s, 50 mW cm^{-2}. Photoinitiating systems: (curve a) Cu1BF4/Iod/EDB (0.72/2.0/2.0 w/w/w%), (curve b) Cu1PF6/Iod/EDB (0.68/2.0/2.0 w/w/w%), (curve c) Cu2BF4/Iod/EDB (0.61/2.0/2.0 w/w/w%), (curve d) Cu2PF6/Iod/EDB (0.63/2.0/2.0 w/w/w%), (curve e) G1/Iod/EDB (0.33/2.0/2.0 w/w/w%) and (curve f) Iod/EDB (2.0/2.0 w/w%).

Interpenetrated Polymer Networks Synthesis

Interpenetrated polymer networks (IPNs) can be synthesized by initiating concomitantly a free radical and a cationic polymerization in order to obtain two chemically different interlaced polymer networks which are not covalently bonded. This reaction could be carried out in one pot by using a photoinitiating system capable of initiating both polymerization in a blend of TMPTA and EPOX (50/50 w/w%), e.g., for a G1-based photoinitiating system [9].

The polymerization of a TMPTA/EPOX blend in the presence of Cu1BF4/Iod/EDB, Cu1PF6/Iod/EDB, Cu2BF4/Iod/EDB or Cu2PF6/Iod/EDB was performed under air using a LED emitting at 405 nm. Again, two photoinitiating systems were used as standards: G1/Iod/EDB and Iod/EDB. Photopolymerization profiles of the IPN synthesis are presented in Figure 16. Among the tested systems, Cu1BF4/Iod/EDB, Cu1PF6/Iod/EDB, Cu2BF4/Iod/EDB, Cu2PF6/Iod/EDB and G1/Iod/EDB exhibited similar polymerization rates and final C=C double bond conversions which are higher than those obtained with the reference Iod/EDB system. Indeed, the formation of the acrylic network with Cu1BF4, Cu1PF6, Cu2BF4, Cu2PF6 or G1 is fast, and it ended in around 50 s with a final conversion around 86%, while the Iod/EDB charge transfer complex leads to a final conversion around 70% after 400 s of irradiation. As for the formation of the epoxy network, each tested three-component photoinitiating system led to a higher final epoxy group conversion than the efficient reference system G1/Iod/EDB and the control system Iod/EDB. During the synthesis of IPNs, an increase in the final conversion of both the C=C double bond and the epoxy group were observed. This improvement could be related to the synergy between the free radical polymerization and the cationic polymerization during the two networks' formation. Indeed, under irradiation, the radical polymerization is at first inhibited by the oxygen in the medium, while the cationic polymerization starts immediately, which induces an increase in the medium viscosity limiting the diffusional oxygen replenishment. The cationic monomer also acts as a diluting agent for the radical polymer network allowing for the achievement of a higher conversion. Indeed, since the formation of the cationic polymer network is generally slower, the cationic monomer delays the gelation phenomenon occurring with the formation of the radical polymer network. The exothermic property of the radical polymerization also tends to boost the cationic polymerization, which is quite temperature sensitive. Thus, the investigated copper complexes Cu1BF4, Cu1PF6, Cu2BF4 and Cu2PF6 are capable of initiating simultaneously the cationic and free radical polymerizations leading to the formation of acrylate/epoxy IPN under visible light. Moreover, these photoinitiating systems led to better results than the corresponding system using the highly efficient photoinitiator G1 (higher final conversion).

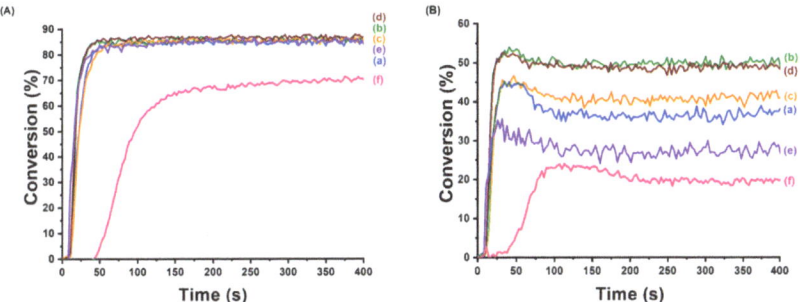

Figure 16. Polymerization profiles (**A**) (acrylate function conversion vs. irradiation time) and (**B**) (epoxide function conversion vs. irradiation time) of TMPTA/EPOX blend (50/50 w/w%) upon irradiation with a LED at 405 nm, under air, sample thickness = 1.4 mm; the irradiation starts at 10 s, 50 mW cm^{-2}. Photoinitiating systems: (curve a) Cu1BF4/Iod/EDB (0.73/2.0/2.0 $w/w/w$%), (curve b) Cu1PF6/Iod/EDB (0.70/2.0/2.0 $w/w/w$%), (curve c) Cu2BF4/Iod/EDB (0.64/2.0/2.0 $w/w/w$%), (curve d) Cu2PF6/Iod/EDB (0.64/2.0/2.0 $w/w/w$%), (curve e) G1/Iod/EDB (0.33/2.0/2.0 $w/w/w$%) and (curve f) Iod/EDB (2.0/2.0 w/w%).

Toward Longer Wavelengths

According to the UV-visible absorption spectra of Cu1BF4, Cu1PF6, Cu2BF4 and Cu2PF6, the formation of acrylate/epoxy IPN using an irradiation wavelength higher than 405 nm seems possible. The polymerization of a TMPTA/EPOX blend (50/50 w/w%) in the presence of Cu1BF4/Iod/EDB, Cu1PF6/Iod/EDB, Cu2BF4/Iod/EDB or Cu2PF6/Iod/EDB, was performed under air using a LED centered at 455 nm or a LED centered at 530 nm. Iod/EDB was used as a standard to ensure that the EDB/Iod charge transfer complex could not initiate the polymerization under this light. Photopolymerization profiles of the IPN synthesis at 455 nm are presented in Figure 17.

Among the tested systems, the four tested three-component photoinitiating systems allowed the formation of acrylate/epoxy IPN at 455 nm. However, performances of the photoinitiating systems are slightly lower under an irradiation at 455 nm than at 405 nm. Indeed, the final C=C double bond conversion is around 84%, the final epoxy group conversion between 35% and 40%. The formation of the acrylate network and the epoxy network with Cu1BF4, Cu1PF6, Cu2BF4 or Cu2PF6 is also slower around 100 s. The lower performance of the photoinitiating system could be explained by the decrease in energy of the irradiation source (LED@455 nm). Moreover, the system EDB/Iod does not initiate any polymerization due to a lack of absorption for the corresponding charge transfer complex at this irradiation wavelength. Therefore, the investigated copper complexes could be used if needed for the polymerization with a LED centered at 455 nm.

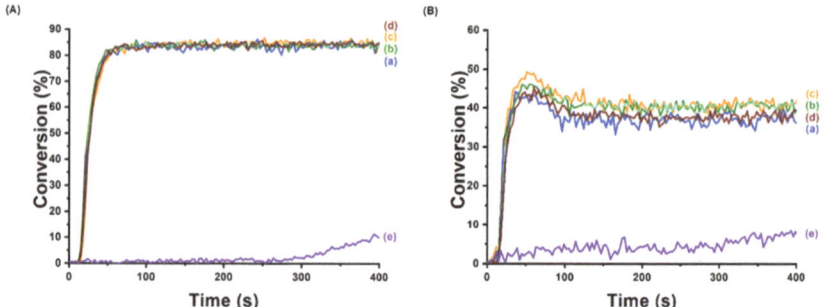

Figure 17. Polymerization profiles (**A**) (acrylate function conversion vs. irradiation time) and (**B**) (epoxide function conversion vs. irradiation time) of TMPTA/EPOX blend (50/50 w/w%) upon irradiation with a LED at 455 nm, under air, sample thickness = 1.4 mm; the irradiation starts at 10 s, 50 mW cm^{-2}. Photoinitiating systems: (curve a) Cu1BF4/Iod/EDB (0.70/2.0/2.0 $w/w/w$%), (curve b) Cu1PF6/Iod/EDB (0.70/2.0/2.0 $w/w/w$%), (curve c) Cu2BF4/Iod/EDB (0.63/2.0/2.0 $w/w/w$%), (curve d) Cu2PF6/Iod/EDB (0.64/2.0/2.0 $w/w/w$%) and (curve e) Iod/EDB (2.0/2.0 w/w%).

Photopolymerization profiles of the IPN synthesized upon irradiation at 530 nm are presented in Figure 18. Among all systems, the four tested three-component photoinitiating systems allowed the formation of acrylate/epoxy IPN at 530 nm. However, performances of the photoinitiating systems are slightly lower under an irradiation at 530 nm than at 455 or 405 nm. Indeed, the final C=C double bond conversion is around 80%, the final epoxy group conversion between 22% and 30%. The formation of the acrylate network and the epoxy network with Cu1BF4, Cu1PF6, Cu2BF4 or Cu2PF6 is also slower around 150 s. The lower performance of the photoinitiating systems can have two origins: (i) the decrease in energy of the irradiation source (LED at 530 nm); and (ii) the lower absorption of Cu1BF4, Cu1PF6, Cu2BF4 or Cu2PF6 at 530 nm compared to 455 or 405 nm. As previously observed at 455 nm, the system EDB/Iod does not initiate any polymerization under irradiation at 530 nm, once again due to the lack of absorption at this wavelength.

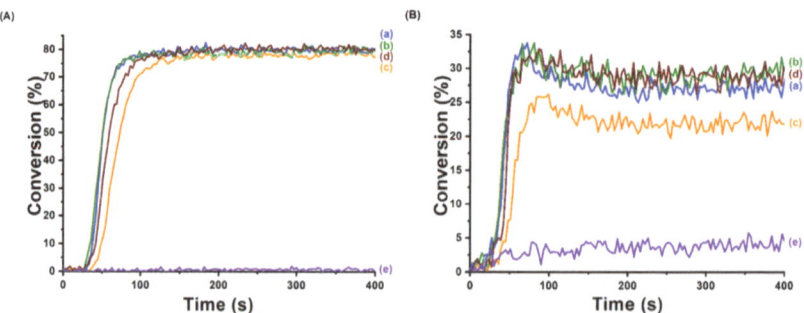

Figure 18. Polymerization profiles (**A**) (acrylate function conversion vs. irradiation time) and (**B**) (epoxide function conversion vs. irradiation time) of TMPTA/EPOX blend (50/50 w/w%) upon irradiation with a LED at 530 nm, under air, sample thickness = 1.4 mm; the irradiation starts at 10 s, 11 mW cm^{-2}. Photoinitiating systems: (curve a) Cu1BF4/Iod/EDB (0.70/2.0/2.0 $w/w/w$%), (curve b) Cu1PF6/Iod/EDB (0.70/2.0/2.0 $w/w/w$%), (curve c) Cu2BF4/Iod/EDB (0.63/2.0/2.0 $w/w/w$%), (curve d) Cu2PF6/Iod/EDB (0.64/2.0/2.0 $w/w/w$%) and (curve e) Iod/EDB (2.0/2.0 w/w%).

Therefore, the investigated copper complexes Cu1BF4, Cu1PF6, Cu2BF4 and Cu2PF6 can be used if needed for the polymerization with a LED centered at 455 nm or 530 nm. Depending on the context of the application, these performances under less harmful irradiations can be really interesting. Despite this possibility, the commonly used LED centered at 405 nm was selected for the rest of this work due to the better polymerization profiles.

Effect of the Concentration of Photoinitiators

Photoinitiating systems based on Cu1BF4, Cu1PF6, Cu2BF4 or Cu2PF6 were efficient, with a relatively high concentration of copper complex (to ensure O.D. = 1, for a 1.4 mm thickness sample). However, a reduction in the copper complex concentration while maintaining the performance could be beneficial for considerations such as leaching safety or cost issues. Moreover, since these formulations do not bleach, a reduction in the copper complex concentration could reduce the intensity of the reddish orange shade of the cured polymers.

Photopolymerization of the TMPTA/EPOX blend (50/50 w/w%) was performed under air using a LED centered at 405 nm, in the presence of the three-component photoinitiating systems Cu1BF4/Iod/EDB, Cu1PF6/Iod/EDB, Cu2BF4/Iod/EDB or Cu2PF6/Iod/EDB with different ratios of copper complex. For each photoinitiating system, two concentrations of copper complex were chosen to ensure O.D. = 1 and O.D. = 0.1. Final conversions for the epoxy and acrylate functions of the tested blends are given in Table 2. For each system, decreases in the final C=C double bond conversion and the final epoxy group conversion were observed while reducing the copper complex concentration. The impact of this reduction is particularly high on the cationic polymerization, while impacting less the free radical polymerization. Noticeably, performances could be maintained, even with a 10-fold reduction in the copper complexes content, tending to confirm our hypothesis on their reactivities as photocatalysts in a photoredox cycle, such as G1.

Table 2. Final conversions for the epoxy and acrylate functions (in percentage) obtained under air (Figure 1). A 4 mm thickness sample for the photopolymerization of TMPTA/EPOX blend (50/50 w/w%) for 400 s exposure to a LED at 405 nm in the presence of different photoinitiating systems.

Photoinitiating System: Cu/Iod (2 w%)/EDB (2 w%)	TMPTA	EPOX
Cu1BF4 (0.73 w%)	86%	37%
Cu1BF4 (0.073 w%)	84%	30%
Cu1PF6 (0.70 w%)	86%	51%
Cu1PF6 (0.069 w%)	82%	30%
Cu2BF4 (0.64 w%)	87%	42%
Cu2BF4 (0.063 w%)	80%	28%
Cu2PF6 (0.64 w%)	86%	49%
Cu2PF6 (0.063 w%)	80%	24%
(reference)	70%	19%

3.2. Copper Complexes with a Ferrocene Derivative Ligand

The ground state absorption spectra of the seven copper complexes with a 1,1'-bis(diphenylphosphino)ferrocene ligand are presented in Figure 19: Cu3BF4, Cu3PF6, Cu4BF4 and Cu4PF6 in acetonitrile; Cu5BF4, Cu5PF6, and Cu6BF4 in dichloromethane. These compounds are characterized by a broad and strong absorption band in the near UV spectral region (350–400 nm) which extends: (i) up to 550 nm for Cu3BF4, Cu3PF6, Cu4BF4, Cu4PF6 and Cu6BF4; (ii) up to 650 nm for the others. The absorption maxima (λ_{max}) and the molar extinction coefficients (ε) for λ_{max} and at the nominal wavelength of the LED (405 nm) are gathered in Table 3. For the seven photoinitiators, the absorption maxima ensure a good overlap with the emission spectrum of the violet LED (centered at 405 nm) used in this work. The seven copper complexes are characterized by rather similar absorption properties, in particular around 405 nm ($\varepsilon_{405\,nm}$ range from 1.0×10^3 to 1.8×10^3 L mol^{-1} cm^{-1}). Moreover, for Cu3BF4, Cu3PF6, Cu4BF4, Cu4PF6, Cu5BF4 and Cu5PF6, the lowest energy transition is redshifted, depending on the electron donating ability of the substituent introduced onto the phenanthroline ligand (Me < Bu < Ph).

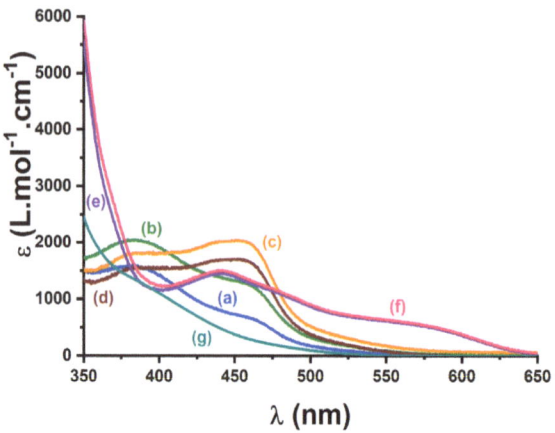

Figure 19. UV-visible absorption spectra of (a) Cu3BF4, (b) Cu3PF6, (c) Cu4BF4, (d) Cu4PF6 in acetonitrile and (e) Cu5BF4, (f) Cu5PF6 (g) Cu6BF4 in dichloromethane.

Table 3. Maximum absorption wavelengths (λ_{max}), extinction coefficients at λ_{max} and at the nominal emission wavelength of the LED (405 nm) for Cu3BF4, Cu3PF6, Cu4BF4 and Cu4PF6 in acetonitrile and for Cu5BF4, Cu5PF6, and Cu6 in dichloromethane.

Compound	λ_{max} (nm)	$\varepsilon_{\lambda max}$ (L mol^{-1} cm^{-1})	ε_{405nm} (L mol^{-1} cm^{-1})
Cu3BF4	380	1.6×10^3	1.3×10^3
Cu3PF6	380	2.1×10^3	1.8×10^3
Cu4BF4	380	1.8×10^3	1.8×10^3
Cu4PF6	380	1.6×10^3	1.5×10^3
Cu5BF4	440	1.6×10^3	1.2×10^3
Cu5PF6	440	1.5×10^3	1.2×10^3
Cu6BF4	380	1.4×10^3	1.0×10^3

To assess the properties of the excited state of the investigated copper complexes, steady state fluorescence analyses were performed. Interestingly, none of the investigated copper complexes were fluorescent. These facts are in line with the finding of Armaroli et al. [13] which reported the non-luminescence of copper complexes containing the 1,1'-*bis*(diphenylphosphino)ferrocene ligand due to a metal ligand charge transfer quenching resulting from a photoinduced intramolecular energy transfer to the ferrocene unit.

The free radical polymerization of TMPTA was performed under air using a LED emitting at 405 nm in the presence of Cu3BF4/Iod/EDB, Cu3PF6/Iod/EDB, Cu4BF4/Iod/EDB, Cu4PF6/Iod/EDB, Cu5BF4/Iod/EDB, Cu5PF6/Iod/EDB or Cu6BF4/Iod/EDB. Again, two photoinitiating systems were used as standards: G1/Iod/EDB and Iod/EDB. Concentration of the different copper complexes was chosen to ensure the same light absorption at 405 nm. Photopolymerization profiles of TMTPA are presented in Figure 20. Among the tested systems, Cu3BF4/Iod/EDB, Cu3PF6/Iod/EDB, Cu4BF4/Iod/EDB, Cu4PF6/Iod/EDB, Cu5BF4/Iod/EDB, Cu5PF6/Iod/EDB and G1/Iod/EDB exhibited similar polymerization rates and final C=C double bond conversions, which are higher than those obtained with the reference Iod/EDB system. Indeed, the reaction with Cu3BF4,

Cu3PF6, Cu4BF4, Cu4PF6, Cu5BF4, Cu5PF6 or G1 was really fast with final conversions around 80%, while the charge transfer complex Iod/EDB could only lead to a final conversion around 65% after 150 s of irradiation. As for the three-component Cu6/Iod/EDB system, the reaction was slower and led to a final conversion around 45% after 400 s. The counter ion, either BF_4^- or PF_6^-, seems not to have an impact on the polymerization. Therefore, Cu3BF4, Cu3PF6, Cu4BF4, Cu4PF6, Cu5BF4 and Cu5PF6 were almost equivalent to G1 and highly efficient to initiate the free radical polymerization. However, none of the three-component systems based on these seven copper complexes were able to initiate a cationic polymerization.

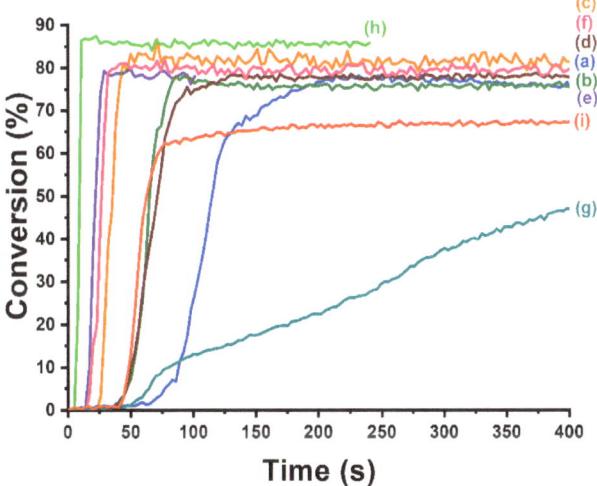

Figure 20. Polymerization profiles (acrylate function conversion vs. irradiation time) of TMPTA upon irradiation with a LED at 405 nm, under air, sample thickness = 1.4 mm; the irradiation starts at 10 s, 50 mW cm^{-2}. Photoinitiating systems: (curve a) Cu3BF4/Iod/EDB (0.50/2.0/2.0 w/w/w%), (curve b) Cu3PF6/Iod/EDB (0.39/2.0/2.0 w/w/w%), (curve c) Cu4BF4/Iod/EDB (0.40/2.0/2.0 w/w/w%), (curve d) Cu4PF6/Iod/EDB (0.50/2.0/2.0 w/w/w%), (curve e) Cu5BF4/Iod/EDB (0.65/2.0/2.0 w/w/w%), (curve f) Cu5PF6/Iod/EDB (0.65/2.0/2.0 w/w/w%), (curve g) Cu6BF4/Iod/EDB (0.68/2.0/2.0 w/w/w%), (curve h) G1/Iod/EDB (0.33/2.0/2.0 w/w/w%) and (curve i) Iod/EDB (2.0/2.0 w/w%).

3.3. Structure/Efficiency Relationship: Role of the Ferrocene Moiety

Using the results obtained above for ferrocene-free compounds vs. ferrocene containing structures, the effect of the iron moiety can be discussed.

3.3.1. Effect on the Panchromatic Behavior

The ground state absorption spectra in dichloromethane of the copper complex comprising 1,1'-*bis*(diphenylphosphino)ferrocene ligand, namely Cu5BF4, and the corresponding ferrocene-free copper complex, Cu1BF4, are presented in Figure 21. A slight increase in the molar extinction coefficient could be observed between 400 and 500 nm for Cu5BF4, which can be confidently assigned to the contribution of the 1,1'-*bis*(diphenylphosphino)ferrocene ligand. However, the molar extinction coefficients of the two complexes were in the similar range: at 405nm $\varepsilon_{Cu1BF4} = 1.0 \times 10^3$ L mol^{-1} cm^{-1} and ε_{Cu5BF4} 1.2×10^3 L mol^{-1} cm^{-1}. Thus, the introduction of the iron moiety in the copper complex does not significantly impact the panchromatic behavior.

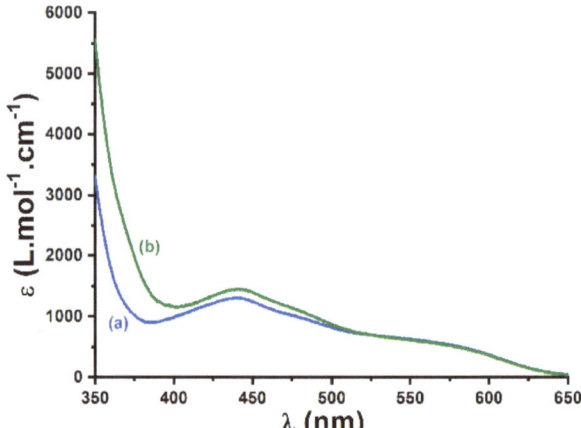

Figure 21. UV-visible absorption spectra of (a) Cu1BF4, (b) Cu5BF4 in dichloromethane.

3.3.2. Effect on the Polymerization Initiating Ability

Radical Polymerization

The free radical polymerization of TMPTA was performed under air using a LED emitting at 405 nm in the presence of Cu1BF4/Iod/EDB or Cu5BF4/Iod/EDB. Again, the photoinitiating system Iod/EDB was used as a standard and the concentration of the copper complex photoinitiator was chosen to ensure the same light absorption at 405 nm for all photoinitiating systems. Photopolymerization profiles of TMTPA are presented in Figure 22. The two systems present similar performances. Notably, the free radical polymerization exhibited fast kinetics in the two cases and a final conversion around 80% could be determined. Thus, the presence of the 1,1'-*bis*(diphenylphosphino)ferrocene ligand in Cu5BF4 does not impact significantly the ability of the system Cu5BF4/Iod/EDB to generate radicals.

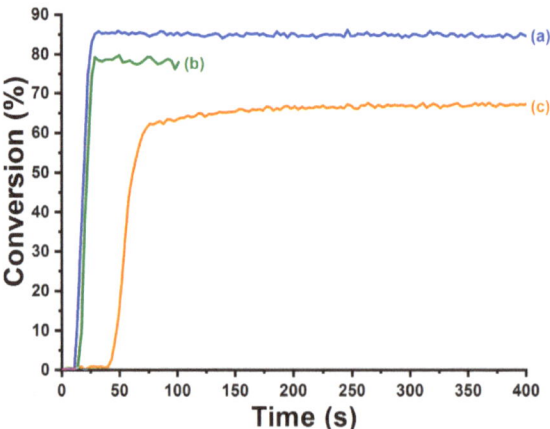

Figure 22. Polymerization profiles (acrylate function conversion vs. irradiation time) of TMPTA upon irradiation with a LED at 405 nm, under air, sample thickness = 1.4 mm; the irradiation starts at 10 s, 50 mW cm^{-2}. Photoinitiating systems: (curve a) Cu1BF4/Iod/EDB (0.73/2.0/2.0 $w/w/w$%), (curve b) Cu5BF4/Iod/EDB (0.64/2.0/2.0 $w/w/w$%), and (curve c) Iod/EDB (2.0/2.0 w/w%).

Cationic Polymerization

The cationic polymerization of EPOX was performed under air using a LED emitting at 405 nm in the presence of Cu1BF4/Iod/EDB or Cu5BF4/Iod/EDB. Again, the photoinitiating system Iod/EDB was used as a standard and the concentration of the copper complex photoinitiator was chosen to ensure the same light absorption at 405 nm. Photopolymerization profiles of EPOX are presented in Figure 23. As stated above, the system Cu1BF4/Iod/EDB was capable of initiating the cationic polymerization and was slightly more efficient than the standard Iod/EDB system. However, the system Cu5BF4/Iod/EDB exhibited almost identical polymerization profiles to those of the reference system. The presence of the iron moiety seems to negatively impact the photoinitiating ability of this system to initiate the cationic polymerization.

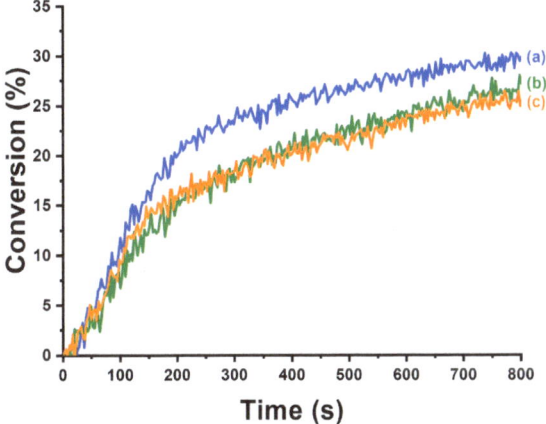

Figure 23. Polymerization profiles (epoxide function conversion vs. irradiation time) of EPOX upon irradiation with a LED at 405 nm, under air, sample thickness = 1.4 mm; the irradiation starts at 10 s, 50 mW cm^{-2}. Photoinitiating systems: (curve a) Cu1BF4/Iod/EDB (0.73/2.0/2.0 $w/w/w$%), (curve b) Cu5BF4/Iod/EDB (0.63/2.0/2.0 $w/w/w$%), and (curve c) Iod/EDB (2.0/2.0 w/w%).

IPN Synthesis

The polymerization of a TMPTA/EPOX blend in the presence of Cu1BF4/Iod/EDB or Cu5BF4/Iod/EDB was performed under air using a LED emitting at 405 nm. Again, the photoinitiating system Iod/EDB was used as a standard and the concentration of the copper complex photoinitiator was chosen to ensure the same light absorption at 405 nm. Photopolymerization profiles of the IPN synthesis are presented in Figure 24. The two systems presented similar performances toward the radical polymerization with a final conversion around 80% but the kinetic with the Cu5BF4/Iod/EDB was slightly slower. For the cationic polymerization, the system Cu1BF4/Iod/EDB could still lead to a high monomer conversion but the system Cu5BF4/Iod/EDB exhibited better performances than the standard Iod/EDB system. Thus, the effects of the presence of the 1,1'-*bis*(diphenylphosphino)ferrocene ligand on the polymerization initiating ability which were evidenced during the free radical polymerization experiments and the cationic polymerization of epoxides are confirmed during the IPN synthesis. However, the detrimental effect for the cationic polymerization seems to be drastically reduced during the polymerization of the TMPTA/EPOX blend.

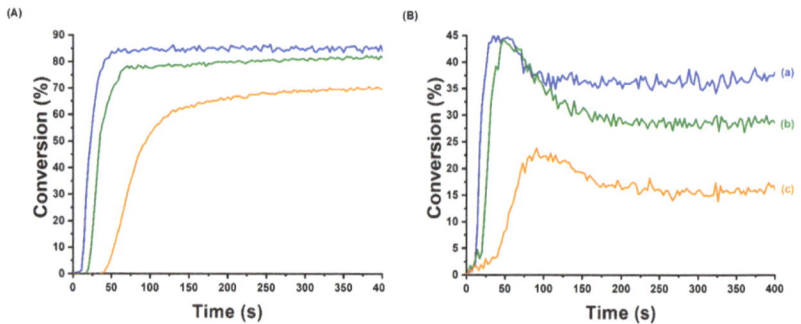

Figure 24. Polymerization profiles (**A**) (acrylate function conversion vs. irradiation time) and (**B**) (epoxide function conversion vs. irradiation time) of TMPTA/EPOX blend (50/50 w/w%) upon irradiation with a LED at 405 nm, under air, sample thickness = 1.4 mm; the irradiation starts at 10 s, 50 mW cm^{-2}. Photoinitiating systems: (curve a) Cu1BF4/Iod/EDB (0.73/2.0/2.0 $w/w/w$%), (curve b) Cu5BF4/Iod/EDB (0.65/2.0/2.0 $w/w/w$%), and (curve c) Iod/EDB (2.0/2.0 w/w%).

4. Conclusions

In the present paper, eleven heteroleptic copper (I) complexes were studied as photoinitiators of polymerization in three-component photoinitiating systems (Cu/Iod/EDB). Among them, ten new copper complexes were designed and synthesized to exhibit panchromatic behavior. Due to the modification of the electron donating substituent on the phenanthroline ligand, interesting redshifted absorption spectra could be obtained which open more possibilities in terms of irradiation wavelengths. The four ferrocene-free copper complexes Cu1BF4, Cu1PF6, Cu2BF4 and Cu2PF6 were able to efficiently initiate both the radical and cationic polymerizations and exhibited similar performances to the benchmark system G1/Iod/EDB based on the outstanding copper complex G1. The formation of acrylate/epoxy IPNs was successfully achieved through the concomitant initiation of both the free radical and the cationic polymerization of 1.4 mm-thick samples under air upon irradiation at 405 nm with a low amount of copper complex. Interestingly, the synthesis of acrylate/epoxy IPN was also possible upon irradiation at 455 nm or at 530 nm. The seven copper complexes containing the 1,1'-*bis*(diphenylphosphino)ferrocene ligand were characterized and were able to initiate efficiently radical polymerizations. Therefore, the iron moiety on the copper complexes did not affect either the panchromatic behavior or the ability to initiate radical polymerization. However, a detrimental effect was observed on their ability to initiate cationic polymerizations. Concerning IPN synthesis, the detrimental effect of the iron moiety, even if less pronounced, is still noticeable. To still improve the safer character of the four iron-free copper complexes, future works will consist of developing copper complexes absorbing in the red or the near-infrared region. New structures of copper complexes will be proposed in forthcoming studies as well as their potential applications in 3D printing.

Author Contributions: Conceptualization, G.N., F.D., C.D. and J.L.; methodology, G.N. and J.L.; validation, A.M., G.N., C.D., B.G., D.G., F.D. and J.L.; formal analysis, A.M., G.N., C.D., B.G., D.G., F.D. and J.L.; investigation, A.M. and G.N.; resources, F.D., D.G. and J.L.; data curation, A.M., G.N., C.D., B.G., D.G., F.D. and J.L.; writing—original draft preparation, A.M., G.N., C.D., B.G., D.G., F.D. and J.L.; writing—review and editing, F.D. and J.L.; visualization, F.D. and J.L.; supervision, F.D. and J.L.; project administration, J.L.; funding acquisition, C.D., F.D., D.G. and J.L. All authors have read and agreed to the published version of the manuscript.

Funding: This research was funded by Aix Marseille University, Université de Haute Alsace, Centre National de la Recherche Scientifique and Agence Nationale de la Recherche (VISICAT project (ANR-17-CE08-0054).

Conflicts of Interest: The authors declare no conflict of interest.

References

1. Kirschner, J.; Szillat, F.; Bouzrati-Zerelli, M.; Becht, J.-M.; Klee, J.E.; Lalevée, J. Iodonium Sulfonates as High-Performance Coinitiators and Additives for CQ-Based Systems: Toward Aromatic Amine-Free Photoinitiating Systems. *J. Polym. Sci. Part A Polym. Chem.* **2019**, *57*, 1664–1669. [CrossRef]
2. Lalevée, J.; Fouassier, J.P.; Graff, B.; Zhang, J.; Xiao, P. Chapter 6: How to Design Novel Photoinitiators for Blue Light. In *Photopolymerisation Initiating Systems*; The Royal Society of Chemistry: London, UK, 2018; pp. 179–199.
3. Mokbel, H.; Anderson, D.; Plenderleith, R.; Dietlin, C.; Morlet-Savary, F.; Dumur, F.; Gigmes, D.; Fouassier, J.-P.; Lalevée, J. Copper Photoredox Catalyst "G1": A New High Performance Photoinitiator for near-UV and Visible LEDs. *Polym. Chem.* **2017**, *8*, 5580–5592. [CrossRef]
4. Ligon, S.C.; Liska, R.; Stampfl, J.; Gurr, M.; Mülhaupt, R. Polymers for 3D Printing and Customized Additive Manufacturing. *Chem. Rev.* **2017**, *117*, 10212–10290. [CrossRef]
5. Mendes-Felipe, C.; Oliveira, J.; Etxebarria, I.; Vilas-Vilela, J.L.; Lanceros-Mendez, S. State-of-the-Art and Future Challenges of UV Curable Polymer-Based Smart Materials for Printing Technologies. *Adv. Mater. Technol.* **2019**, *4*, 1800618. [CrossRef]
6. Lalevée, J.; Blanchard, N.; Tehfe, M.-A.; Peter, M.; Morlet-Savary, F.; Gigmes, D.; Fouassier, J.P. Efficient Dual Radical/Cationic Photoinitiator under Visible Light: A New Concept. *Polym. Chem.* **2011**, *2*, 1986–1991. [CrossRef]
7. Telitel, S.; Lalevée, J.; Blanchard, N.; Kavalli, T.; Tehfe, M.-A.; Schweizer, S.; Morlet-Savary, F.; Graff, B.; Fouassier, J.-P. Photopolymerization of Cationic Monomers and Acrylate/Divinylether Blends under Visible Light Using Pyrromethene Dyes. *Macromolecules* **2012**, *45*, 6864–6868. [CrossRef]
8. Tehfe, M.-A.; Dumur, F.; Xiao, P.; Delgove, M.; Graff, B.; Fouassier, J.-P.; Gigmes, D.; Lalevée, J. Chalcone Derivatives as Highly Versatile Photoinitiators for Radical, Cationic, Thiol–Ene and IPN Polymerization Reactions upon Exposure to Visible Light. *Polym. Chem.* **2013**, *5*, 382–390. [CrossRef]
9. Xiao, P.; Dumur, F.; Zhang, J.; Fouassier, J.P.; Gigmes, D.; Lalevée, J. Copper Complexes in Radical Photoinitiating Systems: Applications to Free Radical and Cationic Polymerization upon Visible LEDs. *Macromolecules* **2014**, *47*, 3837–3844. [CrossRef]
10. Mau, A.; Dietlin, C.; Dumur, F.; Lalevée, J. Concomitant Initiation of Radical and Cationic Polymerisations Using New Copper Complexes as Photoinitiators: Synthesis and Characterisation of Acrylate/Epoxy Interpenetrated Polymer Networks. *Eur. Polym. J.* **2021**, *152*, 110457. [CrossRef]
11. Noirbent, G.; Dumur, F. Recent advances on copper complexes as visible light photoinitiators and (photo)redox initiators of polymerization. *Catalysts* **2020**, *10*, 953. [CrossRef]
12. Xiao, P.; Dumur, F.; Zhang, J.; Gigmes, D.; Fouassier, J.-P.; Lalevée, J. Copper complexes: The effect of ligands on their photoinitiation efficiencies in radical polymerization reactions under visible light. *Polym. Chem.* **2014**, *5*, 6350–6357. [CrossRef]
13. Bonardi, A.H.; Dumur, F.; Grant, T.M.; Noirbent, G.; Gigmes, D.; Lessard, B.H.; Fouassier, J.-P.; Lalevée, J. High Performance Near-Infrared (NIR) Photoinitiating Systems Operating under Low Light Intensity and in the Presence of Oxygen. *Macromolecules* **2018**, *51*, 1314–1324. [CrossRef]
14. Garra, P.; Dietlin, C.; Morlet-Savary, F.; Dumur, F.; Gigmes, D.; Fouassier, J.-P.; Lalevée, J. Photopolymerization Processes of Thick Films and in Shadow Areas: A Review for the Access to Composites. *Polym. Chem.* **2017**, *8*, 7088–7101. [CrossRef]
15. Armaroli, N.; Accorsi, G.; Bergamini, G.; Ceroni, P.; Holler, M.; Moudam, O.; Duhayon, C.; Delavaux-Nicot, B.; Nierengarten, J.-F. Heteroleptic Cu(I) Complexes Containing Phenanthroline-Type and 1,1'-Bis(Diphenylphosphino)Ferrocene Ligands: Structure and Electronic Properties. *Inorg. Chim. Acta* **2007**, *360*, 1032–1042. [CrossRef]
16. Minozzi, C.; Caron, A.; Grenier-Petel, J.-C.; Santandrea, J.; Collins, S.K. Heteroleptic Copper(I)-Based Complexes for Photocatalysis: Combinatorial Assembly, Discovery, and Optimization. *Angew. Chem. Int. Ed.* **2018**, *57*, 5477–5481. [CrossRef] [PubMed]
17. Listorti, A.; Accorsi, G.; Rio, Y.; Armaroli, N.; Moudam, O.; Gégout, A.; Delavaux-Nicot, B.; Holler, M.; Nierengarten, J.-F. Heteroleptic Copper(I) Complexes Coupled with Methano[60]Fullerene: Synthesis, Electrochemistry, and Photophysics. *Inorg. Chem.* **2008**, *47*, 6254–6261. [CrossRef]
18. Lin, J.-T.; Liu, H.-W.; Chen, K.-T.; Cheng, D.-C. Modeling the Kinetics, Curing Depth, and Efficacy of Radical-Mediated Photopolymerization: The Role of Oxygen Inhibition, Viscosity, and Dynamic Light Intensity. *Front. Chem.* **2019**. [CrossRef] [PubMed]
19. Lin, J.-T.; Lalevee, J.; Cheng, D.-C. Kinetics Analysis of Copper Complex Photoredox Catalyst: Roles of Oxygen, Thickness, and Optimal Concentration for Radical/Cationic Hybrid Photopolymerization. *Preprints* **2021**, 2021050597. Available online: https://www.preprints.org/manuscript/202105.0597/v1 (accessed on 10 June 2021).
20. Hu, M.-Y.; He, Q.; Fan, S.-J.; Wang, Z.-C.; Liu, L.-Y.; Mu, Y.-J.; Peng, Q.; Zhu, S.-F. Ligands with 1,10-Phenanthroline Scaffold for Highly Regioselective Iron-Catalyzed Alkene Hydrosilylation. *Nat. Commun.* **2018**, *9*, 221. [CrossRef]
21. Yang, W.; Nakano, T. Synthesis of Poly(1,10-Phenanthroline-5,6-Diyl)s Having a π-Stacked, Helical Conformation. *Chem. Commun.* **2015**, *51*, 17269–17272. [CrossRef]
22. Hebbe-Viton, V.; Desvergnes, V.; Jodry, J.J.; Dietrich-Buchecker, C.; Sauvage, J.-P.; Lacour, J. Chiral Spiro Cu(I) Complexes. Supramolecular Stereocontrol and Isomerisation Dynamics by the Use of TRISPHAT Anions. *Dalton Trans.* **2006**, 2058–2065. [CrossRef] [PubMed]
23. Dietlin, C.; Schweizer, S.; Xiao, P.; Zhang, J.; Morlet-Savary, F.; Graff, B.; Fouassier, J.-P.; Lalevée, J. Photopolymerization upon LEDs: New Photoinitiating Systems and Strategies. *Polym. Chem.* **2015**, *6*, 3895–3912. [CrossRef]

24. Knorn, M.; Rawner, T.; Czerwieniec, R.; Reiser, O. [Copper(Phenanthroline)(Bisisonitrile)]+-Complexes for the Visible-Light-Mediated Atom Transfer Radical Addition and Allylation Reactions. *ACS Catal.* **2015**, *5*, 5186–5193. [CrossRef]
25. Bouzrati-Zerelli, M.; Guillaume, N.; Goubard, F.; Bui, T.-T.; Villotte, S.; Dietlin, C.; Morlet-Savary, F.; Gigmes, D.; Fouassier, J.P.; Dumur, F.; et al. A Novel Class of Photoinitiators with a Thermally Activated Delayed Fluorescence (TADF) Property. *New J. Chem.* **2018**, *42*, 8261–8270. [CrossRef]
26. Garra, P.; Graff, B.; Morlet-Savary, F.; Dietlin, C.; Becht, J.-M.; Fouassier, J.-P.; Lalevée, J. Charge Transfer Complexes as Pan-Scaled Photoinitiating Systems: From 50 Mm 3D Printed Polymers at 405 Nm to Extremely Deep Photopolymerization (31 Cm). *Macromolecules* **2018**, *51*, 57–70. [CrossRef]
27. Wang, D.; Arar, A.; Garra, P.; Graff, B.; Lalevée, J. Charge Transfer Complexes Based on Various Amines as Dual Thermal and Photochemical Polymerization Initiators: A Powerful Tool for the Access to Composites. *J. Polym. Sci.* **2020**, *58*, 811–823. [CrossRef]

Article

Photocatalytic Reduction of CO$_2$ over Iron-Modified g-C$_3$N$_4$ Photocatalysts

Miroslava Edelmannová [1], Martin Reli [1], Kamila Kočí [1], Ilias Papailias [2,*], Nadia Todorova [2], Tatiana Giannakopoulou [2], Panagiotis Dallas [2], Eamonn Devlin [2], Nikolaos Ioannidis [2] and Christos Trapalis [2,*]

Citation: Edelmannová, M.; Reli, M.; Kočí, K.; Papailias, I.; Todorova, N.; Giannakopoulou, T.; Dallas, P.; Devlin, E.; Ioannidis, N.; Trapalis, C. Photocatalytic Reduction of CO$_2$ over Iron-Modified g-C$_3$N$_4$ Photocatalysts. *Photochem* **2021**, *1*, 462–476. https://doi.org/10.3390/photochem1030030

Academic Editor: Marcelo I. Guzman

Received: 24 October 2021
Accepted: 11 November 2021
Published: 13 November 2021

Publisher's Note: MDPI stays neutral with regard to jurisdictional claims in published maps and institutional affiliations.

Copyright: © 2021 by the authors. Licensee MDPI, Basel, Switzerland. This article is an open access article distributed under the terms and conditions of the Creative Commons Attribution (CC BY) license (https://creativecommons.org/licenses/by/4.0/).

[1] Centre for Energy and Environmental Technologies (CEET), Institute of Environmental Technology, VSB-Technical University of Ostrava, 17. Listopadu 15/2172, 70800 Ostrava, Czech Republic; miroslava.edelmannova@vsb.cz (M.E.); martin.reli@vsb.cz (M.R.); kamila.koci@vsb.cz (K.K.)

[2] Institute of Nanoscience and Nanotechnology, NCSR "Demokritos", Patriarchou Gregoriou E & 27 Neapoleos Str., 15341 Agia Paraskevi, Greece; n.todorova@inn.demokritos.gr (N.T.); t.giannakopoulou@inn.demokritos.gr (T.G.); p.dallas@inn.demokritos.gr (P.D.); e.devlin@inn.demokritos.gr (E.D.); n.ioannidis@inn.demokritos.gr (N.I.)

* Correspondence: i.papailias@inn.demokritos.gr (I.P.); c.trapalis@inn.demokritos.gr (C.T.); Tel.: +30-210-650-3347 (I.P.); +30-210-650-3343 (C.T.); Fax: +30-210-651-9430 (I.P. & C.T)

Abstract: Pure g-C$_3$N$_4$ sample was prepared by thermal treatment of melamine at 520 °C, and iron-modified samples (0.1, 0.3 and 1.1 wt.%) were prepared by mixing g-C$_3$N$_4$ with iron nitrate and calcination at 520 °C. The photocatalytic activity of the prepared materials was investigated based on the photocatalytic reduction of CO$_2$, which was conducted in a homemade batch reactor that had been irradiated from the top using a 365 nm Hg lamp. The photocatalyst with the lowest amount of iron ions exhibited an extraordinary methane and hydrogen evolution in comparison with the pure g-C$_3$N$_4$ and g-C$_3$N$_4$ with higher iron amounts. A higher amount of iron ions was not a beneficial for CO$_2$ photoreduction because the iron ions consumed too many photogenerated electrons and generated hydroxyl radicals, which oxidized organic products from the CO$_2$ reduction. It is clear that there are numerous reactions that occur simultaneously during the photocatalytic process, with several of them competing with CO$_2$ reduction.

Keywords: g-C$_3$N$_4$; iron; CO$_2$ reduction; photocatalysis

1. Introduction

The most pressing problems of the present time are undoubtedly the increasing energy requirements that are closely connected to fossil fuel combustion and the increasing concentration of greenhouse gases in the atmosphere. These problems are related, and finding a new clean source of energy is becoming increasingly important. Developed countries are trying to regulate greenhouse gas emissions; for example, the European Union has a long-term goal to reduce its carbon footprint. The first step towards this goal was to reduce greenhouse gas emissions by 20% (from 1990 levels) by 2020. This goal was successfully achieved, and emissions were reduced by 23% in 2018. However, a new legislation was put in place and added to the Paris Agreement. This legislation aims to reduce greenhouse gas emissions by 40% (from 1990 levels) by 2030. According to Energy in figures—Statistical Pocket Book 2019 issued by European Union, world energy production is increasing every year, and it is predicted that it will continue to rise in the years to come [1]. The highest annual increase of world energy production still comes from solid fossil fuels and natural gas compared to renewable or nuclear sources. Fossil fuel reserves are not infinite, and when considering the rate at which the world's energy consumption is increasing, it is clear that more focus has to be placed on finding new sources of energy. The photocatalytic reduction of carbon dioxide (CO$_2$) is currently one of the hottest candidates for this purpose. Not only can the products (CO, CH$_4$) of this reaction be used as an energy source, but the reaction utilizes CO_2, which is the most abundant source of carbon and is also considered to be a waste; therefore, it is not utilized [2]. The photocatalytic community has been studying the photocatalytic reduction of CO_2 for years. While the exact mechanism is not known, many important advances have been made [3,4].

There have been many photocatalysts that have been discovered, and different combinations of these photocatalysts have been tested to determine which combination produces the best photocatalytic reduction of CO_2 over the years [5,6]; however, most attention has focused on TiO_2 [7–9]. TiO_2 has quite large band gap energy, and more importantly, the conduction band potential is not in a convenient position located at around -0.5 V (vs. NHE at pH = 7), which is very close to the required redox potentials of essential reactions such as $CO_2/HCOOH$ (-0.61 V) $CO_2/HCHO$ (-0.48 V), $2H^+/H_2$ (-0.41 V), and CO_2/CH_4 (-0.24 V) [7]. Finding a new photocatalyst is an immediate goal that most of the photocatalytic community is focused on.

Graphitic carbon nitride (g-C_3N_4) has attracted increasing attention from the photocatalytic community, and research groups dealing with the photocatalytic reduction of CO_2 have already noticed its potential [10–12]. Its indirect band gap energy (around 2.7 eV) is especially narrow, and it has a conduction band potential (around -1.4 V) that makes it promising [13,14]. g-C_3N_4 has also high chemical stability, and more importantly, it has much more suitable potential of valence and conduction bands than TiO_2 [15]. The photocatalytic community is trying to shift the reaction into the visible light region, and the band gap energy of 2.7 eV can allow that. This corresponds approximately to a wavelength of 460 nm, which is somewhere on the edge between the blue and green region of visible light.

Even though g-C_3N_4 is a promising photocatalyst, it has a lot of drawbacks, such as low specific surface area, the fast recombination rate of generated charge carriers, and the particle boundary effects that interrupt the delocalization of electrons [16]. The advantages and disadvantages of g-C_3N_4 are summarized and discussed in the following reviews [16,17].

Many research teams have focused their attention to overcome the disadvantages of g-C_3N_4. Various modifications of g-C_3N_4 have been investigated in order to increase photocatalytic activity. For example, the specific surface area of g-C_3N_4 can be significantly enhanced by exfoliation [18,19]. There are several possibilities and techniques that can be used to achieve the exfoliation [20,21]. Another method for enhancing the photocatalytic activity of g-C_3N_4 is metal modification, which is mainly used as an effective approach to tune the electrical and optical properties [22]. Noble metals are the first and obvious choice [10,23]. However, noble metals are expensive, and therefore, other metal modification possibilities have been investigated. One such possibility is zero-valent iron, which is a cost-effective alternative, and g-C_3N_4 with incorporated iron particles has shown the improved absorption of visible light and as well as the lower recombination of charge carriers [24]. Modification caused by iron can also create a synergetic system that combines the Fenton and photocatalytic processes. When iron is present in a trivalent form, Fenton process can proceed [25]. Since there are a lot of lone electron pairs in the g-C_3N_4 heptazine structure, bonding with a foreign metal cation is relatively easy [26].

Although g-C_3N_4 only attracted the attention of the photocatalytic community just a few years ago, numerous papers about various modifications of g-C_3N_4 for the photocatalytic reduction of CO_2 can be found [11,27]. However, to the best of our knowledge, there is only a very limited number of works describing the utilization of g-C_3N_4/Fe for this application. In the present work, iron-modified g-C_3N_4 photocatalysts were synthesized by a simple in situ method. The photocatalysts were characterized and tested for photocatalytic CO_2 reduction where selectivity towards certain products was revealed and discussed.

2. Materials and Methods
2.1. Preparation

Melamine powder 99% was purchased from Alfa Aesar, and iron nitrate 98% was purchased from Chem-Lab.

Bulk g-C$_3$N$_4$ was synthesized by thermal polycondensation of melamine. The conditions were chosen by taking the relevant literature into account [28–31], and the results of our previous work were also considered [32,33]. Typically, 5 g of melamine was put into a covered alumina crucible and was heated in a muffle furnace in air to 520 °C with a rate of 5 °C/min, where it remained for 2 h. The yellow product (yield~2.4 g) was collected and was ground into fine powder. The sample was called CN.

A similar procedure was used for the preparation of the iron-modified g-C$_3$N$_4$ photocatalysts. In detail, 5 g of melamine was mixed with specific amounts of iron nitrate resulting in iron concentrations of 0.1, 0.3, and 1.1 wt.%. The mixture was stirred in 100 mL H$_2$O for 1 h to achieve a homogeneous suspension. After drying at 80 °C in an oven, the powder was put into a covered alumina crucible and was heated in a muffle furnace in air to 520 °C at a rate of 5 °C/min, where it remained for 2 h. The product (yield~2.5 g) was collected and ground into fine powder. The samples were called CN-Fex, where x is the amount (wt.%) of iron.

2.2. Characterization

The crystalline structure, surface area, porosity, chemical composition, and optical characteristics of the samples were thoroughly examined, and the nature of the iron ions was also investigated. Details about the characterization techniques and the equipment that was used are presented in the Supplementary Materials.

2.3. Photocatalytic Test

The photocatalytic reduction of CO$_2$ was performed in a homemade batch-stirred stainless steel reactor (volume 348 mL) with a quartz glass window on the top (Figure S1). A UV 8 W Hg lamp (peak intensity at 365 nm; Ultra-Violet Products Inc., Cambridge, UK) was used as a light source, and it was located over the quartz glass window of the reactor at a specific height so that the intensity on the level of the suspension was 0.833 mW/cm^2.

The reactor was filled with 0.09 g of each photocatalyst and 100 mL of 0.2 M NaOH. The NaOH was added in order to increase the solubility of CO$_2$ in water, thus facilitating the photocatalytic reaction. The suspension was stirred with a magnetic stirrer the entire time to prevent the sedimentation of the photocatalyst. Before the photocatalytic reaction began, the reactor was tightly closed and purged with (He or) CO$_2$. The pH of the suspension decreased during the purging from ~12.5 to ~6.9. A detailed description of the experimental procedure is available in the Supplementary Materials.

The gaseous samples were analyzed in 2 h intervals between the duration 0–8 h, where 0 corresponded to the moment prior to UV irradiation. Samples were collected using a gastight syringe (Hamilton Co., Reno, NV, USA) and were analyzed on a gas chromatograph (Shimadzu Tracera GC-2010 Plus) equipped with a BID detector. Each sample was measured at least three times using the same batch (same photocatalyst, same NaOH) in order to confirm the durability of the photocatalyst and the reproducibility of the experiments (within 5% error). The photon flux and apparent quantum yields (AQY) for individual products in the presence of each photocatalyst were calculated. The photocatalysts were first examined in an inert atmosphere to confirm that no products were formed, and after that the photocatalytic reduction of CO$_2$ was conducted.

3. Results and Discussion

3.1. Crystalline Structure

The XRD patterns of the g-C$_3$N$_4$ photocatalysts are presented in Figure 1. For all of the samples, the two characteristic diffraction peaks of g-C$_3$N$_4$ are observed (JCPDS, PDF #87-1526). The weak peak at 13.1° corresponds to the (100) plane, which is related to the in-plane structural packing motif of the tri-s-triazine units with an interplanar distance of d = 0.675 nm [34,35]. The strong peak at 27.5° corresponds to the (002) plane, which is attributed to the interlayer stacking of aromatic rings with a distance of d = 0.324 nm [36,37]. After the modification, no additional peaks corresponding to iron are observed, which was probably due to the low initial amount of iron nitrate. Furthermore, there are no significant changes to the position or the intensity of the peaks, indicating that the crystalline structure of the materials is preserved [38].

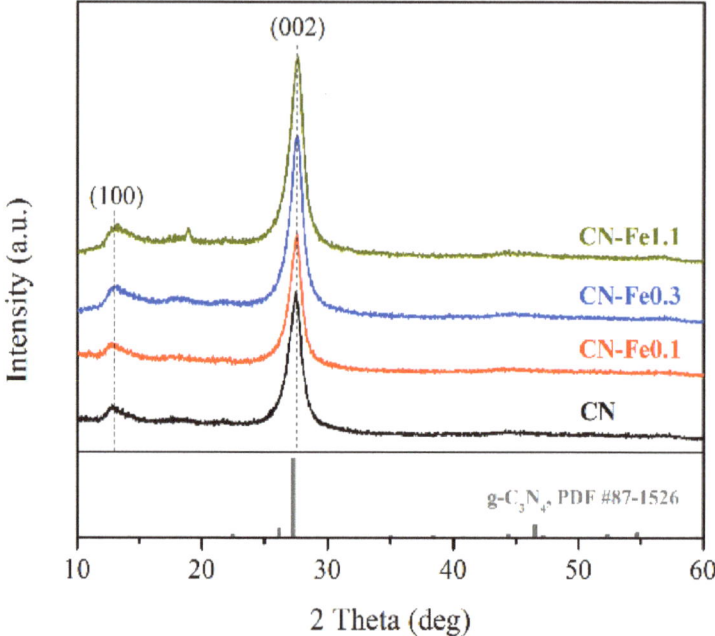

Figure 1. XRD patterns of the g-C_3N_4 photocatalysts.

3.2. Surface Area and Porosity

The N_2 adsorption–desorption isotherms and pore-size distributions of the g-C_3N_4 photocatalysts are displayed in Figure S2, with the characteristic values given in Table 1. All of the samples showed characteristic hysteresis loops of type IV isotherms, which is typical for mesoporous materials [39]. After the modification, the specific surface area of the photocatalysts decreased, which can be attributed to the partial blocking of the g-C_3N_4 pores by the iron. On the other hand, the pore size distribution remained practically unchanged across all of the samples.

Table 1. BET surface area, average pore size, and total pore volume of the g-C_3N_4 photocatalysts.

Sample	BET Surface Area (m^2/g)	Average Pore Size (nm)	Total Pore Volume (cm^3/g)
CN	16	13.7	0.11
CN-Fe0.1	13	19	0.12
CN-Fe0.3	11	18.7	0.11
CN-Fe1.1	7	38.5	0.12

3.3. Chemical Composition

The FT-IR analysis results are shown in Figure 2. For all of the samples, the known characteristic peaks of g-C_3N_4 can be observed. The broad peak at the 3500–3000 cm^{-1} region corresponds to the stretching mode of all of the OH-containing species and N-H stretching from the residual amino groups [40,41]. The shoulder located at 1630 cm^{-1} represents the H-O-H bending mode of all of the water molecules present at the surface of the material. The peaks in the 1628–1232 cm^{-1} region correspond to the characteristic stretching of the C-N heterocycles, including the trigonal N-(C)$_3$ and bridging H-N-(C)$_2$ units [42]. Finally, the sharp absorption peak at 805 cm^{-1} can be attributed to the breathing mode of the triazine units [43]. The FT-IR spectra of the modified materials demonstrate no significant changes. More specifically, there is no observable shift in the peaks, indicating that the chemical bonding of the main g-C$_3$N$_4$ network is unaffected [38].

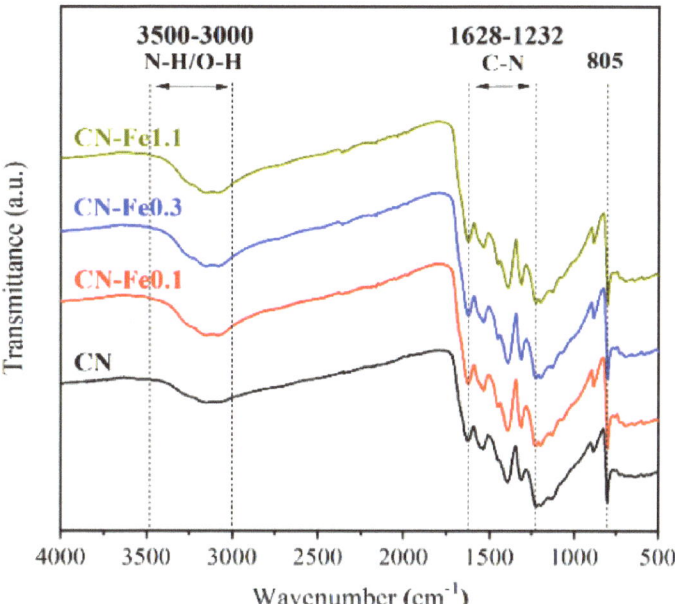

Figure 2. FT-IR spectra of the g-C$_3$N$_4$ photocatalysts.

3.4. Optical Characteristics

The measured diffuse reflectance spectra of the g-C$_3$N$_4$ photocatalysts are shown in the inset of Figure 3. They were used for the construction of the absorption functions $(F \times E)^{1/2} = f(E)$ presented in Figure 4, which allowed the determination of the materials' band gap energy (E_g), as described in Ref. [44]. It is evident that after modification, the light absorption of the materials is increased. However, the E_g only slightly decreased from 2.73 eV for bulk g-C$_3$N$_4$, which is in agreement with the literature [16,17], to 2.71 eV for the modified samples.

The properties of the photogenerated charge carriers were evaluated with fluorescence measurements at λ_{ex} = 365 nm (Figure 4). All of the samples showed a broad band with λ_{em} at around 455 nm, which is in agreement with the absorption edge wavelength measured by UV-vis spectroscopy. Such a signal can be attributed to the band–band PL phenomenon, which mainly results from the n-π* electronic transitions in g-C$_3$N$_4$ [45]. The strongest emission intensity is displayed by bulk g-C$_3$N$_4$, which suggests a faster electron–hole radiative recombination rate. On the other hand, the modified samples show significantly weaker emission intensity, which becomes even more prominent as the iron content increases. This indicates a lower recombination rate and higher charge transfer efficiency, suggesting a potential improvement in the photocatalytic activity of the materials.

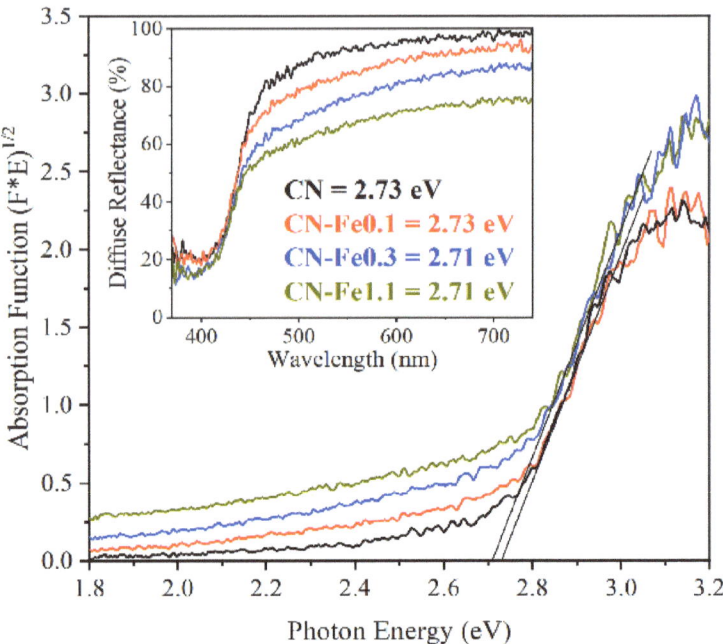

Figure 3. UV-Vis diffuse reflectance (inset) and plots of $(F \times E)^{1/2}$ vs. photon energy of the g-C$_3$N$_4$ photocatalysts.

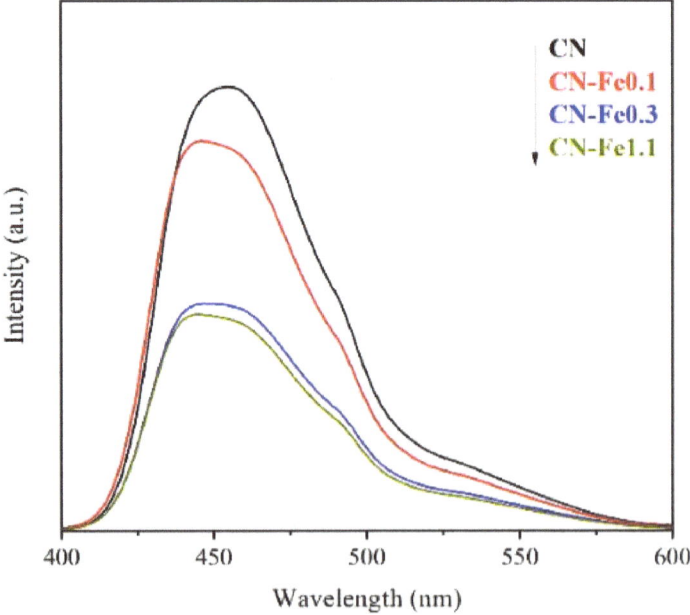

Figure 4. PL spectra of the g-C$_3$N$_4$ photocatalysts.

Photoelectrochemical measurements are one of the characterization techniques that allow the prediction of the photocatalyst behavior. Since the photocatalyst is irradiated under an external potential that has been applied to the working electrode, the recombination of electrons and holes is strongly suppressed. Based on the photocurrent generation (Figure 5), it is clear that the highest amount of charge carriers is produced in the case of bulk g-C$_3$N$_4$. This is connected to a higher specific surface area. However, a higher amount of produced charge carriers does not mean higher photocatalytic activity, especially since bulk g-C$_3$N$_4$ has the highest recombination rate of charge carriers (Figure 4). It is known that the redox potential of Fe^{3+}/Fe^{2+} is below the CB of g-C$_3$N$_4$. After iron modification, the photogenerated electrons could be trapped by the Fe^{2+} sites, leading to the reduced recombination of the photogenerated electron–hole pairs. With a higher iron concentration, more photogenerated electron trapping sites exist, thus leading to the further decrease of the PL intensity.

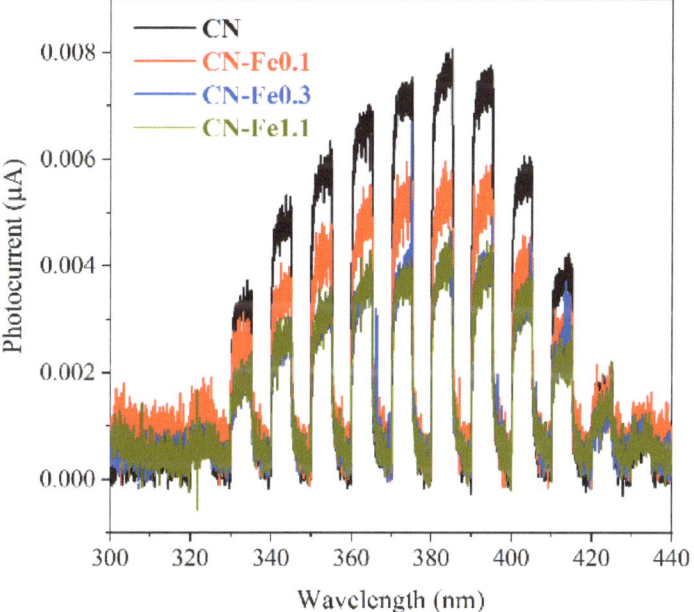

Figure 5. Photocurrent generation of g-C$_3$N$_4$ photocatalysts recorded under 1 V external potential.

3.5. Mössbauer and EPR Spectroscopy

Mössbauer spectroscopy was used to investigate the nature of the iron of the modified g-C$_3$N$_4$ photocatalysts. The 80 K spectra of samples CN-Fe0.3 and CN-Fe1.1 are shown in Figure 6. The most prominent characteristic of both spectra are the two distinct doublets that are characteristic of the Fe^{2+} ions in two different chemical environments [46]. These environments may be associated with the iron that is present in the -C$_3$N$_4$ network gaps or on the layer surface. The spectra also show a lower intensity doublet showing the presence of Fe^{3+} ions. A subspectrum that can be attributed to magnetic Fe$_2$O$_3$ [47], which is most likely due to the reaction of weakly stabilized Fe$^{3+/2+}$ ions with air, is also observed in all of the spectra. At 80 K there is no sign of any subspectral broadening that would suggest the existence of superparamagnetic nanoparticles. Thus, the only crystalline iron-containing phase observed through Mössbauer spectroscopy is that of the Fe$_2$O$_3$ oxide.

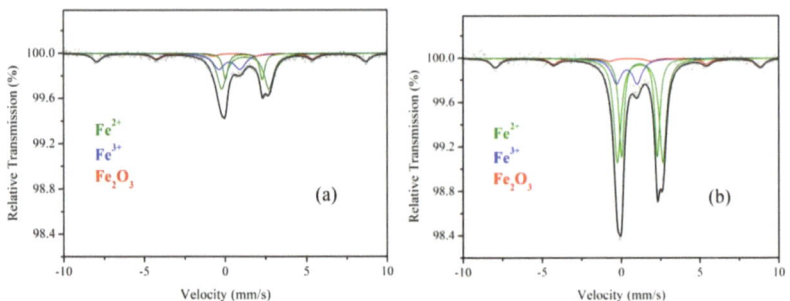

Figure 6. Mössbauer spectra of the samples CN-Fe0.3 (**a**) and CN-Fe1.1 (**b**), where the Fe^{2+} species are in green, the Fe^{3+} species are in blue, and the Fe_2O_3 subspectra are in red.

By comparing the relative spectral areas of the corresponding peaks (Table S2), it is evident that iron in the g-C_3N_4 matrix is mainly present in the form of Fe^{2+} species. This indicates that during melamine polycondensation, the Fe^{3+} ions of the iron nitrate precursor are reduced to Fe^{2+}. It should be also noted that the Mössbauer peaks of the sample CN-Fe0.3 are less intense than those of sample CN-Fe1.1, which confirms the lower amount of Fe present in the photocatalyst. The EPR spectra of the modified samples (Figure S3) display a broad line that is characteristic of iron ion–ion interactions [48]. However, as expected, the Fe^{2+} ions were not visible in the EPR measurements.

Recent studies have revealed that transition metal ions such as Au^{3+}, Ag^+, Cu^{2+}, and Fe^{3+} may be reduced during g-C_3N_4 synthesis at ~550 °C [49,50]. During melamine polycondensation, apart from NH_3, reactive species such as CNH_2, H_2NCN, and CN_2^+ are also released. These species can progressively reduce Fe^{3+} to Fe^{2+} until the formed ions are stabilized in the g-C_3N_4 network [51].

3.6. Photocatalytic Activity

Figure 7 shows the dependence of product yields on the irradiation time (0–8 h). The main products of the photocatalytic reduction of CO_2 are methane (Figure 7a) and carbon monoxide (Figure 7b). However, the hydrogen generated from the photocatalytic water splitting is also present in much higher concentrations (Figure 7c).

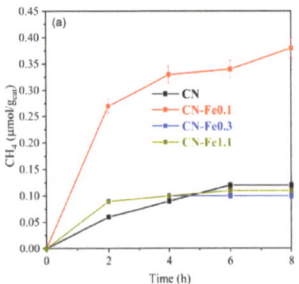
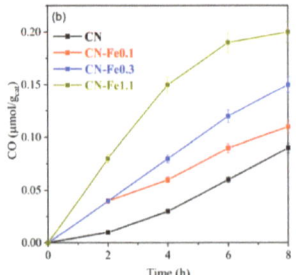
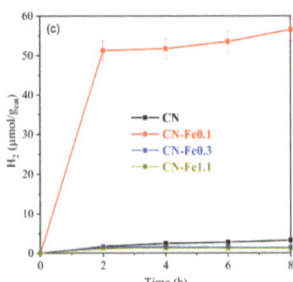

Figure 7. Time dependence of yields of (**a**) CH_4, (**b**) CO, and (**c**) H_2.

Photocatalytic CO_2 reduction is a complex process with a number of reactions occurring simultaneously. The g-C_3N_4 band gap energy around 2.7 eV (Figure 3) corresponds to a wavelength of approximately 460 nm, and since the photocatalytic reduction was conducted under the irradiation of 365 nm, each incident photon should lead to the generation of an electron and hole. The product yields and the apparent quantum yields (AQY) are shown in the Supplementary Materials (Table S1). Among the photocatalysts, the CN-Fe0.1

demonstrated the highest CH$_4$ and H$_2$ AQY at 0.0020% and 0.0733%, respectively. One of the most important properties of a photocatalyst is the CB and VB edge potentials. XPS analysis was used to determine the potential of the valence bands of each photocatalyst (Figure 8).

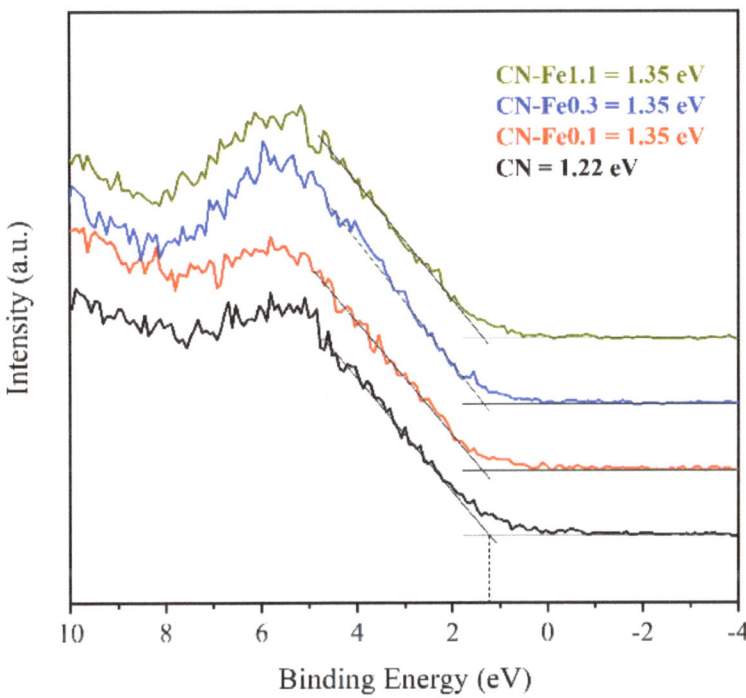

Figure 8. VB edge potentials measured by XPS.

As can be seen, the VB edge potential shifted to more positive values after the modification, which indicates the increased oxidation efficiency of the photogenerated holes [21]. By combining the band gap energy and VB edge potential values, the CB edge potentials of the materials were also calculated. Thus, the electronic band structures of the g-C$_3$N$_4$ photocatalysts were created (Figure 9). As the E_g of the photocatalysts remains practically unchanged after the modification, the CB edge potential shifts to less negative values, indicating a slight decrease in the reduction strength of the photogenerated electrons. However, the potential of CB is still higher than the redox potential that is required for the CO$_2$ reactions (Equations (1)–(5)).

The potentials of valence and conduction bands slightly differ for each g-C$_3$N$_4$ material; however, it is approximately inside the interval -1.5 V (CB potential) and 1.2 V (VB potential) vs. NHE (pH = 7). It is the relatively high reduction strength (negative potential of CB) that makes g-C$_3$N$_4$ such an attractive material.

The potential of CB is negative enough to allow any of the partial redox reactions to proceed (Equations (1)–(5)) [52,53]. However, the direct reduction of a CO$_2$ molecule by a single electron is not possible due to a very negative required potential of -1.9 V (Equation (6)) [2,5,54,55].

$$CO_2 + 2e^- + 2H^+ \rightarrow HCOOH \; E^0 = -0.61 \text{ V} \tag{1}$$

$$CO_2 + 2e^- + 2H^+ \rightarrow CO + H_2O \; E^0 = -0.53 \text{ V} \tag{2}$$

$$CO_2 + 4e^- + 4H^+ \rightarrow HCHO + H_2O \quad E^0 = -0.48 \text{ V} \quad (3)$$

$$CO_2 + 6e^- + 6H^+ \rightarrow CH_3OH + H_2O \quad E^0 = -0.38 \text{ V} \quad (4)$$

$$CO_2 + 8e^- + 8H^+ \rightarrow CH_4 + 2H_2O \quad E^0 = -0.24 \text{ V} \quad (5)$$

$$CO_2 + e^- \rightarrow CO_2^- \quad E^0 = -1.90 \text{ V} \quad (6)$$

All of the redox potentials are stated vs. NHE at pH = 7.

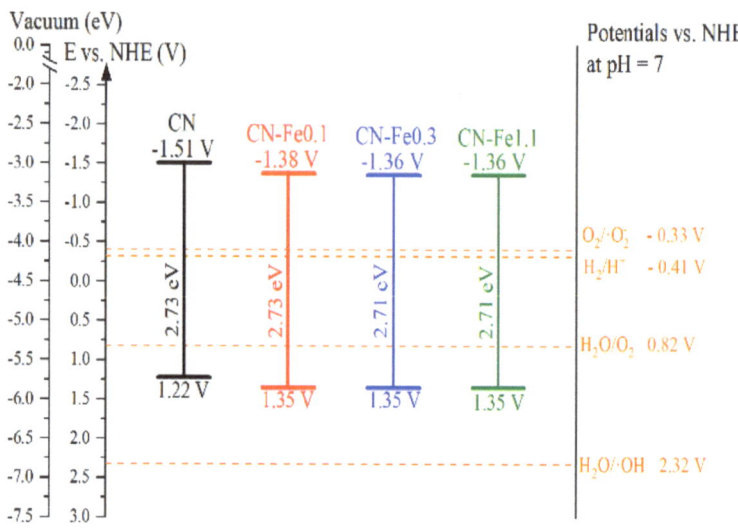

Figure 9. Band structures of the g-C$_3$N$_4$ photocatalysts.

Based on the equations above, it is clear that the photocatalytic reduction of CO$_2$ is a multielectron process that also requires the presence of a hydrogen cation. The necessity of a hydrogen cation is the reason why the reaction has to be conducted in aqueous phase or in the presence of water vapor.

There are two possible ways that water can be oxidized by photogenerated holes toward H$^+$ (Equations (7) and (8)) [56].

$$2H_2O + 4h^+ \rightarrow O_2 + 4H^+ \quad E^0 = +0.82 \text{ V} \quad (7)$$

$$H_2O + h^+ \rightarrow OH + H^+ \quad E^0 = +2.32 \text{ V} \quad (8)$$

It is clear that Equation (8) would be more probable since it requires just one hole and one water molecule; however, its redox potential is much more positive than the VB potential of g-C$_3$N$_4$, and therefore, this reaction cannot proceed in the presence of g-C$_3$N$_4$ material, leaving Equation (7) as the main source of the required H$^+$ ions. This significantly complicates the situation. First of all, the reaction (Equation (7)) is much less probable since it requires two water molecules o and four holes at the same time, but oxygen is also produced along with H$^+$. The reduction of oxygen molecules to a superoxide radical (Equation (9)) is one of the competitive reactions for the reduction of CO$_2$ [27,57].

$$O_2 + e^- \rightarrow O_2^- \quad E^0 = -0.33 \text{ V} \quad (9)$$

The presence of oxygen not only competes with CO$_2$ molecules to be adsorbed on the surface of the photocatalyst but also consumes the necessary electrons that are needed for the reduction.

Another competitive reaction is the reduction of the generated H^+ ions to hydrogen (Equation (10)).

$$2H^+ + 2e^- \rightarrow H_2 \quad E^0 = -0.41 \text{ V} \tag{10}$$

It is this competitive reaction that is responsible for the presence of hydrogen among the detected products (Figure 7c). Methane and carbon monoxide were other detected reaction products that can be generated according to Equations (2) and (5). Unfortunately, the possible products in the liquid phase were below the detection limit, and therefore, their presence could not be confirmed.

It is clear that the photocatalytic reduction of CO_2 is a very complex process that follows several multielectron steps. Nevertheless, the presence of the Fe^{3+}/Fe^{2+} couple (Figure 6) can complicate the situation even more. Many publications say that the presence of iron ions can be beneficial due to the possibility of photo-Fenton process occurring along with the photocatalysis [58–60]. What does this mean in the case of a photocatalytic CO_2 reduction? Part of the photogenerated electrons can directly transfer to Fe^{3+} ions and form Fe^{2+} (Equation (11)).

$$Fe^{3+} + e^- \rightarrow Fe^{2+} \quad E^0 = 0.77 \text{ V} \tag{11}$$

Holes, however, stay in the valence band and can participate in the oxidation reaction. The presence of Fe^{3+} can improve the separation of generated charge carriers; however, the negative side effect is the consumption of electrons needed for the photocatalytic reduction of CO_2; therefore, finding an optimum amount of iron ions is necessary [15].

Since the Fe^{2+} ions are unstable, they easily oxidize back into Fe^{3+} in the presence of oxygen and form either O_2^- (Equation (12)) or H_2O_2 (Equation (13)).

$$Fe^{2+} + O_2 + e^- \rightarrow Fe^{3+} + O_2^- \quad E^0 = -0.05 \text{ V} \tag{12}$$

$$2Fe^{2+} + O_2 + 2H^+ \rightarrow 2Fe^{3+} + H_2O_2 \quad E^0 = +0.68 \text{ V} \tag{13}$$

The above-mentioned equations clearly suggest the consumption of electrons in both redox Fe^{3+}/Fe^{2+} reactions. Higher amounts of iron ions would clearly explain the lower yields of the CO_2 reduction products. On the other hand, the oxygen used in Equations (12) and (13) does not compete with CO_2 molecules needed for adsorption on the surface of the photocatalyst.

In addition, the holes in g-C_3N_4 VB have sufficient potential to oxidize H_2O to H_2O_2 (+0.69 V vs. NHE) [27,61] and the potential of $H_2O_2/\cdot OH$ (1.07 V vs. NHE) [27] is more positive than the redox potential of Fe^{3+}/Fe^{2+} (0.77 V vs. NHE) [27,61,62]; the produced H_2O_2 reacts with Fe^{2+} to produce $\cdot OH$ species. Hydroxyl radical species are very strong oxidizing agents and are very beneficial when the removal of organics in water is the goal. However, the goal is to create organic compounds for the photocatalytic reduction of CO_2. Therefore, a higher amount of Fe would lead to a higher amount of hydroxyl radicals that would be able to oxidize the already generated organic products (CH_4) [27]. This is the reason why the CH_4 yields were lower in the case of photocatalysts containing 0.3 and 1.1 wt.% of Fe (Table S1). Furthermore, when the iron content exceeds an optimal value, the photocatalytic performance decreases, which is possibly due to the competitive capture between the adsorbed CO_2 and Fe^{2+} sites. In this case, an excess of Fe^{2+} sites could trap more photogenerated electrons, leading to fewer electrons being available to react with the adsorbed CO_2 molecules. On the other hand, the higher Fe content in the photocatalyst clearly leads to a higher selectivity toward CO production (Figure 7b). This suggests higher selectivity for CO_2 reduction to CO. Higher amounts of iron ions are incorporated into the g-C_3N_4 lattice and affect its electronic properties, which were confirmed by the valence band position shift after the addition of Fe (Figure 9), forming impurities that lead to an enhanced separation of charge carriers. On the other hand, this means a higher amount of hydroxyl radicals and a higher rate of reverse oxidation for the produced hydrocarbons [63].

4. Conclusions

Iron-modified g-C$_3$N$_4$ photocatalysts were prepared using a simple thermal treatment of melamine mixed with specific amounts of iron nitrate. The XRD and FT-IR measurements did not reveal any iron in the photocatalysts due to its very low concentrations (0.1, 0.3 and 1.1 wt.%). However, its presence in the form of Fe^{2+}, Fe^{3+}, and Fe$_2$O$_3$ was demonstrated by Mössbauer spectroscopy, with Fe^{2+} being the main species.

The addition of iron resulted in a decrease in the photocurrent and specific surface area due to the partial blocking of the g-C$_3$N$_4$ pores by iron. However, this led to the reduced recombination of the photogenerated electron–hole pairs, which is important for high photocatalytic efficiency.

The photocatalytic activity measurements of the prepared materials in the CO$_2$ reduction showed that the photocatalyst with the lowest amount of iron exhibited the biggest methane and hydrogen evolution in comparison with pure g-C$_3$N$_4$ and g-C$_3$N$_4$ with higher iron amounts. Excess iron consumes electrons that are needed for the reduction reactions and generates hydroxyl radicals that oxidize organic products from the CO$_2$ reduction. The iron-modified (0.1 wt.%) g-C$_3$N$_4$ proved to be a promising photocatalyst for CO$_2$ conversion into valuable hydrocarbons.

Supplementary Materials: The following are available online at https://www.mdpi.com/article/10.3390/photochem1030030/s1, Figure S1: The batch reactor used for the photocatalytic reduction of CO$_2$, Figure S2: N$_2$ adsorption-desorption isotherms (a) and pore size distribution curves (b) of the g-C$_3$N$_4$ photocatalysts, Figure S3: EPR spectra of samples CN-Fe0.3 and CN-Fe1.1, Table S1: Recalculated product yields to µmol·s^{-1} and apparent quantum yields (AQY) for each product in the presence of the photocatalysts, Table S2: Mössbauer parameters for samples CN-Fe0.3 and CN-Fe1.1 and corresponding assignments. Isomer Shift (IS), line width (HWHM), Quadrupole Splitting (QS), Hyperfine Magnetic Feld (HMF) and the subspectral area (= relative amount of iron).

Author Contributions: Conceptualization, I.P., C.T., and K.K.; methodology, I.P.; software, M.E.; validation, K.K. and M.R.; formal analysis, M.R.; investigation, N.I., E.D., and P.D.; resources, C.T. and K.K.; data curation, N.T. and T.G.; writing—original draft preparation, M.E., M.R.; writing—review and editing, M.E., M.R., and I.P.; visualization, I.P.; supervision, C.T. and K.K.; project administration, C.T. and K.K.; funding acquisition, C.T. and K.K. All authors have read and agreed to the published version of the manuscript.

Funding: This research was co-financed by the European Union and Greek national funds through the Operational Program Competitiveness, Entrepreneurship and Innovation, under the call RESEARCH-CREATE-INNOVATE (project code: T1EDK-05545) and by EU structural funding in Operational Programme Research Development and Education, project No. CZ.02.1.01/0.0/0.0/16_019/0000853 "IET-ER" and by using Large Research Infrastructure ENREGAT supported by the Ministry of Education, Youth and Sports of the Czech Republic under project No. LM2018098. The authors would also like to acknowledge the Hellenic Foundation for Research and Innovation and the General Secretariat for Research and Innovation for Grant number 1468.

Institutional Review Board Statement: Not applicable.

Informed Consent Statement: Not applicable.

Data Availability Statement: The data presented in this study are available in the article.

Acknowledgments: The authors would like to thank Andreas Karydas and Kalliopi Tsampa from "XRF Laboratory" of Institute of Nuclear and Particle Physics at NCSR "Demokritos" for the XRF measurements.

Conflicts of Interest: The authors declare no conflict of interest.

Sample Availability: Samples of the compounds are not available from the authors.

References

1. European Commission. *EU Energy in Figures-Statistical Pocket Book 2019*; Drukarnia INTERAK SP ZOO: Czarnków, Poland, 2019.
2. Vu, N.N.; Kaliaguine, S.; Do, T.O. Critical Aspects and Recent Advances in Structural Engineering of Photocatalysts for Sunlight-Driven Photocatalytic Reduction of CO$_2$ into Fuels. *Adv. Funct. Mater.* **2019**, *29*, 1901825. [CrossRef]

3. Alkhatib, I.I.; Garlisi, C.; Pagliaro, M.; Al-Ali, K.; Palmisano, G. Metal-organic frameworks for photocatalytic CO_2 reduction under visible radiation: A review of strategies and applications. *Catal. Today* **2020**, *340*, 209–224. [CrossRef]
4. Yuan, L.; Xu, Y.J. Photocatalytic conversion of CO_2 into value-added and renewable fuels. *Appl. Surf. Sci.* **2015**, *342*, 154–167. [CrossRef]
5. Habisreutinger, S.N.; Schmidt-Mende, L.; Stolarczyk, J.K. Photocatalytic reduction of CO_2 on TiO_2 and other semiconductors. *Angew. Chem.* **2013**, *52*, 7372–7408. [CrossRef] [PubMed]
6. Nikokavoura, A.; Trapalis, C. Alternative photocatalysts to TiO_2 for the photocatalytic reduction of CO_2. *Appl. Surf. Sci.* **2017**, *391*, 149–174. [CrossRef]
7. Shehzad, N.; Tahir, M.; Johari, K.; Murugesan, T.; Hussain, M. A critical review on TiO_2 based photocatalytic CO_2 reduction system: Strategies to improve efficiency. *J. CO_2 Util.* **2018**, *26*, 98–122. [CrossRef]
8. Low, J.; Cheng, B.; Yu, J. Surface modification and enhanced photocatalytic CO_2 reduction performance of TiO_2: A review. *Appl. Surf. Sci.* **2017**, *392*, 658–686. [CrossRef]
9. Akhter, P.; Farkhondehfal, M.A.; Hernández, S.; Hussain, M.; Fina, A.; Saracco, G.; Khan, A.U.; Russo, N. Environmental issues regarding CO_2 and recent strategies for alternative fuels through photocatalytic reduction with titania-based materials. *J. Environ. Chem. Eng.* **2016**, *4*, 3934–3953. [CrossRef]
10. Kočí, K.; Van, H.D.; Edelmannová, M.; Reli, M.; Wu, J.C.S. Photocatalytic reduction of CO_2 using Pt/C_3N_4 photocatalyts. *Appl. Surf. Sci.* **2020**, *503*, 144426. [CrossRef]
11. Li, X.; Xiong, J.; Gao, X.; Huang, J.; Feng, Z.; Chen, Z.; Zhu, Y. Recent advances in 3D $g-C_3N_4$ composite photocatalysts for photocatalytic water splitting, degradation of pollutants and CO_2 reduction. *J. Alloys Compd.* **2019**, *802*, 196–209. [CrossRef]
12. Xiang, Q.; Cheng, B.; Yu, J. Graphene-Based Photocatalysts for Solar-Fuel Generation. *Angew. Chem. Int. Ed. Engl.* **2015**, *54*, 11350–11366. [CrossRef] [PubMed]
13. Liu, J. Effect of phosphorus doping on electronic structure and photocatalytic performance of $g-C_3N_4$: Insights from hybrid density functional calculation. *J. Alloys Compd.* **2016**, *672*, 271–276. [CrossRef]
14. Wu, H.Z.; Liu, L.M.; Zhao, S.J. The effect of water on the structural, electronic and photocatalytic properties of graphitic carbon nitride. *Phys. Chem. Chem. Phys.* **2014**, *16*, 3299–3304. [CrossRef] [PubMed]
15. Liu, Q.; Guo, Y.; Chen, Z.; Zhang, Z.; Fang, X. Constructing a novel ternary $Fe(III)/graphene/g-C_3N_4$ composite photocatalyst with enhanced visible-light driven photocatalytic activity via interfacial charge transfer effect. *Appl. Catal. B* **2016**, *183*, 231–241. [CrossRef]
16. Prasad, C.; Tang, H.; Bahadur, I. Graphitic carbon nitride based ternary nanocomposites: From synthesis to their applications in photocatalysis: A recent review. *J. Mol. Liq.* **2019**, *281*, 634–654. [CrossRef]
17. Xiao, M.; Luo, B.; Wang, S.; Wang, L. Solar energy conversion on $g-C_3N_4$ photocatalyst: Light harvesting, charge separation, and surface kinetics. *J. Energy Chem.* **2018**, *27*, 1111–1123. [CrossRef]
18. Praus, P.; Svoboda, L.; Dvorský, R.; Reli, M.; Kormunda, M.; Mančík, P. Synthesis and properties of nanocomposites of WO_3 and exfoliated $g-C_3N_4$. *Ceram. Int.* **2017**, *43*, 13581–13591. [CrossRef]
19. Todorova, N.; Papailias, I.; Giannakopoulou, T.; Ioannidis, N.; Boukos, N.; Dallas, P.; Edelmannová, M.; Reli, M.; Kočí, K.; Trapalis, C. Photocatalytic H_2 Evolution, CO_2 Reduction, and NO_x Oxidation by Highly Exfoliated $g-C_3N_4$. *Catalysts* **2020**, *10*, 1147. [CrossRef]
20. Mohamed, N.A.; Safaei, J.; Ismail, A.F.; Noh, M.F.M.; Arzaee, N.A.; Mansor, N.N.; Ibrahim, M.A.; Ludin, N.A.; Sagu, J.S.; Teridi, M.A.M. Fabrication of exfoliated graphitic carbon nitride, $(g-C_3N_4)$ thin film by methanolic dispersion. *J. Alloys Compd.* **2020**, *818*, 152916. [CrossRef]
21. Papailias, I.; Todorova, N.; Giannakopoulou, T.; Ioannidis, N.; Boukos, N.; Athanasekou, C.P.; Dimotikali, D.; Trapalis, C. Chemical vs thermal exfoliation of $g-C_3N_4$ for NO_x removal under visible light irradiation. *Appl. Catal. B* **2018**, *239*, 16–26. [CrossRef]
22. Wen, J.; Xie, J.; Chen, X.; Li, X. A review on $g-C_3N_4$-based photocatalysts. *Appl. Surf. Sci.* **2017**, *391*, 72–123. [CrossRef]
23. Masih, D.; Ma, Y.; Rohani, S. Graphitic C_3N_4 based noble-metal-free photocatalyst systems: A review. *Appl. Catal. B* **2017**, *206*, 556–588. [CrossRef]
24. Heidarpour, H.; Padervand, M.; Soltanieh, M.; Vossoughi, M. Enhanced decolorization of rhodamine B solution through simultaneous photocatalysis and persulfate activation over Fe/C_3N_4 photocatalyst. *Chem. Eng. Res. Des.* **2020**, *153*, 709–720. [CrossRef]
25. Hu, J.; Zhang, P.; Cui, Y.; An, W.; Liu, L.; Liang, Y.; Yang, Q.; Yang, H.; Cui, W. High-efficiency removal of phenol and coking wastewater via photocatalysis-Fenton synergy over a $Fe-g-C_3N_4$ graphene hydrogel 3D structure. *J. Ind. Eng. Chem.* **2020**, *84*, 305–314. [CrossRef]
26. Ma, T.; Shen, Q.; Xue, B.Z.J.; Guan, R.; Liu, X.; Jia, H.; Xu, B. Facile synthesis of Fe-doped $g-C_3N_4$ for enhanced visible-light photocatalytic activity. *Inorg. Chem. Commun.* **2019**, *107*, 107451. [CrossRef]
27. Sun, Z.; Wang, H.; Wu, Z.; Wang, L. $g-C_3N_4$ based composite photocatalysts for photocatalytic CO_2 reduction. *Catal. Today* **2018**, *300*, 160–172. [CrossRef]
28. Wang, X.; Maeda, K.; Thomas, A.; Takanabe, K.; Xin, G.; Carlsson, J.M.; Domen, K.; Antonietti, M. A metal-free polymeric photocatalyst for hydrogen production from water under visible light. *Nat. Mater.* **2009**, *8*, 76–80. [CrossRef]

29. Zhang, Y.; Pan, Q.; Chai, G.; Liang, M.; Dong, G.; Zhang, Q.; Qiu, J. Synthesis and luminescence mechanism of multicolor-emitting g-C_3N_4 nanopowders by low temperature thermal condensation of melamine. *Sci. Rep.* **2013**, *3*, 1943. [CrossRef]
30. Ge, L. Synthesis and photocatalytic performance of novel metal-free g-C_3N_4 photocatalysts. *Mater. Lett.* **2011**, *65*, 2652–2654. [CrossRef]
31. Baudys, M.; Paušová, Š.; Praus, P.; Brezová, V.; Dvoranová, D.; Barbieriková, Z.; Krýsa, J. Graphitic Carbon Nitride for Photocatalytic Air Treatment. *Materials* **2020**, *13*, 3038. [CrossRef] [PubMed]
32. Papailias, I.; Giannakopoulou, T.; Todorova, N.; Demotikali, D.; Vaimakis, T.; Trapalis, C. Effect of processing temperature on structure and photocatalytic properties of g-C_3N_4. *Appl. Surf. Sci.* **2015**, *358*, 278–286. [CrossRef]
33. Praus, P.; Svoboda, L.; Ritz, M.; Troppová, I.; Šihor, M.; Kočí, K. Graphitic carbon nitride: Synthesis, characterization and photocatalytic decomposition of nitrous oxide. *Mater. Chem. Phys.* **2017**, *193*, 438–446. [CrossRef]
34. Yan, H.; Chen, Y.; Xu, S. Synthesis of graphitic carbon nitride by directly heating sulfuric acid treated melamine for enhanced photocatalytic H_2 production from water under visible light. *Int. J. Hydrogen Energy* **2012**, *37*, 125–133. [CrossRef]
35. Ho, W.; Zhang, Z.; Xu, M.; Zhang, X.; Wang, X.; Huang, Y. Enhanced visible-light-driven photocatalytic removal of NO: Effect on layer distortion on g-C_3N_4 by H_2 heating. *Appl. Catal. B* **2015**, *179*, 106–112. [CrossRef]
36. Zhao, H.; Chen, X.; Jia, C.; Zhou, T.; Qu, X.; Jian, J.; Xu, Y.; Zhou, T. A facile mechanochemical way to prepare g-C_3N_4. *Mater. Sci. Eng. B* **2005**, *122*, 90–93. [CrossRef]
37. Zhao, H.; Yu, H.; Quan, X.; Chen, S.; Zhang, Y.; Zhao, H.; Wang, H. Fabrication of atomic single layer graphitic-C_3N_4 and its high performance of photocatalytic disinfection under visible light irradiation. *Appl. Catal. B* **2014**, *153*, 46–50. [CrossRef]
38. Papailias, I.; Todorova, N.; Giannakopoulou, T.; Ioannidis, N.; Dallas, P.; Dimotikali, D.; Trapalis, C. Novel torus shaped g-C_3N_4 photocatalysts. *Appl. Catal. B* **2020**, *268*, 118733. [CrossRef]
39. Papailias, I.; Todorova, N.; Giannakopoulou, T.; Karapati, S.; Boukos, N.; Dimotikali, D.; Trapalis, C. Enhanced NO_2 abatement by alkaline-earth modified g-C_3N_4 nanocomposites for efficient air purification. *Appl. Surf. Sci.* **2018**, *430*, 225–233. [CrossRef]
40. Zhao, Z.; Sun, Y.; Luo, Q.; Dong, F.; Li, H.; Ho, W.K. Mass-Controlled Direct Synthesis of Graphene-like Carbon Nitride Nanosheets with Exceptional High Visible Light Activity. Less is Better. *Sci. Rep.* **2015**, *5*, 14643. [CrossRef]
41. Zhu, B.; Xia, P.; Ho, W.; Yu, J. Isoelectric point and adsorption activity of porous g-C_3N_4. *Appl. Surf. Sci.* **2015**, *344*, 188–195. [CrossRef]
42. Cao, Y.; Li, Q.; Wang, W. Construction of a crossed-layer-structure MoS_2/g-C_3N_4 heterojunction with enhanced photocatalytic performance. *RSC Adv.* **2017**, *7*, 6131–6139. [CrossRef]
43. Li, Y.; Zhang, H.; Liu, P.; Wang, D.; Li, Y.; Zhao, H. Cross-Linked g-C_3N_4/rGO Nanocomposites with Tunable Band Structure and Enhanced Visible Light Photocatalytic Activity. *Small* **2013**, *9*, 3336–3344. [CrossRef]
44. Giannakopoulou, T.; Todorova, N.; Romanos, G.; Vaimakis, T.; Dillert, R.; Bahnemann, D.; Trapalis, C. Composite hydroxyapatite/TiO_2 materials for photocatalytic oxidation of NO_x. *Mater. Sci. Eng. B* **2012**, *177*, 1046–1052. [CrossRef]
45. Hu, S.; Jin, R.; Lu, G.; Liu, D.; Gui, J. The properties and photocatalytic performance comparison of Fe^{3+}-doped g-C_3N_4 and Fe_2O_3/g-C_3N_4 composite catalysts. *RSC Adv.* **2014**, *4*, 24863–24869. [CrossRef]
46. Oliveira, A.A.S.; Teixeira, I.F.; Ribeiro, L.P.; Lorençon, E.; Ardisson, J.D.; Fernandez-Outon, L.; Macedo, W.A.A.; Moura, F.C.C. Magnetic amphiphilic nanocomposites produced via chemical vapor deposition of CH4 on Fe-Mo/nano-Al2O3. *Appl. Catal. A* **2013**, *456*, 126–134. [CrossRef]
47. Cunha, I.T.; Teixeira, I.F.; Albuquerque, A.S.; Ardisson, J.D.; Macedo, W.A.A.; Oliveira, H.S.; Tristão, J.C.; Sapag, K.; Lago, R.M. Catalytic oxidation of aqueous sulfide in the presence of ferrites (MFe_2O_4, M=Fe, Cu, Co). *Catal. Today* **2016**, *259*, 222–227. [CrossRef]
48. Cuello, N.I.; Elías, V.R.; Winkler, E.; Pozo-López, G.; Oliva, M.I.; Eimer, G.A. Magnetic behavior of iron-modified MCM-41 correlated with clustering processes from the wet impregnation method. *J. Magn. Magn. Mater.* **2016**, *407*, 299–307. [CrossRef]
49. Wang, N.; Han, Z.; Fan, H.; Ai, S. Copper nanoparticles modified graphitic carbon nitride nanosheets as a peroxidase mimetic for glucose detection. *RSC Adv.* **2015**, *5*, 91302–91307. [CrossRef]
50. Li, Z.; Kong, C.; Lu, G. Visible Photocatalytic Water Splitting and Photocatalytic Two-Electron Oxygen Formation over Cu- and Fe-Doped g-C_3N_4. *J. Phys. Chem. C* **2016**, *120*, 56–63. [CrossRef]
51. Bicalho, H.A.; Lopez, J.L.; Binatti, I.; Batista, P.F.R.; Ardisson, J.D.; Resende, R.R.; Lorençon, E. Facile synthesis of highly dispersed Fe(II)-doped g-C_3N_4 and its application in Fenton-like catalysis. *Mol. Catal.* **2017**, *435*, 156–165. [CrossRef]
52. Guzmán, H.; Farkhondehfal, M.A.; Tolod, K.R.; Hernández, S.; Russo, N. Chapter 11-Photo/electrocatalytic hydrogen exploitation for CO_2 reduction toward solar fuels production. In *Solar Hydrogen Production*; Calise, F., D'Accadia, M.D., Santarelli, M., Lanzini, A., Ferrero, D., Eds.; Academic Press: Cambridge, MA, USA, 2019; pp. 365–418.
53. Choi, W. Pure and modified TiO_2 photocatalysts and their environmental applications. *Catal. Surv. Asia* **2006**, *10*, 16–28. [CrossRef]
54. Li, X.; Yu, J.; Jaroniec, M.; Chen, X. Cocatalysts for Selective Photoreduction of CO_2 into Solar Fuels. *Chem. Rev.* **2019**, *119*, 3962–4179. [CrossRef]
55. Viswanathan, B. *Energy Sources: Fundamentals of Chemical Conversion Processes and Applications*; Elsevier Science: Amsterdam, Netherlands, 2016.
56. Wang, Z.; Li, C.; Domen, K. Recent developments in heterogeneous photocatalysts for solar-driven overall water splitting. *Chem. Soc. Rev.* **2019**, *48*, 2109–2125. [CrossRef] [PubMed]

57. Di, L.; Yang, H.; Xian, T.; Chen, X. Enhanced Photocatalytic Degradation Activity of BiFeO$_3$ Microspheres by Decoration with g-C$_3$N$_4$ Nanoparticles. *Mater. Res.* **2018**, *21*, 1–10. [CrossRef]
58. Matos, J.; Arcibar-Orozco, J.; Poon, P.S.; Pecchi, G.; Rangel-Mendez, J.R. Influence of phosphorous upon the formation of DMPO-•OH and POBN-O$_2$•− spin-trapping adducts in carbon-supported P-promoted Fe-based photocatalysts. *J. Photochem. Photobiol. A* **2020**, *391*, 112362. [CrossRef]
59. Matos, J.; Poon, P.S.; Montaña, R.; Romero, R.; Gonçalves, G.R.; Schettino, M.A., Jr.; Passamani, E.C.; Freitas, J.C.C. Photocatalytic activity of P-Fe/activated carbon nanocomposites under artificial solar irradiation. *Catal. Today* **2020**, *356*, 226–240. [CrossRef]
60. Matos, J.; Rosales, M.; Garcia, A.; Nieto-Delgado, C.; Rangel-Mendez, J.R. Hybrid photoactive materials from municipal sewage sludge for the photocatalytic degradation of methylene blue. *Green Chem.* **2011**, *13*, 3431–3439. [CrossRef]
61. Casado, J.; Fornaguera, J. Pilot-Scale Degradation of Organic Contaminants in a Continuous-flow Reactor by the Helielectro-Fenton Method. *Clean* **2008**, *36*, 53–58. [CrossRef]
62. Sayama, K.; Yoshida, R.; Kusama, H.; Okabe, K.; Abe, Y.; Arakawa, H. Photocatalytic decomposition of water into H$_2$ and O$_2$ by a two-step photoexcitation reaction using a WO$_3$ suspension catalyst and an Fe^{3+}/Fe^{2+} redox system. *Chem. Phys. Lett.* **1997**, *277*, 387–391. [CrossRef]
63. Khaing, K.K.; Yin, D.; Xiao, S.; Deng, L.; Zhao, F.; Liu, B.; Chen, T.; Li, L.; Guo, X.; Liu, J.; et al. Efficient Solar Light Driven Degradation of Tetracycline by Fe-EDTA Modified g-C$_3$N$_4$ Nanosheets. *J. Phys. Chem. C* **2020**, *124*, 11831–11843. [CrossRef]

Article

Mono-, Di-, Tri-Pyrene Substituted Cyclic Triimidazole: A Family of Highly Emissive and RTP Chromophores

Daniele Malpicci [1,2], Clelia Giannini [1], Elena Lucenti [2,3], Alessandra Forni [2,3,*], Daniele Marinotto [2,3] and Elena Cariati [1,2,3,*]

[1] Department of Chemistry, Università degli Studi di Milano, Via Golgi 19, 20133 Milano, Italy; daniele.malpicci@unimi.it (D.M.); clelia.giannini@unimi.it (C.G.)
[2] Institute of Chemical Sciences and Technologies "Giulio Natta" (CNR-SCITEC), Via Golgi 19, 20133 Milano, Italy; elena.lucenti@scitec.cnr.it (E.L.); daniele.marinotto@scitec.cnr.it (D.M.)
[3] INSTM RU, Via Golgi 19, 20133 Milano, Italy
* Correspondence: alessandra.forni@scitec.cnr.it (A.F.); elena.cariati@unimi.it (E.C.)

Abstract: The search of new organic emitters is receiving a strong motivation by the development of ORTP materials. In the present study we report on the preparation, optical and photophysical characterization, by both steady state and time resolved techniques, of two pyrene-functionalized cyclic triimidazole derivatives. Together with the already reported mono-substituted derivative, the di- and tri-substituted members of the family have revealed as intriguing emitters characterized by impressive quantum yields in solution and RTP properties in the solid state. In particular, phosphorescence lifetimes increase from 5.19 to 20.54 and 40.62 ms for mono-, di- and trisubstituted compounds, respectively. Based on spectroscopical results and theoretical DFT/TDDFT calculations on the di-pyrene molecule, differences in photophysical performances of the three compounds have been assigned to intermolecular interactions increasing with the number of pyrene moieties appended to the cyclic triimidazole scaffold.

Keywords: fluorescence; organic room temperature phosphorescence; DFT and TDDFT calculations

Citation: Malpicci, D.; Giannini, C.; Lucenti, E.; Forni, A.; Marinotto, D.; Cariati, E. Mono-, Di-, Tri-Pyrene Substituted Cyclic Triimidazole: A Family of Highly Emissive and RTP Chromophores. *Photochem* **2021**, *1*, 477–487. https://doi.org/10.3390/photochem1030031

Academic Editor: Marcelo I. Guzman

Received: 15 October 2021
Accepted: 16 November 2021
Published: 18 November 2021

Publisher's Note: MDPI stays neutral with regard to jurisdictional claims in published maps and institutional affiliations.

Copyright: © 2021 by the authors. Licensee MDPI, Basel, Switzerland. This article is an open access article distributed under the terms and conditions of the Creative Commons Attribution (CC BY) license (https://creativecommons.org/licenses/by/4.0/).

1. Introduction

Organic room temperature phosphorescent materials (ORTP), due to their increased biocompatibility and reduced cost with respect to their organometallic counterparts, are the subject of intense research in the field of functional materials, appearing promising for applications spanning from bioimaging [1,2] to anti-counterfeiting [3,4] and displays [5]. Many strategies have been developed to realize ORTP materials. Among them, π···π stacking interactions [6–8], host–guest systems [9–11], co-assembly based on macrocyclic compounds [12], crystallization [13–15], halogen bonding [16,17], and doping in a polymer matrix [18], can be mentioned. Triimidazo[1,2-*a*:1′,2′-*c*:1″,2″-*e*][1,3,5]triazine, **TT** [7], is the prototype of a family of ORTP materials, showing in the crystalline phase ultralong phosphorescence (up to 1 s) associated with the presence of strong π···π stacking interactions [6]. Introduction of one or multiple halogen atoms on the **TT** scaffold results in a complex excitation dependent photoluminescent behavior including dual fluorescence, molecular and supramolecular phosphorescences, together with ultralong components [8,19–21]. The introduction of one chromophoric fragment, namely 2-fluoropyridine, 2-pyridine and pyrene (**TTPyr$_1$**) has further enriched the photophysical behavior, preserving the solid state ultralong component [22–24]. In particular, the pyrene functionality was chosen due to its favorable emissive properties (i.e., high fluorescence quantum efficiency and good hole-transporting ability) which can be exploited in a large number of applications [25–27]. **TTPyr$_1$** was isolated and characterized in two solvated phases and two not solvated polymorphs. In particular, the two solvated and one polymorph, belonging to centrosymmetric space groups, are obtained at room temperature, while a non-centrosymmetric polymorph can be irreversibly prepared by proper thermal treatment in air starting from the room temperature phases. While in solution the chromophore is characterized by a single fluorescence (quantum yield Φ = 90%, see Table 1), in the solid state intermolecular interactions impart rigidification to the molecule favoring the appearance of a long-lived component whose lifetime and intensity strongly depend on the crystallinity of the sample. Moreover, differently from the room temperature phases, the non-centrosymmetric one displays not only second order nonlinear optical features (SHG ten times that of standard urea), but also an additional fluorescence associated with the presence of two possible almost isoenergetic conformers in its crystals. Additionally, the material in aggregate-state was tested for cellular and bacterial imaging in view of its applications in bioimaging [24].

Table 1. Photophysical parameters of **Pyrene**, **TTPyr$_1$**, **TTPyr$_2$** and **TTPyr$_3$** at 298 K.

		Φ %	λ_{em} (nm)	τ_{av}	k_r (10^7 s^{-1})	k_{nr} (10^7 s^{-1})
Pyrene	DMSO (2.5 × 10^{-6} M)	33.4	374	94.15 ns	0.355	0.707
TTPyr$_1$	DMSO (10^{-5} M)	92	420	2.76 ns	33.3	2.90
	Ground Crystal (TTPyr RT)	54	475 / 514	2.15 ns / 5.19 ms	25.1	21.4
TTPyr$_2$	DMSO (2.5 × 10^{-6} M)	78	419	9.22 ns	8.46	2.39
	Powders	40.2	490 / 528	4.64 ns / 20.54 ms	8.66	12.9
TTPyr$_3$	DMSO (2.5 × 10^{-6} M)	74.4	422	11.16 ns	6.67	2.29
	Powders	36.9	476 / 522	5.12 ns / 40.62 ms	7.21	12.3

Here, we report the synthesis, characterization and photophysical studies of the analogue derivatives with two and three pyrene moieties on the **TT** scaffold, **TTPyr$_2$** and **TTPyr$_3$**, respectively. The photophysical behavior of the solid compounds, comprising one fluorescence and one phosphorescence, is similar to that of **TTPyr$_1$** in the room temperature phases. However, a systematic increase of the phosphorescence lifetime is observed by increasing the number of pyrene groups. Unfortunately, due to the impossibility to get single crystals suitable for X-ray diffraction analysis, an explanation of this phenomenon can be given only at a qualitative level.

2. Materials and Methods

All reagents were purchased from chemical suppliers and used without further purification. Pyrene was crystallized from hot toluene solution before photoluminescence measurements. Triimidazo[1,2-a:1′,2′-c:1″,2″-e][1,3,5]triazine (**TT**) [28], its dibromo- and tribromo-derivatives (namely 3,7-dibromotriimidazo[1,2-a:1′,2′-c:1″,2″-e][1,3,5]triazine, **TTBr$_2$**, and 3,7,11-tribromotriimidazo[1,2-a:1′,2′-c:1″,2″-e][1,3,5]triazine, **TTBr$_3$**) [19] and **TTPyr$_1$** [24] were prepared according to literature procedures.

^{1}H and ^{13}C NMR spectra were recorded on a Bruker AVANCE-400 instrument (400 MHz). Chemical shifts are reported in parts per million (ppm) and are referenced to the residual solvent peak (DMSO, ^{1}H 2.50 ppm, ^{13}C 39.5 ppm; CH$_2$Cl$_2$, ^{1}H 5.32 ppm, ^{13}C 54.0 ppm). Coupling constants (J) are given in hertz (Hz) and are quoted to the nearest 0.5 Hz. Peak multiplicities are described in the following way: s, singlet; d, doublet; m, multiplet.

Mass spectra were recorded on a Thermo Fisher (Thermo Fisher Scientific, Waltham, MA USA) LCQ Fleet Ion Trap Mass Spectrometer equipped with UltiMate™ 3000 HPLC system. UV-Visible spectra were collected by a Shimadzu UV3600 spectrophotometer (Shimadzu Italia S.r.l., Milan, Italy).

Absolute photoluminescence quantum yields were measured using a C11347 (Hamamatsu Photonics K.K). A description of the experimental setup and measurement method can be found in the article of K. Suzuki et al. [29]. For any fixed excitation wavelength, the fluorescence quantum yield Φ is given by:

$$\Phi = \frac{PN(Em)}{PN(Abs)} = \frac{\int \frac{\lambda}{hc} \left[I_{em}^{sample}(\lambda) - I_{em}^{reference}(\lambda) \right] d\lambda}{\int \frac{\lambda}{hc} \left[I_{ex}^{reference}(\lambda) - I_{ex}^{sample}(\lambda) \right] d\lambda}$$

where PN(Em) is the number of photons emitted from a sample and PN(Abs) is the number of photons absorbed by a sample, λ is the wavelength, h is Planck's constant, c is the velocity of light, $I_{em}^{sample}(\lambda)$ and $I_{em}^{reference}(\lambda)$ are the photoluminescence intensities with and without a sample, respectively, $I_{ex}^{sample}(\lambda)$ and $I_{ex}^{reference}(\lambda)$ are the integrated intensities of the excitation light with and without a sample, respectively. PN(Em) is calculated in the wavelength interval [λ_i, λ_f], where λ_i is taken 10 nm below the excitation wavelength, while λ_f is the upper end wavelength in the emission spectrum. The error made was estimated at around 5%.

Steady state emission and excitation spectra and photoluminescence lifetimes were obtained using a FLS 980 (Edinburgh Instrument Ltd., Livingston, United Kingdom) spectrofluorimeter. The steady state measurements were recorded by a 450 W Xenon arc lamp. Photoluminescence lifetime measurements were performed using: Edinburgh Picosecond Pulsed Diode Laser EPL-375, EPLED-300, (Edinburg Instrument Ltd.) and microsecond flash Xe-lamp (60 W, 0.1 ÷ 100 Hz) with data acquisition devices time correlated single-photon counting (TCSPC) and multi-channel scaling (MCS) methods, respectively. Average lifetimes are obtained as:

$$\tau_{av} \equiv \frac{\sum_{n=1}^{m} \alpha_n \tau_n^2}{\sum_{n=1}^{m} \alpha_n \tau_n},$$

where m is the multi-exponential decay number of the fit.

2.1. Synthesis of 3,7-Di(pyren-1-yl)triimidazo[1,2-a:1',2'-c:1",2"-e][1,3,5]triazine (**TT-Pyr$_2$**)

TTPyr$_2$ was prepared by Suzuki coupling between **TTBr$_2$** and pyrene-1-boronic acid (see Scheme 1). In a typical reaction, **TTBr$_2$** (595 mg; 1.67 mmol), pyren-1-ylboronic acid (970 mg, 3.94 mmol), potassium carbonate (1.3 g, 9.42 mmol), Pd(PPh$_3$)$_2$Cl$_2$ (170 mg, 0.24 mmol), water (3 mL) and DMF (25 mL) were transferred inside a 100 mL Schlenk flask equipped with a magnetic stirrer. The system was heated under static nitrogen atmosphere at 130 °C for 12 h. The reaction was then cooled to room temperature, precipitated with water (200 mL) and filtered on a Büchner. The solid crude reaction mixture was further purified by automated flash chromatography on SiO$_2$ with DCM/CH$_3$CN as eluents to give the **TTPyr$_2$** product as a yellow solid (724 mg; Yield 72%; R$_f$ = 0.5 in DCM/CH$_3$CN = 95/5).

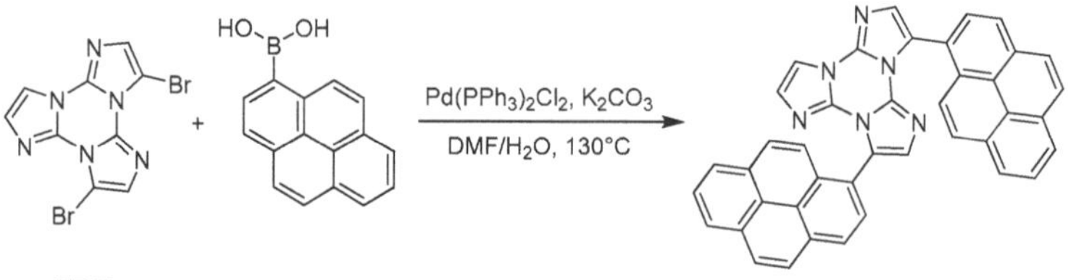

Scheme 1. Synthetic path for the preparation of **TTPyr$_2$**.

NMR data (9.4 T, DMSO-d$_6$, 298 K, δ, ppm): ^{1}H NMR 8.46–8.07 (18H, m), 8.01 (1H, d, J = 1.5 Hz), 7.53 (1H, s), 6.98 (1H, s), 6.88 (1H, d, J = 1.5 Hz); ^{13}C NMR 136.7, 136.6, 135.9, 131.4, 131.3, 130.9, 130.7, 130.4, 130.3, 129.8, 129.5, 128.8, 128.4, 128.2, 128.1, 128.0, 127.7, 127.5, 127.3, 127.2, 126.4, 125.7, 125.6, 125.4, 124.3, 124.2, 124.1, 123.7, 123.6, 123.5, 110.9.

MS (ESI-positive ion mode): m/z: 599 [M + H]$^+$.

2.2. Synthesis of 3,7,11-Tri(pyren-1-yl)triimidazo[1,2-a:1',2'-c:1",2"-e][1,3,5]triazine (**TT-Pyr₃**)

TTPyr₃ was prepared by Suzuki coupling between **TTBr₃** and pyrene-1-boronic acid (see Scheme 2). In a typical reaction, **TTBr₃** (400 mg; 0.91 mmol), pyren-1-ylboronic acid (702 mg, 2.85 mmol), potassium carbonate (1.3 g, 9.19 mmol), Pd(PPh$_3$)$_2$Cl$_2$ (64 mg, 0.09 mmol), water (2 mL) and DMF (10 mL) were transferred inside a 100 mL Schlenk flask equipped with a magnetic stirrer. The system was heated under static nitrogen atmosphere at 130 °C for 12 h. The reaction was then cooled to room temperature, precipitated with water (200 mL) and filtered on a Büchner. The solid crude reaction mixture was further purified by automated flash chromatography on SiO$_2$ with Hexane/Et$_2$O/DCM as eluents to give the **TTPyr₃** product as a pale yellow solid (435 mg; Yield 60%; R$_f$ = 0.5 in Hexane/Et$_2$O/DCM = 60/20/20).

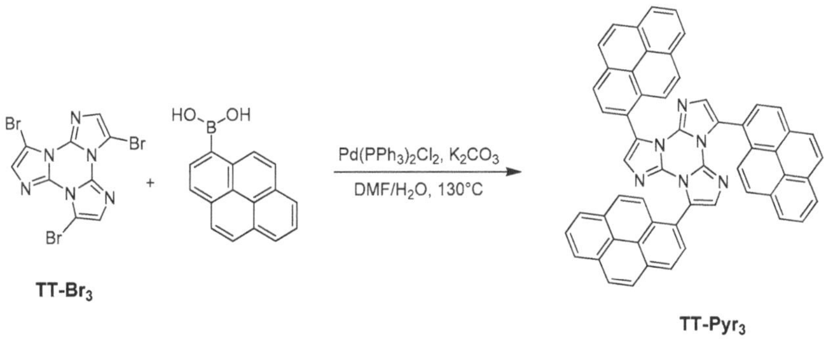

Scheme 2. Synthetic path for the preparation of **TTPyr₃**.

NMR data (9.4 T, CD$_2$Cl$_2$, 300 K, δ, ppm): ^{1}H NMR 8.34–8.06 (27H, m), 7.04 (3H, s); ^{13}C NMR 136.9, 132.7, 131.9, 131.4, 130.3, 129.0, 128.8, 127.9, 126.9, 126.3, 126.1, 125.6, 125.1, 124.7, 123.7.

MS (ESI-positive ion mode): m/z: 799 [M + H]$^+$.

2.3. Computational Details

Geometry optimization of **TTPyr₂** has been performed at the ωB97X/6-311++G(d,p) level of theory [30], in agreement with the computational protocol previously developed to study **TT** [7] and its derivatives, in particular **TTPyr₁** [24]. The adopted functional was in fact demonstrated to be able to accurately describe ground and excited states properties of **TT** and its derivatives, besides dispersive intermolecular interactions which are essential to interpret the multi-faceted properties of this family of compounds. Calculations on the tri-pyrene derivative have not be performed due to the required high computational costs. All calculations have been performed with Gaussian 16 [31].

3. Results

3.1. Synthesis and Molecular Structures

TTPyr₂ and **TTPyr₃** were prepared by Suzuki coupling between pyrene-1-boronic acid and the corresponding di- and tri-brominated **TT** precursors (see Schemes 1 and 2). The new pyrene derivatives were characterized by ^{1}H and ^{13}C NMR spectroscopy (Figures S1–S5) and mass spectrometry (Figures S3 and S6), samples with the purity grade required for photophysical characterization were obtained by repeated crystallizations.

In order to get insight on the structural features of the investigated compounds, DFT calculations were performed on the smaller derivative, **TTPyr₂** (see optimized structures in Figure 1). Owing to the lack of the crystal structure of this compound, suitable models were built up starting from structural information previously obtained on **TTPyr₁** by both X-ray diffraction analysis and theoretical calculations [24]. For this latter compound, crystallizing in different phases and polymorphs, two possible conformations were detected, which correspond to different orientation of pyrene with respect to the **TT** moiety as denoted by the (H)C1-C2-C10-C11(H) torsion angle τ (C2-C10 being the bond connecting **TT** with pyrene). In particular, the room temperature phases display τ values equal to about 129° (RT conformation), while the polymorph obtained at high temperature shows τ equal to 49° (HT conformation) [24]. DFT geometry optimization of the two conformations leads to different almost isoenergetic minima (having τ = 110.3 and 67.4°, respectively) in a very flat region. The HT conformation is only 0.6 kcal/mol more stable than the RT one, suggesting that both conformations are expected to be present in solution. To build up the **TTPyr$_2$** model, an additional pyrene moiety was added, in the proper position, to both the RT and HT optimized geometries of **TTPyr$_1$** imposing the same RT or HT conformation. The resulting optimized structures (see Figure 1), denoted respectively as 'RT/RT' and 'HT/HT' conformations, display C1–C2–C10–C11 and C1′–C2′–C10′–C11′ torsion angles equal to 112.3 and −67.2° (RT/RT) and 67.4 and 67.3° (HT/HT conformation), the latter being more stable by 0.54 kcal/mol. Of course, also the mixed 'RT/HT' conformation is envisaged to be present in solution, but no significant differences in the electronic levels are expected with respect to those computed by TDDFT for the RT/RT and HT/HT conformations, which are themselves very similar (see Table S1).

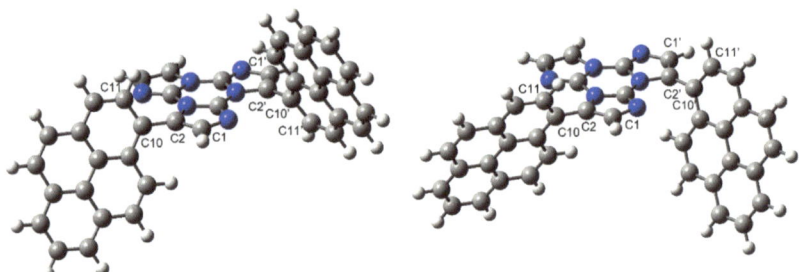

Figure 1. ωB97X/6-311++G(d,p) optimized geometry of **TTPyr$_2$** in the RT/RT (**left**) and HT/HT (**right**) conformations (hydrogen, carbon and nitrogen atoms in white, grey and blue, respectively).

Owing to the twisted arrangement of **TT** and pyrene in the investigated systems, a Quantum Theory of Atoms in Molecules (QTAIM) topological analysis [32] on the **TTPyr$_1$** and **TTPyr$_2$** electron density was performed to detect the possible presence of intramolecular interactions between the two moieties. In all cases, a bond critical point (bcp) between a **TT** nitrogen atom (N3) and the pyrene C10 carbon atom, located at about 3.18 Å from each other, has been found ($\rho_{bcp} \sim 0.05$ e Å^{-3}, see Figures S7 and S8 for the **TTPyr$_1$** and **TTPyr$_2$** molecular graphs). Owing to the nature of the involved atoms, such N···C bond paths cannot be considered as classical non-covalent interactions [33], though contributing to stabilize the molecule due to the privileged electron-exchange channel associated with the bond path [34].

Interestingly, the molecular graphs substantially change when considering the optimized geometries of **TTPyr$_1$** and **TTPyr$_2$** in the excited state (see Figures S9 and S10). The relaxed S$_1$ geometries are characterized by lower molecular twisting than the ground state one. In particular, for **TTPyr$_1$** τ varies from 110.3 to 138.5° (RT conformation) and from 67.4 to 41.5° (HT conformation), respectively, with a concomitant shortening of the C2-C10 bond (from 1.478 to 1.435 Å), implying a greater conjugation between the **TT** and pyrene moieties compared with the ground state [24]. For **TTPyr$_2$**, on the other hand, only one pyrene unit undergoes a substantial relaxation with respect to **TT**, τ varying from 112.3 to 139.0° (RT/RT conformation) and from 67.3 to 40.8° (HT/HT conformation). The corresponding C2–C10 bond shortens from 1.477 to 1.422 and 1.433 Å (RT/RT and HT/HT conformation, respectively). For the other pyrene unit, τ shows only slight variations (from −67.2 to −64.6° in the RT/RT conformation and from 67.4 to 64.0° in the HT/HT conformation) and the C2′–C10′ bond remains almost unvaried. These results indicate that, as determined for **TTPyr$_1$**, also **TTPyr$_2$** shows in the excited state a greater conjugation, which is however extended on **TT** and only one pyrene unit. The associated molecular graphs (see Figures S9 and S10) reveal, for both conformations of **TTPyr$_1$** and **TTPyr$_2$**, the formation of an intramolecular C–H···N hydrogen bond between the **TT** and pyrene moieties. This interaction stabilizes the two conformations in the excited state allowing to reduce the molecular twisting.

3.2. Photophysical Studies

TTPyr$_2$ and **TTPyr$_3$** have been photophysically characterized both in solution and in solid state by steady state and time resolved spectroscopy. The results are reported in Table 1 together with those previously obtained for the room temperature non-solvated species of **TTPyr$_1$** and those of pyrene molecule in the same conditions.

Diluted DMSO solutions of the three compounds show two absorptions with peaks at 257, 268, 279 nm (ES$_2$) and 332, 347 nm (ES$_1$) respectively, with molar absorption coefficients, ε, computed at 347 nm, equal to 34,894, 46,408 and 84,443 M^{-1} cm^{-1} for **TTPyr$_1$**, **TTPyr$_2$** and **TTPyr$_3$**, respectively. Such bands, even though strongly red shifted and broadened with respect to the parent pyrene moiety (see Figure 2), are ascribed to transitions of main pyrene character, as previously suggested based on the similarity with the absorption spectrum of pyrene itself in the same conditions and theoretical calculations [24]. Looking at the HOMO and LUMO energies of pyrene, **TTPyr$_1$**, and **TTPyr$_2$**, it is found that, while the HOMO energy is almost constant along the series, the LUMO one significantly decreases (by 0.17 or 0.19 eV according to the conformation) from pyrene to **TTPyr$_1$**, and remains almost unvaried going to **TTPyr$_2$**.

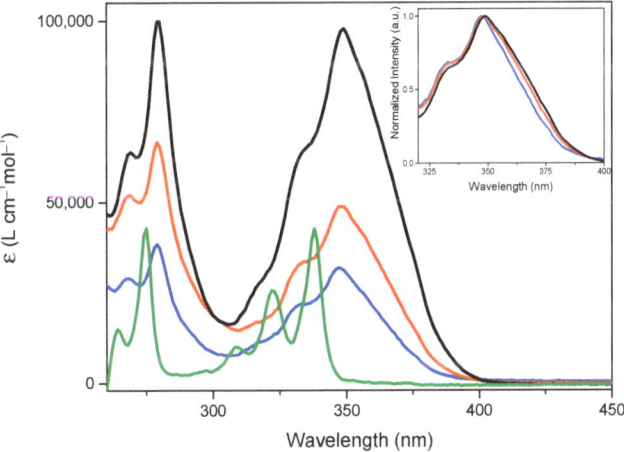

Figure 2. Molar absorption coefficient for 2.5 10^{-6} M DMSO solution of **TTPyr$_1$** (blue line), **TTPyr$_2$** (red line), **TTPyr$_3$** (black line) and pyrene (green line). Inset: expansion of 350 nm region for normalized spectra of **TTPyr$_1$** (blue line), **TTPyr$_2$** (red line) and **TTPyr$_3$** (black line).

For the three compounds, perfectly overlapped emission spectra, consisting in an intense very broad single fluorescent emission at about 420 nm, are measured (see Figure 3). The three compounds display impressively high quantum yields when compared with pyrene itself (Φ = 33.4%) in the same conditions (see Table 1), suggesting that the **TT** moiety successfully suppresses the ACQ (aggregation caused quenching) phenomena affecting pyrene fluorescence [35–37]. Specifically, Φ equal to 92, 78 and 74% have been measured for **TTPyr$_1$**, **TTPyr$_2$**, **TTPyr$_3$**, respectively, with lifetime, τ, equal to 2.76 [24], 9.22 and 11.16 ns (Figures S11 and S14), resulting in k$_r$ values 33.3, 8.46 and 6.67 × 10^7 s^{-1} and k$_{nr}$ 2.90, 2.39

and 2.29×10^7 s^{-1}, respectively (see Table 1). The observed quantum yields are similar or higher than those reported in the literature for other mono-, di- and tri-pyrene substituted families. For example, DCM solutions of di- and tri-pyrene derivatives of carbazole and mono-, di- and tri-pyrene functionalized adamantane display Φ equal to 94 and 72% [32] and 17, 19 and 19% [38] respectively.

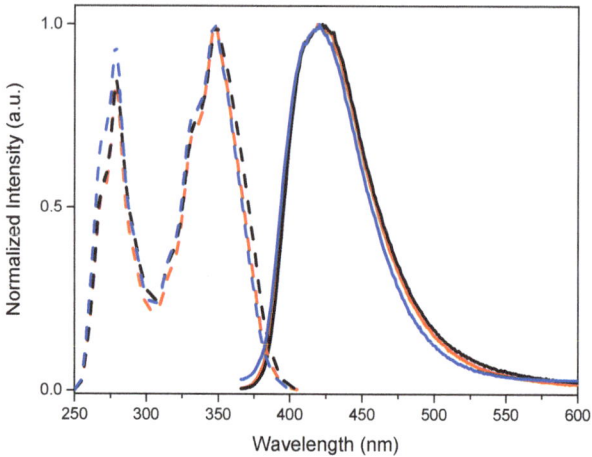

Figure 3. Normalized emission (λ_{exc} = 350 nm, solid lines) and excitation (λ_{em} = 420 nm, dashed lines) spectra of 2.5×10^{-6} M DMSO solutions at 298 K of **TTPyr$_1$** (blue line), **TTPyr$_2$** (red line) and **TTPyr$_3$** (black line).

To justify the slight decrease of quantum yield for **TTPyr$_2$** and **TTPyr$_3$** with respect to **TTPyr$_1$**, close inspection of optical properties has been performed. When better focusing on the 347 nm absorption of the three compounds (ES$_1$, see inset of Figure 2), besides the already mentioned macroscopic red shift and broadening with respect to pyrene, a slight red shift and widening of the band from **TTPyr$_1$** to **TTPyr$_2$** and **TTPyr$_3$** can also be recognized. Based on DFT/TDDFT calculations, such band results from an envelope of more transitions (see Table S1 and Figures S20 and S21 for the simulated absorption spectra). In particular, two transitions are computed at low energy for pyrene (at 301 and 294 nm with oscillator strength f = 0.0003 and 0.347, respectively) and **TTPyr$_1$** (at 304 and 302 nm with f = 0.060 and 0.498, respectively), with a bathochromic displacement of the stronger transition and intensification of the low energy one, justifying the experimental red shift and broadening of **TTPyr$_1$** ES$_1$. Going to **TTPyr$_2$**, four transitions are computed in the same region (at 306, 304, 303 and 301 nm with f = 0.380, 0.101, 0.066 and 0.445, respectively, see Table S1 and Figures S20 and S21), implying further broadening of the associated band with very minor red shift with respect to **TTPyr$_1$**. It is as well to be taken into account that more conformations of similar energy are expected in solution, due to the flatness of the potential energy surface associated with the rotation of pyrene with respect to TT [24]. As reported in Table S1, such conformations are characterized by the same position of the transitions but slightly different oscillator strengths, further contributing to the band broadening. Similar results are expected for the tri-pyrene derivative, which was not submitted to theoretical calculations owing to the computational costs required by the further increased number of conformational degrees of freedom. Geometry optimization of S$_1$ for both conformations of **TTPyr$_2$** leads, as determined for **TTPyr$_1$**, to a reduced molecular twisting (see Figures S9 and S10) with a large energy separation from the higher excited states and approximately the same energy position (λ = 372 nm), oscillator strength (0.693) and orbital composition (HOMO-LUMO with 93% weight) as found for **TTPyr$_1$** (λ = 370 nm, f = 0.698, HOMO-LUMO with weight = 94%). Based on the strict similarity between **TTPyr$_1$** and **TTPyr$_2$** at molecular level, minor ACQ phenomena should be taken into account for **TTPyr$_2$**

and **TTPyr₃**. A similar result was previously obtained for pyrene-functionalized carbazole derivatives, with fluorescence quantum yield decreasing by increasing the number of pyrene moieties [39].

The three compounds are good emitters also in the solid state (see Table 1), where the emission is implemented by the presence, as already reported for **TTPyr₁** [24], of one phosphorescent component in addition to the fluorescent band (at 475, 490 and 476 nm for **TTPyr₁**, **TTPyr₂**, **TTPyr₃**, respectively, see Figure 4). The long lived component (at 514, 528 and 522 nm, with lifetimes equal to 5.19 [24], 20.54 and 40.62 ms for **TTPyr₁**, **TTPyr₂**, **TTPyr₃** respectively, see Figures S13 and S16) can be selectively activated by exciting at low energy (480 nm). Based on the conclusions drawn for **TTPyr₁** [24], even for **TTPyr₂** and **TTPyr₃** such long-lived emission could be explained by (i) easy singlet-to-triplet intersystem crossing due to almost overlapped singlet and triplet energy levels (see Figure S20), and (ii) interchromophoric interactions in the solid state which inhibit the molecular motions reducing the thermal non-radiative dissipation. Unfortunately, differently from **TTPyr₁** where such interactions have been deeply analyzed for the different isolated and crystallized phases, for **TTPyr₂** and **TTPyr₃** no single crystals suitable for X-ray diffraction analysis could be obtained. However, based on the longer lifetimes of **TTPyr₃**, a high number of interactions are expected for this latter with respect to the other members of the family. Such interactions are not efficacious for producing crystals due to the large number of conformational degrees of freedom (as expected, even though at lower extent, for **TTPyr₂**).

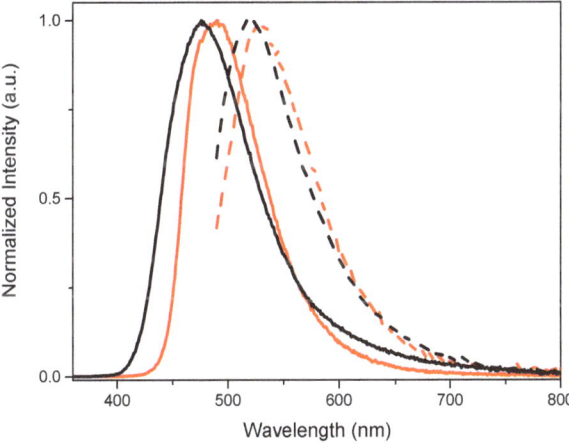

Figure 4. Normalized emission spectra (λ_{exc} = 350 nm solid line, λ_{exc} = 480 nm dashed line) at 298 K of powders of **TTPyr₂** (red) and **TTPyr₃** (black).

4. Conclusions

In conclusion, two new intriguing chromophores, **TTPyr₂** and **TTPyr₃**, have been prepared and characterized. Their photophysical behavior fully analyzed in solution and in solid state has been compared with that of parent **TTPyr₁** revealing a trend depending on the number of pyrene moieties appended to the **TT** scaffold. In particular, the fluorescence quantum yield in solution (much higher than that of pyrene itself) slightly decreases by increasing the number of pyrene units, while, in the solid state, the phosphorescent component, visible for all compounds at room temperature, possesses lifetimes increasing in the same order due to stabilizing interchromophoric interactions.

Supplementary Materials: The following are available online at https://www.mdpi.com/article/10.3390/photochem1030031/s1, **Figure S1.** 1H NMR spectrum and expanded region of **TTPyr2** (400 MHz, DMSO-d6), **Figure S2.** 13C NMR spectrum and expanded region of **TTPyr2** (400 MHz, DMSO-d6), **Figure S3.** LC-MS profile of **TTPyr2, Figure S4.** 1H NMR spectrum and expanded region of **TTPyr3** (400 MHz, CD2Cl2), **Figure S5.** 13C NMR spectrum and expanded region of **TTPyr3** (400 MHz, CD2Cl2), **Figure S6.** LC-MS profile of **TTPyr3, Figure S7.** Molecular graphs of **TTPyr1** in ground state RT (left) and HT (right) conformation with bond paths, bond critical points (green circles) and ring critical points (red circles), **Figure S8.** Molecular graphs of **TTPyr2** in ground state RT/RT (left) and HT/HT (right) conformation with bond paths, bond critical points (green circles) and ring critical points (red circles), **Figure S9.** Molecular graphs of **TTPyr1** in excited state RT (left) and HT (right) conformation with bond paths, bond critical points (green circles) and ring critical points (red circles), **Figure S10.** Molecular graphs of **TTPyr2** in excited state RT/RT (left) and HT/HT (right) conformation with bond paths, bond critical points (green circles) and ring critical points (red circles), **Figure S11.** Lifetime measurement (exc = 374 nm, em = 420 nm) of **TTPyr2** in DMSO at 298 K, **Figure S12.** Lifetime measurement (exc = 375 nm, em = 490 nm) of **TTPyr2** powders at 298 K, **Figure S13.** Lifetime measurement (exc = 340 nm, em = 530 nm) of **TTPyr2** powders at 298 K, **Figure S14.** Lifetime measurement (exc = 374 nm, em = 420 nm) of **TTPyr3** in DMSO at 298 K, **Figure S15.** Lifetime measurement (exc = 300 nm, em = 470 nm) of **TTPyr3** powders at 298 K, **Figure S16.** Lifetime measurement (exc = 340 nm, em = 520 nm) of **TTPyr3** powders at 298 K, **Figure S17.** Lifetime measurement (exc = 374 nm, em = 394 nm) of **Pyrene** in DMSO at 298 K, **Figure S18.** Photographs of **TTPyr2** (left) and **TTPyr3** (right) solutions under UV light OFF (left) or UV light ON (right, exc = 366 nm), **Figure S19.** Photographs of powders of **TTPyr2** (left) and **TTPyr3** (right) under UV light OFF (left) or UV light ON (right, exc = 366 nm). **Table S1.** Excitation energies (nm), oscillator strength (f) and composition of the first singlet transitions computed for pyrene, **TT-Pyr1** (HT and RT conformations) and **TT-Pyr2** (HT/HT and RT/RT conformations), **Figure S20.** Electronic levels computed for pyrene, **TTPyr1** and **TTPyr2** at molecular level. In blue are reported the singlet levels with oscillator strength f 0.001 and the corresponding values of f (see detailed information in Table S1), **Figure S21.** Simulated absorption spectra of pyrene (top), **TTPyr1**, RT conformation (middle) and **TTPyr2**, RT/RT conformation (bottom) at B97X/6-311++G(d,p) level of theory, resulting from convolution of the singlet excitation energies with 0.1 eV of half-bandwidth (singlet levels plotted as blue sticks according to their oscillator strength), **Figure S22.** Plots of the B97X/6-311++G(d,p) MOs mainly involved in the lowest energy transitions of **TT-Pyr1** in RT conformation (isosurfaces value 0.02), **Figure S23.** Plots of the B97X/6-311++G(d,p) MOs mainly involved in the lowest energy transitions of **TT-Pyr2** in RT/RT conformation (isosurfaces value 0.02). **Table S1.** Excitation energies, oscillator strength and composition of the first singlet transitions computed for pyrene, **TT-Pyr$_1$** and **TT-Pyr$_2$**.

Author Contributions: Conceptualization, E.C., A.F. and D.M. (Daniele Malpicci); Investigation, D.M. (Daniele Malpicci), D.M. (Daniele Marinotto) and E.L.; Validation, C.G.; Quantum-mechanical calculations, A.F.; Writing the original draft, E.C. and A.F.; Writing—Review and editing, E.C., A.F. and E.L; Funding acquisition, E.C., A.F. and C.G. All authors have read and agreed to the published version of the manuscript.

Funding: This research received no external funding.

Institutional Review Board Statement: Not applicable.

Informed Consent Statement: Not applicable.

Data Availability Statement: Not applicable.

Acknowledgments: The use of instrumentation purchased through the Regione Lombardia-Fondazione Cariplo joint SmartMatLab Project is gratefully acknowledged.

Conflicts of Interest: The authors declare no conflict of interest.

References

1. Qin, W.; Zhang, P.; Li, H.; Lam, J.W.Y.; Cai, Y.; Kwok, R.T.K.; Qian, J.; Zheng, W.; Tang, B.Z. Ultrabright red AIEgens for two-photon vascular imaging with high resolution and deep penetration. *Chem. Sci.* **2018**, *9*, 2705–2710. [CrossRef] [PubMed]
2. Zhi, J.; Zhou, Q.; Shi, H.; An, Z.; Huang, W. Organic Room Temperature Phosphorescence Materials for Biomedical Applications. *Chem. Asian J.* **2020**, *15*, 947–957. [CrossRef]

3. Gu, L.; Wu, H.; Ma, H.; Ye, W.; Jia, W.; Wang, H.; Chen, H.; Zhang, N.; Wang, D.; Qian, C.; et al. Color-tunable ultralong organic room temperature phosphorescence from a multicomponent copolymer. *Nat. Commun.* **2020**, *11*, 944. [CrossRef]
4. Lei, Y.; Dai, W.; Guan, J.; Guo, S.; Ren, F.; Zhou, Y.; Shi, J.; Tong, B.; Cai, Z.; Zheng, J.; et al. Wide-Range Color-Tunable Organic Phosphorescence Materials for Printable and Writable Security Inks. *Angew. Chem. Int. Ed.* **2020**, *59*, 16054–16060. [CrossRef]
5. Hirata, S.; Totani, K.; Kaji, H.; Vacha, M.; Watanabe, T.; Adachi, C. Reversible Thermal Recording Media Using Time-Dependent Persistent Room Temperature Phosphorescence. *Adv. Opt. Mater.* **2013**, *1*, 438–442. [CrossRef]
6. An, Z.; Zheng, C.; Tao, Y.; Chen, R.; Shi, H.; Chen, T.; Wang, Z.; Li, H.; Deng, R.; Liu, X.; et al. Stabilizing triplet excited states for ultralong organic phosphorescence. *Nat. Mater.* **2015**, *14*, 685–690. [CrossRef] [PubMed]
7. Lucenti, E.; Forni, A.; Botta, C.; Carlucci, L.; Giannini, C.; Marinotto, D.; Previtali, A.; Righetto, S.; Cariati, E. H-Aggregates Granting Crystallization-Induced Emissive Behavior and Ultralong Phosphorescence from a Pure Organic Molecule. *J. Phys. Chem. Lett.* **2017**, *8*, 1894–1898. [CrossRef] [PubMed]
8. Lucenti, E.; Forni, A.; Botta, C.; Carlucci, L.; Giannini, C.; Marinotto, D.; Pavanello, A.; Previtali, A.; Righetto, S.; Cariati, E. Cyclic Triimidazole Derivatives: Intriguing Examples of Multiple Emissions and Ultralong Phosphorescence at Room Temperature. *Angew. Chem. Int. Ed.* **2017**, *56*, 16302–16307. [CrossRef] [PubMed]
9. Ma, X.-K.; Liu, Y. Supramolecular Purely Organic Room-Temperature Phosphorescence. *Acc. Chem. Res.* **2021**, *54*, 3403–3414. [CrossRef] [PubMed]
10. Kabe, R.; Adachi, C. Organic long persistent luminescence. *Nature* **2017**, *550*, 384–387. [CrossRef]
11. Li, D.; Lu, F.; Wang, J.; Hu, W.; Cao, X.-M.; Ma, X.; Tian, H. Amorphous Metal-Free Room-Temperature Phosphorescent Small Molecules with Multicolor Photoluminescence via a Host–Guest and Dual-Emission Strategy. *J. Am. Chem. Soc.* **2018**, *140*, 1916–1923. [CrossRef]
12. Zhang, Z.; Xu, W.-W.; Xu, W.-S.; Niu, J.; Sun, X.; Liu, Y. A Synergistic Enhancement Strategy for Realizing Ultralong and Efficient Room-Temperature Phosphorescence. *Angew. Chem. Int. Ed.* **2020**, *59*, 18748–18754. [CrossRef]
13. Hayduk, M.; Riebe, S.; Voskuhl, J. Phosphorescence Through Hindered Motion of Pure Organic Emitters. *Chem. A Eur. J.* **2018**, *24*, 12221–12230. [CrossRef]
14. Baroncini, M.; Bergamini, G.; Ceroni, P. Rigidification or interaction-induced phosphorescence of organic molecules. *Chem. Commun.* **2017**, *53*, 2081–2093. [CrossRef]
15. Sun, L.; Zhu, W.; Yang, F.; Li, B.; Ren, X.; Zhang, X.; Hu, W. Molecular cocrystals: Design, charge-transfer and optoelectronic functionality. *Phys. Chem. Chem. Phys.* **2018**, *20*, 6009–6023. [CrossRef] [PubMed]
16. Bolton, O.; Lee, K.; Kim, H.-J.; Lin, K.Y.; Kim, J. Activating Efficient Phosphorescence from Purely Organic Materials by Crystal Design. *Nat. Chem.* **2011**, *3*, 205–210. [CrossRef]
17. Shi, H.; An, Z.; Li, P.-Z.; Yin, J.; Xing, G.; He, T.; Chen, H.; Wang, J.; Sun, H.; Huang, W.; et al. Enhancing Organic Phosphorescence by Manipulating Heavy-Atom Interaction. *Cryst. Growth Des.* **2016**, *16*, 808–813. [CrossRef]
18. Lin, Z.; Kabe, R.; Nishimura, N.; Jinnai, K.; Adachi, C. Organic Long-Persistent Luminescence from a Flexible and Transparent Doped Polymer. *Adv. Mater.* **2018**, *30*, 1803713. [CrossRef] [PubMed]
19. Lucenti, E.; Forni, A.; Botta, C.; Carlucci, L.; Colombo, A.; Giannini, C.; Marinotto, D.; Previtali, A.; Righetto, S.; Cariati, E.M. The Effect of Bromo Substituents on the Multifaceted Emissive and Crystal-Packing Features of Cyclic Triimidazole Derivatives. *ChemPhotoChem* **2018**, *2*, 801–805. [CrossRef]
20. Lucenti, E.; Forni, A.; Botta, C.; Giannini, C.; Malpicci, D.; Marinotto, D.; Previtali, A.; Righetto, S.; Cariati, E.M. Intrinsic and Extrinsic Heavy-Atom Effects on the Multifaceted Emissive Behavior of Cyclic Triimidazole. *Chem. A Eur. J.* **2019**, *25*, 2452–2456. [CrossRef] [PubMed]
21. Giannini, C.; Forni, A.; Malpicci, D.; Lucenti, E.; Marinotto, D.; Previtali, A.; Carlucci, L.; Cariati, E. Room Temperature Phosphorescence from Organic Materials: Unravelling the Emissive Behaviour of Chloro-Substituted Derivatives of Cyclic Triimidazole. *Eur. J. Org. Chem.* **2021**, *2021*, 2041–2049. [CrossRef]
22. Previtali, A.; Lucenti, E.; Forni, A.; Mauri, L.; Botta, C.; Giannini, C.; Malpicci, D.; Marinotto, D.; Righetto, S.; Cariati, E. Solid State Room Temperature Dual Phosphorescence from 3-(2-Fluoropyridin-4-yl)triimidazo[1,2-a:1′,2′-c:1″,2″-e][1,3,5]triazine. *Molecules* **2019**, *24*, 2552. [CrossRef]
23. Lucenti, E.; Forni, A.; Previtali, A.; Marinotto, D.; Malpicci, D.; Righetto, S.; Giannini, C.; Virgili, T.; Kabacinski, P.; Ganzer, L.; et al. Unravelling the intricate photophysical behavior of 3-(pyridin-2-yl)triimidazotriazine AIE and RTP polymorphs. *Chem. Sci.* **2020**, *11*, 7599–7608. [CrossRef] [PubMed]
24. Previtali, A.; He, W.; Forni, A.; Malpicci, D.; Lucenti, E.; Marinotto, D.; Carlucci, L.; Mercandelli, P.; Ortenzi, M.A.; Terraneo, G.; et al. Tunable Linear and Nonlinear Optical Properties from Room Temperature Phosphorescent Cyclic Triimidazole-Pyrene Bio-Probe. *Chem. A Eur. J.* **2021**. [CrossRef] [PubMed]
25. Figueira-Duarte, T.M.; Müllen, K. Pyrene-Based Materials for Organic Electronics. *Chem. Rev.* **2011**, *111*, 7260–7314. [CrossRef]
26. De Silva, T.P.D.; Youm, S.G.; Fronczek, F.R.; Sahasrabudhe, G.; Nesterov, E.E.; Warner, I.M. Pyrene-Benzimidazole Derivatives as Novel Blue Emitters for OLEDs. *Molecules* **2021**, *26*, 6523. [CrossRef] [PubMed]
27. Yang, J.; Qin, J.; Ren, Z.; Peng, Q.; Xie, G.; Li, Z. Pyrene-Based Blue AIEgen: Enhanced Hole Mobility and Good EL Performance in Solution-Processed OLEDs. *Molecules* **2017**, *22*, 2144. [CrossRef]
28. Schubert, D.M.; Natan, D.T.; Wilson, D.C.; Hardcastle, K.I. Facile Synthesis and Structures of Cyclic Triimidazole and Its Boric Acid Adduct. *Cryst. Growth Des.* **2011**, *11*, 843–850. [CrossRef]

29. Suzuki, K.; Kobayashi, A.; Kaneko, S.; Takehira, K.; Yoshihara, T.; Ishida, H.; Shiina, Y.; Oishi, S.; Tobita, S. Reevaluation of absolute luminescence quantum yields of standard solutions using a spectrometer with an integrating sphere and a back-thinned CCD detector. *Phys. Chem. Chem. Phys.* **2009**, *11*, 9850–9860. [CrossRef]
30. Chai, J.-D.; Head-Gordon, M. Systematic optimization of long-range corrected hybrid density functionals. *J. Chem. Phys.* **2008**, *128*, 084106. [CrossRef]
31. Frisch, M.J.; Trucks, G.W.; Schlegel, H.B.; Scuseria, G.E.; Robb, M.A.; Cheeseman, J.R.; Scalmani, G.; Barone, V.; Petersson, G.A.; Nakatsuji, H.; et al. *Gaussian 16*; Revsion. A. 03; Gaussian Inc.: Wallingford, CT, USA, 2016.
32. Bader, R.F.W. *Atoms in Molecules: A Quantum Theory*; Clarendon Press: Oxford, UK, 1990.
33. Bader, R.F.W. Bond Paths Are Not Chemical Bonds. *J. Phys. Chem. A* **2009**, *113*, 10391–10396. [CrossRef]
34. Pendás, A.M.; Francisco, E.; Blanco, M.A.; Gatti, C. Bond Paths as Privileged Exchange Channels. *Chem. A Eur. J.* **2007**, *13*, 9362–9371. [CrossRef] [PubMed]
35. Winnik, F.M. Photophysics of preassociated pyrenes in aqueous polymer solutions and in other organized media. *Chem. Rev.* **1993**, *93*, 587–614. [CrossRef]
36. Lee, S.H.; Kim, S.H.; Kim, S.K.; Jung, J.H.; Kim, J.S. Fluorescence Ratiometry of Monomer/Excimer Emissions in a Space-Through PET System. *J. Org. Chem.* **2005**, *70*, 9288–9295. [CrossRef] [PubMed]
37. Suzuki, I.; Ui, A.M.; Yamauchi, A. Supramolecular Probe for Bicarbonate Exhibiting Anomalous Pyrene Fluorescence in Aqueous Media. *J. Am. Chem. Soc.* **2006**, *128*, 4498–4499. [CrossRef]
38. Wrona-Piotrowicz, A.; Makal, A.; Zakrzewski, J. Triflic Acid-Promoted Adamantylation and tert-Butylation of Pyrene: Fluorescent Properties of Pyrene-Decorated Adamantanes and a Channeled Crystal Structure of 1,3,5-Tris(pyren-2-yl)adamantane. *J. Org. Chem.* **2020**, *85*, 11134–11139. [CrossRef] [PubMed]
39. Kotchapradist, P.; Prachumrak, N.; Tarsang, R.; Jungsuttiwong, S.; Keawin, T.; Sudyoadsuk, T.; Promarak, V. Pyrene-functionalized carbazole derivatives as non-doped blue emitters for highly efficient blue organic light-emitting diodes. *J. Mater. Chem. C* **2013**, *1*, 4916–4924. [CrossRef]

Article

Solution and Solid-State Optical Properties of Trifluoromethylated 5-(Alkyl/aryl/heteroaryl)-2-methyl-pyrazolo[1,5-*a*]pyrimidine System

Felipe S. Stefanello [1], Jean C. B. Vieira [1], Juliane N. Araújo [1], Vitória B. Souza [2], Clarissa P. Frizzo [1], Marcos A. P. Martins [1], Nilo Zanatta [1], Bernardo A. Iglesias [2,*] and Helio G. Bonacorso [1,*]

[1] Núcleo de Química de Heterociclos (NUQUIMHE), Departamento de Química, Universidade Federal de Santa Maria, Santa Maria 97105-900, RS, Brazil; felipestefa@gmail.com (F.S.S.); jeanbauer96@gmail.com (J.C.B.V.); julianenascimentoaraujo2@gmail.com (J.N.A.); clarissa.frizzo@gmail.com (C.P.F.); marcos.nuquimhe@gmail.com (M.A.P.M.); nilo.zanatta@ufsm.br (N.Z.)
[2] Laboratório de Bioinorgânica e Materiais Porfirínicos, Departamento de Química, Universidade Federal de Santa Maria, Santa Maria 97105-900, RS, Brazil; vitoriabarbosadesouza@gmail.com
* Correspondence: bernardopgq@gmail.com (B.A.I.); helio.bonacorso@ufsm.br (H.G.B.)

Abstract: This paper describes the photophysical properties of a series of seven selected examples of 5-(alkyl/aryl/heteroaryl)-2-methyl-7-(trifluoromethyl)pyrazolo[1,5-*a*]pyrimidines (3), which contain alkyl, aryl, and heteroaryl substituents attached to the scaffolds of 3. Given the electron-donor groups and -withdrawing groups, the optical absorption and emission in the solid state and solution showed interesting results. Absorption UV–Vis and fluorescence properties in several solvents of a pyrazolo[1,5-*a*]pyrimidines series were investigated, and all derivatives were absorbed in the ultraviolet region despite presenting higher quantum emission fluorescence yields in solution and moderate emission in the solid state. Moreover, the solid-state thermal stability of compounds **3a–g** was assessed using thermogravimetric analysis. The thermal decomposition profile showed a single step with almost 100% mass loss for all compounds **3**. Additionally, the values of $T_{0.05}$ are considerably low (72–187 °C), especially for compound **3a** (72 °C), indicating low thermal stability for this series of pyrazolo[1,5-*a*]pyrimidines.

Keywords: pyrazoles; pyrimidines; pyrazolo[1,5-*a*]pyrimidines; photophysical properties

1. Introduction

According to the Web of Science, there have been about nine hundred publications on photophysical properties and organic compounds in the last five years [1]. This shows the importance of synthesizing organic compounds with these photophysical characteristics, which have drawn considerable attention and have been widely used in industrial and scientific fields [2].

For many organic molecules to exhibit outstanding photophysical properties, in most cases, a combination of factors is required, which are related mainly to their structural properties. These properties may involve the polarization of the chemical scaffolds due to the presence of electron-donating (EDG) and electron-withdrawing groups (EWG) [3], chain arrangements, and conformations (stereochemistry) [4–6], as well as the presence of charge-transfer bands, such as *intramolecular charge transfer* transitions (ICT) [7,8].

In this regard, *N*-heterocyclic skeletons present many classes of compounds that exhibit photophysical properties [9–12]. One such class is the pyrazolo[1,5-*a*]pyrimidines that have π-extended electronic systems by two planar fused rings with three nitrogen atoms of different electronic atom nature [13]; in fact, given its structural diversity, numerous studies have highlighted its importance in materials science [14–19].

For these reasons, this study sought to evaluate and study, for the first time, the photophysical properties of pyrazolo-pyridimine derivatives, more specifically, the compounds

named 5-(alkyl/aryl/heteroaryl)-2-methyl-7-(trifluoromethyl)pyrazolo[1,5-a]pyrimidines (3), where the synthetic approaches have already been mostly described in the literature [13,20,21], although there is still a lack of studies on the absorption and emission properties of these derivatives, both in solution and in the solid state. Given this context, UV–Vis absorption analysis and steady-state fluorescence emission properties, both in liquid and the solid state, will be discussed and studied. Furthermore, the solvent polarity on absorption and emission effects and the thermal stability in the solid state will also be discussed and presented (Scheme 1).

Scheme 1. Summary of this study: synthesis and photophysical properties of 5-(alkyl/aryl/heteroaryl)-2-methyl-7-(trifluoromethyl)pyrazolo[1,5-a]pyrimidines (3).

2. Materials and Methods

2.1. General

Unless otherwise indicated, all common reagents and solvents were used as obtained from commercial suppliers without further purification. The ^{1}H, ^{13}C, and NMR spectra were acquired on a Bruker Avance III 600 MHz (3a–g) spectrometer for one-dimensional experiments with 5 mm sample tubes at 298 K and digital resolution of 0.01 ppm in CDCl$_3$ as the solvent, using TMS as the internal reference, and the atoms numbering according to Figure 1. All spectra can be found in the Supplementary Information (Figures S1–S8). All melting points were determined using coverslips on a Microquímica MQAPF-302 apparatus and are uncorrected. The HRMS analyses were performed on a hybrid high-resolution and high-accuracy (5 mL L^{-1}) micrOTOF-Q mass spectrometer (Bruker Scientifics, Billerica, MA, USA) at Caxias do Sul University (Brazil).

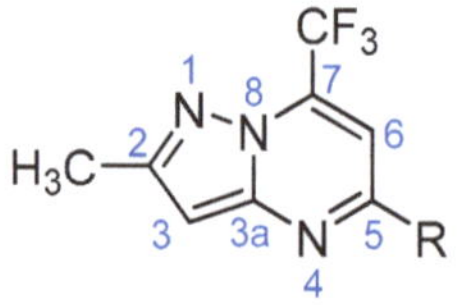

Figure 1. Atom numbering for NMR chemical shifts assignment of 3a–g.

2.2. Synthetic Procedures

General procedure was used for the synthesis of 5-(alkyl/aryl/heteroaryl)-2-methyl-7-(trifluoromethyl)pyrazolo[1,5-a]pyrimidines (3a–g).

According to Frizzo and collaborators [13], a solution of 3-amino-5-methyl-1H-pyrazole (1.0 mmol, 0.097 mg) (2) in acetic acid (5 mL) was added to a magnetically stirred solution of the respective 4-(alkyl/aryl)-4-methoxy-1,1,1-trifluoroalk-3-en-2-ones (1.0 mmol) (1a–g), also diluted in acetic acid (5 mL). The mixture was stirred at 80 °C for 16 h. After the reaction time (TLC), the products 3a–g were extracted with chloroform (3 × 10 mL), washed with distilled water (3 × 10 mL), and dried over anhydrous magnesium sulfate. The chloroform was removed in a rotary evaporator under reduced pressure and the respective compounds 3a–g were purified by recrystallization from ethanol.

2.2.1. 2,5-Dimethyl-7-(trifluoromethyl)pyrazolo[1,5-a]pyrimidine (3a)

Yellow solid, yield 50%, m.p. 52–53 °C. Literature [20] (Yield 87%, oil)

^{1}H NMR (600 MHz, CDCl$_3$) δ (ppm): 6.96 (s, 1H, H-6), 6.52 (s, 1H, H-3), 2.67 (s, 3H, CH$_3$), 2.57 (s, 3H, CH$_3$).

2.2.2. 2-Methyl-5-phenyl-7-(trifluoromethyl)pyrazolo[1,5-a]pyrimidine (**3b**)

Yellow solid, yield 85%, m.p. 123–124 °C. Literature [20] (Yield 82%, m.p. 123–124 °C).
^{1}H NMR (600 MHz, CDCl$_3$) δ (ppm): 8.30–8.00 (m, 2H, Ph), 7.75–7.52 (m, 3H, H-6/Ph), 6.68 (s, 1H, H-3), 2.62 (s, 3H, CH$_3$).

2.2.3. 5-(4-Methoxyphenyl)-2-methyl-7-(trifluoromethyl)pyrazolo[1,5-a]pyrimidine (**3c**)

Yellow solid, yield 60%, m.p. 182–183 °C. Literature [21]
^{1}H NMR (600 MHz, CDCl$_3$) δ (ppm): 8.09 (d, J = 8.6 Hz, 2H. Ph), 7.50 (s, 1H, H-6), 7.06 (d, J = 8.6 Hz, 1H, Ph), 6.62 (s, 1H, H-3), 3.92 (s, 3H, OCH$_3$), 2.60 (s, 3H, CH$_3$).

2.2.4. 5-(4-Fluorophenyl)-2-methyl-7-(trifluoromethyl)pyrazolo[1,5-a]pyrimidine (**3d**)

Yellow solid, yield 98%, m.p.155–156 °C. Literature [20] (Yield 96%, m.p. 141–144 °C).
^{1}H NMR (600 MHz, CDCl$_3$) δ (ppm): 8.12 (dd, J = 8.9, 5.3 Hz, 2H, Ph), 7.50 (s, 1H, H-6), 7.24 (t, J = 8.6 Hz, 2H, Ph), 6.66 (s, 1H, H-3), 2.61 (s, 3H, CH$_3$).

2.2.5. 5-(4-Bromophenyl)-2-methyl-7-(trifluoromethyl)pyrazolo[1,5-a]pyrimidine (**3e**)

Yellow solid, yield 70%, m.p. 171–173 °C. Literature [13,20] (Yield 86%, m.p. 171–173 °C).
^{1}H NMR (600 MHz, CDCl$_3$) δ (ppm): 8.00 (d, J = 8.6 Hz, 2H, Ph), 7.69 (d, J = 8.6 Hz, 2H, Ph), 7.50 (s, 1H, H-6), 6.68 (s, 1H, H-3), 2.62 (s, 3H, CH$_3$).

2.2.6. 2-Methyl-5-(4-nitrophenyl)-7-(trifluoromethyl)pyrazolo[1,5-a]pyrimidine (**3f**)

Orange solid, yield 80%, m.p. 223–224 °C.
^{1}H NMR (600 MHz, CDCl$_3$) δ (ppm): 8.41 (d, J = 8.6 Hz, 2H, Ph), 8.32 (d, J = 8.8 Hz, 2H, Ph), 7.60 (s, 1H, H-6), 6.77 (s, 1H, H-3), 2.65 (s, 3H, CH$_3$).
^{13}C{^{1}H} NMR (150 MHz, DMSO-d$_6$) δ (ppm): 157.8 (C-2), 152.1 (C-5), 150.2 (Ph), 149.1 (C-3a), 142.0 (Ph), 134.1 (q, J = 37.1 Hz, C-7), 128.1 (Ph), 124.2 (Ph), 119.4 (q, J = 274.8 Hz, CF$_3$), 102.4 (d, J = 4.2 Hz, C-6), 98.7 (C-3), 14.9 (CH$_3$).
HRMS (ESI): (M + H): Calcd. for C$_{14}$H$_{10}$F$_3$N$_4$O$_2$ = 323.0756; Found: 323.0759.

2.2.7. 2-Methyl-5-(2-thienyl)-7-(trifluoromethyl)pyrazolo[1,5-a]pyrimidine (**3g**)

Yellow solid, yield 72%, m.p. 155–156 °C. Literature [20] (Yield 89%, m.p. 152–154 °C).
^{1}H NMR (600 MHz, CDCl$_3$) δ (ppm): 7.73 (d, J = 3.6 Hz, 1H, thienyl), 7.58 (d, J = 5.0 Hz, 1H, thienyl), 7.41 (s, 1H, H-6), 7.19 (t, J = 4.4 Hz, 1H, thienyl), 6.60 (s, 1H, H-3), 2.59 (s, 3H, CH$_3$).

2.3. Photophysical Measurements

2.3.1. Photophysical Measurements in Solution

Electronic UV–Vis analysis of compounds **3a–g** in several solvents with distinct polarity (CH$_3$CN, CHCl$_3$, THF, toluene, EtOH, and DMSO) were measured using a Shimadzu UV2600 spectrophotometer (data interval, 1.0 nm, and slit 1.0 mm). Steady-state fluorescence emission spectra of derivatives **3a–g** in the same solutions were measured with a Horiba Jobin Yvon FluoroMax 4 Plus spectrofluorometer (slit 5.0 mm; Em/Exc) and corrected according to the manufacturer's instructions. Fluorescence quantum yield (Φ_f; in %) values of compounds **3a–g** were determined by comparing the corrected fluorescence spectra with that of standard 9,10-diphenylanthracene (DPA) in CHCl$_3$ solution (Φ_f = 65%, λ_{exc} = 375 nm) according to the current literature [12,22–24].

2.3.2. Photophysical Measurements in the Solid State

For the absorption and UV–Vis measurements in the solid state, derivatives **3a–g** were treated as powder, and the baseline in the solid state was obtained using a barium sulphate standard (BaSO$_4$; Wako Company®, Richmond, VA, USA). The diffuse reflectance spectra

(DRUV) were measured using an integrating sphere attachment on a Shimadzu UV-2600 spectrophotometer in the 250–700 nm range.

The fluorescence emission spectra in the solid state were measured in the 300–700 nm range using the Horiba Yvon-Jobin Fluoromax Plus (Em/Exc; slit 5.0 mm) instrument. Fluorescence quantum yields (Φ_f) in the solid state were determined by comparing the integrated area to the corrected fluorescence spectrum of compounds with the integrated area to the corrected fluorescence spectrum of a standard compound (in this case, sodium ascorbate — $\Phi_f = 55\%$), as reported elsewhere [23].

Fluorescence lifetimes in the solid state of related compounds were recorded using the time-correlated single-photon counting (TCSPC) method with DeltaHub controller and Horiba spectrofluorometer. Data were processed with the DAS6 and Origin® 8.5 software (Northampton, MA, USA) using mono-exponential fitting of raw data. NanoLED (1.0 MHz; pulse width < 1.2 ns; 284 nm excitation wavelength) was used as a source of excitation.

2.4. Thermogravimetric Analysis

Thermogravimetric analyses (TGA) were performed using a TGA Q5000 instrument (TA Instruments Inc., New Castle, DE, USA) at a heating rate of 10 °C min^{-1}, from 40 °C to 600 °C under a N_2 flux of 25 mL min^{-1}. The masses were approximately 1 mg for all samples. Data analysis was performed using the OriginPro 8.5 software (Northampton MA, USA). The confirmation of calibration of apparatus before analysis was done with $CaC_2O_4 \cdot H_2O$ (99.9%).

Differential scanning calorimetry (DSC) analyses were carried out using a Q2000 DSC calorimeter (TA Instruments, New Castle, DE, USA) equipped with an RCS refrigeration accessory and with N_2 as purge gas (50 mL min^{-1}). The heating rate used was 5 °C min^{-1}. The calibration of instruments in standard DSC mode was verified with indium (99.99%). The masses of the samples (1–5 mg) were weighed on a Sartorius balance (M500P) with a precision of ±0.001 mg. All samples were subjected to three heating–cooling cycles, as follows: 25 to 250 °C.

3. Results

3.1. Synthesis and Structural Characterization

The precursors 4-alkoxy-4-(alkyl/aryl/heteroaryl)-1,1,1-trifluoroalk-3-en-2-ones (**1a–g**) were first synthesized through the trifluoracetylation of enol ethers and acetals according to the literature procedures [25–33]. The 3-amino-5-methyl-1*H*-pyrazole precursor **2** was acquired from a commercial supplier (Sigma-Aldrich, São Paulo, Brazil).

The method employed to synthesize the 5-(alkyl/aryl/heteroaryl)-2-methyl-7-(trifluoromethyl)pyrazolo[1,5-*a*]pyrimidines (**3a–e, 3g**) has already been described elsewhere [13]. The compounds (**3a–e**) and (**3g**) were obtained in 50–98% yields (Scheme 2), which showed the appearance of air-stable yellow-orange solids [13,20,21]. The compound 2-methyl-5-(4-nitrophenyl)-7-(trifluoromethyl)pyrazolo[1,5-*a*]pyrimidine (**3f**) has yet to be described in the literature, and it was obtained in 80% yield after recrystallization (also from ethanol).

All products were fully characterized with ^{1}H NMR and the melting point showed spectral data typical for these compounds and also in agreement with the literature [13,20,21]. Until now, an unpublished compound (**3f**) was also characterized by ^{1}H- and ^{13}C NMR and HRMS. For instance, in the NMR chemical shifts assignment, compound **3f** presented a chemical shift at 7.60 ppm at the ^{1}H NMR spectrum, which was assigned to the pyrimidine H-6; a signal at 6.77 ppm was assigned to the pyrazole H-3, a signal at 2.65 ppm referred to the unique methyl substituent, and a signal at 8.41 and 8.32 ppm was assigned to the p-phenyl substituted aromatic ring. The same compound **3f** showed chemical shifts in the ^{13}C{^{1}H} NMR spectrum as a singlet at 157.8 (C-2), 152.1 (C-5), 149.1 (C-3a), 102.4 (C-6) and 98.7 (C-3) ppm, and a quartet for C-7 and CF_3 group appearing at 134.1 ppm with J = 37.1 Hz and 119.4 ppm with J = 274.8 Hz, respectively, due to the ^{13}C–^{19}F scalar coupling.

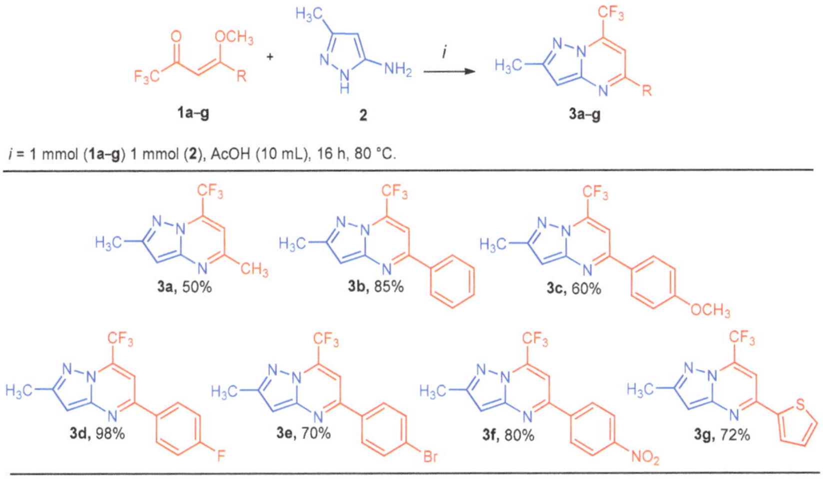

Scheme 2. Synthesis of 5-(alkyl/aryl/heteroaryl)-2-methyl-7-(trifluoromethyl)pyrazolo[1,5-*a*]pyrimidines (3).

3.2. Photophysical Properties of Pyrazolo[1,5-a]pyrimidines (3a–g)

3.2.1. Solution Analysis

Regarding the photophysical properties of pyrazolo derivatives **3a–g**, the photophysical properties of all compounds in different solvent polarities (toluene, CHCl$_3$, CH$_3$CN, THF, EtOH, and DMSO) were analyzed. For exemplification purposes, the spectral profile of derivative 3b in all solvents studied is illustrated in Figure 2, and the absorption parameters of compounds are listed in Table 1; all UV–Vis absorption spectra are listed in the Supplementary Information (Figures S9–S14).

In general, all derivatives showed electronic transition bands in the UV region and can be attributed to $\pi \rightarrow \pi^*$ and $n \rightarrow \pi^*$ type transitions, which are characteristics of this type of heterocyclic and aromatic skeleton, according to the literature [12,23,24,34]. As seen in Figure 2, the derivatives studied show a similar absorption behavior according to the nature of the solvent. Additionally, by analyzing the UV–Vis spectra in the ground state of related compounds, small changes according to the solvent property are also observed, and some spectral changes occur due to the presence of electron-donor or -acceptor substituents (Table 1).

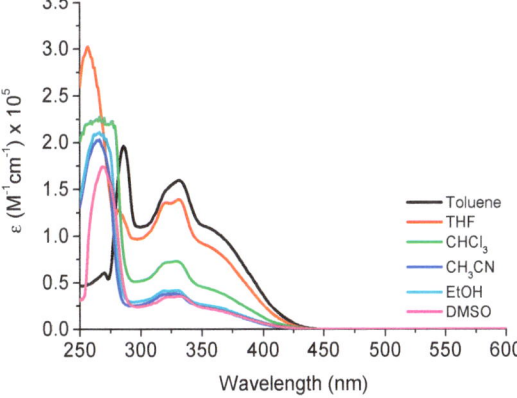

Figure 2. Comparative UV-Vis absorption spectra in several solvents of compound **3b**.

Table 1. Photophysical data analysis of derivatives 3a–g in different solvents.

Compound	Solvent [a]	λ_{abs}, nm (ε; $M^{-1}cm^{-1}$)	λ_{em}, nm (QY, %) [b]	SS (nm/cm^{-1}) [c]
3a	CHCl$_3$	274 (15670); 309 (4850)	501 (67.0)	192/12,400
	THF	273 (19870); 308 (4505)	473 (87.0)	165/11,325
	Toluene	283 (16120); 309 (8870)	492 (92.0)	183/12,040
	CH$_3$CN	306 (12550); 337 (7150)	511 (67.0)	174/10,105
	EtOH	305 (17500); 338 (9690)	456 (55.0)	118/7655
	DMSO	307 (9995); 342 (5120)	441 (86.0)	99/6565
3b	CHCl$_3$	266 (22520); 325 (7320); 357 (sh)	504 (79.0)	147/8170
	THF	256 (30195); 324 (13535); 362 (sh)	507 (78.0)	145/7900
	Toluene	286 (19550); 331 (15950); 362 (sh)	508 (88.0)	146/7940
	CH$_3$CN	265 (20210); 324 (3800); 360 (sh)	506 (71.0)	146/8015
	EtOH	265 (21035); 325 (4175); 365 (sh)	514 (71.0)	149/7940
	DMSO	269 (17410); 328 (3570); 362 (sh)	521 (91.0)	159/8430
3c	CHCl$_3$	290 (13420); 338 (5785); 367 (sh)	500 (77.0)	133/7250
	THF	256 (18080); 295 (23610); 339 (12985); 368 (sh)	500 (75.0)	132/7170
	Toluene	294 (22040); 338 (19095); 365 (sh)	501 (84.0)	136/7435
	CH$_3$CN	287 (14085); 324 (3800); 360 (sh)	554 (63.0)	194/9730
	EtOH	289 (11740); 337 (5285); 367 (sh)	487 (56.0)	120/6715
	DMSO	269 (10020); 327 (2010); 364 (sh)	509 (88.0)	145/7825
3d	CHCl$_3$	269 (15780); 331 (2870); 370 (sh)	504 (79.0)	134/7185
	THF	256 (17380); 288 (8800); 326 (9640); 362 (sh)	509 (77.0)	147/7980
	Toluene	289 (18920); 336 (6020); 369 (sh)	510 (87.0)	141/7490
	CH$_3$CN	266 (20720); 325 (4070); 359 (sh)	516 (73.0)	157/8475
	EtOH	267 (17185); 328 (3030); 360 (sh)	514 (73.0)	154/8320
	DMSO	269 (15720); 327 (3125); 359 (sh)	521 (93.0)	162/8660
3e	CHCl$_3$	276 (17465); 330 (3800); 364 (sh)	507 (79.0)	143/7750
	THF	257 (9915); 291 (9720); 329 (8275); 365 (sh)	511 (78.0)	146/7830
	Toluene	290 (22170); 334 (15485); 362 (sh)	510 (87.0)	148/8015
	CH$_3$CN	272 (10370); 327 (2195); 364 (sh)	516 (73.0)	152/8090
	EtOH	274 (20375); 327 (4695); 366 (sh)	514 (73.0)	148/7865
	DMSO	275 (17760); 330 (4285); 366 (sh)	524 (90.0)	158/8240
3f	CHCl$_3$	299 (10955); 344 (sh)	496 (64.0)	152/8910
	THF	271 (10545); 296 (12640); 342 (sh)	524 (73.0)	182/10,155
	Toluene	297 (11850); 346 (sh); 382 (sh)	520 (89.0)	138/6950
	CH$_3$CN	295 (11625); 341 (sh); 373 (sh)	545 (70.0)	172/8460
	EtOH	292 (17630); 340 (sh); 373 (sh)	419 (24.0)	46/2940
	DMSO	257 (14555); 301 (16145); 347 (sh)	558 (95.0)	211/10,900
3g	CHCl$_3$	280 (19115); 337 (8370); 375 (sh)	481 (63.0)	106/5875
	THF	257 (15280); 304 (20765); 336 (19370); 378 (sh)	511 (75.0)	133/6885
	Toluene	290 (17520); 345 (8170); 375 (sh)	510 (86.0)	135/7060
	CH$_3$CN	276 (11520); 343 (5715); 372 (sh)	517 (73.0)	145/7540
	EtOH	278 (18050); 343 (8620); 371 (sh)	496 (56.0)	125/6790
	DMSO	281 (21175); 346 (11215); 375 (7770)	520 (89.0)	145/7435

[a] Dielectric constant (ε) and refractive index (η): toluene (ε = 2.38; η = 1.4969), THF (ε = 7.50; η = 1.4072), CHCl$_3$ (ε = 4.81; η = 1.4459), CH$_3$CN (ε = 36.6; η = 1.3441), EtOH (ε = 24.5; η = 1.3614), and DMSO (ε = 46.7; η = 1.4793); [b] Excited at lower transition band and using 9,10-diphenylanthracene (DPA) in chloroform as standard (λ_{exc} = 375 nm; Φ_f = 0.65); [c] Stokes shifts: $\Delta\lambda = \lambda_{em} - \lambda_{abs} = 1/\lambda_{abs} - 1/\lambda_{em}$; sh = sholuder.

By comparing the electronic effect of the substituent on the aromatic moiety (3c—OCH$_3$ and 3f—NO$_2$ units), very subtle shifts can be observed in the other solvents investigated, revealing that there is no significant change in the ground state (Table 1).

Regarding fluorescent emission properties, derivatives 3a–g were investigated in the same solvent polarities used in the UV-Vis analysis, and the data regarding the emission peaks (λ_{em}), quantum fluorescence yield (QY), and Stokes shifts (SS) are presented in Table 1. The normalized fluorescence emission spectra of derivatives in all solvents are presented in the Supplementary Information (Figures S15–S20). Regarding the fluorescence lifetime measurements of the derivatives in solution, time-resolved measurements were not made because the proper NanoLED source was unsuitable for this analysis.

In general, derivatives **3a–g** have emission bands located in the blue to green range. As with the UV–Vis absorption analysis, compound **3b** was chosen as an example, and the fluorescence emission spectra in all solvents and natural/UV light solution photography are listed in Figure 3. According to the spectra in Figure 3c, the solvent polarity does not show any significant changes in the emission peaks of compound **3c**. As for compound **3f** (containing NO_2 group), more visible changes are observed, mainly in the protic medium (Supplementary Information—Figure S24). We can attribute this to a difference in the stabilization of the structures in the excited state, primarily in the presence of electron-withdrawing groups and the secondary H-bonding interactions in ethanol solution.

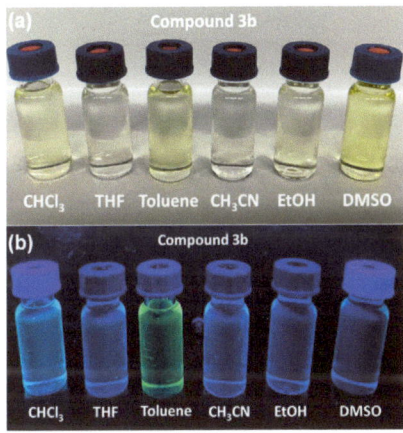

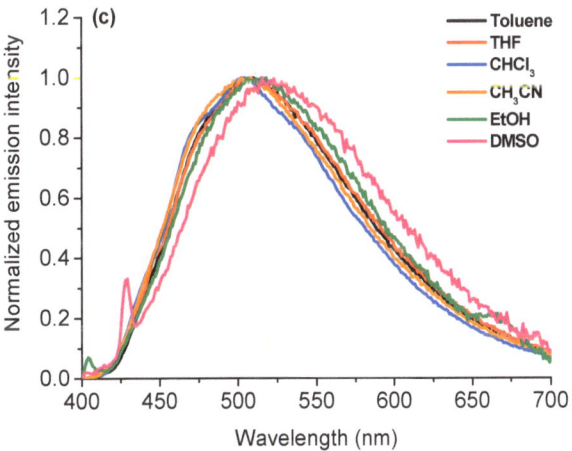

Figure 3. (a) Solutions in natural light, (b) solutions in UV_{365nm} light irradiation, and (c) comparative steady-state fluorescence emission spectra in several solvents of compound **3b**.

As for the Φ_f values, the compounds presented higher QYs; this may be associated with a greater stabilization and solvation of these molecules in the singlet excited state (Table 1) and dependence on the substituent electronic property. Finally, moderate to large SS were observed for all derivatives in the solvents studied, and this can be attributed to the vibrational relaxation or dissipation and solvent reorganization, which can decrease the separation of the energy levels of the ground and excited states (Table 1).

3.2.2. Aggregation-Induced Emission Behavior

In a generalized way, the aggregation-induced emission (AIE) phenomenon describes the behavior of a molecule that shows dim or no emission in dilute solution but much-enhanced emission in aggregates or the solid state [35,36]. The fluorescence emission behaviors of the selected compounds **3b**, **3c**, and **3f** were examined in the THF-H_2O mixture (0–90% water fraction) to confirm the possibility of AIE characteristics. Studied compounds emit a blue to green region under a UV lamp with 365 nm in THF solution (Figure 3). All fluorescence emission spectra in the THF-H_2O mixture of compounds **3c** and **3f** are listed in the Supplementary Information (Figures S23 and S24).

Interestingly, the fluorescence emission of derivatives **3b**, **3c**, and **3f** is sensitive to solvent polarity; thus, we aimed to explore their emission behavior in THF as an aprotic water-miscible solvent. The emission responses of compound **3b** upon adding different amounts of water to THF solution is presented in Figure 4. With the increase of water content (0–90% v/v), a great decrease in the emission peak intensities was observed, and the fluorescence intensity as a function of water content showed a slightly bathochromic shift. Tigreros and co-workers previously described similar behavior in a study with pyrazolo derivatives containing a triphenylamine substituent [16]. Thus, the AIE properties were not observed, and this decrease in the emission intensities of derivatives can be directly attributed to an aggregation phenomenon (J- or H-aggregate types) as the water fraction increases. Consequently, this result demonstrates that (trifluoromethyl)pyrazolo-based probes can act as possible fluorescent sensors for small amounts of acid or protic molecules.

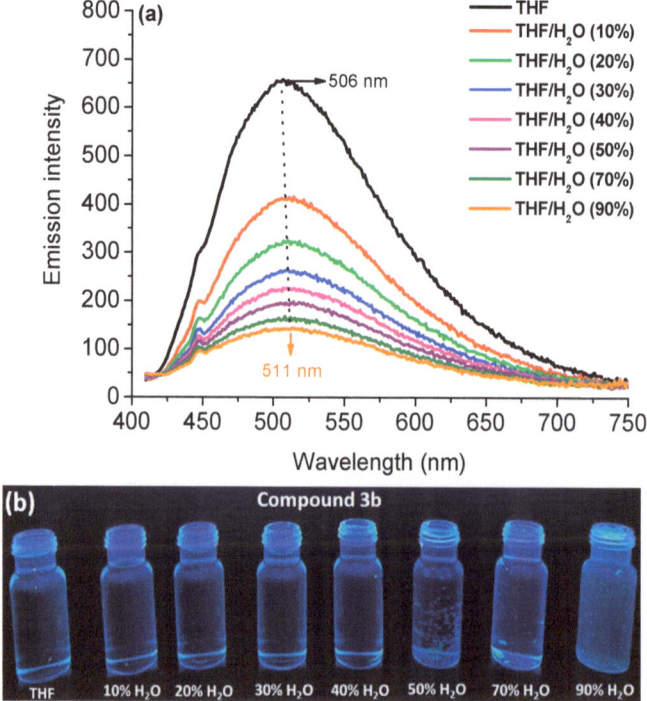

Figure 4. (a) Fluorescence emission spectra of compound **3b** and (b) photograph of compound **3b** solutions in THF/water mixture with different water fractions (0–90%) under a UV lamp (365 nm).

3.2.3. Solid-State Analysis—First Evidences

A solid-state absorption and fluorescence emission spectroscopy analysis in powder was performed as the (trifluoromethyl)-pyrazolo derivatives **3a–g** present fluorescence emission in the solid state. The reflectance spectra of the compounds revealed similar absorption peaks compared to the solution study, in which we observed the broadening of the absorption bands (Supplementary Information; Figure S25).

The fluorescence emission data of derivatives **3a–g** in the solid state are listed in Table 2, and all spectra are presented in Figure 5a. Thus, compared to the spectra in solution, the derivatives presented emission peaks very close to the values obtained in organic solvents (Table 1). The variations in emission peaks observed in the solid state can be attributed to a change in the molecular arrangement in the absence of the solvent, which may be favored by π-π stacking interactions. The QY values observed in the solid state for derivatives **3a–g** are smaller than those observed in the solution, which may be directly related to the solid-state arrangement.

Table 2. Photophysical data analysis of derivatives **3a–g** in the solid state.

Compound	λ_{abs}, nm	λ_{em} nm (QY,%) [a]	SS (nm/cm^{-1}) [b]	τ_f, ns (χ^2) [c]	k_r (10^8 s^{-1}) [d]	k_{nr} (10^8 s^{-1}) [e]
3a	261, 338, 417	493 (29.0)	76/3700	3.50 ± 0.44 (1.131901)	0.83	2.03
3b	286, 338, 425	493 (23.0)	68/3245	3.03 ± 0.59 (1.143012)	0.76	2.55
3c	283, 335, 425	485 (21.0)	60/2910	8.62 ± 0.37 (1.091558)	0.24	0.92
3d	294, 340, 427	509 (28.0)	82/3770	3.00 ± 0.45 (1.051683)	0.93	2.40
3e	290, 428	483 (24.0)	55/2660	6.62 ± 0.48 (1.131901)	0.36	1.15
3f	283, 335, 427	542 (29.0)	115/4970	1.36 ± 0.82 (1.151343)	2.13	5.22
3g	268, 331, 433	507 (24.0)	74/3370	6.14 ± 0.52 (0.919048)	0.39	1.23

[a] Excitation at a less-energy absorption peak using sodium salicylate as standard (Φ_f = 55%); [b] Stokes shifts: $\Delta\lambda = \lambda_{em} - \lambda_{abs} = 1/\lambda_{abs} - 1/\lambda_{em}$; [c] Using excitation by NanoLED source at 284 nm; [d,e] Determined by [23].

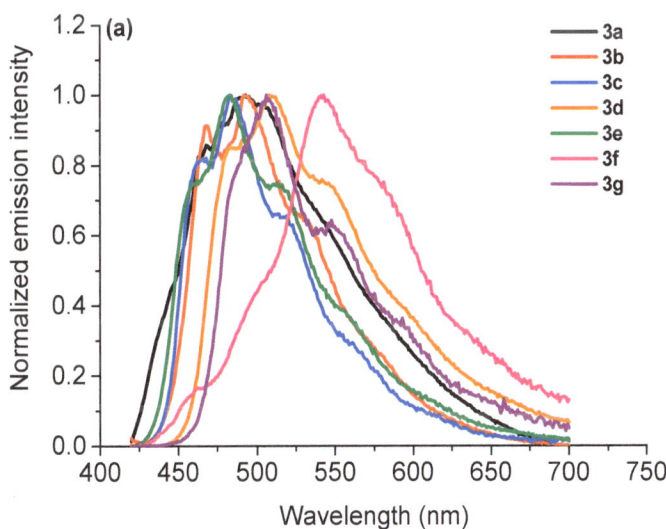

Figure 5. *Cont.*

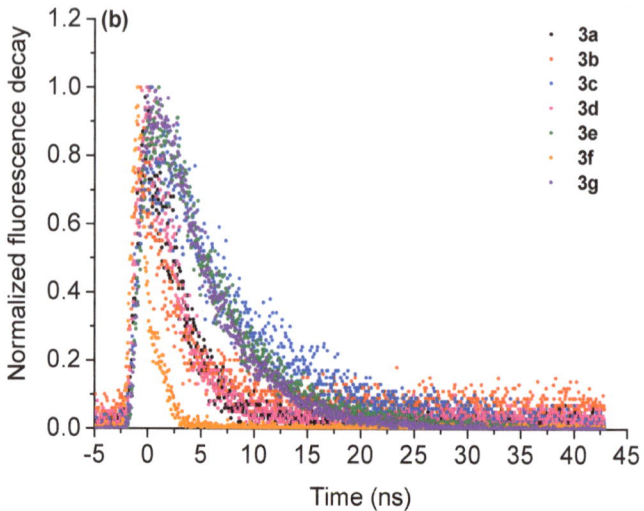

Figure 5. (a) Normalized steady-state fluorescence emission spectra of compounds **3a–g** in the solid state and (b) normalized fluorescence decay of compounds **3a–g** in the solid state when excited by a NanoLED source at 284 nm.

Compared with the solution study, solid-state fluorescence lifetime measurements were conducted, and lifetime decay plots and the τ_f, radiative (k_r) and non-radiative (k_{nr}) values for derivatives **3a–g** are presented in Figure 5b and Table 2, respectively. It is possible to note a variation in the τ_f values according to the electronic nature of the molecule, which is attributed to the non-influence of the solvent in the excited state and a greater ordering of the molecules in the solid state (Table 2). In addition, we can evidence a decrease in the radiative (k_r) rates with an increase in the non-radiative (k_{nr}) rates, and this is probably evidenced by a relaxation of the vibrational levels of the molecules and restricted motion.

3.3. Thermal Stability in the Solid State

The solid-state thermal stability of compounds **3a–g** was accessed using TGA, and the results are summarized in Table 3, where $T_{0.05}$ expresses the temperature at which 5.0% of mass loss occurred and T_d is the temperature of maximum decomposition rate (i.e., the peak of the derivative curve). The order of thermal stability was established in terms of $T_{0.05}$ as follows: **3a** < **3b** < **3d** < **3e** < **3g** < **3c** < **3f**. The TGA curves for compounds **3a**, **3e**, and **3f** are presented in Figure 6, and the other results, including DSC/TGA/DTG curves for compounds **3b** and **3d**, are shown in the Supplementary Information (Figures S26–S33). It is possible to note from Figure 6 and the other curves that the thermal decomposition occurs in a single step with almost 100% of mass loss. Additionally, the values of $T_{0.05}$ are considerably low, especially for compound **3a**, indicating low thermal stability for this series of pyrazolo[1,5-a]pyrimidines. Regarding $T_{0.05}$ and structure relations, no direct correspondence between molar masses and thermal stability was observed for the entire series. More important than the molar mass of the compounds was the nature of the R substituent. Nonetheless, more detailed explanations for the observed order of thermal stability would require further analysis. From the values of $T_{0.05}$ in Table 3 and the melting temperatures of compounds **3a–g**, it is worth noticing that the majority of the compounds presented considerable mass loss (5%) below their melting point, narrowing possible applications to the solid state.

Table 3. Results of the TGA analysis.

Compound	$T_{0.05}$ (°C)	T_d (°C)	Mass Loss (%)
3a	72	110	99
3b	117	161	97
3c	171	200	99
3d	134	169	96
3e	147	187	97
3f	187	230	99
3g	151	187	98

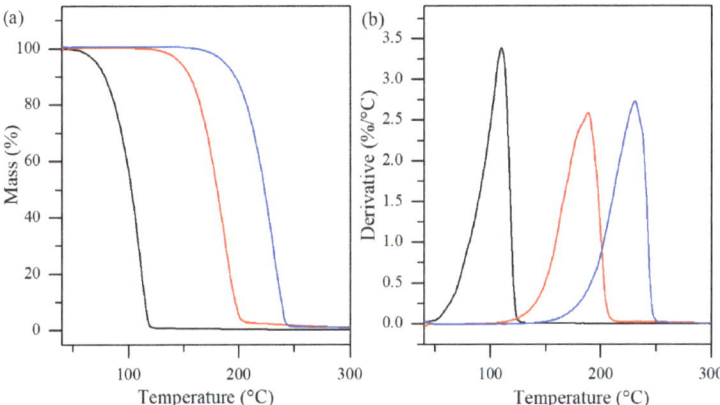

Figure 6. TGA (a) and derivative (b) curves for compounds 3a (black line), 3e (red line), and 3f (blue line).

4. Conclusions

The synthesis in yields of 50–98% and photophysical behavior of a series of seven examples of 5-(alkyl/aryl/heteroaryl)-substituted 2-methyl-7-(trifluoromethyl)pyrazolo[1,5-a]pyrimidine core (3) was achieved, where one new compound (3f) was obtained and fully structurally characterized. The optical properties in solution and the solid state of this geminated system 3 were also successfully investigated. In the photophysical evaluation of the molecules, transition bands were observed in the UV region, and moderate to higher values in the quantum fluorescence yields for the derivatives 3a–g. Regarding the solvent polarity variation, the changes vary according to the electronic nature of the molecules evaluated in the presence or absence of the substituent. Furthermore, photophysical analysis in the solid state and AIE phenomena were also evaluated. For this series of pyrazolo[1,5-a]pyrimidines, regarding $T_{0.05}$ and structure relations, no direct correspondence between molar masses and thermal stability was observed for the entire series. Additionally, it is worth noticing that most of the compounds presented considerable mass loss (5%) below their melting point, narrowing possible applications to the solid state.

Supplementary Materials: The following supporting information can be downloaded at: https://www.mdpi.com/article/10.3390/photochem2020024/s1. Reference [37] is cited in the supplementary materials.

Author Contributions: Conceptualization, H.G.B. and B.A.I.; methodology, H.G.B., B.A.I. and C.P.F.; validation and formal analysis, F.S.S., J.N.A., V.B.S., B.A.I., J.C.B.V., C.P.F. and N.Z.; investigation, writing—original draft preparation, writing—review and editing, F.S.S., H.G.B., B.A.I., C.P.F., M.A.P.M. and N.Z.; supervision and funding acquisition, H.G.B., B.A.I. and M.A.P.M. All authors have read and agreed to the published version of the manuscript.

Funding: This research received no external funding.

Informed Consent Statement: Not applicable.

Acknowledgments: The authors would like to thank the following entities: The Coordination for Improvement of Higher Education Personnel—CAPES (Finance Code 001) for the fellowships and the National Council for Scientific and Technological Development—CNPq: proc. no 305.379/2020-8 and 403.134/2021-8 (H.G. Bonacorso); proc. no. 409.150/2018-5, 305.458/2021-3, 403.210/2021-6 and 304.711/2018-7 (B.A. Iglesias), and the Research Support Foundation of the State of Rio Grande do SulFAPERGS: proc. no. 17/2551-0001275-5 (H.G. Bonacorso) and 21/2551-0002114-4 (B.A. Iglesias) for the financial support. We would also like to thank Atlas Assessoria Linguística for language editing.

Conflicts of Interest: There are no conflict of interest to declare.

References

1. Searched as a Topic the Words Organic Compounds and Fluorescent Properties in the Last 5 Years. Available online: https://www.webofscience.com/wos/woscc/citation-report/f531ee1e-c0ea-42bf-90a4-e37ad40ddb7c-2901a49f (accessed on 8 March 2022).
2. Klymchenko, A.S. Solvatochromic and fluorogenic dyes as environment-sensitive probes: Design and biological applications. *Acc. Chem. Res.* **2017**, *50*, 366–375. [CrossRef] [PubMed]
3. Tigreros, A.; Ortiz, A.; Insuasty, B. Effect of π-conjugated linkage on photophysical properties: Acetylene linker as the better connection group for highly solvatochromic probes. *Dye. Pigm.* **2014**, *111*, 45–51. [CrossRef]
4. Wang, K.; Xiao, H.; Qian, L.; Han, M.; Wu, X.; Guo, Z.; Zhan, H. Diversified AIE and mechanochromic luminescence based on carbazole derivative decorated dicyanovinyl groups: Effects of substitution sites and molecular packing. *CrystEngComm* **2020**, *22*, 2166–2172. [CrossRef]
5. Cao, X.; Li, Y.; Han, Q.; Gao, A.; Wang, B.; Chang, X.; Hou, J.T. Design of large π-conjugated α-cyanostilbene derivatives as colorimetric sensors for volatile acids and organic amine gases. *J. Mater. Chem. C* **2020**, *8*, 4058–4064. [CrossRef]
6. Yu, Y.; Fan, Y.; Wang, C.; Wei, Y.; Liao, Q.; Li, Q.; Li, Z. Phenanthroimidazole derivatives with minor structural differences: Crystalline polymorphisms, different molecular packing, and totally different mechanoluminescence. *J. Mater. Chem. C* **2019**, *7*, 13759–13763. [CrossRef]
7. Li, G.Y.; Han, K.L. The sensing mechanism studies of the fluorescent probes with electronically excited state calculations. *WIREs Comput. Mol. Sci.* **2018**, *8*, e1351. [CrossRef]
8. Kournoutas, F.; Fihey, A.; Malval, J.P.; Spangenberg, A.; Fecková, M.; Le Poul, P.; Katan, C.; Robin-Le Guen, F.; Bureš, F.; Achelle, S.; et al. Branching effect on the linear and nonlinear optical properties of styrylpyrimidines. *Phys. Chem. Chem. Phys.* **2020**, *22*, 4165–4176. [CrossRef]
9. Gherasim, C.; Airinei, A.; Tigoianu, R.; Craciun, A.M.; Danac, R.; Nicolescu, A.; Lucian, D.; Mangalagiu, I.I. Synthesis and photophysical insights of new fused N-heterocyclic derivatives with isoquinoline skeleton. *J. Mol. Liq.* **2020**, *310*, 113196. [CrossRef]
10. Kappenberg, Y.G.; Stefanello, F.S.; Zanatta, N.; Martins, M.A.P.; Nogara, P.A.; Rocha, J.B.T.; Tisoco, I.; Iglesias, B.A.; Bonacorso, H.G. Hybridized 4-trifluoromethyl-(1,2,3-triazol-1-yl)quinoline system: Synthesis, photophysics, selective DNA/HSA bio-interactions and molecular docking. *ChemBioChem* **2022**, *23*, e202100649. [CrossRef]
11. Rocha, I.O.; Kappenberg, Y.G.; Rosa, W.C.; Frizzo, C.P.; Zanatta, N.; Martins, M.A.P.; Tisoco, I.; Iglesias, B.A.; Bonacorso, H.G. Photophysical, photostability, and ROS generation properties of new trifluoromethylated quinoline-phenol Schiff bases. *Beilstein J. Org. Chem.* **2021**, *17*, 2799–2811. [CrossRef]
12. Silva, B.; Stefanello, F.S.; Feitosa, S.C.; Frizzo, C.P.; Martins, M.A.P.; Zanatta, N.; Iglesias, B.A.; Bonacorso, H.G. Novel 7-(1H-pyrrol-1-yl)spiro[chromeno[4,3-b]quinoline-6,10-cycloalkanes]: Synthesis, cross-coupling reactions, and photophysical properties. *New J. Chem.* **2021**, *45*, 4061–4070. [CrossRef]
13. Frizzo, C.P.; Martins, M.A.P.; Marzari, M.R.B.; Campos, P.T.; Claramunt, R.M.; Garcıa, M.A.; Sanz, D.; Alkorta, I.; Elguero, J. Structural Studies of 2-Methyl-7-substituted Pyrazolo[1,5-a]Pyrimidines. *J. Heterocycl. Chem.* **2010**, *47*, 1259–1268. [CrossRef]
14. Stefanello, F.S.; Kappenberg, Y.G.; Araujo, J.N.; Zorzin, S.F.; Martins, M.A.P.; Zanatta, N.; Iglesias, B.A.; Bonacorso, H.G. Trifluoromethyl-substituted aryldiazenyl-pyrazolo[1,5-a]pyrimidin-2-amines: Regioselective synthesis, structure, and optical properties. *J. Fluor. Chem.* **2022**, *255–256*, 109967. [CrossRef]
15. Tigreros, A.; Aranzazu, S.; Bravo, N.; Zapata-rivera, J.; Portilla, J. Pyrazolo[1,5-a]pyrimidines-based fluorophores: A comprehensive theoretical-experimental study. *RSC Adv.* **2020**, *10*, 39542–39552. [CrossRef] [PubMed]
16. Tigreros, A.; Macías, M.; Portilla, J. Photophysical and crystallographic study of three integrated pyrazolo[1,5-a]pyrimidine—Triphenylamine systems. *Dye. Pigm.* **2021**, *184*, 108730. [CrossRef]
17. Castillo, J.C.; Tigreros, A.; Portilla, J. 3-Formylpyrazolo[1,5-a]pyrimidines as key intermediates for the preparation of functional fluorophores. *J. Org. Chem.* **2018**, *83*, 10887–10897. [CrossRef]
18. Tigreros, A.; Castillo, J.; Portilla, J. Cyanide chemosensors based on 3-dicyanovinylpyrazolo[1,5-a]pyrimidines: Effects of peripheral 4-anisyl group substitution on the photophysical properties. *Talanta* **2020**, *215*, 120905. [CrossRef]
19. Arias-Gómez, A.; Godoy, A.; Portilla, J. Functional Pyrazolo[1,5-a]pyrimidines: Current Approaches in Synthetic Transformations and Uses As an Antitumor Scaffold. *Molecules* **2021**, *26*, 2708. [CrossRef]

20. Buriol, L.; München, T.S.; Frizzo, C.P.; Marzari, M.R.B.; Zanatta, N.; Bonacorso, H.G.; Martins, M.A.P. Resourceful synthesis of pyrazolo[1,5-*a*]pyrimidines under ultrasound irradiation. *Ultrason. Sonochem.* **2013**, *20*, 1139–1143. [CrossRef]
21. Goldfarb, D.S. Method for Altering the Lifespan of Eukaryotic Organisms. US Patent 2009/0163545 A1, 25 June 2009.
22. Rosa, W.C.; Rocha, I.O.; Schmitz, B.F.; Rodrigues, M.B.; Martins, M.A.P.; Zanatta, N.; Acunha, T.V.; Iglesias, B.A.; Bonacorso, H.G. Synthesis and photophysical properties of trichloro(fluoro)-substituted 6-(3-oxo-1-(alk-1-en-1-yl)amino)coumarins and their 2,2-Difluoro-2*H*-1,3,2-oxazaborinin-3-ium-2-uide heterocycles. *J. Fluor. Chem.* **2020**, *238*, 109614. [CrossRef]
23. Santos, G.C.; Rocha, I.O.; Stefanello, F.S.; Copetti, J.P.P.; Tisoco, I.; Martins, M.A.P.; Zanatta, N.; Frizzo, C.P.; Iglesias, B.A.; Bonacorso, H.G. Investigating ESIPT and donor-acceptor substituent effects on the photophysical and electrochemical properties of fluorescent 3,5-diaryl-substituted 1-phenyl-2-pyrazolines. *Spectrochim. Acta Part A Mol. Biomol. Spectrosc.* **2022**, *269*, 120768. [CrossRef] [PubMed]
24. Stefanello, F.S.; Kappenberg, Y.G.; Ketzer, A.; Franceschini, S.Z.; Salbego, P.R.S.; Acunha, T.V.; Nogara, P.A.; Rocha, J.B.T.; Martins, M.A.P.; Zanatta, N.; et al. New 1-(Spiro[chroman-2,1′-cycloalkan]-4-yl)-1*H*-1,2,3-Triazoles: Synthesis, QTAIM/MEP analyses, and DNA/HSA-binding assays. *J. Mol. Liq.* **2021**, *324*, 114729. [CrossRef]
25. Effenberger, F. Synthese und Reaktionen von 1,4-Bis-athoxymethylen-butandion-(2.3). *Chem. Ber.* **1965**, *98*, 2260–2265. [CrossRef]
26. Effenberger, F.; Maier, R.; Schönwälder, K.-H.; Ziegler, T. Die Acylierung von Enolethern mit reaktiven Carbonsäure-chloriden. *Chem. Ber.* **1982**, *115*, 2766–2782. [CrossRef]
27. Hojo, M.; Masauda, R.; Sakaguchi, S.; Takagawa, M. A convenient synthetic method for β-alkoxy-and β-phenoxyacrylic acids and 3, 4-dihydro-2*H*-pyran-5-and 2, 3-dihydrofuran-4-carboxylic acids. *Synthesis* **1986**, *12*, 1016–1017. [CrossRef]
28. Bonacorso, H.G.; Martins, M.A.P.; Bittencourt, S.R.T.; Lourega, R.V.; Zanatta, N.; Flores, A.F.C. Trifluoroacetylation of unsymmetrical ketone acetals. A convenient route to obtain alkyl side chain trifluoromethylated heterocycles. *J. Fluor. Chem.* **1999**, *99*, 177–182. [CrossRef]
29. Hojo, M.; Masuda, R.; Kokuryo, Y.; Shioda, H.; Matsuo, S. Electrophilic substitutions of olefinichydrogens II. Acylation of vinyl ethers and N-vinyl amides. *Chem. Lett.* **1976**, *5*, 499–502. [CrossRef]
30. Colla, A.; Martins, M.A.P.; Clar, G.; Krimmer, S.; Fischer, P. Trihaloacetylated enol ethers-general synthetic procedure and heterocyclic ring closure reactions with hydroxylamine. *Synthesis* **1991**, *6*, 483–486. [CrossRef]
31. Flores, A.F.C.; Brondani, S.; Zanatta, N.; Rosa, A.; Martins, M.A.P. Synthesis of 1,1,1-trihalo-4-methoxy-4-[2-heteroaryl]-3-buten-2-ones, the corresponding Butan-1,3-dione and azole derivatives. *Tetrahedron Lett.* **2002**, *43*, 8701–8705. [CrossRef]
32. Nenajdenko, V.G.; Balenkova, E.S. Preparation of α,β-unsaturated trifluromethylketones and their application in the synthesis of heterocycles. *ARKIVOC* **2011**, *i*, 246–328. [CrossRef]
33. Siqueira, G.M.; Flores, A.F.C.; Clar, G.; Zanatta, N.; Martins, M.A.P. Sintese de β-aril-β-metóxiviniltrialometilcetonas: Acilação de cetais. *Quim. Nova* **1994**, *17*, 24–26.
34. Rosa, W.C.; Rocha, I.O.; Schmitz, B.F.; Martins, M.A.P.; Zanatta, N.; Tisoco, I.; Iglesias, B.A.; Bonacorso, H.G. 4-(Trifluoromethyl) coumarin-fused pyridines: Regioselective synthesis and photophysics, electrochemical, and antioxidative activity. *J. Fluor. Chem.* **2021**, *248*, 109822. [CrossRef]
35. Zhao, Z.; Zhang, H.; Lam, J.W.Y.; Tang, B.Z. Aggregation-Induced Emission: New Vistas at the Aggregate Level. *Angew. Chemie Int. Ed.* **2020**, *59*, 9888–9907. [CrossRef]
36. Mei, J.; Hong, Y.; Lam, J.W.Y.; Qin, A.; Tang, Y.; Tang, B.Z. Aggregation-induced emission: The whole is more brilliant than the parts. *Adv. Mater.* **2014**, *26*, 5429–5479. [CrossRef] [PubMed]
37. Bristow, W.T.; Webb, K.S. Intercomparison study on accurate mass measurement of small molecules in mass spectrometry. *J. Am. Soc. Mass Spectrom.* **2003**, *14*, 1086–1098. [CrossRef]

Article

Enhanced Photoluminescence of Electrodeposited Europium Complex on Bare and Terpyridine-Functionalized Porous Si Surfaces

Min Hee Joo [1,2], So Jeong Park [1], Hye Ji Jang [1], Sung-Min Hong [1,2], Choong Kyun Rhee [1] and Youngku Sohn [1,2,*]

[1] Department of Chemistry, Chungnam National University, Daejeon 34134, Korea; wnalsgml4803@naver.com (M.H.J.); jsjs5921@naver.com (S.J.P.); gpwldndud@naver.com (H.J.J.); qwqe212@naver.com (S.-M.H.); ckrhee@cnu.ac.kr (C.K.R.)
[2] Department of Chemical Engineering and Applied Chemistry, Chungnam National University, Daejeon 34134, Korea
* Correspondence: youngkusohn@cnu.ac.kr; Tel.: +82-42-821-6548

Abstract: The trivalent Eu(III) ion exhibits unique red luminescence and plays an significant role in the display industry. Herein, the amperometry electrodeposition method was employed to electrodeposit Eu(III) materials on porous Si and terpyridine-functionalized Si surfaces. The electrodeposited materials were fully characterized by scanning electron microscopy, X-ray diffraction crystallography, Fourier-transform infrared spectroscopy, and X-ray photoelectron spectroscopy. Photoluminescence (PL) spectroscopy revealed that PL signals were substantially increased upon deposition on porous Si surfaces. PL signals were mainly due to direct excitation and charge-transfer-indirect excitations before and after thermal annealing, respectively. The as-electrodeposited materials were of a Eu(III) complex consisting of OH, H_2O, NO_3^-, and CO_3^{2-} groups. The complex was transformed to Eu_2O_3 upon thermal annealing at 700 °C. The electrodeposition on porous surfaces provide invaluable information on the fabrication of thin films for displays, as well as photoelectrodes for catalyst applications.

Keywords: porous Si; photoluminescence; europium; lanthanide; electrodeposition; terpyridine; functionalization

Citation: Joo, M.H.; Park, S.J.; Jang, H.J.; Hong, S.-M.; Rhee, C.K.; Sohn, Y. Enhanced Photoluminescence of Electrodeposited Europium Complex on Bare and Terpyridine-Functionalized Porous Si Surfaces. Photochem 2021, 1, 38–52. https://doi.org/10.3390/photochem1010004

Academic Editor: Marcelo I. Guzman

Received: 5 March 2021
Accepted: 3 April 2021
Published: 6 April 2021

Publisher's Note: MDPI stays neutral with regard to jurisdictional claims in published maps and institutional affiliations.

Copyright: © 2021 by the authors. Licensee MDPI, Basel, Switzerland. This article is an open access article distributed under the terms and conditions of the Creative Commons Attribution (CC BY) license (https://creativecommons.org/licenses/by/4.0/).

1. Introduction

Porous silicon (PS) is known to show unique physicochemical properties and a high surface area [1–5]. The pores have commonly been prepared by electrochemical anodization process in a hydrofluoric acid (HF) solution by varying many experimental parameters such as concentration, organic-solvent additive, time, and applied potential [6,7]. The properties have consequently been determined by the finally achieved pore sizes and depths. The application areas are very wide, and include sensors, catalysts, solar cells, and charge storages [8–17]. For sensors, Ramírez-González et al. controlled the thickness of the PS layer and tested ethanol-sensing performance by measuring conductivity. They showed that the thickness is an important factor for a target-sensing gas [14]. Dwivedi et al. prepared TiO_2 nanotube-decorated PS, and showed that the heterojunction exhibited a selective ethanol-sensing performance (with sub-ppm level down to 0.5 ppm) and a good linear response at low detection level [15]. For catalyst applications, Kim et al. used PS nanoparticles (NPs) for embedding Pd (~12 nm) and superparamagnetic γ-Fe_2O_3 (~7 nm) NPs [16]. They demonstrated that the nanocomposites showed an excellent catalytic performance for the reduction of 4-nitrophenol to 4-aminophenol by $NaBH_4$. For solar cells, Sundarapura et al. prepared an anodic SiO_2 layer on a PS surface, and demonstrated that the solar-cell efficiency was significantly improved [13]. For energy storage, Ortaboy et al. synthesized MnO_x-decorated carbonized porous Si nanowire, and showed that the PS nanowire-based pseudocapacitor electrode had a specific capacitance of 635 F g^{-1}, areal power of 100 mW cm^{-2}, and energy of 0.46 mW h cm^{-2} [17]. They also reported a power density of 25 kW kg^{-1} and an energy density of 261 W h kg^{-1} at 0.2 mA/cm^2, and a large potential window of 3.6 V. Surface functionalization on Si has also widely been demonstrated to widen the application areas and performances [18–20]. Yaghoubi et al. prepared a PS by the electrochemical etching method and modified with lectins via amino-silane functionalization followed by glutaaldehyde incubation [20]. They demonstrated that a lectin-conjugated PS showed good biosensing performances for label-free and real-time detection of *Escherichia coli* and *Staphylococcus aureus* by reflectometric interference Fourier transform spectroscopy.

In the present study, the electrodeposition method was employed to embed electrodeposited materials inside the pores of a PS and to examine new characteristics arising from the newly developed PS materials. In addition, surface functionalization was performed to further modify the surface property on electrodeposition. The Eu(III) ion was selected because it exhibits strong red luminescence and depends on local environment [21]. Therefore, luminescence profiles were examined, and consequently, dramatically enhanced luminescence characteristics was observed. The newly reported results provide valuable information on the development of thin-film display materials, as well as thin films for other application areas, such as catalyst electrodes.

2. Materials and Methods

Chemicals of Eu(III) nitrate hexahydrate (99.9%, Alfa Aesar, Karlsruhe, Germany), sodium perchlorate (NaClO$_4$, ≥98.0%, Sigma-Aldrich, St. Louis, MO, USA), (3-aminopropyl) trimethoxysilane (APTMS, 97%, Sigma-Aldrich), and 2,2':6',2''-terpyridine-4'-carboxylic acid (Tpy-COOH, 95%, Alfa Aesar) were used as received without further purification. An Si wafer (<100> B-doped p-type, Siltron Inc., Gumi, Korea) was cut to a size of 20 mm × 20 mm and precleaned with isopropyl alcohol and deionized water, repeatedly. The precleaned Si wafer was dipped in a solution of H$_2$O:H$_2$O$_2$:NH$_4$OH (5:1:1 v/v) for 15 min, then dipped again in a 2% HF solution for 30 min, and dried under an infrared lamp. Gold (Au) sputtering was performed on the Si wafer at a current of 5 mA for 90 s, and then thermal treated at 700 °C for 2 h in a furnace. For the fabrication of a porous Si (PS), the Au-deposited Si wafer was dipped in a 5% HF solution for 48 h, and then cleaned with deionized water. In the second etching step, the pre-etched Si wafer was used as an anode. A carbon rod was used as a cathode with an applied potential of 30.0 V for 1 h in an etching solution (ethanol:deionized water:HF = 2:1:1 molar ratio). Au nanoparticle residues on the Si wafer were further removed by dipping in an aqua regia solution for 4 h. The removal of Au was checked by X-ray photoelectron spectroscopy. For the functionalization of terpyridine ligand, the finally obtained cleaned porous Si (PS) substrate was initially dipped in a warm (60 °C) H$_2$O/NH$_4$OH/H$_2$O$_2$ (5:3:1, v/v) solution to make an OH-terminated surface. After that, for the surface functionalization with amino-silane, the OH-terminated PS substrate was dipped for 12 h in a 1% APTMS/methanol solution, then fully washed with pure methanol, and dried under nitrogen gas stream. The amino-silane functionalized substrate was then dipped in a 1 mM Tpy-COOH/DMSO solution for 6 h to finally functionalize terpyridine ligand on the surface. The terpyridine-functionalized substrate was abbreviated as PS-Si-Tpy.

For the electrochemical experiments of cyclic voltammetry and amperometry, a conventional three-electrode system was employed using a Potentiostat/Galvanostat (WPG100 model, WonATech Co., Ltd., Seoul, Korea). A Pt wire (0.5 mm) and Ag/AgCl electrodes were used as the counter and reference electrodes, respectively. Working electrodes were bare Si, bare PS, and terpyridine-functionalized PS (PS-Si-Tpy) substrates. A 0.1 M NaClO$_4$ solution was used as a supporting electrolyte with 10 mM Eu(III) ions. For electrodeposition of Eu on the three different substrates, the amperometry was performed at an applied potential of −1.8 V (vs. Ag/AgCl) for 5 h in a 10 mM Eu(III)/ 0.1 M NaClO$_4$ solution. The Eu-electrodeposited samples were further examined before and after thermal annealing at 700 °C.

For the surface morphologies of PS, Eu-electrodeposited PS and Eu-electrodeposited PS-Si-Tpy, a field-emission scanning electron microscope (model Hitachi S-4800 FE-SEM, Hitachi Ltd., Tokyo, Japan) was employed to take surface images. For the analysis of crystal phase formation, X-ray diffraction (XRD) patterns were obtained using a X-ray diffractometer (model MiniFlex II, Rigaku Corp., Tokyo, Japan, CNU chemistry core-facility) equipped with a Cu K_α X-ray light source. For crystal models, the VESTA software program (ver. 3.5.7) was used. To examine the major functional groups and the complex formation of the electrodeposited Eu materials before and after thermal annealing, FT-IR spectra were taken using a Nicolet iS 10 FT-IR spectrometer (Thermo Scientific Korea, Seoul, Korea) with an attenuated total reflection (ATR) mode. To examine photoluminescence (PL) characteristics of the Eu-electrodeposited Si, PS, and PS-Si-Tpy samples, a Sinco FS-2 fluorescence spectrometer (Sinco, Seoul, Korea) was used to obtain emission and excitation spectra, and their corresponding 2D and 3D photoluminescence (PL) counter maps scanned at various excitation wavelengths. The PL decay curves were taken using a FluoroLog 3 spectrometer (Horiba-Jobin Yvon, Kyoto, Japan) with a pulsed laser diode 374 nm ($\pm$10), a pulsed nano LED 264 nm ($\pm$10), and a single-photon-counting photomultiplier tube. X-ray photoelectron spectroscopy (XPS) was employed to confirm the surface oxidation states using a Thermo-VG Scientific K-alpha$^+$ spectrometer (Thermo VG Scientific, Waltham, MA, USA) equipped with a hemispherical energy analyzer and an Al K_α X-ray source.

3. Results

3.1. Surface Functionalization

The surface-functionalization process on a porous Si (PS) surface is depicted in Figure 1. As described in the experimental section, a PS substrate was initially terminated by OH groups. The surface was then functionalized with amino-silane. The amino group reacted with -COOH, forming a peptide bond [22,23]. Consequently, a terpyridine ligand was functionalized on the surface. The Lewis basic ligand strongly interacted with Lewis acidic metal cations such as the Eu(III) ion in the present study [22,23].

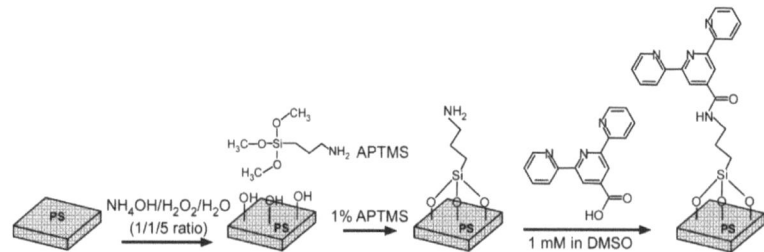

Figure 1. Surface functionalization process on a porous Si surface.

3.2. Surface Morphology of Porous Si and Electrodeposited Samples

Figure 2A,B display cyclic voltammetry (CV) profiles in 0.1 M NaClO$_4$ and 10 mM Eu(III)/0.1 M NaClO$_4$ electrolytes over PS and PS-Si-Tpy electrodes, respectively. A current increase (starting from -1.0 V in the negative-going potential) was commonly observed, attributed to a hydrogen-production reaction [22,23]. Upon addition of 10 mM Eu(III) ion, the current densities were enhanced by 1.5$\times$ and 1.7$\times$ at -1.8 V (vs. Ag/AgCl) over the PS and PS-Si-Tpy electrodes, respectively, due to the reduction/complexation current. For the functionalized electrode, the current density of hydrogen production current was relatively diminished. Therefore, the morphology after electrodeposition was expected to be different from that over PS, as discussed below. Figure 2a–d show the SEM images of Si after the first and second etching steps. As seen in Figure 2a, the Si surface was not fully etched by dipping in an HF solution. Furthermore, Au particles were still present on the surface. After the second etching step by electrochemistry, followed by Au removal, the SEM images showed submicron size pores well distributed on the surface shown in

Figure 2b. In the SEM image (Figure 2c) of the PS surface upon electrodeposition of Eu, clustered small particles were observed, and thin-film morphology was observed inside the pores. Upon thermal annealing at 700 °C, the surface was markedly changed, and small dots appeared to be embedded inside the pores. In the SEM image (Figure 2d) of the PS-Si-Tpy surface upon electrodeposition of Eu, thick films were appeared to be formed on the surface. The pores were fully covered by the electrodeposited materials. Upon thermal annealing of the sample, aggregated particles were found to be formed on the surface.

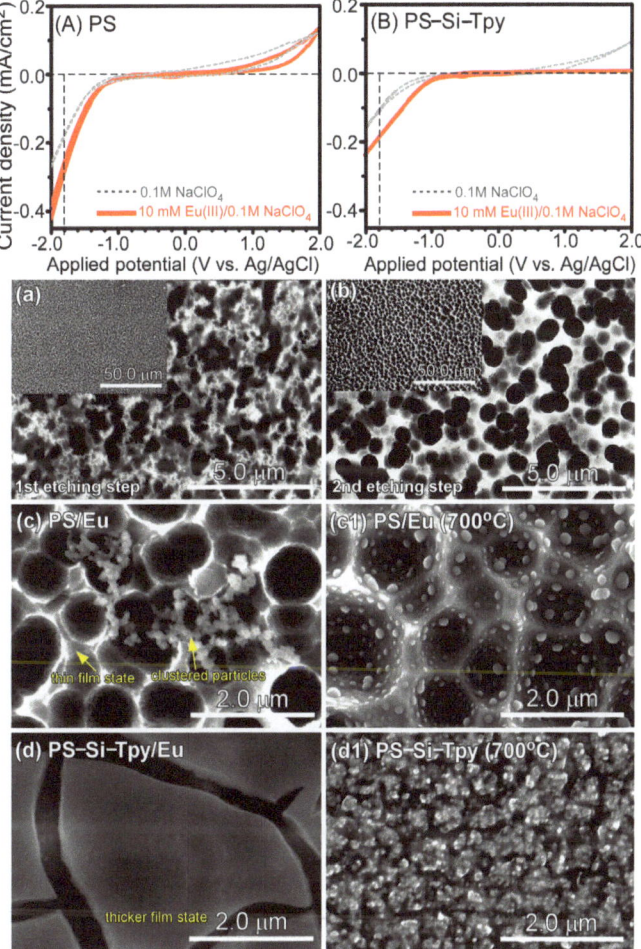

Figure 2. Cyclic voltammetry profiles of bare PS (**A**) and PS-Si-Tpy (**B**) in blank 0.1 M NaClO$_4$ (gray dotted line) and 0.1 M Eu(III)/0.1 M NaClO$_4$ (thick red line) electrolytes; SEM images of Si (**a**) after the first etching step, porous Si (**b**) after the second etching step, Eu-electrodeposited PS before (**c**) and after (**c1**) thermal annealing; and Eu-electrodeposited PS-Si-Tpy before (**d**) and after (**d1**) thermal annealing.

3.3. Crystal Phases of the As-Electrodeposited and Thermal-Annealed Samples

Figure 3 displays the XRD patterns of PS-Si-Tpy, Eu-electrodeposited PS-Si-Tpy, and thermal (700 °C) annealed Eu-electrodeposited PS-Si-Tpy samples. Before electrodeposition, the XRD signals were mainly due to bare Si [11]. Upon Eu deposition, new XRD peaks were clearly observed at 2θ = 10.1°, 20.1°, 28.0°, and 49.8°. Interestingly, these XRD peak positions were in good consistency with those of the $Ln_2(OH)_x(NO_3)_y(SO_4)_z \cdot nH_2O$ complex [24], further discussed below. Joo et al. also reported similar XRD profiles for a series of lanthanide complexes on carbon and Ni substrates prepared by amperometry electrodeposition [25,26]. For the thermal-treated sample, several XRD peaks were observed at 2θ = 28.5°, 32.9°, 47.3°, and 56.1°, a good match with the (222), (004), (044), and (226) crystal planes of cubic phase (Ia-3) Eu_2O_3 (ref. no 98-004-0472) [21]. The XRD peak at 2θ = 28.5° was stronger than other peaks and corresponded to the (222) crystal plane. The unit cell structure and the structure projections of the (222) and (440) crystal planes are shown in the inset of Figure 3.

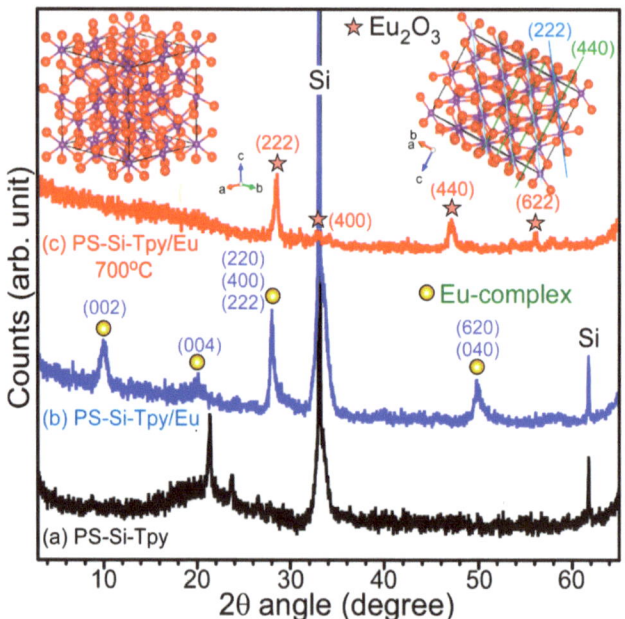

Figure 3. XRD profiles of porous Si (**a**), Eu-electrodeposited PS-Si-Tpy (**b**), and thermal-annealed PS-Si-Tpy/Eu (**c**), and the unit cell of Eu_2O_3 and the structure projections of the (222) and (440) crystal planes.

3.4. FT-IR of the As-Electrodeposited and Thermal-Annealed Samples

To confirm the major functional groups of a complex deduced from the XRD patterns, FT-IR spectra were taken and shown in Figure 4. Compared with the FT-IR profiles of PS (Figure 4a) and PS-Si-Tpy (Figure 4b) substrates, the substrates after Eu-electrodeposition showed strong transmittance FT-IR signals, and the two FT-IR profiles (Figure 4(a1,b1)) were observed to be very similar. This indicated that the complex formation was the same, although the corresponding morphologies (Figure 2) were different. A very strong and broad peak was observed around 3570 cm^{-1}, attributed to O–H stretching vibrations [21,25,26]. The O–H bending vibration appeared around 1620 cm^{-1}. A strong vibrational peak around 1100 cm^{-1} indicated a presence of ClO_4^- species trapped (or complexed) in the complex [25–27]. The peaks around 1350 cm^{-1} were attributed to the stretching modes of the CO_3^{2-} group [25,26]. The vibrational modes of the NO_3^- group also were found around this position [24]. Additional vibrational modes of CO_3^{2-} group were observed at 815 cm^{-1} and 935 cm^{-1}, confirming the presence of CO_3^{2-} group in the complex. The peak around 620 cm^{-1} could be assigned to a Eu–O vibrational mode [21,25,26]. On the basis of the FT-IR profiles, the complex of the electrodeposited materials possibly consisted of Eu, OH, H_2O, NO_3^-, and CO_3^{2-} groups. Upon thermal annealing at 700 °C, most of the peaks disappeared. The O-H vibrational peaks were weakly observed, as expected. Two major peaks were observed at 1395 cm^{-1} and 1500 cm^{-1}, and were assigned to chemisorbed CO_2 species [21].

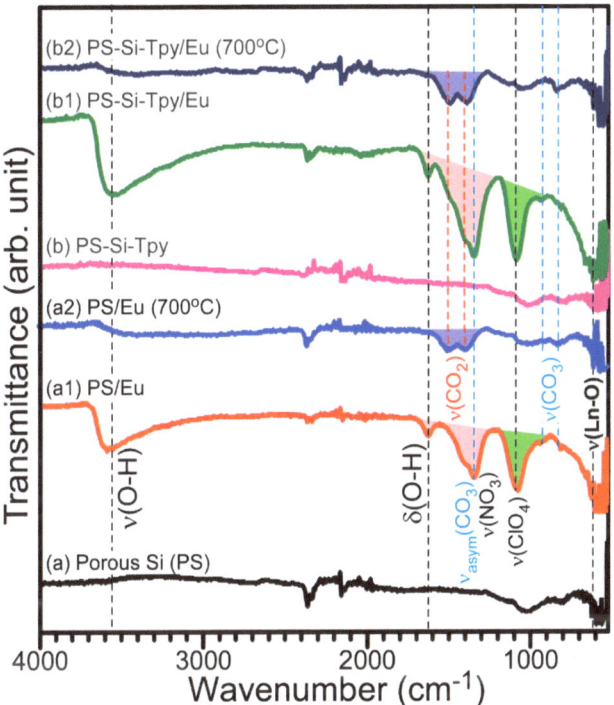

Figure 4. Transmittance FT-IR spectra of porous Si (**a**), Eu-deposited PS (**a1**), thermal-annealed PS/Eu (**a2**), terpyridine-functionalized PS (**b**), PS-Si-Tpy/Eu (**b1**), and thermal-annealed PS-Si-Tpy/Eu (**b2**).

3.5. Eu-Complexation with the Monolayer Terpyridine Ligand

Complexation of the Eu(III) ion with the functionalized terpyridine ligand was examined. This complexation occurred on the monolayer level, not on a bulk state by electrodeposition. A PS-Si-Tpy substrate was dipped in a 10 mM Eu(III)/0.1 M NaClO$_4$ solution for 1 h, gently rinsed with deionized water, and dried under a nitrogen gas stream. Photoluminescence spectroscopy and X-ray photoelectron spectroscopy were employed to confirm the complexation between the Eu(III) ion and the terpyridine group. For PS-Si-Tpy-Eu(III), Figure 5a displays the emission spectra taken at excitation wavelengths of 395 nm and 300 nm. The 395 nm corresponds to the $^5L_6 \leftarrow {}^7F_0$ direct excitation for Eu(III) energy levels [21]. Under this excitation, a broad peak was observed around 612 nm, attributed to the $^5D_0 \to {}^7F_2$ transition of the Eu(III) ion [21]. Under 300 nm excitation, two peaks were observed around 590 nm and 612 nm, assigned to the $^5D_0 \to {}^7F_1$ and $^5D_0 \to {}^7F_2$ transitions of the Eu(III) ion, respectively. The corresponding 2D and 3D photoluminescence contour maps (Figure 5(a1,a2)) show a more closely spaced region (or stronger signals) under 300 nm. This indicated that the Eu(III) emission was mainly excited by an indirect charge-transfer process from porous Si support to the Eu(III) site via the functionalized group.

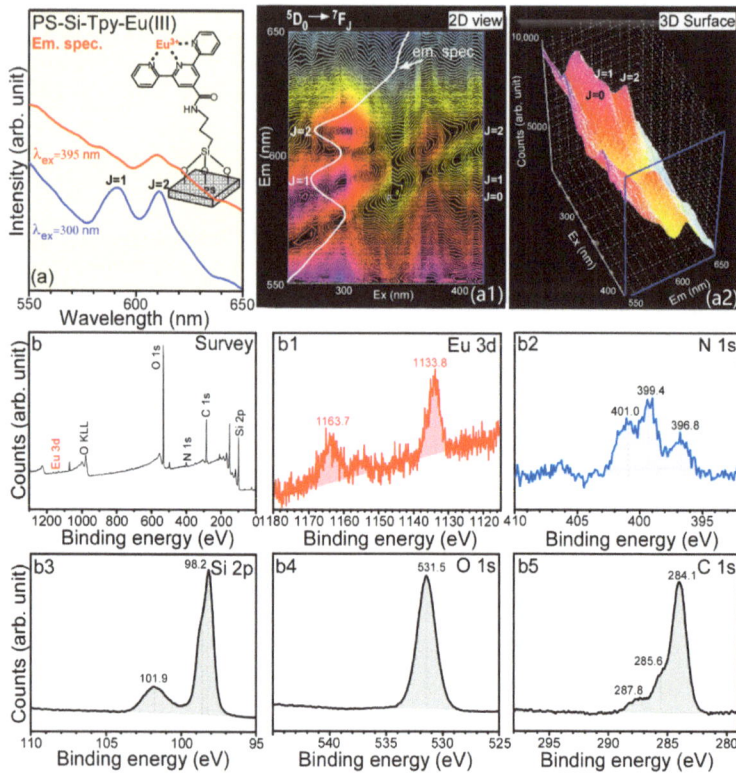

Figure 5. Emission spectra (**a**), corresponding 2D (**a1**), and 3D (**a2**) photoluminescence imaging profiles of Eu(III) coordinated with the functionalized terpyridine; and survey (**b**), Eu 3d (**b1**), N 1s (**b2**), Si 2p (**b3**), O 1s (**b4**), and C 1s (**b5**) XPS profiles.

Figure 5(b–b5) display the survey, Eu 3d, N 1s, Si 2p, O 1s, and C 1s XPS profiles for PS-Si-Tpy-Eu(III). Strong signals of Si, C, and O elements were expected from the PS and the functionalized molecules. The weak Eu 3d signal was due to Eu(III) complexed with the terpyridine ligand. Eu $3d_{5/2}$ and Eu $3d_{3/2}$ peaks were observed at 1133.8 eV and 1163.7 eV, and were attributed to the Eu(III) oxidation state [21,26]. Si 2p peaks were observed around 98.2 and 101.9 eV, and were attributed to Si^0 and Si-O species, respectively [11]. A strong O 1s peak at 531.5 eV was mainly due to Si-O species. Three N 1s peaks were observed, and were tentatively assigned to N of the amide bond (at 396.8 eV), and N of the singly (401.0 eV) and doubly (399.4 eV) coordinated terpyridine groups, respectively. The C 1s peaks at 284.1 eV, 285.6 eV, and 287.8 eV were attributed to C-C, C-N, and C-O species, respectively.

3.6. Photoluminescence of Electrodeposited Eu on the Si Substrate

Figure 6(a,a1) display the excitation and emission spectra for the Eu-electrodeposited Si substrate with no pores. The corresponding 2D and 3D photoluminescence contour maps are shown in Figure 6(c,c1), respectively. For the excitation spectra at emission wavelengths of 610 nm and 590 nm, weak peaks were observed around 280 nm and 395 nm, respectively. For the emission spectrum at an excitation wavelength of 280 nm, a broad peak was observed between 350 nm and 650 nm. The PL signals of Eu(III) ions were significantly observed between 550 nm and 720 nm. For the emission spectrum at 395 nm, weak signals appeared around 589 nm, 614 nm, 650 nm, and 690 nm, and were attributed to the transitions from the excited $^5D_0$ level to the lower $^7F_1$, $^7F_2$, $^7F_3$, and $^7F_4$ levels, respectively [21,23].

Upon thermal annealing at 700 °C, the PL signals (Figure 6(b,b1)) were increased, due to a change in the crystal phase and a decrease in the PL quenching centers. In the excitation spectrum at an emission wavelength of 610 nm, the several peaks at 365 nm, 382 nm, 395 nm, and 405 nm were assigned to the transitions from the ground $^7F_0$ level to the upper $^5D_4$, $^5G_J/^5L_7$, $^5L_6$, and $^5D_3$ levels, respectively [21,23]. A much stronger (2×) and broader peak was observed around 280 nm, commonly attributed to a charge-transfer ($O^{2-} \to Eu^{3+}$) excitation [21,23]. For the excitation spectrum at 590 nm, a broad peak around 280 nm was mainly observed. For the emission spectrum at an excitation wavelength of 280 nm, two sharp peaks appeared at 588 nm and 611 nm, and were attributed to the transitions from the excited $^5D_0$ level to the lower $^7F_1$ and $^7F_2$ levels, respectively [21,23]. On the other hand, for the emission spectrum at 395 nm, the peak at 611 nm was mainly observed, and the peak at 588 nm was drastically diminished. The corresponding 2D and 3D photoluminescence contour maps taken at various excitation wavelengths are shown in Figure 6(d,d1), respectively.

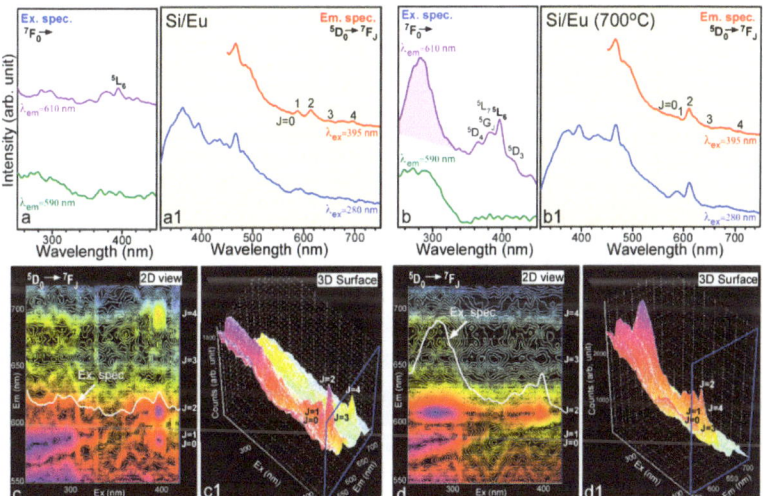

Figure 6. Excitation (**a,b**) and emission (**a1,b1**) spectra for electrodeposited Eu on Si (**left panel**) and thermal-annealed Si/Eu (**right panel**), and their 2D (**c,c1**) and 3D (**d,d1**) photoluminescence imaging profiles.

3.7. Photoluminescence of Electrodeposited Eu on the Porous Si Substrate

Figure 7(a,a1) display the excitation and emission spectra for a porous Si substrate after Eu electrodeposition, respectively. Compared with the PL signals in Figure 6, the PL intensity was strongly enhanced by about 11×. In the excitation spectrum at an emission wavelength of 612 nm, several peaks were observed around 295 nm, 320 nm, 364 nm, 381 nm, 395 nm, and 415 nm, and were attributed to the transitions from the $^7F_0$ level to the upper $^5F_4$, $^5H_5$, $^5D_4$, $^5G_J/^5L_7$, $^5L_6$, and $^5D_3$, levels, respectively [21,23]. The peak at 395 nm was major. In the excitation spectrum at 590 nm, the sharper peaks were somewhat weakened, while the peak around 290 nm was increased. For the emission spectra at an excitation wavelength of 290 nm, no characteristics of Eu(III) PL signals were significantly observed. However, the emission spectrum at 395 nm showed sharp PL signals around 575 nm, 591 nm, 612 nm, 650 nm, and 697 nm, attributed to the transitions from the excited $^5D_0$ level to the lower $^7F_1$, $^7F_2$, $^7F_3$, and $^7F_4$ levels, respectively, as mentioned above [21,23]. The corresponding 2D and 3D photoluminescence contour maps are shown in Figure 7(c,c1), respectively.

Upon thermal annealing at 700 °C, the PL profiles (Figure 7(b,b1)) were drastically changed, and the intensity was also substantially increased by about 29×. One major change was found in the excitation spectra at excitation wavelengths of 614 nm and 590 nm. In the excitation spectrum at 614 nm, the charge-transfer peak at 295 nm was observed to be about 5.1× stronger than the peak at 395 nm. In the excitation spectrum at 590 nm, the charge-transfer peak was mainly observed, and no transition signals of Eu(III) ions were significantly observed between 350 nm and 450 nm. In the emission spectrum at 295 nm, two major peaks were observed at 594 nm ($^5D_0 \rightarrow {}^7F_1$) and 613 nm ($^5D_0 \rightarrow {}^7F_2$). On the other hand, in the emission spectrum at 395 nm, a peak at 614 nm ($^5D_0 \rightarrow {}^7F_2$) was mainly observed. The peak at 697 nm ($^5D_0 \rightarrow {}^7F_4$) was observed to be substantially decreased upon thermal annealing. The corresponding 2D and 3D photoluminescence contour maps are shown in Figure 7(d,d1), respectively.

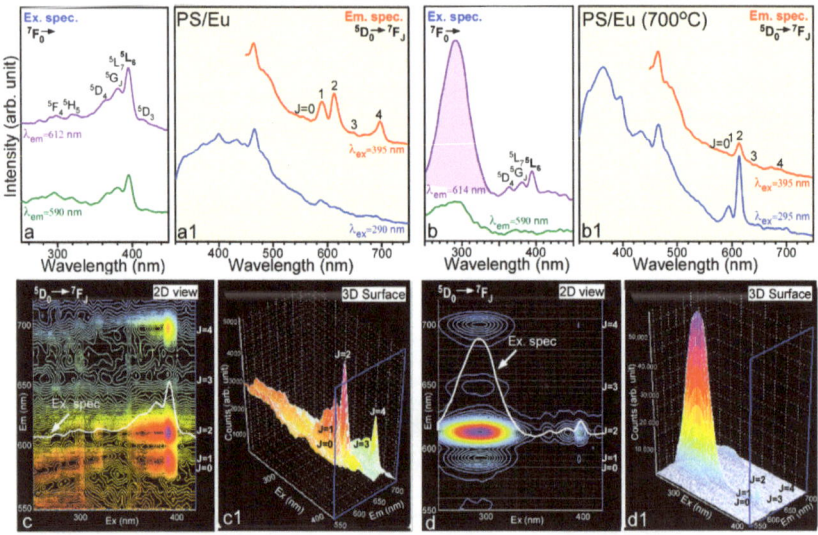

Figure 7. Excitation (**a,b**) and emission (**a1,b1**) spectra for electrodeposited Eu on PS (**left panel**) and thermal-annealed PS/Eu (**right panel**), and their 2D (**c,c1**) and 3D (**d,d1**) photoluminescence imaging profiles.

3.8. Photoluminescence of Electrodeposited Eu on the PS-Si-Tpy Substrate

Figure 8(a,a1) displays excitation and emission spectra for the PS-Si-Tpy substrate after Eu electrodeposition, respectively. The PL profiles were quite similar to those of the PS substrate after Eu electrodeposition shown in Figure 7. One difference is that the PL intensity was weaker (by about 0.2×) than those shown in Figure 7. The peak positions and the corresponding assignments were the same, as discussed above. The corresponding 2D and 3D photoluminescence contour maps are shown in Figure 8(c,c1), respectively. Upon thermal annealing at 700 °C, the PL profiles (Figure 8(b,b1)) were also drastically changed, and the intensity was also substantially increased, as shown in Figure 7. The peak positions and profiles were also very similar to those found in Figure 7(b,b1). In the excitation spectrum at 614 nm, the charge-transfer peak at 295 nm was observed to be about 7.8× stronger than the peak at 395 nm. The corresponding 2D and 3D photoluminescence contour maps are shown in Figure 8(d,d1), respectively.

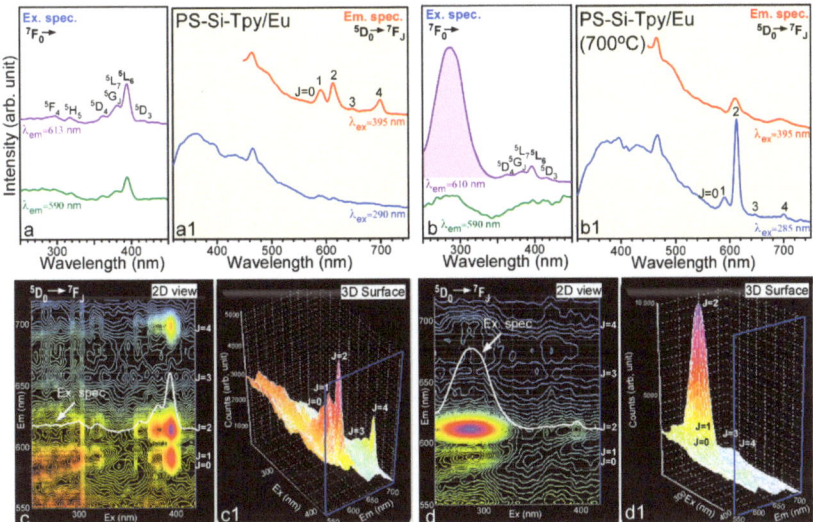

Figure 8. Excitation (**a,b**) and emission (**a1,b1**) spectra for electrodeposited Eu on terpyridine-functionalized PS (**left panel**) and thermal-annealed PS-Si-Tpy/Eu (**right panel**), and their 2D (**c,c1**) and 3D (**d,d1**) photoluminescence imaging profiles.

3.9. Photoluminescence Decay Kinetics

Figure 9 displays photoluminescence decay profiles at an emission wavelength of 610 nm, corresponding to the $^5D_0 \rightarrow {}^7F_2$ transition. The fitting parameters are summarized in Table 1 using a third-order exponential decay function [23]. The average lifetimes at an excitation wavelength of 374 nm were estimated to be 18.7 ns and 23.4 ns for the PS-Si-Tpy/Eu and thermal-annealed PS-Si-Tpy/Eu samples, respectively. The as-electrodeposited sample showed a shorter lifetime. The lifetime at an excitation wavelength of 264 nm was much shorter, at 15.4 ns.

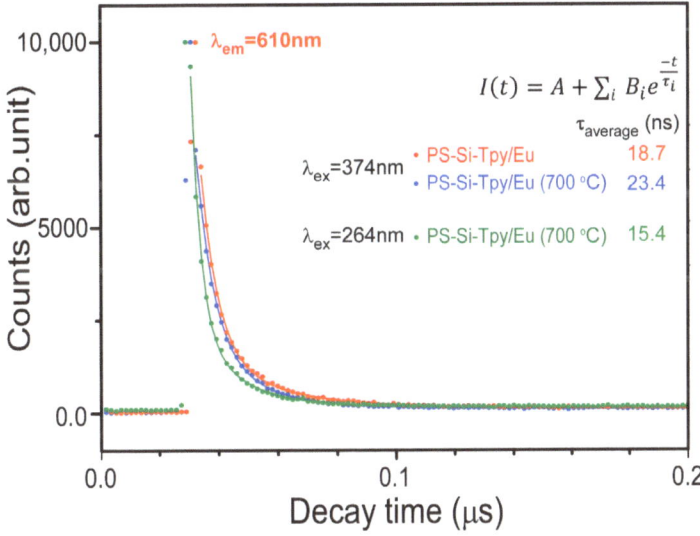

Figure 9. Photoluminescence decay times for 610 nm emission signals at excitation wavelengths of 264 nm and 374 nm for the PS-Si-Tpy/Eu and thermal-annealed PS-Si-Tpy/Eu samples.

Table 1. Fitting parameters of the PL decay curves for 610 nm emission signals at excitation wavelengths of 264 nm and 374 nm for the PS-Si-Tpy/Eu and thermal-annealed PS-Si-Tpy/Eu samples.

Parameters	PS-Si-Tpy/Eu (374 nm)	PS-Si-Tpy/Eu-700 °C (374 nm)	PS-Si-Tpy/Eu-700 °C (264 nm)
τ_1 (ns)	19.58	25.34	14.75
τ_2 (ns)	798.36	867.749	1102.64
τ_3 (ns)	6.18	5.94	3.19
B_1	1558.68	1360.836	2288.291
B_2	83.2513	98.52946	76.09427
B_3	5271.567	4926.381	6620.526
R_1	0.2254	0.2131	0.2547
R_2	0.0120	0.0154	0.0085
R_3	0.7625	0.7715	0.7368
A	34.77629	37.63854	93.39494
χ^2	1.066872	1.102067	1.091194
$\tau_{average}$ (ns)	18.74 (±0.20)	23.37 (±0.24)	15.44 (±0.24)

Note: $I(t) = A + \sum_i B_i e^{\frac{-t}{\tau_i}}$, $R_i = B_i / (\sum_{i=1}^{3} B_i)$, $\tau_{average} = \sum_{i=1}^{3} R_i \cdot \tau_i$, R = the relative ratio factor.

3.10. Surface Chemical States of the Electrodeposited Materials

Figure 10 displays the survey, Eu 3d, and O 1s XPS profiles of the PS-Eu and PS-Si-Tpy-Eu before and after thermal annealing. The survey spectra commonly showed the elements of Eu (Eu 3d, Eu 4s, Eu 4p, and Eu 4d), O (O 1s), and C (C 1s), as expected. For the as-electrodeposited samples before thermal annealing, additional peaks were observed at binding energies (BEs) of 407 eV and 208 eV, and were attributed to N of NO_3^- and Cl of ClO_4^-, respectively [23,25,26]. These two groups were observed in the FT-IR spectra (Figure 4). Upon thermal annealing, these two elements disappeared. For the Eu 3d XPS profiles before thermal annealing, the Eu $3d_{5/2}$ and Eu $3d_{3/2}$ peaks were observed at 1133.8 eV and 1163.5 eV, respectively, with a spin-orbit splitting energy of 29.7 eV, which was attributed to the Eu(III) oxidation state [21,28]. No indication of the Eu(II) oxidation state was found. Upon thermal annealing, the Eu $3d_{5/2}$ and Eu $3d_{3/2}$ peaks were observed at 1135.0 eV and 1164.8 eV, respectively, with a spin-orbit splitting energy of 29.8 eV, which was attributed to the Eu(III) oxidation state. Interestingly, two additional peaks were found around 1126 eV and 1156 eV; these were attributed to the Eu $3d_{5/2}$ and Eu $3d_{3/2}$ peaks of the Eu(II) oxidation state. For the O 1s XPS profiles before thermal annealing, two peaks could be assigned at 530.5 eV and 532.0 eV; these were attributed to the Eu-OH and C-O/ClO_4 species, respectively [23,25,26]. Upon thermal annealing, three peaks could be assigned at 528.2 eV, 530.2 eV and 532.2 eV; these were attributed to lattice oxygen, defect oxygen, and surface O-H/H_2O species, respectively [23,25,26]. The dominant O 1s peak at 532.2 eV could be due to the hygroscopic nature of Eu_2O_3 [21]. The defect-related peak was plausibly due to the existence of two oxidation states of Eu(III) and Eu(II) ions after thermal annealing.

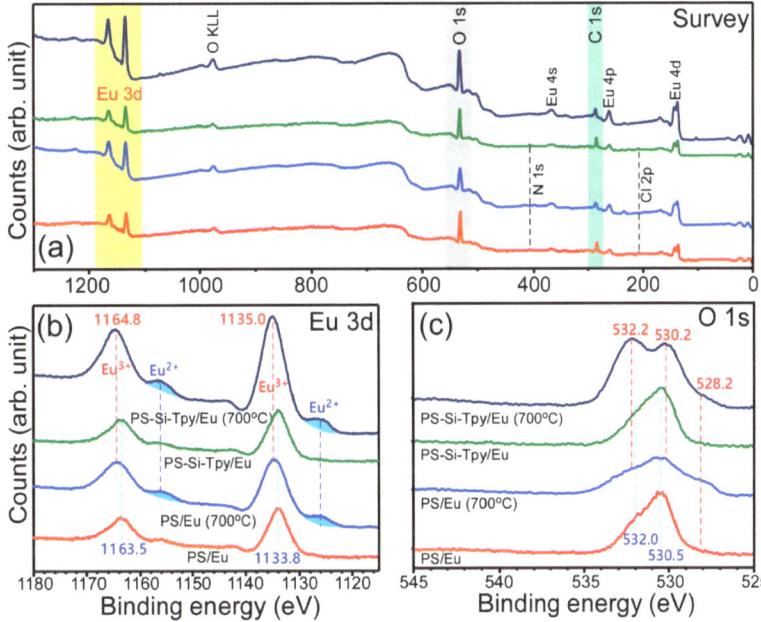

Figure 10. Survey (**a**), Eu 3d (**b**), and O 1s (**c**) XPS profiles of PS-Eu and PS-Si-Tpy-Eu before and after thermal annealing.

4. Discussion

Eu electrodeposition on both PS and terpyridine-functionalized PS (PS-Si-Tpy) were observed to be successful. It was initially assumed that light absorption and emission are both increased when Eu species were deposited on a PS with a high surface area. Moreover, when it was applied to an electrode catalyst material, the catalytic surface area was increased, and finally, the catalytic performance was enhanced. On the basis of the CV profiles, the currents at the applied potential were due to hydrogen evolution reaction and Eu complexation process. Thereby, the morphologies over the two different surfaces were expected to be different. SEM images (Figure 2) confirmed Eu was deposited inside the pores of PS. On the PS-Si-Tpy substrate, the film that was formed was thicker, and the pores were fully covered by the film. The uniform film formation on the PS-Si-Tpy electrode could be because the terpyridine ligand initially interacted with the Eu(III) ion that may initiate uniform electrodeposition. The interaction between Eu(III) ion and the terpyridine group was evidenced by PL and XPS (Figure 5). The Eu(III) emission characteristics were observed in the PL profiles at both direct and indirect excitation wavelengths. The PL intensity was stronger at an indirect (charge-transfer) excitation wavelength of 300 nm, compared with that at a direct excitation. This indicated that the Eu(III) ion was well bound with the terpyridine group. Thereby, a charge-transfer effect was observed to be efficient. The higher N 1s BE position also indicated an interaction between Eu(III) ions and three N atoms of the terpyridine group.

The materials were observed to be electrodeposited as a form of $Eu_2(OH)_x(NO_3)_y(CO_3)_z \cdot nH_2O$ complex [24–26]. One of several experimental evidences was the vibrational modes of H_2O/OH, CO_3^{2-}, and NO_3^- species in the FT-IR spectra (Figure 4). Another evidence was the XRD profile (Figure 3) of the corresponding sample. Wu et al. reported a very similar XRD profile for a $Ln_2(OH)_5(SO_4)_x(NO_3)_y \cdot nH_2O$ complex. On the basis of the literature, when SO_4^{2-} is replayed by CO_3^{2-}, we expect a similar complex formation, as well as a similar XRD profile. At a fixed potential during electrodeposition, an electric double layer (positive and negative ions) is commonly formed. Therefore, in the

21. Kang, J.-G.; Jung, Y.; Min, B.-K.; Sohn, Y. Full characterization of Eu(OH)$_3$ and Eu$_2$O$_3$ nanorods. *Appl. Surf. Sci.* **2014**, *314*, 158–165. [CrossRef]
22. Park, S.J.; Joo, M.H.; Hong, S.-M.; Rhee, C.K.; Kang, J.-G.; Sohn, Y. Electrochemical Eu(III)/Eu(II) Behaviors and Recovery over Terpyridyl-Derivatized Modified Indium Tin Oxide Electrode Surfaces. *Chem. Eng. J.* **2021**, *15*, 128717. [CrossRef]
23. Park, S.J.; Joo, M.H.; Hong, S.-M.; Kang, J.-G.; Rhee, C.K.; Lee, S.W.; Sohn, Y. Electrochemical Eu(III) behaviors and Eu oxysulfate recovery over terpyridine-functionalized indium tin oxide electrode. *Inorg. Chem. Front.* **2020**, *8*, 1175–1188. [CrossRef]
24. Wu, X.; Li, J.-G.; Zhu, Q.; Liu, W.; Li, J.; Li, X.; Sun, X.; Sakka, Y. One-step freezing temperature crystallization of layered rare-earth hydroxide (Ln$_2$(OH)$_5$NO$_3$·nH$_2$O) nanosheets for a wide spectrum of Ln (Ln = Pr–Er, and Y), anion exchange with fluorine and sulfate, and microscopic coordination probed via photoluminescence. *J. Mater. Chem. C* **2015**, *3*, 3428–3437. [CrossRef]
25. Joo, M.H.; Park, S.J.; Hong, S.M.; Rhee, C.K.; Sohn, Y. Electrochemical recovery and behaviors of rare earth (La, Ce, Pr, Nd, Sm, Eu, Gd, Tb, Dy, Ho, Er, Tm, and Yb) ions on Ni sheets. *Materials* **2020**, *13*, 5314. [CrossRef]
26. Joo, M.H.; Park, S.J.; Hong, S.-M.; Rhee, C.K.; Kim, D.; Sohn, Y. Electrodeposition and Characterization of Lanthanide Elements on Carbon Sheets. *Coatings* **2021**, *11*, 100. [CrossRef]
27. Chen, Y.; Zhang, Y.-H.; Zhao, L.-Z. ATR-FTIR spectroscopic studies on aqueous LiClO$_4$, NaClO$_4$, and Mg(ClO$_4$)$_2$ solutions. *Phys. Chem. Chem. Phys.* **2004**, *6*, 537–542. [CrossRef]
28. NIST X-ray Photoelectron Spectroscopy Database, Version 4.1 National Institute of Standards and Technology, Gaithersburg. 2012. Available online: http://srdata.nist.gov/xps/ (accessed on 2 April 2021).
29. Millero, F.J.; Magdalena Santana-Casiano, J.; Gonzalez-Davila, M. The formation of Cu(II) complexes with carbonate and bicarbonate ions in NaClO$_4$ solutions. *J. Solut. Chem.* **2010**, *39*, 543–558. [CrossRef]
30. Sohn, Y. Structural and spectroscopic characteristics of terbium hydroxide/oxide nanorods and plates. *Ceram. Int.* **2014**, *40*, 13803–13811. [CrossRef]

Review

A Photochemical Overview of Molecular Solar Thermal Energy Storage

Alberto Gimenez-Gomez [1,†], Lucien Magson [1,†], Beatriz Peñin [1], Nil Sanosa [1], Jacobo Soilán [1], Raúl Losantos [1,2,*] and Diego Sampedro [1,*]

[1] Department of Chemistry, Centro de Investigación en Síntesis Química (CISQ), Universidad de La Rioja, Madre de Dios, 53, 26006 Logroño, Spain
[2] ITODYS, Université Paris Cité and CNRS, 75006 Paris, France
* Correspondence: raul.losantosc@unirioja.es (R.L.); diego.sampedro@unirioja.es (D.S.)
† These authors contributed equally to this work.

Abstract: The design of molecular solar fuels is challenging because of the long list of requirements these molecules have to fulfil: storage density, solar harvesting capacity, robustness, and heat release ability. All of these features cause a paradoxical design due to the conflicting effects found when trying to improve any of these properties. In this contribution, we will review different types of compounds previously suggested for this application. Each of them present several advantages and disadvantages, and the scientific community is still struggling to find the ideal candidate suitable for practical applications. The most promising results have been found using norbornadiene-based systems, although the use of other alternatives like azobenzene or dihydroazulene cannot be discarded. In this review, we primarily focus on highlighting the optical and photochemical aspects of these three families, discussing the recently proposed systems and recent advances in the field.

Keywords: MOST; solar energy storage; solar fuels; norbornadiene; azobenzene; dihydroazulene

1. Introduction

Energy generation and storage has become one of the major challenges in our society and are especially relevant for industry [1,2]. The current energy demand is continuously rising [3] each year by 1.3%, and this progression is expected to last at least until 2040 [4], even considering that many industries worldwide have been affected by COVID-19. According to the International Energy Agency, buildings are responsible for almost 30% of energy consumption and account for 28% of CO_2 emissions [4,5]. To avoid the environmental impact from conventional energy sources, the use of renewable electricity needs to augment considerably. However, we are not yet able to avoid our dependence on fossil fuels. Consequently, significant efforts to find better alternatives to generate and store energy are under exploration. This is especially relevant for solar energy use and storage [6], which has been envisioned as an abundant, clean, and promising energy source.

Using natural photosynthesis as a working model for solar energy use, scientists are designing and preparing chemical systems capable of capturing and storing solar energy. Nowadays, different alternatives to make use of sunlight are under research, including direct use of photonic solar power and heating capacity of solar radiation. The variability in solar income is a very significant drawback to solar energy, as the power of both types of energy (photonic and heating) is not constant during the four seasons of the year [7] and strongly depends on the weather and geographical factors. This non-constant power supply unequivocally demands a storage solution, which should allow wider usability under conditions such as night or winter. Consequently, different methodologies have been developed to exploit solar power such as underground solar energy storage (USES) and molecular solar thermal (MOST) systems.

The USES system mechanism consists of the storage of sun energy underground during summer months using a pile [8,9]. There are four basic types of USES systems: hot-water-thermal storage, borehole thermal storage, aquifer thermal storage, and water gravel pit storage [10]. This mechanism requires a plant of quite large dimensions, making it quite difficult to use this technique once the building has already been built [11]. Similarly, in these approaches, the thermal insulation requirements usually imply a challenge for long-term storage.

On the other hand, MOST technology has become a promising candidate to capture and store solar energy in a sustainable and efficient manner. These systems have been expanded significantly in the last decades [7], even though the first idea dates a while back [12]. The MOST approach is based on the storage of solar energy as chemical energy using a photoactive molecule, which, after being exposed to sunlight, isomerizes into a metastable high-energy photoisomer [13]. The release of chemical energy as heat can be performed during the back conversion step using an external stimulus (Figure 1a) either through heat, through a catalyst, or electrochemically [14].

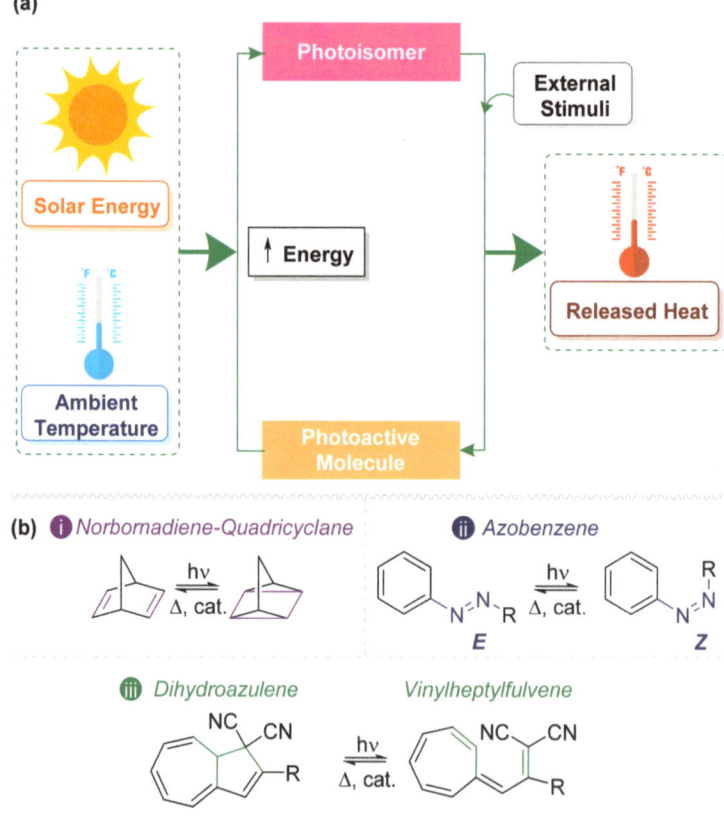

Figure 1. (**a**) Concept of the MOST system [15]. (**b**) Photoswitches most used in solar energy storage: (i) norbornadiene–quadricyclane, (ii) E/Z–azobenzene, and (iii) dihydroazulene–vinylheptafulvene.

Along the years, a large number of systems have been proposed as MOST candidates. There are at least six major requirements for a practical system, which makes the design of successful candidates a challenging task; this will be described further in the next section. Until now, none of the previously proposed compounds completely fulfils this list of requirements. Thus, the design and preparation of new alternatives to MOST technology is still a hot topic. In the following sections, the most relevant efforts to prepare suitable compounds according to these requirements will be discussed. For the scope of this review, we will describe the three types of systems that have been mainly used within the current framework of MOST technology: norbornadiene, azobenzene, and dihydroazulene derivatives [15–17] (Figure 1b). In addition, we will focus on the central role of the optical properties in the storage of solar energy.

2. Requisites for MOST Systems: Optical Properties

The development of new MOST candidates is a challenging task, which has gained more attention and visibility in the last decades owing to the expectations raised by this technology. As mentioned above, the ideal compound for a MOST device is still unknown, so many different strategies combining experimental and computational tools are being used to assist in the molecular design.

In order to maximize the efficiency of MOST systems, their optimization has received a great deal of attention; in light of this, the design of a suitable photoswitch must meet a large number of objectives to fulfil the ideal MOST system. Thus, the design criterion of a MOST system is subjected to several parameters involving both engineering and chemical challenges [18].

The first key step in the molecular solar thermal energy storage system is the absorption of light by the parent molecule, which undergoes a reversible photoisomerization reaction to its corresponding metastable isomer. This photoisomer should be stable enough to store the chemical potential for varying periods of time, depending on the envisioned application. Then, this stored energy should be released when and where required, in the form of heat. For the initial photoisomerization part of the MOST cycle to be successful, the photoswitch pair needs to fulfil a long list of features, in some cases even contradictory [3]:

- A large difference in the free energies of the parent molecule and its photoisomer, with a minimal increase in the molecular weight to maximize energy density [19,20].
- A moderately large kinetic barrier for back conversion [21,22].
- An absorption spectrum of the photoactive molecule matching the solar spectrum [22,23].
- A high quantum yield ($\Phi = 1$) for the photogeneration of the metastable photoisomer.
- A photochemically inactive (or non-absorbing) photoisomer.
- Negligible degradation of both the photoactive molecule and its photoisomer after multiple cycles, especially moving towards higher temperatures.

These are the major requirements that should be optimized to improve the performance of any potential candidate. Furthermore, when considering a MOST molecule in an integrated device, the use of environmentally friendly compounds and solvents is desired to minimize the risks in case of leaching or losses to its surroundings. In this review, we will focus on the optical properties. Thus, a more detailed explanation of some of the requirements related to the photochemistry of these compounds will be presented: the absorption spectrum (solar match), photoreaction efficiency (quantum yield), and energy storage capacity.

2.1. Solar Match

The photochemical reaction from a low (parent molecule, photoactive molecule) to a high energy configuration (photoisomer) is the central focus of the photochemical part of the MOST cycle. For this transformation to take place, the main requirement is the absorption of energy supply from the sun. Hence, the ideal MOST systems should absorb the maximum number of photons in the UV and visible range of the solar spectrum, 300–700 nm, which corresponds to the maximum intensity of sunlight. Ideally, the absorption spectrum of the lower-energy isomer should overlap with the most intense region of the solar energy window (solar match) and preferably in the solar spectrum range between 340–540 nm, where the solar radiation is relatively high [22]. Moreover, it is desirable that no absorption overlap between the initial isomer and photoisomer exists to avoid a non-desirable photon absorption competition between the two states.

Most of the basic cores of the photoswitches reported to date do not show wavelengths going far beyond 350 nm (for instance, parent norbornadiene has a maximum at 236 nm and some ruthenium derivatives absorb at 350) [24]. This is a significant drawback as the solar photons' flux at wavelengths below 330 nm is quite low. In this regard, both experimental [25,26] and computational [27] progress has been made in providing functionalized photoswitches that absorb at larger wavelengths. The most used and successful chemical strategy to shift the absorption of MOST compounds toward higher wavelengths is by creating a 'push–pull' effect through the introduction of donor–acceptor substituents and increasing the molecule's conjugated π-system. Preferably, low molecular weight electron donor and acceptor groups are prominent targets for generating relevant photoswitches, as they cause a lower impact on the energy density (affected by both the difference in energy between isomers and the molecular weight). However, beyond the stored energy, the chemical modification of photoswitches may also negatively affect other relevant properties, making it clear that the optimization of a set of molecules for MOST applications is extremely challenging.

2.2. Quantum Yield

Once the parent molecule absorbs a photon from sunlight, the excitation from the ground state (S_0) to the excited state (S_n) takes place. Subsequently, a certain number of molecules will undergo photo-conversion, but a fraction could undergo relaxation, returning to their initial state. To quantify the fraction of molecules effectively performing the photoconversion from the photoactive molecule to the photoisomer, the quantum yield is measured. This dimensionless number provides the probability of a parent molecule to furnish the metastable high-energy photoisomer per absorbed photon. From an efficiency perspective, the photo-conversion that leads to the high isomer needs to be as high as possible, being close to unity if possible. This should allow for an efficient conversion of solar energy into chemical energy, hence avoiding other competitive processes such as radiative, non-radiative, or quenching.

2.3. Storage Energy Density

While it is not strictly a photochemical property, another crucial concern in MOST systems is the energy storage. MOST technology is designed for generating the greatest possible increase in temperature after releasing the stored chemical energy in the photoisomer as heat. In this way, the key property to achieve this goal is the enthalpy difference (ΔH) between photoisomers. This means that, the bigger the energy difference between the (not charged) metastable photoisomer and its parent state, the larger the energy storage density that will be accumulated in the system. As a rule of thumb, MOST systems should provide at least 0.3 MJ/kg to be of practical use, prior to the subsequent heat release. Then, heat could be released using an external stimulus like a thermal increment or via a catalyst. Thus, the photoisomer should not undergo back-conversion quickly at room temperature in order to store the energy for hours, days, or months (storage time) depending on the target application. Even if this review is focused on the photochemical aspects of the MOST technology, it is also relevant to mention that alternative cooling and heating methods are available. For instance, the use of phase transition materials and water adsorption in zeolites has been already commercialized [28,29].

3. Photoswitches Used in MOST Technology

As a brief introduction to the state-of-the-art of the historic development of MOST candidates, very different sets of families were considered at some point. However, most of them were discarded at a relatively early stage because of some practical reasons or flaws. Considering the most explored molecular systems, the main parts of them are photoswitches, as explained above. This is partly because of the considerable overlap between the requirements for photoswitches in general and the compounds used in the MOST concept (absorption, high quantum yields, photostability, and robustness). Historically, the compounds designed to be used in MOST systems can be grouped into two main types [2,16], depending on the photochemical transformation that takes place. In this sense, the mechanism of the photochemical process transforming sunlight into chemical energy can be an isomerization or a cycloaddition. Other types of photochemically induced intramolecular rearrangements were also studied. As an example of some more complex rearrangements, organometallic diruthenium fulvalene's have also been considered [30].

According to the photochemical transformation involved, the systems based on an isomerization are typical examples of molecular photoswitches, like stilbenes [31], azobenzenes, retinal-based photoswitches [32], or other less-known families like hydantoins [33]. The main problem behind using traditional cis-trans photoswitches is the typical small energy gap between the two isomers, producing a small amount of energy storage. This problem has been overcome by two different strategies. Firstly, stabilization of the E-isomer usually occurs when increasing electronic delocalization. Secondly, with a destabilization of the Z-isomer attributed to vicinal groups, steric interactions are incurred. Combining those strategies, some stilbene derivatives could be designed to reach an energy storage of 100 kJ/mol higher than the original unsubstituted stilbene molecule, reaching 105 kJ/mol [34]. Comparably, following the same strategies, retinal-like systems were postulated for this application too, with more modest energy storage capacities [35].

The employment of systems based on a photochemical cycloaddition typically has better properties in terms of energy storage, but their optical properties (absorption spectra) are usually less tuneable as absorption usually lies in the high-energy region of the UV spectra. The main exponent of this approach is the norbornadiene (NBD)–quadricyclane (QC) couple [36], which has been studied since the 1980s and nowadays is a focus of most of the efforts from the community. One of the first proposals using cycloaddition reactions was the use of anthracene derivatives, thanks to their well-known intermolecular [4 + 4] cycloaddition. These compounds also present some problems, as the absorption usually occurs below 300 nm, meaning a low efficiency exposed to solar radiation. This was partially solved by adding (a) bridge group(s) to link two anthracene moieties, but in this case, the efficiency decreased drastically [37]. Another cycloaddition system used is the pair based on dihydroazulene (DHA) and vinylheptafulvene (VHF) [38,39]. Unfortunately, the parent compounds in this couple present a small energy difference between isomers, plus the tunability of the optical properties has already been exhaustively explored [40].

Other systems such as ruthenium fulvalene complexes have also been proposed and studied but have been discarded for practical applications because of the low efficiency and high preparation costs [30,41].

In summary, many molecular systems have been studied along the years as potential MOST candidates. In the following, we will focus our attention on the three most promising families of MOST molecules to date, namely, norbornadiene, azobenzenes, and dihydroazulenes. These three families combine relatively good (or tuneable) properties and are synthetically attainable.

3.1. Norbornadiene/Quadricyclane Couple

Among the previously mentioned MOST systems under investigation, the most profoundly explored is without a doubt the NBD to QC isomerization. Even if the foundations of the MOST concept did not begin with the NBD/QC photoswitch, it is nowadays the main area of research in the field. These molecules have reached energy density values close to the maximum energy density limit of a solar thermal battery at 1 MJ/kg [42]. In contrast, the absorption of unsubstituted NBD is within the UVC range (less than 267 nm) and does not overlap with the solar spectrum, which begins at 340 nm [26].

The ideal absorption scenario for molecular solar thermal energy storage systems is to use solar radiation, which reaches the Earth's surface at high intensities [43]. Thus, targeting a photoisomerization induced reaction in the 350–450 nm range is highly desirable. In designing new NBD/QC molecules, the difference in the absorption maxima between the NBD and QC molecules needs to be large enough to minimize spectral overlap [25,44], which could diminish the incident radiation flux reaching the photoactive isomer. A significant advantage of NBDs over other photoswitches (i.e., azobenzenes) is that, upon absorption, a blue-shift usually occurs. Thus, the QC molecule tends to absorb outside of the visible spectrum, which purposely mismatches the solar spectrum and, consequently, the presence of photostationary states is circumvented. In turn, this avoids the need for engineering adjustments such as band pass filters.

The long list of requirements for an optimal molecule for MOST applications has caused the design of new molecules with increased complexity. For instance, two principal chemical strategies have been applied in the shifting of the absorption of the parent molecule to higher wavelengths [2]. One of them involves adding electron-donating or electron-withdrawing substituent groups to create a 'push–pull' effect, while the other involves increasing the system's extent of π-conjugation [45]. Some representative examples are shown in Table 1 and Figure 2.

Table 1. Molar mass, λ_{max}; enthalpy of isomerization ($\Delta H_{isomerization}$); and energy density of NBD derivatives, which are shown in Figure 2 [2].

NBD	Molar Mass (g/mol)	λ_{max} (nm)	$\Delta H_{isomerization}$ (kJ/mol)	Energy Density (kJ/kg)
Unsubstituted	92	236	113	1228
1	244	308	96	393
2	274	309	97	354
3	342	318	98	287
4	299	350	97	324
5	355	365	102	287
6	217	331	-	-
7	247	355	-	-
8	223	340	-	-
9	260	398	103	396
10	193	309	122	632
11	223	326	89	397
12	288	380	91	315
13	356	359	183	514
14	356	334	99	278
15	256	362	-	-
16	308	350	-	-
17	308	308	173	562
18	495	336	-	-

The first method is typically performed by adding substituents with electron-rich phenyl rings to red shift the absorption to higher wavelengths. A potential drawback caused by this modification is an increased spectral overlap between the NBD and QC isomers [46]. Concordantly, solely changing the acceptor group from a cyano to a trifluoroacetyl group red shifts the absorbance by 100 nm, although it considerably shortens the lifetimes of QC [47].

In the second strategy, the use of NBD dimers in MOST systems is being explored, although the origin of this molecular cooperativity remains to be fully understood. Extending the π-conjugation by linking two NBD units through an electron-rich aromatic unit minimizes the impact of molecular weight increase as two units are considered and two photons may be absorbed. Unfortunately, shorter wavelengths of absorption are required for the second photoisomerization process from QC/NBD to QC/QC, ultimately decreasing the quantum yield. Moreover, both NBD moieties should be photoisomerized as NBD/QC is far more labile than QC/QC [24].

Another method to try to improve the performance of the NBD/QC couple is the use of alternative deactivation channels. In this sense, thermally activated delayed fluorescent (TADF) molecules undergo an excitation to the lowest-lying singlet state, relax to the triplet state, and finally can be thermally converted back to the singlet state (otherwise known as a reversible intersystem crossing) and relax from this singlet excited state. TADF molecules have the lowest-lying excitation band ideally situated in the perfect range of the solar spectrum required for MOST systems, at lower wavelengths than typical phosphorescent molecules (480–550 nm) and higher wavelengths than fluorescent molecules (330–380 nm). TADF molecules were originally built around a benzophenone core structure, with the addition of carbazole and diphenylamine moieties being common, wherein keto groups are in a co-planar position relative to aromatic moieties [48]. This co-planarity effect enhances the overlap between n and π^* states, creating a very small singlet–triplet energy gap. A new promising approach was attempted linking thermally activated delayed fluorescence molecules like phenoxazine–triphenyltriazine (PXZ-TRZ) to NBD moieties, where red-shifted absorption peaks were in the 400–430 nm range [49]. Moreover, NBD molecules were substituted onto these structures with an increased conjugation as a possible solution to maximize the low-energy storage density of a single NBD unit.

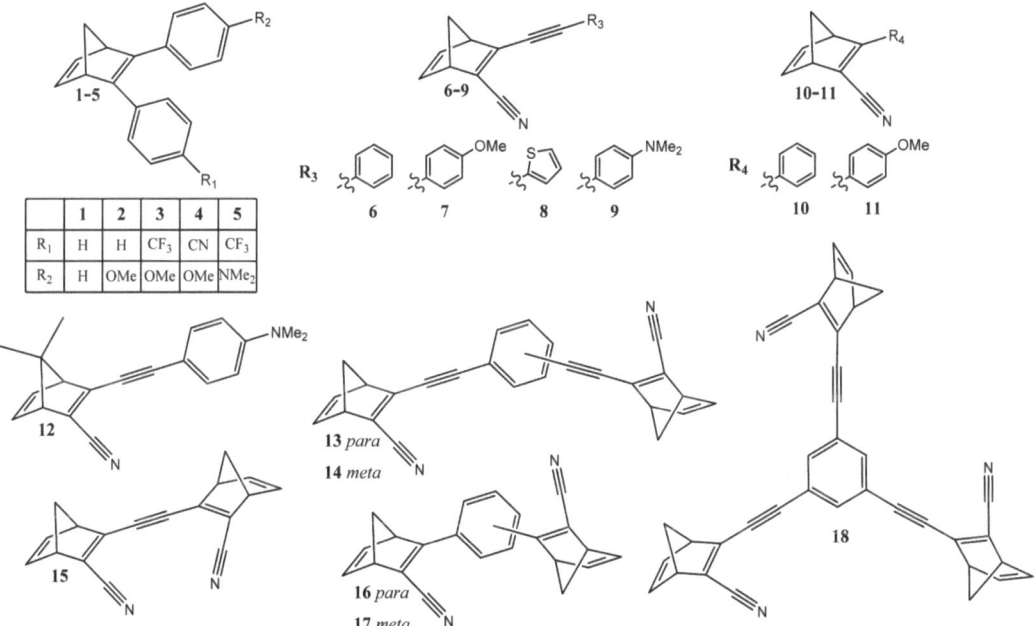

Figure 2. List of NBD derivatives with absorption maxima in the 300–400 nm range.

Despite the copious publications using the NBD/QC photoswitch, the optimal system has not been devised yet. The prime-substituted NBD has a red-shifted absorption of 59 nm, yet it has an increase in molar mass of 131 g/mol, inevitably reducing the energy density by 13.3 kJ/mol. The energy density of unsubstituted NBDs to the present date still outcompetes the prime-substituted NBD by a margin of 13% [50]. To implement an NBD/QC MOST photoswitch that absorbs in the 350–450 nm region into practical applications will entail a compromise of 20 kJ/mol of storage energy for a better fitting with the solar spectrum, especially considering that, when moving to highly absorbing materials like NBD, a major fraction of light will nonetheless be converted to heat. Controlling thermal fluxes at the surface is not only required to keep the photoisomer stable, but it is also a safety mechanism essential in energy storage devices like batteries to prohibit overheating under working conditions. Utilizing waste heat by coupling a heat exchanger to the final device will prevent local temperature extremes and maintain a truly closed MOST system [44].

As mentioned previously, it is also crucial to control the efficiency of the photochemical process, which starts with the absorption of a photon by NBD. The incoming photon can induce an electronic transition from the minimum in the ground state to the S_1 state, causing a [2 + 2] intramolecular cycloaddition [51]. Under natural solar irradiation conditions, this photoconversion does not take place in the parent NBD, as just a few photons below 300 nm arrive at sea level. Therefore, the unsubstituted NBD/QC couple is chemically inactive to sunlight. Furthermore, in the event that the NBD absorbs a proper solar radiation, just a few molecules will undergo photoisomerization because of the quantum yield of this system being quite low (ϕ = 0.05). In consideration of these requirements, the NBD/QC system has been chemically modified in order to increase the quantum yield [18] as well as red-shift the wavelength of absorption.

As shown in Figure 3, the introduction of donor–acceptor groups increases the quantum yield and makes the system capable of absorbing natural solar irradiation above 400 nm. Thus, when the incoming photon is absorbed by NBD, the photoisomerization transformation starts through a S_0–S_1 transition. The molecules will then undergo relaxation in the excited state potential energy surface to reach the minimum energy conical intersection point (MECI) S_0/S_1, leading to the photoisomer QC [24]. In the heat release step, the thermodynamic driving force of the process (ΔH_{isom} = 372 kJ/mol) pushes QC in forming the less-strained geometry (NBD), based on the cleavage of two single bonds of the four-membered ring and the transformation of the remaining bonds to double bonds, as seen in Figure 4. In this stage, the reaction releases a high amount of energy because of the high-standard enthalpy of the reversion of QC to NBD. This energy also depends on the gravimetric energy density (MJ/kg), hence systems with smaller molecular weights will hold comparatively greater energy densities. Thus, cyano-substituted NBD derivatives have been computational and experimentally studied as they show attractive energy densities (0.4–0.6 MJ/kg) [25].

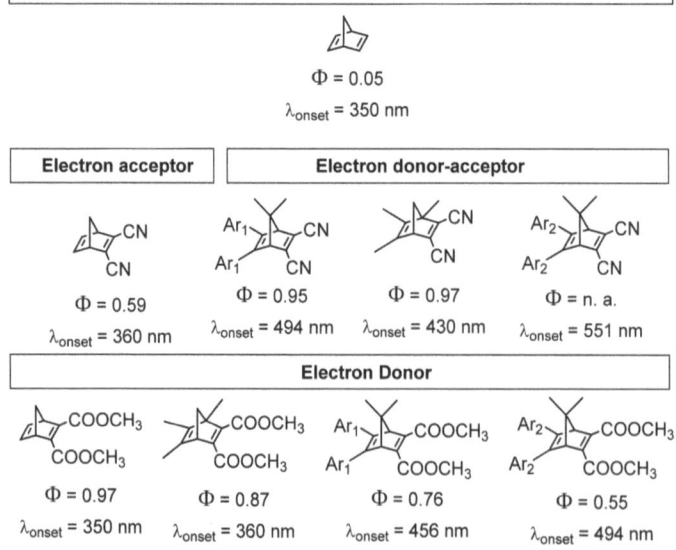

Figure 3. Summary of the effect of electron donor–acceptor groups on the quantum yield and on the onset of the absorption wavelength. Ar_1 means a phenyl group and Ar_2 means a p-methoxyphenyl group. Data collected from [18,52–54].

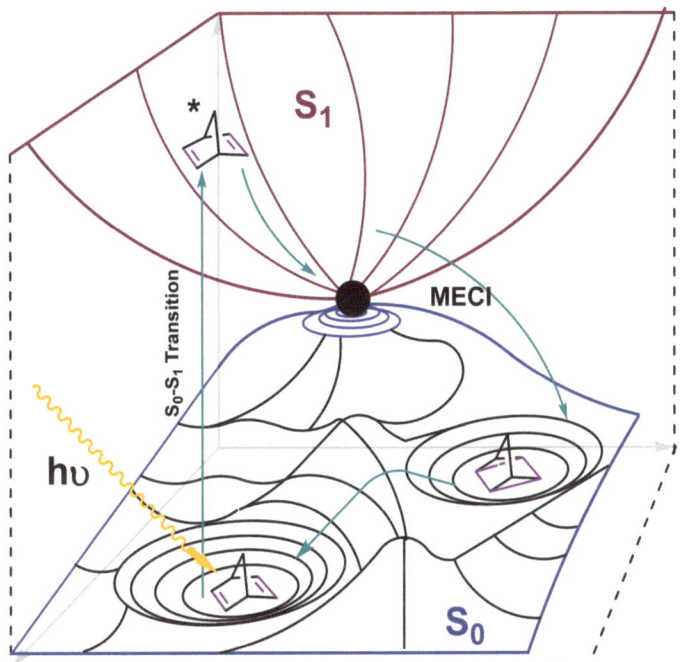

Figure 4. Approximate 3D illustration of a ground-state (S$_0$, blue) and excited (S$_1$, red) energy landscape depicting the whole photoisomerization mechanism.

Control of the back conversion reaction is crucial in providing energy at the right time. It is well known in these systems that the optimization of solar match and storage lifetime at the same time is challenging, because one property is improved at the expense of the other. However, a recent novel strategy defies this, wherein NBD molecules have an improved solar match, yet the storage time remains untouched. Chiefly, via the introduction of substituents in the *ortho* position, a substantial increase in back conversion ΔS* occurs, which comes from the steric interference of the side group of the parent molecule [55].

3.2. Azobenzene Photoswitches

Azobenzene is a chemical compound based on diazene (HN=NH), where both hydrogens are substituted by a phenyl ring [56,57]. Azobenzene can be found either in a *cis* or *trans* conformation [58]. The *trans* to *cis* isomerization can be triggered by different stimuli such as irradiation with ultraviolet (UV) light [59], mechanical force [60], or an electrostatic stimulus [61,62]. Contrastingly, the *cis* to *trans* isomerization can be observed in dark conditions when applying heat owing to the thermodynamic stability of the *trans* isomer, although it can also be driven by visible light (see Figure 5a). The *trans* conformation assumes a planar structure with C$_{2h}$ symmetry [63,64], with a maximum distance of 9.0 Å between the most distant positions of the aromatic rings. On the other hand, the *cis* one adopts a non-planar conformation with C$_2$ symmetry [65,66] and the end-to-end distance is reduced to 5.5 Å. The *cis*-isomer is not planar as a result of steric hindrance, and this causes the π-electron clouds of the aromatic rings to face each other, leading to a small dipolar moment in the molecule (μ ~3D) [67]. This ring's disposition is also reflected in the proton nuclear magnetic resonance (^{1}H-NMR) spectra. The shielding effect produced by the anisotropy of the π-electron cloud in the *cis*-isomer affects the signals, thus making them appear at a higher shift compared with the *trans*-isomer.

The absorbance spectrum of the *trans*-azobenzene molecule typically presents two absorption bands (see Figure 5b), corresponding to the electronic transitions n–π* and π–π*. The electronic transition π–π* produces a strong band in the UV region and is also present in analogous-carbon systems such as stilbene [68]. The n–π* electronic transition presents a different band, which is far less intense, in the visible region. This latter transition is produced by the nitrogen's lone pair of electrons [69], which generates a completely different photoisomerization process compared with its analogous carbon system, the stilbene. The *trans*–*cis* isomerization process is usually followed by a color change towards more intense tonalities. The absorbance spectra of both isomers are differentiated by the following aspects: (i) *trans*-isomer: the absorption band related to the π–π* electronic transition is very intense, with a molar extinction coefficient (ε) around 2–3 × 10^4 M^{-1} cm^{-1}. On the other hand, the band corresponding to the n–π* electronic transition is much weaker (ε ~400 M^{-1} cm^{-1}), as this transition is forbidden by the symmetry rules in this isomer. (ii) *Cis*-isomer: the absorption band related to the π–π* electronic transition is hipsochromically shifted and its intensity diminishes greatly (ε ~7–10 × 10^{-3} M^{-1} cm^{-1}) with respect to the *trans*-isomer. On the other hand, the band corresponding to the n–π* electronic transition (380–520 nm) is allowed in this isomer, thus significantly increasing the intensity of this band (ε ~1500 M^{-1} cm^{-1}). These optical properties can be greatly modified upon substitution and this fact has been exploited to match the requirements of MOST technology.

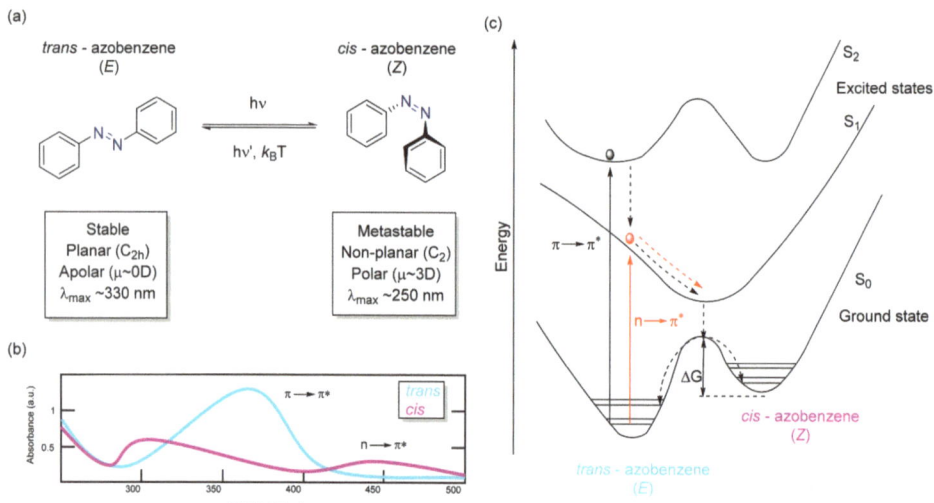

Figure 5. (a) Azobenzene isomer structures and an overview of properties. (b) Absorbance spectrum scheme of *trans* and *cis*-azobenzene isomers with its main bands assigned to electronic transitions. (c) Schematic view of lower energy levels and pathways of the azobenzene isomers, adapted from Georgiev et al. [70].

Although azobenzene can adopt *trans* and *cis* conformations, the former is more stable at the electronic ground state by roughly 58.6 kJ/mol (0.6 eV) [71]; this is explained by the lack of delocalization and a distorted configuration of the *cis*-form in comparison with the *trans*-isomer. The experimentally measured energy barrier between *trans* and *cis* conformations is about 96.2 kJ/mol (1.0 eV) [72]; thus, in dark conditions and at room temperature, the predominant species is the *trans* one (Figure 5c). As explained above, the energy difference between isomers will have a strong effect on the stored energy density. This value could be also affected by substitution, but in general, smaller values of energy density could be obtained using azobenzene compared with other MOST molecules such as NBD/QC.

The differences in spectroscopical properties for the *cis* and *trans* isomers and the distribution of excited state electronic levels allow them to undergo isomerization upon irradiation. This photochemical interconversion usually ends up providing a varying mixture of *cis* and *trans* isomers on the photostationary states (PSS). The ratio of isomers in the mixture depends on the substitution of the azobenzene core and the wavelength of irradiation. In turn, the excitation wavelength at which this process takes place depends on the nature of the substituents on the aryl groups of the azobenzene; although, usually, in the *trans* to *cis* isomerization, this process is promoted by wavelengths around 320–380 nm, while exposition to 400–450 nm wavelengths typically favors *cis* to *trans* isomerization. This reversion can also be induced thermally. Although both photochemical conversions take place in the order of picoseconds, the *cis* to *trans* thermal relaxation is in the order of seconds or even hours and can be tuned by substitution.

Although several mechanistic studies have focused on the isomerization mechanisms of azobenzene, in fact, the effect of the substituents on the phenyl rings as well as the influence of other parameters yields a complex mechanism [73–75]. In fact, several mechanisms of the isomerization of the azobenzene and its derivatives can take place depending on the nature of the substituents and the reaction conditions (see Figure 6): (i) The inversion of one of the N–C bonds corresponds to an in-plane inversion of the NNC angle between the azo bond and the first carbon of the benzene ring (angle Φ), reaching 180°, while the angle of CNNC remains fixed at 0°. (ii) Through the rotation of the N=N double bond. This mechanism is similar to the one in the stilbene isomerization [73]. The rotational pathway starts with the breakage of the N=N p-bond, thus becoming an N–N bond. After that, there is an out-of-plane rotation of the CNNC dihedral (φ) angle, while the CNN angle remains at 120°. (iii) Through the concerted inversion [76], where there is a simultaneous increase in the NNC angles up to 180°. (iv) Lastly, through inversion-assisted rotation mechanism, which implies changes in both CNNC and NNC angles at the same time.

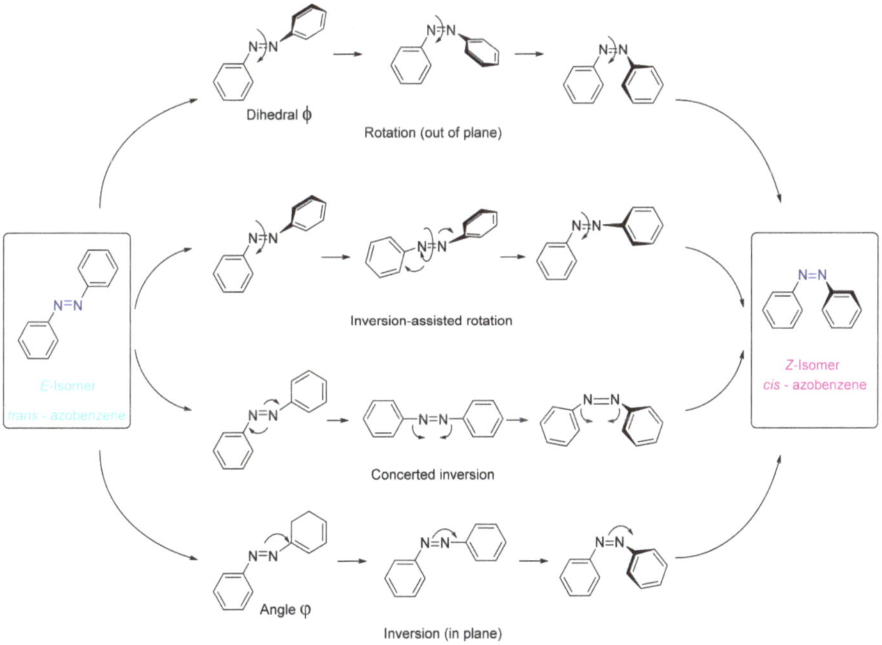

Figure 6. Schematic view of four different isomerization mechanisms of azobenzene described in the literature.

Nevertheless, recent computational studies [75] using ab initio methodology including dynamic electron correlation have precisely depicted the topography of the potential energy surfaces of the S_1 (n–π*) and S_2 (π–π*) energy levels. The inclusion of the dynamic correlation makes this study more reliable as the topography of the PES can be highly altered when it is not included, thus providing contradictory data because of changes in the location of the minima and shape of the reaction paths. A high-level computational study [75] at the CASSCF/CASPT2 (Complete Active Space Self Consistent Field/CAS second-order perturbation theory) level, reoptimizing some of the critical points of the excited states at the MS-CASPT2 level, was performed. The mechanisms described fit into the different reaction pathways depending on the isomer excited and the excited state reached (see Figure 7). The inversion mechanism is the most likely pathway for the *cis* to *trans* thermal-back isomerization, but for the *trans* to *cis* case, depending on the excitation, different paths are available.

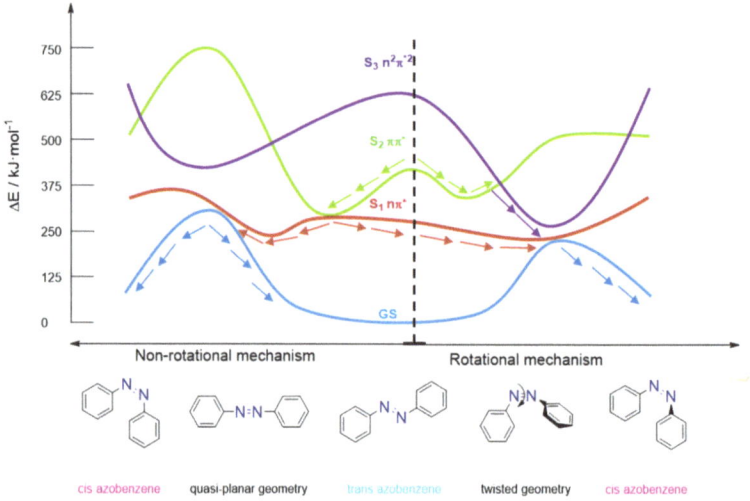

Figure 7. Different electronic states and paths involved in the free-rotation and restricted rotation of azobenzene derivatives. Adapted from Reguero et al. [75].

When an azobenzene is promoted to the S_1 state and rotation is not hindered by substituents, a CNNC rotation will take place, reaching the S_1 potential energy surface minimum. An available conical intersection between this excited state and the ground state is located close in terms of geometrical distortion and energy, usually above by just 4–8 kJ/mol. Therefore, the population is funneled to the ground-state surface, yielding photoisomerization. For the excitation to S_2, an energy degenerated region is usually present, thus leading to the same rotational pathway discussed for S_1. However, alternative pathways can lead to photoisomerization back to the starting material. All of these processes produce differences in the quantum yield observed [76] depending on the excited state reached and the geometries allowed. Consequently, the PSS composition and the photoisomerization quantum yield become largely dependent on the substitution of the phenyl rings of the azobenzene and the specific irradiation wavelength used.

As shown in previous paragraphs, substitution has a major impact on the properties related to the MOST concept. Azobenzene derivatives can be classified into three different groups depending on the energetic order of its electronic states (π–π* and n–π*). This order is mainly dependent on the electronic nature of the azobenzene aromatic rings and their substitution pattern. The three classes are usually labelled as follows (see Figure 8a): (i) Azobenzene-type molecules, in which the electronic nature of the phenyl rings is analogous to that existing in the unsubstituted azobenzene (Ph-N=N-Ph). They present a strong π–π* band in the UV region and a weak n–π* band in the yellow visible region. (ii) Aminoazobenzenes, which can be *ortho-* or *para*-substituted with an electron-donating group (EDG), [(*para-* or *ortho-*(EDG)-C₆H₄-N=N-Ar)]. The π–π* and n–π* bands are very close or even overlap in the UV/Vis region; they are typically orange. (iii) Pseudo stilbenes, which have an electron-donating and an electron-withdrawing group (EWG) at the *para* position of the phenyl rings, thus being a push–pull system [(*p-*(EDG)-C₆H₄-N=N-C₆H₄-*p-*(EWG)]. They are typically red and the absorption band corresponding to the π–π* electronic transition undergoes a bathochromic shift, overlapping bands or even changing the order of appearance with the corresponding band of the n–π* electronic transition.

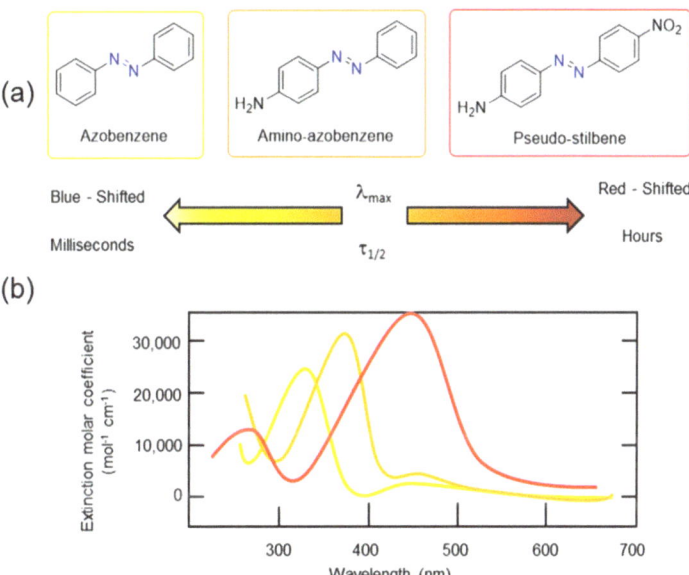

Figure 8. (a) Azobenzene classification by Rau et al. [77] helps describe the optical and physical properties of this family of compounds (i.e., their color). (b) Typical example of absorption spectra of the different azobenzene derivatives.

Pseudo-stilbenes have a highly asymmetric electron distribution alongside the molecule, which turns into a large molecular dipole and anisotropic optical properties. In addition, this class also presents the best photo-response, which is mainly caused by an overlapped absorption spectrum of the E and Z forms, thus reaching a mixed photo-stationary state where the *trans* and *cis* isomers are continuously interconverting. Therefore, for pseudo-stilbenes, a single visible light wavelength can induce both forward and reverse isomerization. Thermal relaxation from *cis* to *trans* spans from milliseconds to seconds.

On the other hand, in the other two classes of compounds, the absorption spectra usually do not overlap, meaning that two different wavelengths are needed to switch between the *cis* and *trans* forms, which is ideal for photoswitchable materials. Particularly, azobenzene-type molecules are proven to isomerize back from *cis* into *trans* isomers very slowly through thermal relaxation when bulky substituents are introduced into the structure, thus providing a thermal relaxation time from *cis* to *trans* isomers of hours for the azobenzene-type molecules and minutes for the amino-azobenzene-type molecules. Recently, it has even been proven that some stable *cis*-isomers can be isolated using the proper solvent, which leads to a stable two-state photoswitchable system.

Even if the optical properties of azobenzenes discussed in previous paragraphs are quite interesting for MOST applications, the applicability of these compounds is somewhat limited by the energy density. The energy difference of a typical azobenzene is usually around 40 kJ/mol, which does not outcompete other systems. In addition, spectral overlap and the increased molecular weight to obtain excellent optical properties have led to a relative decrease in the interest of these compounds [16]. However, the broad tunability and the synthetic availability of different compounds achieved in the last two decades have reactivated their use as MOST candidates. Additionally, the wider scope obtained by azoheteroarenes, in which one of two phenyl rings are modified by a heterocyclic ring, can exponentially increase its possibilities by tuning some of the relevant properties.

For instance, we have already discussed that the photosiomerization quantum yield should be as high as possible [78,79]. Thus, this parameter has been subject to extensive research. The effect of various solvents, temperatures, and rigid environments was explored. Zimmerman et al. reported that n–π* quantum yield is more efficient than π–π* in both directions of conversion (*cis-trans*/*trans-cis*) and, in accordance with Birnbaum and Style, the absorption in the *cis-trans* conversion is more efficient (*cis-trans*: 0.39, while *trans-cis*: 0.33) [80–82]. This should be carefully considered when designing MOST applications based on azobenzenes. On the other hand, the ability of azobenzenes to absorb light in the visible region to induce the photoisomerization has been considered a clear advantage of these compounds [83].

The poor performance of azobenzenes with respect to energy storage has been addressed differently. In 2014, Grossman and Kolpak performed a series of density functional theory (DFT) computational studies of the azobenzene coupled with carbon nanotubes (CNTs) (Figure 9a). The addition of nanotubes to azobenzene increases the rigidity and conformational hindrance of the structures, so the energy storage per azobenzene increased up to 30% and the storage lifetimes almost reached the unit, in addition to increased fatigue resistance (a half-life ($t_{1/2}$) of 33 h was observed for the *cis* isomer) [84].

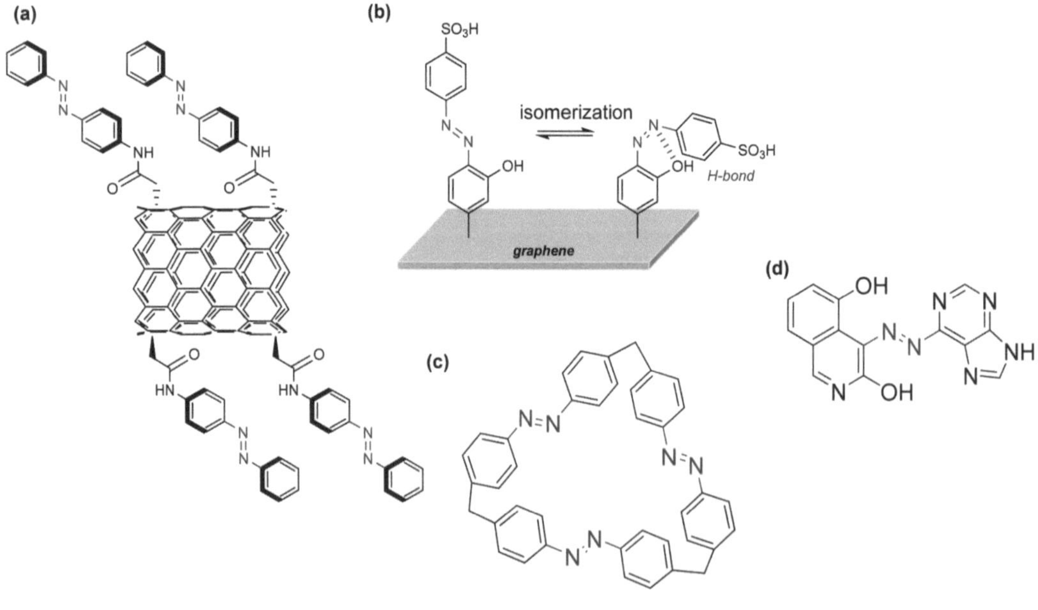

Figure 9. (**a**) Azobenzene coupled with carbon nanotubes. (**b**) (Z)-isomer stabilized by the substitution of the azobenzene. (**c**) Azobenzene macrocycles. (**d**) Azoheteroarene derivative.

Later, the same authors extended their DFT study, incorporating azobenzenes functionalized with carbon-based templates such as graphene, fullerene, and β-carotene, identifying various potential molecules that have improved properties. Upon modification, the *cis*-isomer is stabilized by intramolecular hydrogen bonds (Figure 9b), leading to long-term storage lifetimes ($t_{1/2}$ = 5408 h) [85,86]. In view of these results, different experimental groups carried out the synthesis of the photoswitches, obtaining good results [87,88]. Specifically, they obtained molecules with 11–12 carbon atoms per azobenzene and the hydrogen bond stabilization of the azobenzenes was confirmed by NMR, FT-IR, and DFT studies.

In a different approach, the addition of azobenzene to macrocycles [89] has also been suggested to improve energy storage capabilities [90]. The use of rings formed by azobenzenes connected through a suitable linker agent increase the barrier of the back reaction (Figure 9c). In this approach, the formation of the macrocycle contributes to increasing the rigidity of the system. In addition, it is possible to add functional groups to the macrocycle, as done in the graphene, allowing to establish hydrogen bonds with the aim of obtaining longer lifetimes for the energy storage.

Even if azobenzene-functionalized CNTs have increased the energy density, these systems cannot be deposited into uniform films. To avoid this problem, an azopolymer was prepared to act as solid-state solar-thermal fuel (STF) [91]. This novel design allows to prepare uniform films. This device allowed to increase the temperature of the heat generated in the back conversion by up to 180 °C [86]. Following this breakthrough, different studies have been conducted on azopolymers [92,93] because of the ease of application of the photoswitches in large areas. Additionally, different studies attempted to combine the rigidity obtained by coupling a polymer and the addition of functional groups to generate non-covalent interactions to the azobenzene, with the aim of stabilizing the azopolymers. Non-covalent interactions have also demonstrated an increase in the lifetime of the stored energy, such as π–π stacking and hydrogen bonds. Taking advantage of these interactions, a new mechanism to synthesize azopolymers controlling the thickness and film morphology by electrodeposition was described. Unfortunately, when the substitution of the polymers is modified, the energy storage is greatly affected [94].

It is known that preparing azobenzenes with very bulky substituents increases the rigidity and, therefore, the lifetime of energy storage. However, different and simpler systems continued to be sought in parallel. In 2017, Grossman and co-workers demonstrated that it was not necessary to use templates or even polymers to achieve SFT materials with high-energy density and thermal stability. In this study, they synthesized various azobenzenes substituted with bulky aromatic rings, obtaining an enthalpy difference between the *cis* and *trans* isomers comparable to the unsubstituted azobenzene. A critical factor to improve the azobenzene properties was to distort the molecule to avoid planarity, demonstrating that using small molecules made it possible to generate thin films with excellent charging and cycling properties [95].

Lately, a novel approach implying the use of azoheteroarene photoswitches has been explored. In various experimental and computational studies, it was discovered that changing the type of heteroaromatic ring or the position of the heteroatom with respect to the azo group had a crucial effect on the photoswitching properties [35,96]. Using this approach, it was possible to prepare molecules with a half-life of days or even years, providing an excellent alternative to use these compounds in a MOST system. The increase in the half-lifetime in the photoswitches is due to the formation of an intramolecular hydrogen bond affecting the energy difference between isomers, and thus the energy storage (Figure 9d). However, not all studied azoheteroarenes absorb in the visible region, making their use for energy storage problematic. Again, the long list of features that a system should fulfil to be of practical use in the MOST technology makes it extremely difficult to select the ideal candidate. However, the ever-growing list of available azoheteroarenes turns these compounds into excellent candidates to absorb and store sunlight in MOST devices.

3.3. Dihdroazulene–Vinylheptafluvene (DHA–VHF)

Another system widely studied in the context of MOST applications is the dihdroazulene/vinylheptafulvene (DHA/VHF) couple. While these compounds have some practical issues that hamper their applied use, they also feature some interesting properties. The optical properties of this system imply a DHA absorption of ca. 350 nm and a band at

480 nm for VHF, which also includes an isosbestic point at 400 nm with a very low overlap between both species. The photoisomerization quantum yield is around 0.6, which could be considered a good value for practical applications, although clearly lower than the better examples of NBD/QC, but by far better in comparison with NBD (0.05) or azobenzene (0.33). However, the main drawback for this couple is the short half-life time of parent VHF, being only 10 h at 25 °C. This should rule out its use for a long-term storage application, but others implying daily cycles (as charging during the day and releasing at night) can be considered as an alternative. On the other hand, it has shown some promising features in certain application tests. For instance, it has been used on solar and fluidic devices to provide an efficiency of up to 0.13% in non-laboratory experiments, considering the total solar income harvested. In addition, DHA/VHF also has a very high robustness against degradation, featuring less than 0.01% degradation in 70 irradiation back conversion cycles in toluene solutions [2,17]. Another drawback of this system is the low-energy storage capacity of DHA/VHF, which present an energy difference of ca. 28 kJ/mol, which is quite modest in comparison with NBD derivatives.

The general energy landscape of this couple can be seen in Figure 10. Upon light absorption, DHA isomerizes to VHF through a photoinduced carbon–carbon ring-opening reaction. In the excited state, DHA planarizes, allowing the stereochemical conditions to open the ring. This is favored by an increased steric hindrance due to the electronic reorganization. After the photochemical step, two possible isomers can be obtained. The initially formed *cis*-VHF can further evolve to the thermodynamically stable *trans*-VHF [38], with both species being in equilibrium through a small energy barrier [97]. This behaviour can facilitate an early back reversion to DHA, decreasing the lifetime of the photoisomer and affecting the energy density at the same time. In addition, the enthalpy of the photoreaction was measured to be ca. 35.2 kJ/mol in a vacuum [2].

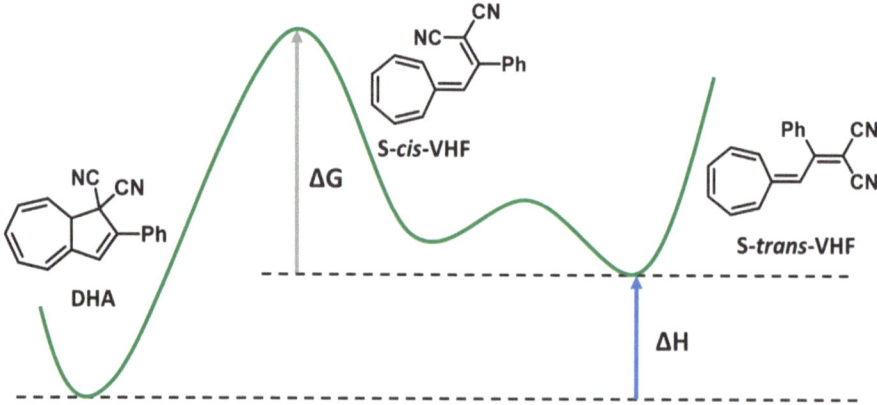

Figure 10. Dihydroazulene (DHA)/vinylhetafulvene (VHF) couple and potential energy landscape. Adapted from Nielsen et al. [40].

Another relevant point is the large solvent effect that has been observed for the DHA/VHF couple. The photoreaction from DHA to VHF is faster in polar solvents than in apolar ones [98]. In contrast, the robustness through different irradiation conditions decreases notably in polar solvents (0.18–0.28%) compared with in apolar ones like toluene (0.01%). From these studies, some differences in the kinetic parameters have been invoked to explain the stabilization of charge-separated species, which acts as an intermediate during the photochemical transformation [39,98]. On the other hand, DHA is 10 times more soluble in toluene (apolar and aprotic) than in ethanol (polar and protic) and the photoisomerization quantum yield is larger (0.6) in toluene than in ethanol (0.5), being intermediate in acetonitrile (0.55). To sum up these effects, the more polar the solvent, the better the kinetic parameters, but, in contrast, an additional enhancement of system robustness is found with apolar solvents.

Along the years, many different studies were performed to better understand and improve the DHA capabilities, aiming at a better performance. The effect of some vicinal chemical modifications over the photoisomerization from DHA to VHF was improved in the 90s, usually by adding some donor or acceptor electronic groups on *para* positions of the phenyl ring. This yields in a moderate modification of the photoreaction quantum yield, ranging from 0.6 to 0.3 when increasing the electron-withdrawing strength of the substituent in the *para* position [98].

More recently, some extensive computational analysis was performed for the screening of interesting new candidates. In this work [40], an impressively wide number of chemical modifications has been performed on DHA systems to tune some of their properties. The most relevant and effective ones are summarized in Figure 11, whereby the main interest was the substitutions in positions 1–3 and 7, according to the numbering seen below for DHA systems.

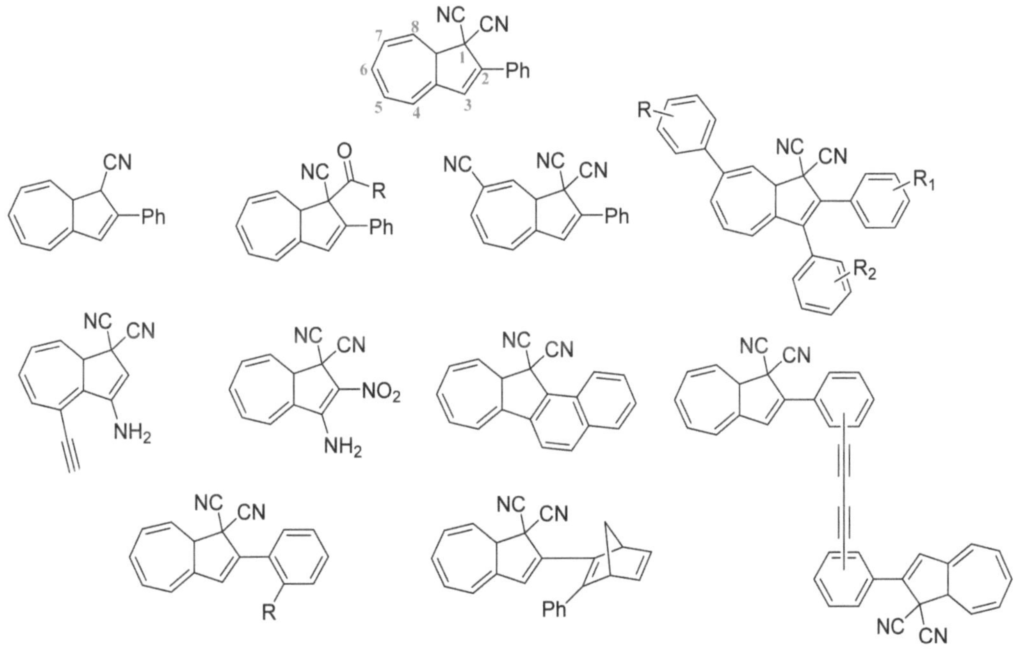

Figure 11. Numbered positions and more relevant modifications in DHA structure, yielding some type of improvement.

One of the first modifications was to check the importance of the CN groups in position 1 and, as can be expected according to the mechanism, the elimination of this strong electron-withdrawing group causes a lack of photoswitching capacity, proving that the strong electrophilic behavior of C1 is crucial in the charge transfer intermediate on the photochemical transformation. As the elimination lacks the photoisomerization, some interesting attempts to replace one of the CN with other EWGs were carried out [99]. To finalize, the substitution of one CN by an ethoxycarbonyl or carbamoyl group can maintain the photoswitching ability and enhance the energy storage capacity, increasing it by 0.05–0.1 MJ/kg on average, being a relevant increment, also considering the higher molecular weight. As counterpoint, the predicted ΔG is lower than that found in base DHA, facilitating the back conversion. Moreover, with these systems, the solvent dependence seems to be slightly lower, even being quite representative.

The modulation on the stability of VHF and the back reversion speed can be controlled by the modification of positions 2, 3, and 7; herein, the effect of donor- and electron-withdrawing groups can drastically change some properties.

The inclusion of a donor group at C2 and an acceptor group at C3 or C7 have some impact, increasing the lifetime of these derivatives [100,101]. The addition of electron-withdrawing groups (EWGs), like cyano, in position 7 can increase the half-life of VHF by up to six times, in some cases reaching an exceptionally long-lived VHF in acetonitrile. This behavior can be explained by the stabilization of the VHF form, increasing the energy needed to initiate the back-reaction, but, in contrast, having a low impact on the energy storage capacity.

The inclusion of an amino group in position 3 yields a hydrogen bond between the amino and the electron-withdrawing groups (CN). This seems to stabilize the DHA system, increasing the storage energy, also blocking VHF in the s-*cis*-VHF conformer. This means that the s-*cis*-VHF becomes the most stable isomer of VHF. Thus, geometrically, it is more similar to the intermediate and, as expected, the back-reaction barrier becomes even lower. This behavior reduces its usability because of the fast reversion achieved despite the higher energy stored. In contrast, the addition of a donor group (amino) on position 3 and an acceptor group (NO_2) at C2 produces an increase in the ΔG of the back-reaction, also lengthening the lifetime of the VHF form [40].

Some other modifications, like the condensed effect of adding a 9-anthryl group in position 2 and 3, were found to increase the storage density to 0.38 MJ/kg, but this facilitates the VHF-to-DHA ring closure in a few seconds [46].

The addition of bulky groups on the *ortho* position of the phenyl group in C2 can stabilize the *cis* conformer of VHF, increasing the energetic barrier of the back-reaction [102]. The quantum yields of photoisomerization for these ortho substituted derivatives range in acetonitrile close to DHA (0.55) values, thus improving the lifetime without perturbing the photoisomerization rate. This effect is only with an iodine atom; if the size of the added group increases, this effect is maximized, reaching a 60 times longer half-life and quantum yields slightly higher than DHA of ca. 0.6–0.7.

The last modification that we will comment about is the preparation of multi-switcher devices; on these, the combination of two DHA moieties, connected through a bisacetylene bridge, can differ too much depending on the substitution in *ortho*, *meta*, and *para*, ranging from an increase in the lifetime from a few hours to a couple of days [102].

In addition, combinations with other MOST candidates were attempted, offering some promising results, especially the increase in storage density due to the charge of two molecules, and in case of DHA-NBD, the combined absorption is notably red-shifted and the spectral overlap is decreased [103]. Moreover, the use of an azobenzene derivative together with phenylene bridge bis-DHA moieties, all included in a macrocyclic structure, can yield a modification of the DHA isomerization because of the predominant azobenzene isomerization found. This can limit the possibilities of combining azobenzenes and DHA moieties [104].

A principal outlook towards the use of DHA systems presents a few advantages, such as efficient isomerization and good optical properties, as well as a series of drawbacks including the modest energy storage and short lifetimes.

4. Conclusions

In this review, we have covered the three more relevant families of compounds that are under investigation as MOST systems, norbornadiene, azobenzene, and dihydroazulene derivatives. Every set of compounds has its own drawbacks and strengths, which should be carefully considered for any specific application. However, it is also relevant to recognize that we are still far from finding the ideal MOST candidate. The mentioned improvements in the molecular design constitute the basis for the future development of these systems. In addition, the use of molecular modelling and machine learning strategies can provide a fast and valuable input in this way [105]. A general increase in the performance of the molecules used for solar energy storage is required before this technology could provide an alternative and efficient way of harvesting and storing solar energy, as well as its use and release on demand. In the near future of MOST devices, the exploitation of hybrid strategies (multijunction devices) is the more promising field to improve the overall performance.

Author Contributions: Bibliographic search and writing—original draft preparation, A.G.-G., L.M., N.S., B.P. and J.S.; writing—review and editing, R.L. and D.S.; supervision, R.L. and D.S.; funding acquisition, D.S. All authors have read and agreed to the published version of the manuscript.

Funding: This research was funded by the Spanish Ministerio de Ciencia e Innovación, grant number PID2021-126075NB-I00.

Institutional Review Board Statement: Not applicable.

Informed Consent Statement: Not applicable.

Data Availability Statement: Not applicable.

Acknowledgments: A.G.-G., L.M., N.S., and B.P. thank the European Union's H2020 research and innovation program under grant agreement No 951801. R.L. thanks Universidad de La Rioja and Ministerio de Universidades for his Margarita Salas grant.

Conflicts of Interest: The authors declare no conflict of interest.

References

1. de Amorim, W.S.; Valduga, I.B.; Ribeiro, J.M.P.; Williamson, V.G.; Krauser, G.E.; Magtoto, M.K.; de Andrade Guerra, J.B.S.O. The nexus between water, energy, and food in the context of the global risks: An analysis of the interactions between food, water, and energy security. *Environ. Impact Assess. Rev.* **2018**, *72*, 1–11. [CrossRef]
2. Sun, C.-L.; Wang, C.; Boulatov, R. Applications of Photoswitches in the Storage of Solar Energy. *ChemPhotoChem* **2019**, *3*, 268–283. [CrossRef]
3. Wang, Z.; Erhart, P.; Li, T.; Zhang, Z.-Y.; Sampedro, D.; Hu, Z.; Wegner, H.A.; Brummel, O.; Libuda, J.; Nielsen, M.B.; et al. Storing energy with molecular photoisomers. *Joule* **2021**, *5*, 3116–3136. [CrossRef]
4. Lee, D.S.; Fahey, D.W.; Skowron, A.; Allen, M.R.; Burkhardt, U.; Chen, Q.; Doherty, S.J.; Freeman, S.; Forster, P.M.; Fuglestvedt, J.; et al. The contribution of global aviation to anthropogenic climate forcing for 2000 to 2018. *Atmos. Environ.* **2021**, *244*, 117834. [CrossRef]
5. IEA. *Global Status Report for Buildings and Construction*; International Energy Agency: Paris, France, 2019.
6. Rabaia, M.K.H.; Abdelkareem, M.A.; Sayed, E.T.; Elsaid, K.; Chae, K.J.; Wilberforce, T.; Olabi, A.G. Environmental impacts of solar energy systems: A review. *Sci. Total Environ.* **2021**, *754*, 141989. [CrossRef]
7. Xu, X.; Wang, G. Molecular Solar Thermal Systems towards Phase Change and Visible Light Photon Energy Storage. *Small* **2022**, *18*, 2107473. [CrossRef]
8. Ma, Q.; Wang, P.; Fan, J.; Klar, A. Underground solar energy storage via energy piles: An experimental study. *Appl. Energy* **2022**, *306*, 118042. [CrossRef]
9. Ma, Q.; Wang, P. Underground solar energy storage via energy piles. *Appl. Energy* **2020**, *261*, 114361. [CrossRef]
10. Wu, D.; Kong, G.; Liu, H.; Jiang, Q.; Yang, Q.; Kong, L. Performance of a full-scale energy pile for underground solar energy storage. *Case Stud. Therm. Eng.* **2021**, *27*, 101313. [CrossRef]
11. Wang, H.; Qi, C. Performance study of underground thermal storage in a solar-ground coupled heat pump system for residential buildings. *Energy Build.* **2008**, *40*, 1278–1286. [CrossRef]
12. Ciamician, G. The Photochemistry of the Future. *Science* **1912**, *36*, 385–394. [CrossRef] [PubMed]
13. Moth-Poulsen, K. *Organic Synthesis and Molecular Engineering*; Nielsen, M.B., Ed.; John Wiley & Sons, Inc.: Hoboken, NJ, USA, 2014; pp. 179–196.
14. Moth-Poulsen, K.; Ćoso, D.; Börjesson, K.; Vinokurov, N.; Meier, S.K.; Majumdar, A.; Vollhardt, K.P.C.; Segalman, R.A. Molecular solar thermal (MOST) energy storage and release system. *Energy Environ. Sci.* **2012**, *5*, 8534–8537. [CrossRef]
15. Zhang, Z.Y.; He, Y.; Wang, Z.; Xu, J.; Xie, M.; Tao, P.; Ji, D.; Moth-Poulsen, K.; Li, T. Photochemical Phase Transitions Enable Coharvesting of Photon Energy and Ambient Heat for Energetic Molecular Solar Thermal Batteries That Upgrade Thermal Energy. *J. Am. Chem. Soc.* **2020**, *142*, 12256–12264. [CrossRef] [PubMed]
16. Lennartson, A.; Roffey, A.; Moth-Poulsen, K. Designing photoswitches for molecular solar thermal energy storage. *Tetrahedron Lett.* **2015**, *56*, 1457–1465. [CrossRef]
17. Vlasceanu, A.; Broman, S.L.; Hansen, A.S.; Skov, A.B.; Cacciarini, M.; Kadziola, A.; Kjaergaard, H.G.; Mikkelsen, K.V.; Nielsen, M.B. Solar Thermal Energy Storage in a Photochromic Macrocycle. *Chem.—A Eur. J.* **2016**, *22*, 10796–10800. [CrossRef]
18. Yoshida, Z.-i. New molecular energy storage systems. *J. Photochem.* **1985**, *29*, 27–40. [CrossRef]
19. Gur, I.; Sawyer, K.; Prasher, R. Searching for a Better Thermal Battery. *Science* **2012**, *335*, 1454–1455. [CrossRef]

20. Bren, V.A.; Dubonosov, A.D.; Minkin, V.I.; Chernoivanov, V.A. Norbornadiene–quadricyclane—An effective molecular system for the storage of solar energy. *Russ. Chem. Rev.* **1991**, *60*, 451–469. [CrossRef]
21. Philippopoulos, C.; Marangozis, J. Kinetics and efficiency of solar energy storage in the photochemical isomerization of norbornadiene to quadricyclane. *Ind. Eng. Chem. Prod. Res. Dev.* **1984**, *23*, 458–466. [CrossRef]
22. Börjesson, K.; Lennartson, A.; Moth-Poulsen, K. Efficiency Limit of Molecular Solar Thermal Energy Collecting Devices. *ACS Sustain. Chem. Eng.* **2013**, *1*, 585–590. [CrossRef]
23. Fei, L.; Yin, Y.; Yang, M.; Zhang, S.; Wang, C. Wearable solar energy management based on visible solar thermal energy storage for full solar spectrum utilization. *Energy Storage Mater.* **2021**, *42*, 636–644. [CrossRef]
24. Mansø, M.; Petersen, A.U.; Wang, Z.; Erhart, P.; Nielsen, M.B.; Moth-Poulsen, K. Molecular solar thermal energy storage in photoswitch oligomers increases energy densities and storage times. *Nat. Commun.* **2018**, *9*, 1945. [CrossRef] [PubMed]
25. Quant, M.; Lennartson, A.; Dreos, A.; Kuisma, M.; Erhart, P.; Börjesson, K.; Moth-Poulsen, K. Low Molecular Weight Norbornadiene Derivatives for Molecular Solar-Thermal Energy Storage. *Chem.—A Eur. J.* **2016**, *22*, 13265–13274. [CrossRef] [PubMed]
26. Gray, V.; Lennartson, A.; Ratanalert, P.; Börjesson, K.; Moth-Poulsen, K. Diaryl-substituted norbornadienes with red-shifted absorption for molecular solar thermal energy storage. *Chem. Commun.* **2014**, *50*, 5330–5332. [CrossRef]
27. Kuisma, M.; Lundin, A.; Moth-Poulsen, K.; Hyldgaard, P.; Erhart, P. Optimization of Norbornadiene Compounds for Solar Thermal Storage by First-Principles Calculations. *ChemSusChem* **2016**, *9*, 1786–1794. [CrossRef]
28. Solmuş, İ.; Kaftanoğlu, B.; Yamalı, C.; Baker, D. Experimental investigation of a natural zeolite–water adsorption cooling unit. *Appl. Energy* **2011**, *88*, 4206–4213. [CrossRef]
29. Liu, Y.; Leong, K.C. Numerical modeling of a zeolite/water adsorption cooling system with non-constant condensing pressure. *Int. Commun. Heat Mass Transf.* **2008**, *35*, 618–622. [CrossRef]
30. Vollhardt, K.P.C.; Weidman, T.W. Synthesis, structure, and photochemistry of tetracarbonyl(fulvalene)diruthenium. Thermally reversible photoisomerization involving carbon-carbon bond activation at a dimetal center. *J. Am. Chem. Soc.* **1983**, *105*, 1676–1677. [CrossRef]
31. Schwerzel, R.E.; Klosterman, N.E.; Kelly, J.R.; Hillenbrand, L.J. Catalytic Extraction of Stored Solar Energy from Photochemicals. U.S. Patent 4105014/1978, 1978.
32. Blanco-Lomas, M.; Martínez-López, D.; Campos, P.J.; Sampedro, D. Tuning of the properties of rhodopsin-based molecular switches. *Tetrahedron Lett.* **2014**, *55*, 3361–3364. [CrossRef]
33. Martinez-Lopez, D.; Yu, M.L.; Garcia-Iriepa, C.; Campos, P.J.; Frutos, L.M.; Golen, J.A.; Rasapalli, S.; Sampedro, D. Hydantoin-based molecular photoswitches. *J. Org. Chem.* **2015**, *80*, 3929–3939. [CrossRef]
34. Bastianelli, C.; Caia, V.; Cum, G.; Gallo, R.; Mancini, V. Thermal isomerization of photochemically synthesized (Z)-9-styrylacridines. An unusually high enthalpy of Z→E conversion for stilbene-like compounds. *J. Chem. Soc. Perkin Trans. 2* **1991**, *22*, 679–683. [CrossRef]
35. Losantos, R.; Sampedro, D. Design and Tuning of Photoswitches for Solar Energy Storage. *Molecules* **2021**, *26*, 3796. [CrossRef] [PubMed]
36. Orrego-Hernández, J.; Hölzel, H.; Wang, Z.; Quant, M.; Moth-Poulsen, K. Norbornadiene/Quadricyclane (NBD/QC) and Conversion of Solar Energy. In *Molecular Photoswitches*; Wiley: Hoboken, NJ, USA, 2022; pp. 351–378.
37. Jones, G.; Reinhardt, T.E.; Bergmark, W.R. Photon energy storage in organic materials—The case of linked anthracenes. *Sol. Energy* **1978**, *20*, 241–248. [CrossRef]
38. Wang, Z.; Udmark, J.; Börjesson, K.; Rodrigues, R.; Roffey, A.; Abrahamsson, M.; Nielsen, M.B.; Moth-Poulsen, K. Evaluating Dihydroazulene/Vinylheptafulvene Photoswitches for Solar Energy Storage Applications. *ChemSusChem* **2017**, *10*, 3049–3055. [CrossRef] [PubMed]
39. Broman, S.L.; Brand, S.L.; Parker, C.R.; Petersen, M.Å.; Tortzen, C.G.; Kadziola, A.; Kilså, K.; Nielsen, M.B. Optimized synthesis and detailed NMR spectroscopic characterization of the 1,8a-dihydroazulene-1,1-dicarbonitrile photoswitch. *ARKIVOC* **2011**, *2011*, 51–67. [CrossRef]
40. Koerstz, M.; Christensen, A.S.; Mikkelsen, K.V.; Nielsen, M.B.; Jensen, J.H. High throughput virtual screening of 230 billion molecular solar heat battery candidates. *PeerJ Phys. Chem.* **2021**, *3*, e16. [CrossRef]
41. Börjesson, K.; Ćoso, D.; Gray, V.; Grossman, J.C.; Guan, J.; Harris, C.B.; Hertkorn, N.; Hou, Z.; Kanai, Y.; Lee, D.; et al. Exploring the Potential of Fulvalene Dimetals as Platforms for Molecular Solar Thermal Energy Storage: Computations, Syntheses, Structures, Kinetics, and Catalysis. *Chem. Eur. J.* **2014**, *20*, 15587–15604. [CrossRef]
42. Kucharski, T.J.; Tian, Y.; Akbulatov, S.; Boulatov, R. Chemical solutions for the closed-cycle storage of solar energy. *Energy Environ. Sci.* **2011**, *4*, 4449–4472. [CrossRef]
43. International, A. *A International in Standard Tables for Reference Solar Spectral Irradiances: Direct Normal and Hemispherical on 37° Tilted Surface*; ASTM International: West Conshohocken, PA, USA, 2012.
44. Jorner, K.; Dreos, A.; Emanuelsson, R.; El Bakouri, O.; Galván, I.F.; Börjesson, K.; Feixas, F.; Lindh, R.; Zietz, B.; Moth-Poulsen, K.; et al. Unraveling factors leading to efficient norbornadiene–quadricyclane molecular solar-thermal energy storage systems. *J. Mater. Chem. A* **2017**, *5*, 12369–12378. [CrossRef]

45. Liu, X.; Xu, Z.; Cole, J.M. Molecular Design of UV–vis Absorption and Emission Properties in Organic Fluorophores: Toward Larger Bathochromic Shifts, Enhanced Molar Extinction Coefficients, and Greater Stokes Shifts. *J. Phys. Chem. C* **2013**, *117*, 16584–16595. [CrossRef]
46. Skov, A.B.; Broman, S.L.; Gertsen, A.S.; Elm, J.; Jevric, M.; Cacciarini, M.; Kadziola, A.; Mikkelsen, K.V.; Nielsen, M.B. Aromaticity-Controlled Energy Storage Capacity of the Dihydroazulene-Vinylheptafulvene Photochromic System. *Chem.—A Eur. J.* **2016**, *22*, 14567–14575. [CrossRef] [PubMed]
47. Petersen, A.U.; Hofmann, A.I.; Fillols, M.; Mansø, M.; Jevric, M.; Wang, Z.; Sumby, C.J.; Müller, C.; Moth-Poulsen, K. Solar Energy Storage by Molecular Norbornadiene–Quadricyclane Photoswitches: Polymer Film Devices. *Adv. Sci.* **2019**, *6*, 1900367. [CrossRef] [PubMed]
48. Bas, E.E.; Ulukan, P.; Monari, A.; Aviyente, V.; Catak, S. Photophysical Properties of Benzophenone-Based TADF Emitters in Relation to Their Molecular Structure. *J. Phys. Chem. A* **2022**, *4*, 473–484. [CrossRef]
49. Meng, F.-Y.; Chen, I.-H.; Shen, J.-Y.; Chang, K.-H.; Chou, T.-C.; Chen, Y.-A.; Chen, Y.-T.; Chen, C.-L.; Chou, P.-T. A new approach exploiting thermally activated delayed fluorescence molecules to optimize solar thermal energy storage. *Nat. Commun.* **2022**, *13*, 797. [CrossRef]
50. Wang, Z.; Roffey, A.; Losantos, R.; Lennartson, A.; Jevric, M.; Petersen, A.U.; Quant, M.; Dreos, A.; Wen, X.; Sampedro, D.; et al. Macroscopic heat release in a molecular solar thermal energy storage system. *Energy Environ. Sci.* **2019**, *12*, 187–193. [CrossRef]
51. Dubonosov, A.D.; Bren, V.A.; Chernoivanov, V.A. Norbornadiene–quadricyclane as an abiotic system for the storage of solar energy. *Russ. Chem. Rev.* **2002**, *71*, 917–927. [CrossRef]
52. Harel, Y.; Adamson, A.W.; Kutal, C.; Grutsch, P.A.; Yasufuku, K. Photocalorimetry. 6. Enthalpies of isomerization of norbornadiene and of substituted norbornadienes to corresponding quadricyclenes. *J. Phys. Chem.* **1987**, *91*, 901–904. [CrossRef]
53. Dilling, W.L. Intramolecular Photochemical Cycloaddition of Nonconjugated Olefins. *Chem. Rev.* **1966**, *66*, 373–393. [CrossRef]
54. Sadao, M.; Yoshinobu, I.; Zen-ichi, Y. Photochromic Solid Films Prepared by Doping with Donor–Acceptor Norbornadienes. *Chem. Lett.* **1987**, *16*, 195–198.
55. Jevric, M.; Petersen, A.U.; Mansø, M.; Kumar Singh, S.; Wang, Z.; Dreos, A.; Sumby, C.; Nielsen, M.B.; Börjesson, K.; Erhart, P.; et al. Norbornadiene-Based Photoswitches with Exceptional Combination of Solar Spectrum Match and Long-Term Energy Storage. *Chem.—A Eur. J.* **2018**, *24*, 12767–12772. [CrossRef]
56. Mitscherlich, E. Ueber das Stickstoffbenzid. *Ann. Der Phys.* **1834**, *108*, 225–227. [CrossRef]
57. Noble, A. III. Zur Geschichte des Azobenzols und des Benzidins. *Justus Liebigs Ann. Der Chem.* **1856**, *98*, 253–256. [CrossRef]
58. Hartley, G.S. The Cis-form of Azobenzene. *Nature* **1937**, *140*, 281. [CrossRef]
59. Durr, H.; Bouas-Laurent, H. *Photochromism: Molecules and Systems*; Elsevier Science: Amsterdam, The Netherlands, 2003.
60. Turanský, R.; Konôpka, M.; Doltsinis, N.L.; Štich, I.; Marx, D. Switching of functionalized azobenzene suspended between gold tips by mechanochemical, photochemical, and opto-mechanical means. *Phys. Chem. Chem. Phys.* **2010**, *12*, 13922–13932. [CrossRef] [PubMed]
61. Henzl, J.; Mehlhorn, M.; Gawronski, H.; Rieder, K.H.; Morgenstern, K. Reversible cis-trans isomerization of a single azobenzene molecule. *Angew. Chem.-Int. Ed.* **2006**, *45*, 603–606. [CrossRef]
62. Tong, X.; Pelletier, M.; Lasia, A.; Zhao, Y. Fast cis-trans isomerization of an azobenzene derivative in liquids and liquid crystals under a low electric field. *Angew. Chem.-Int. Ed.* **2008**, *47*, 3596–3599. [CrossRef]
63. Crecca, C.R.; Roitberg, A.E. Theoretical study of the isomerization mechanism of azobenzene and disubstituted azobenzene derivatives. *J. Phys. Chem. A* **2006**, *110*, 8188–8203. [CrossRef]
64. Brown, C.J. A refinement of the crystal structure of azobenzene. *Acta Crystallogr.* **1966**, *21*, 146–152. [CrossRef]
65. Mostad, A.; Romming, C. A refinement of the crystal structure of cis-azobenzene. *Acta Chem. Scand.* **1971**, *25*, 3561–3568. [CrossRef]
66. Gagliardi, L.; Orlandi, G.; Bernardi, F.; Cembran, A.; Garavelli, M. A theoretical study of the lowest electronic states of azobenzene: The role of torsion coordinate in the cis–trans photoisomerization. *Theor. Chem. Acc.* **2004**, *111*, 363–372. [CrossRef]
67. Sudesh, G.; Road, T.B.; Sciences, P.; Green, B. Photochemistry of Azobenzene-Containing Polymers. *Chem. Rev.* **1989**, *89*, 1915–1925.
68. Sension, R.J.; Repinec, S.T.; Szarka, A.Z.; Hochstrasser, R.M. Femtosecond laser studies of the cis-stilbene photoisomerization reactions. *J. Chem. Phys.* **1993**, *98*, 6291–6315. [CrossRef]
69. Hamm, P.; Ohline, S.M.; Zinth, W. Vibrational cooling after ultrafast photoisomerization of azobenzene measured by femtosecond infrared spectroscopy. *J. Chem. Phys.* **1997**, *106*, 519–529. [CrossRef]
70. Georgiev, A.; Bubev, E.; Dimov, D.; Yancheva, D.; Zhivkov, I.; Krajčovič, J.; Vala, M.; Weiter, M.; Machkova, M. Synthesis, structure, spectral properties and DFT quantum chemical calculations of 4-aminoazobenzene dyes. Effect of intramolecular hydrogen bonding on photoisomerization. *Spectrochim. Acta Part A Mol. Biomol. Spectrosc.* **2017**, *175*, 76–91. [CrossRef] [PubMed]
71. Schulze, F.W.; Petrick, H.J.; Cammenga, H.K.; Klinge, H.Z. Thermodynamic properties of the structural analogues benzo[c]-cinnoline, trans-azobenzene and cis-azobenzene. *Z. Phys. Chem. Neue Fol.* **1997**, *1*, 107. [CrossRef]
72. Brown, E.V.; Granneman, G.R. Cis-Trans Isomerism in the Pyridyl Analogs of Azobenzene. A Kinetic and Molecular Orbital Analysis. *J. Am. Chem. Soc.* **1975**, *97*, 621–627. [CrossRef]
73. Tamai, N.; Miyasaka, H. Ultrafast Dynamics of Photochromic Systems. *Chem. Rev.* **2000**, *100*, 1875–1890. [CrossRef]

74. Casellas, J.; Bearpark, M.J.; Reguero, M. Excited-State Decay in the Photoisomerisation of Azobenzene: A New Balance between Mechanisms. *ChemPhysChem* **2016**, *17*, 3068–3079. [CrossRef]
75. Rau, H. Further evidence for rotation in the π,π^* and inversion in the n,π^* photoisomerization of azobenzenes. *J. Photochem.* **1984**, *26*, 221–225. [CrossRef]
76. Fujino, T.; Arzhantsev, S.Y.; Tahara, T. Femtosecond Time-Resolved Fluorescence Study of Photoisomerization of trans-Azobenzene. *J. Phys. Chem. A* **2001**, *105*, 8123–8129. [CrossRef]
77. Rau, H. *Photochromism: Molecules and Systems*, 1st ed.; Elsevier: Amsterdam, The Netherlands, 1990; pp. 165–192.
78. Kunz, A.; Heindl, A.H.; Dreos, A.; Wang, Z.; Moth-Poulsen, K.; Becker, J.; Wegner, H.A. Intermolecular London Dispersion Interactions of Azobenzene Switches for Tuning Molecular Solar Thermal Energy Storage Systems. *ChemPlusChem* **2019**, *84*, 1145–1148. [CrossRef] [PubMed]
79. Wang, Z.; Losantos, R.; Sampedro, D.; Morikawa, M.-a.; Börjesson, K.; Kimizuka, N.; Moth-Poulsen, K. Demonstration of an azobenzene derivative based solar thermal energy storage system. *J. Mater. Chem. A* **2019**, *7*, 15042–15047. [CrossRef]
80. Shigeru, Y.; Hiroshi, O.; Osamu, T. The cis-trans Photoisomerization of Azobenzene. *Bull. Chem. Soc. Jpn.* **1962**, *35*, 1849–1853.
81. Zimmerman, G.; Chow, L.-Y.; Paik, U.-J. The Photochemical Isomerization of Azobenzene1. *J. Am. Chem. Soc.* **1958**, *80*, 3528–3531. [CrossRef]
82. Ladányi, V.; Dvořák, P.; Al Anshori, J.; Vetráková, Ľ.; Wirz, J.; Heger, D. Azobenzene photoisomerization quantum yields in methanol redetermined. *Photochem. Photobiol. Sci.* **2017**, *16*, 1757–1761. [CrossRef]
83. Mahimwalla, Z.; Yager, K.G.; Mamiya, J.-i.; Shishido, A.; Priimagi, A.; Barrett, C.J. Azobenzene photomechanics: Prospects and potential applications. *Polym. Bull.* **2012**, *69*, 967–1006. [CrossRef]
84. Kolpak, A.M.; Grossman, J.C. Azobenzene-Functionalized Carbon Nanotubes As High-Energy Density Solar Thermal Fuels. *Nano Lett.* **2011**, *11*, 3156–3162. [CrossRef]
85. Kucharski, T.J.; Ferralis, N.; Kolpak, A.M.; Zheng, J.O.; Nocera, D.G.; Grossman, J.C. Templated assembly of photoswitches significantly increases the energy-storage capacity of solar thermal fuels. *Nat. Chem.* **2014**, *6*, 441–447. [CrossRef]
86. Wu, S.; Butt, H.-J. Solar-Thermal Energy Conversion and Storage Using Photoresponsive Azobenzene-Containing Polymers. *Macromol. Rapid Commun.* **2020**, *41*, 1900413. [CrossRef]
87. Feng, W.; Luo, W.; Feng, Y. Photo-responsive carbon nanomaterials functionalized by azobenzene moieties: Structures, properties and application. *Nanoscale* **2012**, *4*, 6118–6134. [CrossRef]
88. Feng, Y.; Liu, H.; Luo, W.; Liu, E.; Zhao, N.; Yoshino, K.; Feng, W. Covalent functionalization of graphene by azobenzene with molecular hydrogen bonds for long-term solar thermal storage. *Sci. Rep.* **2013**, *3*, 3260. [CrossRef] [PubMed]
89. Norikane, Y.; Kitamoto, K.; Tamaoki, N. Novel crystal structure, cis-trans isomerization, and host property of meta-substituted macrocyclic azobenzenes with the shortest linkers. *J. Org. Chem.* **2003**, *68*, 8291–8304. [CrossRef]
90. Durgun, E.; Grossman, J.C. Photoswitchable Molecular Rings for Solar-Thermal Energy Storage. *J. Phys. Chem. Lett.* **2013**, *4*, 854–860. [CrossRef] [PubMed]
91. Zhitomirsky, D.; Cho, E.; Grossman, J.C. Solid-State Solar Thermal Fuels for Heat Release Applications. *Adv. Energy Mater.* **2016**, *6*, 1502006. [CrossRef]
92. Lv, J.-A.; Liu, Y.; Wei, J.; Chen, E.; Qin, L.; Yu, Y. Photocontrol of fluid slugs in liquid crystal polymer microactuators. *Nature* **2016**, *537*, 179–184. [CrossRef] [PubMed]
93. Jeong, S.P.; Renna, L.A.; Boyle, C.J.; Kwak, H.S.; Harder, E.; Damm, W.; Venkataraman, D. High Energy Density in Azobenzene-based Materials for Photo-Thermal Batteries via Controlled Polymer Architecture and Polymer-Solvent Interactions. *Sci. Rep.* **2017**, *7*, 17773. [CrossRef] [PubMed]
94. Zhitomirsky, D.; Grossman, J.C. Conformal Electroplating of Azobenzene-Based Solar Thermal Fuels onto Large-Area and Fiber Geometries. *ACS Appl. Mater. Interfaces* **2016**, *8*, 26319–26325. [CrossRef] [PubMed]
95. Cho, E.N.; Zhitomirsky, D.; Han, G.G.D.; Liu, Y.; Grossman, J.C. Molecularly Engineered Azobenzene Derivatives for High Energy Density Solid-State Solar Thermal Fuels. *ACS Appl. Mater. Interfaces* **2017**, *9*, 8679–8687. [CrossRef]
96. Calbo, J.; Weston, C.E.; White, A.J.P.; Rzepa, H.S.; Contreras-García, J.; Fuchter, M.J. Tuning Azoheteroarene Photoswitch Performance through Heteroaryl Design. *J. Am. Chem. Soc.* **2017**, *139*, 1261–1274. [CrossRef]
97. Saritas, K.; Grossman, J.C. Accurate Isomerization Enthalpy and Investigation of the Errors in Density Functional Theory for Dihydroazulene/Vinylheptafulvene Photochromism Using Diffusion Monte Carlo. *J. Phys. Chem. C* **2017**, *121*, 26677–26685. [CrossRef]
98. Goerner, H.; Fischer, C.; Gierisch, S.; Daub, J. Dihydroazulene/vinylheptafulvene photochromism: Effects of substituents, solvent, and temperature in the photorearrangement of dihydroazulenes to vinylheptafulvenes. *J. Phys. Chem.* **1993**, *97*, 4110–4117. [CrossRef]
99. Christensen, O.; Nielsen, L.E.; Johansen, J.; Holk, K.; Udmark, J.; Nielsen, M.B.; Cacciarini, M.; Mikkelsen, K.V. Density Functional Theory Study of Carbamoyl-Substituted Dihydroazulene/Vinylheptafulvene Derivatives and Solvent Effects. *J. Phys. Chem. C* **2022**, *126*, 4815–4825. [CrossRef]
100. Broman, S.L.; Jevric, M.; Nielsen, M.B. Linear Free-Energy Correlations for the Vinylheptafulvene Ring Closure: A Probe for Hammett σ Values. *Chem.—A Eur. J.* **2013**, *19*, 9542–9548. [CrossRef] [PubMed]
101. Kilde, M.D.; Hansen, M.H.; Broman, S.L.; Mikkelsen, K.V.; Nielsen, M.B. Expanding the Hammett Correlations for the Vinylhepta-fulvene Ring-Closure Reaction. *Eur. J. Org. Chem.* **2017**, *2017*, 1052–1062. [CrossRef]

102. Ranzenigo, A.; Cordero, F.M.; Cacciarini, M.; Nielsen, M.B. ortho-Substituted 2-Phenyldihydroazulene Photoswitches: Enhancing the Lifetime of the Photoisomer by ortho-Aryl Interactions. *Molecules* **2021**, *26*, 6462. [CrossRef]
103. Schøttler, C.; Vegge, S.K.; Cacciarini, M.; Nielsen, M.B. Long-Term Energy Storage Systems Based on the Dihydroazulene/Vinylheptafulvene Photo-/Thermoswitch. *ChemPhotoChem* **2022**, *6*, e202200037. [CrossRef]
104. Abedi, M.; Pápai, M.; Henriksen, N.E.; Møller, K.B.; Nielsen, M.B.; Mikkelsen, K.V. Theoretical Investigation on the Control of Macrocyclic Dihydroazulene/Azobenzene Photoswitches. *J. Phys. Chem. C* **2019**, *123*, 25579–25584. [CrossRef]
105. Wang, Z.; Hölzel, H.; Moth-Poulsen, K. Status and challenges for molecular solar thermal energy storage system based devices. *Chem. Soc. Rev.* **2022**. Advance Article. [CrossRef]

Perspective

Photochemistry of Metal Nitroprussides: State-of-the-Art and Perspectives

Paula M. Crespo [1], Oscar F. Odio [2,*] and Edilso Reguera [1,*]

[1] Instituto Politécnico Nacional, Centro de Investigación en Ciencia Aplicada y Tecnología Avanzada, U. Legaria, Ciudad de México 11500, Mexico; moncbarrera@gmail.com
[2] CONACyT-Instituto Politécnico Nacional, Centro de Investigación en Ciencia Aplicada y Tecnología Avanzada, U. Legaria, Ciudad de México 11500, Mexico
* Correspondence: odiochacon@gmail.com (O.F.O.); edilso.reguera@gmail.com (E.R.)

Abstract: This contribution summarizes the current state in the photochemistry of metal nitroprussides, which is dominated by the electronic structure of the nitrosyl group. From the combination of p orbitals of the nitrogen and oxygen atoms in the NO^+ ligand, a π^*NO molecular orbital of relatively low energy is formed, which has π^*_{2px} and π^*_{2py} character. This is a double degenerate orbital. When the nitrosyl group is found coordinated to the iron atom in the nitroprusside ion, the availability of that low energy π^*NO orbital results in light-induced electronic transitions from the iron atom d_{xy}, d_{xz} and d_{yz} orbitals, $2b_2$ (xy) → 7e (π^*NO) and 6e (xz,yz) → 7e (π^*NO), which are observed at 498 and 394 nm, respectively. These light-induced transitions and the possibility of NO isomer formation dominate the photochemistry of metal nitroprussides. In this feature paper, we discuss the implications of such transitions in the stability of coordination compounds based on the nitroprusside ion in the presence of water molecules for both 3D and 2D structures, including the involved degradation mechanisms. These photo-induced electronic transitions modify the physical and functional properties of solids where the nitroprusside ion forms part of their structure and appear as an opportunity for tuning their magnetic, electrical, optical and as energy-applied materials, for instance. This contribution illustrates these opportunities with results from some recently reported studies, and possible research subjects, even some not explored, are mentioned.

Keywords: photochemistry of metal nitroprussides; stability of transition metal nitroprussides; meta-stable states in the nitroprusside ion; photo-isomerization; photochemistry

Citation: Crespo, P.M.; Odio, O.F.; Reguera, E. Photochemistry of Metal Nitroprussides: State-of-the-Art and Perspectives. *Photochem* **2022**, *2*, 390–404. https://doi.org/10.3390/photochem2020027

Academic Editor: Marcelo I. Guzman

Received: 15 April 2022
Accepted: 28 May 2022
Published: 31 May 2022

Publisher's Note: MDPI stays neutral with regard to jurisdictional claims in published maps and institutional affiliations.

Copyright: © 2022 by the authors. Licensee MDPI, Basel, Switzerland. This article is an open access article distributed under the terms and conditions of the Creative Commons Attribution (CC BY) license (https://creativecommons.org/licenses/by/4.0/).

1. Introduction

Metal nitroprussides are salts of the pentacyanonitrosylferrate(II) ion, $[Fe(CN)_5NO]^{2-}$, which were first reported in 1849 by Playfair [1]. Of these six ligands, only the five CNs at their N ends are available to bond a metal center (T); the NO^+ moiety remains unlinked at its O end [2,3]. Such salts can be formed with alkaline, alkaline earth [4], transition [5–7] and even rare-earth metals [8] and can even form 2D structures when the axial CN remains unlinked at its N end, resulting in the formation of layered coordination polymers [9]. Certain organic ligands can disrupt the CN_{Ax}-T bond to occupy the metal axial coordination sites [10–13].

These organic molecules behave as a pillar between adjacent layers. Since the metal nitroprussides discovery, many studies related to their physical, chemical and electronic properties have been performed and were recently reviewed by Reguera et al. [14]. One of the most relevant properties is related to the hypotensive action of the sodium nitroprusside, NaNP, which was demonstrated in 1929 and made the NaNP useful in cases of severe hypertension [15]. This feature is due to the presence and release of the NO moiety when the NaNP is dispersed in an aqueous solution and receives illumination.

Metal nitroprussides can absorb light through three mechanisms [16]: (1) metal-ligand-charge transfer (MLCT); (2) band-to-band transitions; (3) d-d transitions when an appropriate transition metal (T) is involved. The photochemistry of metal nitroprussides is dominated by the first mechanism, related to the photoactive properties of the nitrosyl group. It is known that the MLCT is the main process in the photochemistry of transition metal complexes [17], and its excitation depends on the ligands, particularly when these last ones participate in a strong metal-ligand π-bonding, such a charge transfer is possible.

Moreover, these transitions can produce an intermolecular redox reaction when the excited states are promoted from the metal d orbitals to the orbitals (π, π*, σ and σ*) of the ligand. Many papers have been published regarding the photochemistry of NaNP, and only a handful have been made about other nitroprusside salts. Transition metal nitruprussides, form a family of coordination polymers with demonstrated potential applications for small molecules separation and storage, battery materials, sensors and actuators, information storage, photo-catalysis, in healthcare and other areas [14].

Nevertheless, their stability in the presence of water molecules and under the light incidence, two usually coinciding conditions in a normal environment, remains poorly documented and only recently has been studied [18,19]. This contribution discusses the accumulated information by our research group on the photochemistry of that family of materials, in conditions close to that expected during their applications. The light-induced electronic transitions Fe → π*NO modifies the electronic structure of metal nitroprussides and their physical properties. This has the potential for applications through materials with light-driven functional properties. This subject is also considered in this feature paper.

2. Photochemistry of the Nitroprusside Block

Although the metal nitroprussides were synthesized in 1848, the crystal structure of the sodium nitroprusside was determined a few decades later [2]. From this last study, we now know that the NP ion has approximately C_{4v} symmetry, being all the C-N distances similar to other cyanide complexes and that the N-O distance is similar to the one found in NOX, where X is a halogen. However, the Fe-NO distance is short enough to suggest a triple bond between the Fe and N atoms, where the two π-bonds lay on the antibonding π*(NO) and the σ-bond goes to an empty d-orbital from the iron atom [20]. The work of Manoharan et al., using SCCC-MO calculations, gave valuable information about the ordering of the energy levels on the nitroprusside ion [21]. The resulting energy is shown in Figure 1. The highest-filled molecular orbitals, HOMO, in the ground state (GS) of the NP ion are represented as $(6e)^2(2b_2)^2$, where the former has d_{xz} and d_{yz} character but also contains about a quarter of π*(NO) character, and the second one has around 85% d_{xy} character, with the remaining 14% being π(CN) and 2% π*(CN) contributions.

On the other hand, the lowest-unfilled molecular orbital, LUMO, is labeled as 7e and is around 73% of π*(NO) character, completed with small contributions from the metal, d_{xy}, d_{xz}, d_{yz} and contributions from the σ(CN), π(CN) and π*(CN) orbitals. Because of the π*(NO) has a high contribution to the HOMO 6e, which has mainly metal character, a π-back bonding interaction between the Fe atom and the NO moiety can be inferred. This leads to a significant decrease in the electron density on the iron atom, reducing its π-back bonding interaction with the CN ligands, and in consequence, the electron density found at their 5σ orbital lowers.

This remains well documented from XPS spectra of metal nitroprussides [22]. The large contribution of the π*(NO) orbital to the 7e molecular orbital is probed by the frequency for the ν(NO) stretching band in the IR spectrum, which decreases when the $2b_2$ (d_{xy}) → 7e (π*NO) charge transfer increases [20]. For sodium nitroprusside, this transition is observed at 498 nm, which explains the red-purple color of crystals of that sodium salt. For transition metal nitroprussides, crystals of a varied diversity of colors (and absorption bands) are formed related to the combination of light absorption via MLCT and d-d transitions for the metal linked at the N end of the CN ligands [23]. Such d-d transitions have a minor impact on the photochemistry of metal nitroprussides.

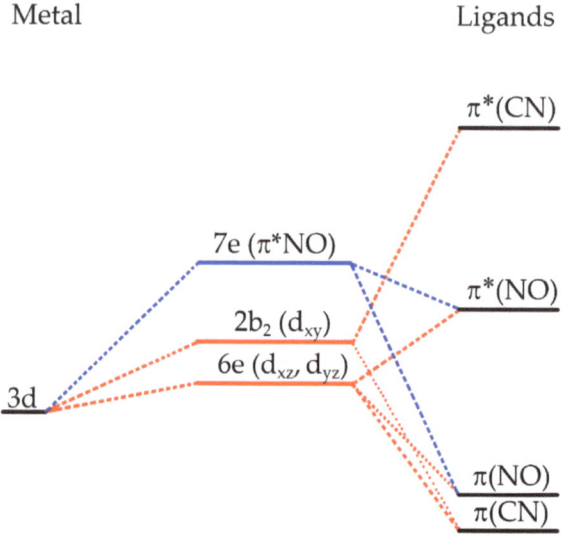

Figure 1. Ordering of the MO of NP, as proposed by Manoharan et al. [21]. The LUMO is represented in blue and the HOMO in red.

In 1977, Hauser et al. [24,25] observed a light-induced metastable state recording Mössbauer spectra at low temperature while a single crystal of NaNP was irradiated with a blue-green laser beam. With this technique, these authors found a spectrum that was different from the one obtained for the ground state (GS) measured without the laser beam irradiation. The spectrum collected during the irradiation was only visible at a certain angle.

Moreover, the irradiation must be continuous to produce this new state, so they ruled out the possibility of the new state being generated by light absorption or inelastic scattering. After these pioneer works, Zöllner et al. [26] studied by DSC (differential scanning calorimetry) measurements the thermal behavior of an irradiated sample of NaNP and discovered two long-lived metastable states (MS1 and MS2), the second decaying at a lower temperature; also, they found that the production of the MS states is exhibited in both single crystals or powders and even in other nitroprusside salts.

Subsequent experimental and theoretical studies [27–31] have enlighted a consistent picture on the subject: the light-induced metastable states are linkage isomers of the nitroprusside block, one of which has the O end of the NO bonded to the iron atom, the isonitrosyl η^1-O (MS1), and the other one has the NO side-bounded, η^2-NO (MS2) [32] (Figure 2).

These two metastable states result from the irradiation with a light source in the range of λ = 350 to 580 nm while the sample is maintained underneath 100 K [33]; however, when the temperature is raised to 165 K, these MS decay to the GS again, and thus it is evident that the population of these metastable states depends on the polarization direction and wavelength of the light beam used [34]. Furthermore, these two states can be converted into one another by irradiating the sample with light of appropriate energy, 620–850 nm (Figure 2). The metastable states can also be de-excited by raising the temperature or irradiating the sample [26].

Further studies, recording SQUID (superconducting quantum interference device) magnetic data for nickel nitroprusside, Ni[Fe(CN)$_5$NO]·xH$_2$O, at 5 K, irradiating the sample with light from an Ar$^+$ laser, λ = 475 nm, revealed the appearance of magnetic order at low temperature. Such cooperative magnetic interaction disappears when the sample is warmed above 200 K [35,36]. Such behavior was explained as due to the light-driven electron transfer from the iron atom to the NO ligand, resulting in the appearance of two new paramagnetic centers in the solid, on the iron atom and the NO ligand.

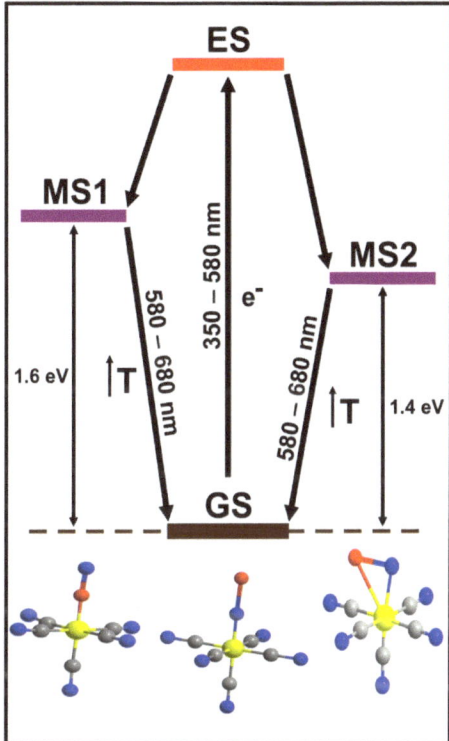

Figure 2. Excitation of isomeric transitions for the NO$^+$ ligand by absorption in the UV–vis spectral region.

The observed cooperative magnetic order results from their superexchange interaction with the two unpaired electrons on the Ni atom. The authors reported the appearance of a new IR band at 1821 cm^{-1}, close to the frequency (1831 cm^{-1}) corresponding to the MS1 of sodium nitroprusside [37]. This suggests that the photo-induced magnetic order in nickel nitroprusside is related to a light excited state (ES), which then decays to the MS1 and MS2 states for the NO group (Figure 2). Such interpretation is congruent with the corresponding Mössbauer spectra recorded at low temperature (300 mK), which identified the presence of Fe(III) species in the irradiated sample [8]. This provides a conclusive clue to understanding the photochemistry of metal nitroprussides and their relationship with the MLCT in this ion.

From the nature of the HOMO and LUMO orbitals described above, these metastable states are considered as a d-π*(NO) one-electron excitation, however, Güdel proposed that the nature of the MS cannot be due to one electron configuration, so the MS can also be produced by a d-d transition added to the dominant MLCT [33]. These states have also been found in other nitrosyl complexes with transition metals when irradiated within the 350–580 nm wavelength range, either in 3D or 2D structures [26,38].

3. Photodecomposition: A Consequence of the Photoinduced Metal-Ligand-Charge-Transfer

3.1. Photodegradation Studies on Sodium Nitroprusside (NaNP)

The photolysis of NaNP aqueous solution has been the subject of intense studies due to its importance in medical issues as a potent hypotensive agent. In 1963 Mitra et al. [39] reported that a solution of NaNP is acidified when irradiated with an unfiltered mercury discharge lamp; this effect was reversible provided the irradiation was for a short time; however, under prolonged irradiation, irreversible changes and secondary reactions appeared to take place. The authors proposed that the changes in the NaNP solution were triggered by the photolysis of the NP block, which releases the NO ligand as NO⁺. Subsequent detailed studies covering spectroscopic and analytic techniques, mostly those from the groups of Wolfe, Stockel and de Oliveira [40–42], agree with a general pathway presented in Scheme 1.

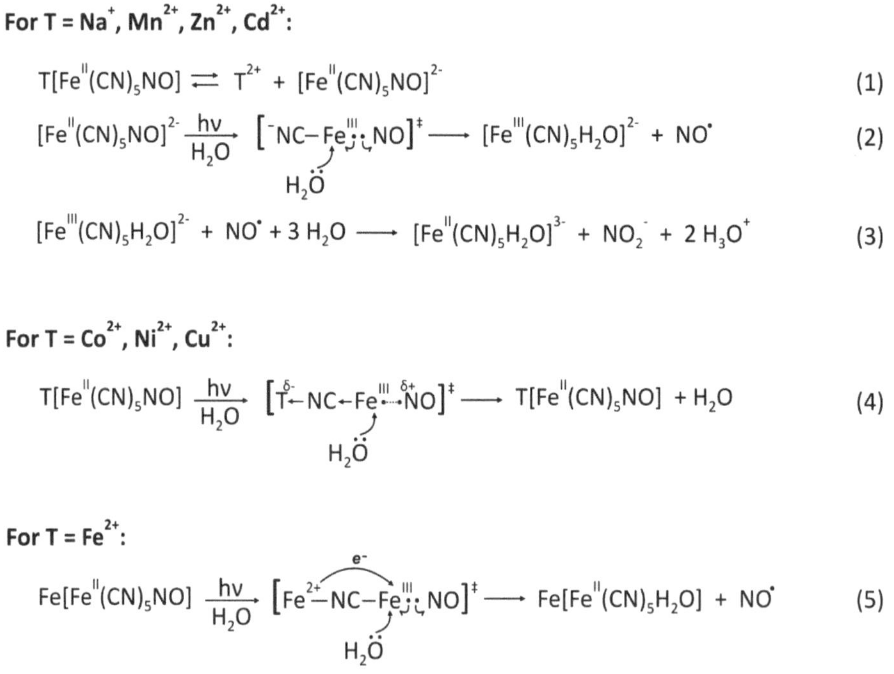

For T = Na^+, Mn^{2+}, Zn^{2+}, Cd^{2+}:

$$T[Fe^{II}(CN)_5NO] \rightleftharpoons T^{2+} + [Fe^{II}(CN)_5NO]^{2-} \tag{1}$$

$$[Fe^{II}(CN)_5NO]^{2-} \xrightarrow[H_2O]{h\nu} [^-NC-Fe^{III}\cdots NO]^{\ddagger} \longrightarrow [Fe^{III}(CN)_5H_2O]^{2-} + NO^{\bullet} \tag{2}$$

$$[Fe^{III}(CN)_5H_2O]^{2-} + NO^{\bullet} + 3H_2O \longrightarrow [Fe^{II}(CN)_5H_2O]^{3-} + NO_2^- + 2H_3O^+ \tag{3}$$

For T = Co^{2+}, Ni^{2+}, Cu^{2+}:

$$T[Fe^{II}(CN)_5NO] \xrightarrow[H_2O]{h\nu} [T^{\delta-}-NC-Fe^{III}\cdots NO^{\delta+}]^{\ddagger} \longrightarrow T[Fe^{II}(CN)_5NO] + H_2O \tag{4}$$

For T = Fe^{2+}:

$$Fe[Fe^{II}(CN)_5NO] \xrightarrow[H_2O]{h\nu} [Fe^{2+}-NC-Fe^{III}\cdots NO]^{\ddagger} \longrightarrow Fe[Fe^{II}(CN)_5H_2O] + NO^{\bullet} \tag{5}$$

Scheme 1. General photodegradation pathways for transition metal nitroprussides as a function of the external metal (T).

The photodegradation of the nitroprusside ion in solution begins with the formation of the ES due to the intra-molecular electronic transition 6e → 7e resulting from the photon absorption in the blue region (c.a. 400 nm). As we noted before, orbital 6e has a marked bonding character reflecting the π-back bonding interaction of the Fe(II)-NO bond, whereas orbital 7e has a predominant π*(NO) character; hence, after the 6e → 7e transition, the resulting Fe(III)-NO bond is significantly weakened respect to the ground state since now the interaction is only of σ nature.

Thus, such destabilizing effect can give rise to the heterolytic rupture of the Fe-NO bond, leading to the release of radical NO and the formation of $[Fe^{III}(CN)_5H_2O]^{2-}$ ion after the incorporation of a water molecule to complete the coordination sphere of the central Fe(III) atom (step **2** in Scheme 1).

The effect of the solvent appears to be important in assisting the photo-degradation process since experiments performed in neat conditions [18] prevent the NO release, likely due to the cage effect. The NO release has been documented not only for nitroprusside aqueous solutions but also for nitroprusside dispersed in several solvents [41,43].

The formation of $[Fe^{III}(CN)_5H_2O]^{2-}$ ion and the release of NO to the solution are followed by several secondary photo- and redox-reactions that account for the observed acidification of the solutions and also yield as byproducts new species like $[Fe^{II}(CN)_5H_2O]^{3-}$, NO_2^-, NO_3^-, CN^- and Prussian blue (PB). The extent of these reactions depends on the experimental conditions during the photodegradation test, as are the intensity and duration of the irradiation, sample concentration, pH, and level of oxygen dissolved. The $[Fe^{II}(CN)_5H_2O]^{3-}$ species can be formed by either thermal- and photo-reduction of the primary $[Fe^{III}(CN)_5H_2O]^{2-}$ ion [41,42,44], or by a redox reaction between this last species and the released NO molecules, leading to NO^+ cation that forms promptly NO_2^- (step 3 in Scheme 1). In addition, NO can also be consumed in aerobic oxidation with the O_2 dissolved in the solution to form NO_2^- and NO_3^- anions [45].

The key role of the 6e → 7e electronic transition is highlighted when studies are performed using monochromatic radiation [40,46]; hence, several works have reported that irradiating the nitroprusside solutions at the wavelength corresponding to the $2b_2$ → 7e transition (near 500 nm), no release of the nitrosyl radical is detected. The explanation lies in the fact that the $2b_2$ orbital is almost entirely composed of the iron d_{xy} orbital, i.e., it has a marked nonbonding character; therefore, the promotion of an electron to the $\pi^*(NO)$ orbital does not influence significantly the strength of the Fe-NO bond.

An interesting question relies on whether the NO release competes with the linkage metastable isomers MS1 and MS2, or it happens from the metastable states as a two-step consecutive mechanism; this is a difficult task because photodegradation is usually observed after prolonged irradiation times at room temperature and in solution phase, i.e., under experimental conditions for which MS1 and MS2 decay rapidly and cannot be detected with steady-state techniques.

Up to now, there has been no consensus regarding this issue. Several works have postulated the consecutive mechanism [46–48] on the theoretical grounds that the structure of MS1 and MS2 could favor the Fe-NO bond rupture [49,50]. On the other hand, the work of Lynch et al. [37] supports the competitive mechanism for NaNP solutions at room temperature based on the picosecond transient Infrared Spectroscopy technique, with which they were able to trace simultaneously the appearance of MS1, MS2 and NO species.

3.2. Effect of T in the photodegradation of $T[Fe(CN)_5NO] \cdot xH_2O$

The extent of the nitroprusside photolysis depends on the efficiency of the electronic reorganization within the ES; conditions that favor a strong Fe(III)-NO σ interaction provoke a fast deactivation of the ES, preventing bond rupture; on the contrary, as the σ interaction becomes weaker, the life-time of the ES is expected to be longer, thus, increasing the probabilities for Fe-NO bond rupture [46]. As can be inferred, there is room for controlling the photolysis of nitroprusside-containing materials by tuning the electronic properties of the internal block.

Based on this hypothesis, our research group tested the photo-stability of a series of aqueous 3D transition metal nitroprusside dispersions in order to study the influence of the external cation (T) bonded to the N ends of the $T[Fe(CN)_5NO] \cdot xH_2O$ structure, where T = Mn, Fe, Co, Ni, Cu, Zn and Cd [18]. For this endeavor, we subjected the aqueous dispersions to an intense white light lamp and analyzed the resulting products by FT-IR and Mössbauer Spectroscopies along with potentiometric measurements to track the pH variations during the photoreactions.

Since the ligands in the nitroprusside block are active in infrared absorption, giving well-localized narrow and intense bands, FTIR is a valuable technique to sense the changes involving them; at the same time, small variations in the coordination of the internal Fe atom can be probed with Mössbauer spectroscopy. Thus, these techniques allow for tracing the loss of the NO ligand from the nitroprusside block and a facile identification of the degradation products $[Fe^{III}(CN)_5H_2O]^{2-}$ and $[Fe^{II}(CN)_5H_2O]^{3-}$.

Our results evidenced two limiting situations. On one hand, when T = Mn and Zn, the 3D compounds completely dissolve in water, and the recovered materials (by rotoevaporating the solutions) show the typical spectroscopic signs encountered for NaNP photodegradation, including a significant decrease of the solution pH. On the other hand, for T = Co, Ni and Cu, the compounds are practically insoluble, which permits the sample recovery by centrifugation; in these cases, the pH does not vary throughout the reaction, and the recovered solids do not display spectroscopic variations respect to the original materials.

The case of the Cd compound presents an intermediate behavior: the insoluble fraction remains unaltered and the soluble fraction is photodegraded. From these results, it is possible to rationalize the influence of T on the photodegradation tendency: 3D nitroprusside of metal cations with low polarizing power are easily ionizable in water, which liberates the internal building block and thus increases the electronic density in the N end of the CN ligands (step **1** in Scheme 1); this charge is then available for the Fe(III) atom to stabilize the photo-excited state, allowing longer a life-time that favors the rupture of the Fe(III)-NO bond (step **2** in Scheme 1).

In contrast, in the nitroprussides where T forms stronger T-NC bonds due to the significant polarization of the electronic density in the CN 5σ orbital toward the cation, the available electronic density in the nitroprusside block is lower; as a consequence, the σ interaction Fe(III) ← NO in the ES becomes stronger and leads to rapid quenching of the ES, frustrating the release of the nitrosyl radical from the nitroprusside block (step **4** in Scheme 1).

The case of the FeNP is quite interesting because contrary to the expected behavior owing to its polarizing power (intermediate between Mn and Co) not only the soluble but also the insoluble fraction presents spectroscopic footprints of degradation, which are now limited to the $[Fe^{II}(CN)_5H_2O]^{3-}$ species only. In addition, a careful analysis of the Mossbauer spectrum at 5 K suggests the presence of Fe^{3+} cation in a high spin configuration that is external to the building block.

All the experimental evidence points to the formation of the mixed-valence compound $Fe[Fe^{II}(CN)_5H_2O]$, similar to PB. Inspired by this finding, we performed a series of experiments where several 3D TNP powder samples were extensively irradiated, with FeNP as the only compound that showed traces of photodegradation, expressed as the occurrence of the mixed-valence compound.

This result not only indicates the assisting role of the water in the NO release from the nitroprusside block but also demonstrates the unusual sensitivity of the FeNP toward photodegradation, which is clearly related with the PB-like species. In the initial ES, the configuration Fe^{2+}–NC–Fe(III)–NO leads to the more stable Fe^{3+}–NC–Fe(II)–NO by a metal-to-metal electron transfer from the external HS Fe^{2+} to the internal LS Fe(III) mediated by the CN ligand, leading to a more stable configuration that significantly stabilizes the ES (step **5** in Scheme 1).

3.3. Effect of Pyridine (Py) in the Photodegradation of $T(Py)_2[Fe(CN)_5NO]$

To obtain further insights on the influence of the nitroprusside electronic structure on the material photostability, we prepared and evaluated a series of 2D transition metal nitroprusside derivatives containing pyridine (Py) as pillar molecules: $T(Py)_2[Fe(CN)_5NO]$ [19]. Effectively, the treatment of the 3D materials with an excess of the organic molecule drives the inclusion of it in the coordination sphere of the external metal as axial ligands, which liberates the axial cyanide ligand of the nitroprusside block and transforms the initial hydrated 3D structure into an anhydrous 2D structure [11].

As in the previous study, we tested the photostability of aqueous dispersions containing these 2D compounds as a function of the external metal T. For the sake of comparison, experiments in dark conditions were first performed. The most striking feature was the partial or complete recovery of the 3D phase due to the loss of the Py molecules from the metal coordination sphere; the 3D phase was detected by FTIR, and the free Py molecules can be sensed due to the basification of the dispersed medium.

The 2D → 3D transformation is more prominent as the polarizing power of T decreases; for instance, while for the Ni compound the 3D phase is barely detected, for the Mn and Zn counterparts, the whole samples are recovered in the soluble fraction as the 3D phases. For the rest of the samples, both phases coexist with a 2D/3D composition that is consistent with the cation polarizing power order. Using the Cu compound as a model, we were able to tune the extent of Py losses by varying the ionic force of the medium, since high salt concentration prevents the substitution of Py by water molecules in the coordination sphere of T.

With this in mind, we then performed the analogous experiments under white light illumination. The recorded degradation effects for the case of metals with low polarizing power (Mn, Zn and Fe) were similar to those encountered during the photostability study of the 3D TNP. This was expected because of the fast 3D formation at the initial stages of the experiment; the pH behavior is consistent with this notion: it increases suddenly due to Py release from the nitroprusside structure and then begins to decrease as a consequence of the photolysis reactions.

The analysis of the data corresponding to the compounds containing cations of higher polarizing power demonstrated that, contrary to the 3D TNP materials, a certain level of photodegradation can be detected together with the ubiquitous 3D phase. In order to rationalize this finding, we analyzed the NO release as a function of the ionic force of the medium for the Cu compound. The obtained results indicated that NO losses are more extensive as the ionic force becomes higher; at the same time, the same direct relation can be found between the ionic force and the extent of Py losses, which is the inverse behavior seen during the in-darkness experiments. Therefore, it seems that not only does NO and Py release go hand-in-hand under illuminating conditions but also, more importantly, NO release triggers the Py release.

The key factor for understanding the situation relies in the differential structural factor that appears in the nitroprusside block on 2D formation: the free axial cyanide ligand (CN_{axial}). For the cases where T is a low polarizing cation, the fast Py release provokes the sudden formation of the 3D phase, which becomes ionized, and this process liberates the nitroprusside block from the influence of T; therefore, the free CN_{axial} initially encountered in the 2D structure does not have the opportunity to influence the photodegradation process. Now well, the picture is different when the 2D structure contains a metal with high polarizing power.

As it was explained, in this case, T retains the Py molecules in its coordination sphere, thus, retarding the 2D → 3D transformation; in other words, the CN_{axial} of the nitroprusside block remains unbonded to metal centers. This allows the CN_{axial} to play part in the electronic rearrangement that takes place after photoexcitation (step **1** in Scheme 2); in fact, since now the available electronic density within the nitroprusside block is higher with respect to the case when all the CN ligands are forming T-NC bonds, the extra stabilization of the Fe(III) atom in the ES favors the rupture of the Fe-NO bond (step **2** in Scheme 2).

That explains why the 2D materials are more sensitive to photodegradation that the 3D counterparts when T = Co, Cu and Ni. This proposed pathway is consistent with the direct proportionality between the ionic force of the medium and the extent of NO release: as the ionic force increases, the dielectric constant of the medium decreases, causing the excess electronic density around the free CN_{axial} to be less dispersed throughout the medium, i.e., more available for stabilizing the photo-induced ES in the nitroprusside block.

Yet, what about the observed parallel behavior between NO and Py loses? Again, the free CN_{axial} plays a fundamental role. The release of the NO ligand produces $[Fe^{III}(CN)_5H_2O]^{2-}$ residues throughout the structure; the extensive electronic rearrangement that takes place within the aquopentacyanide complex provokes the CN_{axial} to bear higher electronic density with respect to the case of the nitroprusside complex. As a consequence, the CN_{axial} ligand becomes more nucleophilic and could displace one Py molecule by attacking the metal center of an adjacent layer, thus, forming a local 3D substructure (step 3 in Scheme 2).

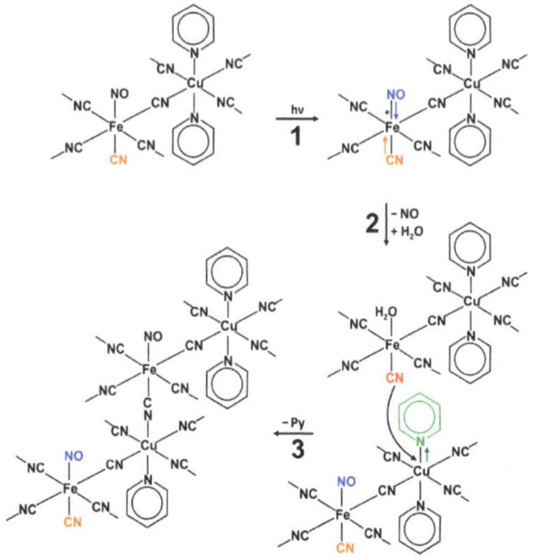

Scheme 2. Photodegradation pathways for the Co, Ni and CuPyNP samples.

Hence, it follows that, as the photolysis events are more frequent, the release of Py molecules from the structure becomes more extensive. Clearly, the extent of the pillar's loss depends on its nucleophilicity. Thus, when Py is substituted by a more nucleophilic derivative, such as 4-methoxypyridine [51], it is apparent that the rate of NO loss increases respect to the rate of molecule loss.

Altogether, several useful conclusions can be drawn about the potential of transition metal nitroprussides to tune the photodegradation response, which are summarized in Table 1 for the studied compounds. For instance, the series of 3D TNP materials demonstrate that there is a clear gap in the photo sensibility as a function of the metal nature. However, the introduction of pillared molecules appears to open up possibilities for finer tuning; in this regard, the variation of other structural properties (the nucleophilicity of the organic ligand) or the medium conditions (the dielectric constant) could also be helpful. Controlling the NO release in photo-sensible systems is crucial in the design of functional materials with biomedical applications [52,53].

Table 1. The photostability of the aqueous dispersions for the studied 3D and 2D transition metal nitroprussides *.

T	T[Fe(CN)$_5$NO]·xH$_2$O (3D)	T(Py)$_2$[Fe(CN)$_5$NO] (2D)
Mn	Low	Low
Fe	Low	Low
Co	High	Medium
Ni	High	High
Cu	High	Medium
Zn	Low	Low
Cd	Medium	-

* Experimental conditions: illuminating source: white-light lamp (8×10^4 lux); sample concentration: 0.5 g/L; reaction volume: 100 mL; reaction time: 4 h. See other details in references [18,19].

4. Role of the Photo-Induced Charge Transfer in the Physical and Functional Properties

The relatively high stability (lifetime $\tau > 10^7$ s below 160 K) for the photo-induced metastable states (MS1 and MS2) together with the GS state (Figure 2) represent three available possible light-driven configurations in the nitroprusside ion to modify the physical and functional properties of materials based on that building block. The existence of such long-lifetime meta-stable states stimulated the study of crystals of sodium, barium and guanidinium nitroprussides as materials for holographic information storage in the past [54].

In addition, the orientation of the Fe-NO group can be tuned through a controlled $2b_2$ (xy) → $7e$ (π*NO) charge transfer within the GS, enlarging such possibilities for nitroprusside-based materials. That last effect is observed, for instance, during the 3D to 2D structural transition, which is probed as a progressive decrease in the frequency of the ν(NO) stretching vibration but without reaching the frequency values corresponding to the meta-stable states [55].

The above-discussed photo-induced magnetic order in nickel nitroprusside is a typical example of light-driven physical properties in metal nitroprusside taking advantage of the MLCT by photon absorption in a certain energy region [35,36]. Such a charge transfer leads to the appearance of an unpaired electron on the iron atom and, at the same time, to a reduction in the electron density on this atom.

This last effect reduces the charge density accumulated at the CN 5σ orbital; however, the Ni(2+) has a sufficiently high polarizing power to participate in a strong superexchange magnetic interaction with the iron atom, thereby, resulting in the reported photo-induced magnetic order at low temperature. The unpaired electrons in the Ni and iron atoms are found in orthogonal orbitals, e_g and t_{2g}, respectively, and the magnetic interaction between them has a ferrimagnetic character (Figure 3A). This is a typical example of a cooperative physical property induced by the sample illumination.

For the Mo analog, $Cs_{1.1}Fe_{0.95}[Mo(CN)_5NO]\cdot 4H_2O$, the photo-switchable ion conduction was documented [56]. In the crystal structure of that material, six molecular blocks $(Mo(CN)_5NO)$ are arranged in such a way that their six unbridged NO groups are oriented toward the center of a cube forming a cavity of ca. 5 Å. One of the water molecules is found coordinated to the iron atom, which behaves as a Lewis acid, inducing the release of a proton from the coordinated water molecule.

The remaining water molecules form a chain of hydrogen-bonding interactions where the coordinated water molecule and the O end of the NO group participate. Such a network of hydrogen-bonding interactions supports fast ionic conduction, 1×10^{-3} Scm^{-1}, for the released proton, via its hopping between water molecules (Grotthuss mechanism), in the non-irradiated sample. When the sample is irradiated with a light of 532 nm, the ionic conductivity decreases to 6×10^{-5} Scm^{-1}, which is accompanied by a decrease in the frequency of the ν(NO) vibration.

Such changes were ascribed to a photo-isomerization of the NO ligand, breaking the chain of hydrogen-bonding interactions and, from this fact, the observed reduction in the ionic conductivity (Figure 3B); once the irradiation ceases, the conductivity progressively recovers the initial value [56]. More recently, the same research group has explored the use of a dysprosium-containing nitroprusside complex as a reversible photo-switchable nonlinear-optical crystal [57]. Upon irradiation at 473 nm, there is a notable increase in the second harmonic generation signal due to the large hyper-polarizability of the –ON+ group belonging to the linkage isomer MS1, which is depleted if the crystal is brought back to its ground state by irradiating at 804 nm.

Recently, the formation of a salt of the nitroprusside ion with polar organic cation, e.g., dimethylammonium ($Me_2NH_2^+$), was evaluated as a route to obtain light-tunable ferroelectric materials [58]. That organic cation has a permanent dipole moment of 2.2 Debye. Such hybrid solids are obtained by evaporation of an aqueous solution of sodium nitroprusside and the organic cation as hydrochloride. From the obtained solid, crystals of (Me$_2$NH$_2$)Na[Fe(CN)$_5$NO] are separated and then submitted to a variable-temperature structural study, in the 298 to 458 K range, complemented with DSC data.

This material undergoes a reversible structural transition that is easily detected in the recorded DSC curves as exothermic/endothermic peaks at ca. 423 and 408 K. These thermal effects and the XRD powder patterns provide a conclusive clue on the occurrence of an order-disorder phase transition. The structure for the phase below the transition temperature (T_C) corresponds to a polar space group $Pna2_1$, where the organic cation behaves as a charge balancing cation occupying the cavities in a framework formed by Na[Fe(CN)$_5$NO] units, interacting with the N ends of the ligands through hydrogen bonds.

Once the material is heated above T_C, the structure changes to a centrosymmetric space group ($Pnam$) related to a certain rotational disorder for the organic cation. Such structural change has relevant implications for the material physical properties. Below T_C, it behaves as a ferroelectric material, and the observed structural change results in a first-order ferroelectric phase transition. Below Tc, all the organic cation dipole moments appear ordered (Figure 3C). According to the recorded IR spectra before and under irradiation with light of 532 nm, in this material, both MS1 and MS2 meta-stable states can be excited.

From this fact, the authors advanced that such a photo-induced effect opens the possibility to obtain multiple-states ferroelectric materials with attractive potential applications. An analog study was reported for (Me$_2$NH$_2$)K[Fe(CN)$_5$NO] with the additional attractive that its authors report a uniaxial positive thermal expansion below T_C but above the transition temperature, the three axes elongate without a notable anisotropy [59]. Considering the photo-isomerization in the nitroprusside building block, its effect on both the ferroelectric state and the atypical thermal expansion is anticipated but not documented.

Recently, 2D ferrous nitroprussides pillared with organic ligands have emerged as a series of hybrid inorganic-organic solids with thermally-induced spin-crossover (SCO) behavior [12,13,60]. Such an unusually large number of compositions with SCO was ascribed to the role of the NO group as an electron buffer to modulate the electron density found at CN 5σ orbital in the equatorial CN ligands [60]. This allows the tuning of the local crystal field sensed by the iron atom to have a 10Dq value appropriate to stabilize the low spin (LS) state ($e_g^0 t_{2g}^6$) on the sample cooling.

If the sample in the LS is irradiated with light of a wavelength appropriate to induce a Fe to NO charge transfer, the electron density in that molecular orbital (5σ) lowers, diminishing the value of 10Dq, and the spin state will be perturbed, favoring the LS to HS (high spin) transition. That effect has not been studied; however, it is expected. In that case, the sample irradiation would be a route to commute between the LS and HS states with consequent potential applications (Figure 3D).

In the interlayer region of 2D transition metal nitroprussides, the NO group is found as a dangling ligand with a negative charge density at its O end. This is an appropriate adsorption site for small quadrupolar molecules, e.g., H$_2$, N$_2$, CO$_2$ and CS$_2$. The presence of such guest molecules in that region can be probed as a certain frequency shift for the ν(CN) stretching band accompanied by a change in its splitting [55]. These effects could be amplified through the sample illumination, within the GS for the NO ligand.

Figure 3 illustrates the main mechanisms involved in the light-driven functional properties related to the photo-induced NO isomerization discussed above. The study of molecular materials and those of hybrid inorganic-organic nature is an emerging research area, and many examples where the nitroprusside ion forms part of these materials could be reported in the near future. In such materials, the photo-induced isomerization of the NO ligand appears like a route for tuning their physical and functional properties.

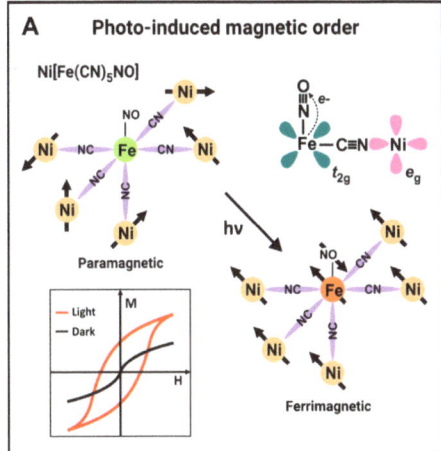
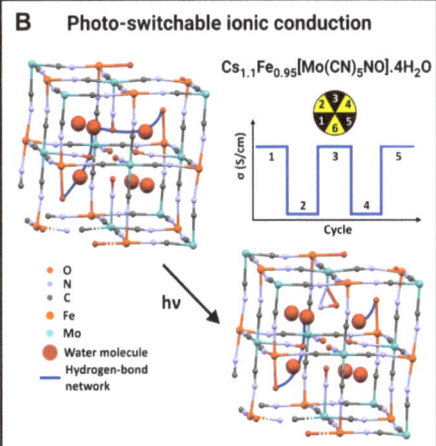
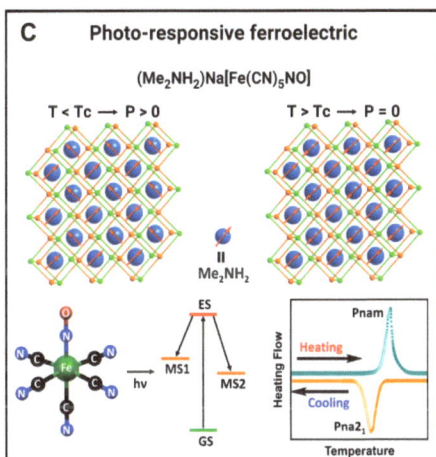
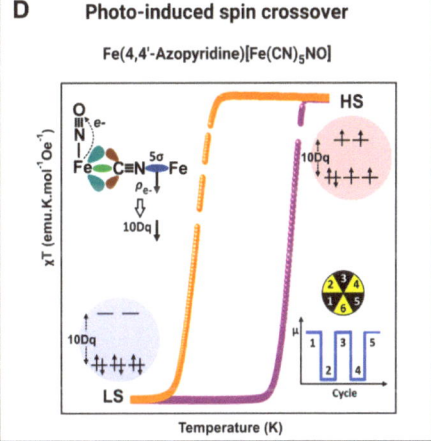

Figure 3. Photo-driven functional properties of materials based on the nitroprusside ion. Images prepared from the reported results in: (**A**) Ref. [36], (**B**) Ref. [56], (**C**) Ref. [58] and (**D**) Ref. [60].

5. Concluding Remarks

Nitroprussides exhibit interesting physical and functional properties—particularly the coordination polymers formed with transition metal ions. Such properties have been recently summarized and include their potential application for small molecule separation and storage, battery materials, sensors, actuators, information storage, optical materials, photo-catalysis, healthcare, etc. In the presence of water molecules, including those found in their framework as structural or absorbed water, we discussed the light absorption and induced photo-isomerization results in the liberation of the photo-active nitrosyl group with the consequent material degradation.

In this paper, the understanding of the degradation mechanisms in the presence of light and water received particular attention. In the absence of water molecules, the photo-isomerization of the NO group appears as a reversible process that can be used to modulate the physical and functional properties of metal nitroprussides. The photo-chemistry of the nitroprusside ion in other solvents remains poorly documented, and considering the potential application of materials based on this building block, this is a subject that deserves to be considered in further studies. Molecular and hybrid inorganic–organic solids are opening new opportunities to prepare materials where different physical properties are combined to obtain novel functional materials.

This will be an active research area for metal nitroprussides in the future for applications in spintronic, photo-optical devices, sensors and actuators and smart materials with the ability to respond to external agents as observed in solids with a spin-crossover effect. In this contribution, some of these potentialities are briefly summarized, and others are anticipated. This feature paper was prepared with the intention of helping to understand the photochemistry of nitroprusside-based materials and how to control the favorable and unfavorable effects.

Author Contributions: P.M.C., O.F.O. and E.R. had a similar level of contribution to its preparation, including the design, writing, artwork preparation and revision. All authors have read and agreed to the published version of the manuscript.

Funding: This research received no external funding.

Acknowledgments: The authors thank LNCAE (Laboratorio NaciAlmacenamiento de Energía) for the access to its experimental facility. The preparation of this contribution was partially supported by the project SECITI/185/2021.

Conflicts of Interest: The authors have no conflict of interest. The information contained in this manuscript has a basic character and concerns new scientific knowledge. The data provided by the authors through this manuscript are available on request.

References

1. Playfair, L., XXXIII. On the nitroprussides, a new class of salts. *London Edinb. Dublin Philos. Mag. J. Sci.* **1850**, *36*, 271–283. [CrossRef]
2. Manoharan, P.T.; Hamilton, W.C. The Crystal Structure of Sodium Nitroprusside. *Inorg. Chem.* **1963**, *2*, 1043–1047. [CrossRef]
3. Enemark, J.; Feltham, R. Principles of structure, bonding, and reactivity for metal nitrosyl complexes. *Coord. Chem. Rev.* **1974**, *13*, 339–406. [CrossRef]
4. Punte, G.; Rigotti, G.; Rivero, B.E.; Podjarny, A.D.; Castellano, E.E. Structure of calcium nitroprusside tetrahydrate. *Acta Crystallogr. Sect. B* **1980**, *36*, 1472–1475. [CrossRef]
5. Mullica, D.F.; Tippin, D.B.; Sappenfield, E.L. Synthesis, Spectroscopic Studies and X-Ray Crystal Structure Analysis of Cobalt Nitroprusside, Co[Fe(CN)5NO]·5H2O. *J. Coord. Chem.* **1991**, *24*, 83–91. [CrossRef]
6. Balmaseda, J.; Reguera, E.; Gomez, A.; Roque, J.; Vazquez, C.; Autie, M. On the Microporous Nature of Transition Metal Nitroprussides. *J. Phys. Chem. B* **2003**, *107*, 11360–11369. [CrossRef]
7. Rodríguez-Hernández, J.; Reguera, L.; Lemus-Santana, A. Silver nitroprusside: Atypical coordination within the metal nitroprussides series. *Inorg. Chim. Acta* **2015**, *428*, 51–56. [CrossRef]
8. Rusanov, V.; Stankov, S.; Ahmedova, A.; Trautwein, A. Determination of the Mössbauer parameters of rare-earth nitroprussides: Evidence for new light-induced magnetic excited state (LIMES) in nitroprussides. *J. Solid State Chem.* **2009**, *182*, 1252–1259. [CrossRef]
9. Reguera, E.; Gómez, A.; Hernández, J.R. Unique coordination in metal nitroprussides: The structure of Cu[Fe(CN)5NO]?2H2O and Cu[Fe(CN)5NO]. *J. Chem. Crystallogr.* **2004**, *34*, 893–903. [CrossRef]
10. Avila, Y.; Rodríguez-Hernández, J.; Crespo, P.M.; González, M.; Reguera, E. 2D ferrous nitroprussides stabilized through organic molecules as pillars: Preparation, crystal structure and related properties. *J. Coord. Chem.* **2021**, *74*, 695–713. [CrossRef]
11. Ávila, Y.; Osiry, H.; Plasencia, Y.; Torres, A.E.; González, M.; Lemus-Santana, A.A.; Reguera, E.; Plasencia, Y.; Torres, E.A. From 3D to 2D Transition Metal Nitroprussides by Selective Rupture of Axial Bonds. *Chem. Eur. J.* **2019**, *25*, 11327–11336. [CrossRef] [PubMed]
12. Avila, Y.; Crespo, P.; Plasencia, Y.; Mojica, H.; Hernández, J.R.; Reguera, E. Intercalation of 3X-pyridine with X = F, Cl, Br, I, in 2D ferrous nitroprusside. Thermal induced spin transition in Fe(3F-pyridine)2[Fe(CN)5NO]. *J. Solid State Chem.* **2020**, *286*, 121293. [CrossRef]
13. Plasencia, Y.; Avila, Y.; Rodríguez-Hernández, J.; Ávila, M.; Reguera, E. Thermally induced spin transition in Fe(pyrazine)[Fe(CN)5NO]. *J. Phys. Chem. Solids* **2021**, *150*, 109843. [CrossRef]
14. Reguera, L.; Avila, Y.; Reguera, E. Transition metal nitroprussides: Crystal and electronic structure, and related properties. *Coord. Chem. Rev.* **2021**, *434*, 213764. [CrossRef]
15. Butler, A.R.; Glidewell, C. Recent chemical studies of sodium nitroprusside relevant to its hypotensive action. *Chem. Soc. Rev.* **1987**, *16*, 361–380. [CrossRef]

16. Böer, K.W.; Pohl, U.W. *Band-to-Band Transitions*; Böer, K.W., Pohl, U., Eds.; Semiconductor Physics, Springer International Publishing: Cham, Switzerland, 2014; pp. 1–29.
17. Vogler, A. Photoreactivity of metal-to-ligand charge transfer excited states. *Coord. Chem. Rev.* **1998**, *177*, 81–96. [CrossRef]
18. Crespo, P.; Odio, O.; Ávila, Y.; Perez-Cappe, E.; Reguera, E. Effect of water and light on the stability of transition metal nitroprussides. *J. Photochem. Photobiol. A Chem.* **2021**, *412*, 113244. [CrossRef]
19. Crespo, P.M.; Odio, O.F.; Ávila, Y.; Reguera, E. Effect of water and light on the stability of Pyridine pillared 2D transition metal nitroprussides. *J. Photochem. Photobiol. A Chem.* **2022**. submitted. [CrossRef]
20. Swinehart, J. The nitroprusside ion. *Coord. Chem. Rev.* **1967**, *2*, 385–402. [CrossRef]
21. Manoharan, P.T.; Gray, H.B. Electronic Structure of Nitroprusside Ion. *J. Am. Chem. Soc.* **1965**, *87*, 3340–3348. [CrossRef]
22. Cano, A.; Lartundo-Rojas, L.; Shchukarev, A.; Reguera, E. Contribution to the coordination chemistry of transition metal nitroprussides: A cryo-XPS study. *New J. Chem.* **2019**, *43*, 4835–4848. [CrossRef]
23. Inoue, H.; Yanagisawa, S. Bonding nature of coordination polymers, KM [Co (CN) 6]. *Keio Eng. Rep.* **1972**, *25*, 1–11.
24. Hauser, U.; Oestreich, V.; Rohrweck, H.D. On optical dispersion in transparent molecular systems. *Z. Für Phys. A At. Nucl.* **1977**, *280*, 17–25. [CrossRef]
25. Hauser, U.; Oestreich, V.; Rohrweck, H.D. On optical dispersion in transparent molecular systems. *Z. Für Phys. A At. Nucl.* **1978**, *284*, 9–19. [CrossRef]
26. Zöllner, H.; Krasser, W.; Woike, T.; Haussühl, S. The existence of light-induced long-lived metastable states in different $X_n[Fe(CN)_5NO]\cdot yH_2O$ crystals, powders and solutions. *Chem. Phys. Lett.* **1989**, *161*, 497–501. [CrossRef]
27. Pressprich, M.R.; White, M.A.; Vekhter, Y.; Coppens, P. Analysis of a metastable electronic excited state of sodium nitroprusside by X-ray crystallography. *J. Am. Chem. Soc.* **1994**, *116*, 5233–5238. [CrossRef]
28. Morioka, Y.; Takeda, S.; Tomizawa, H.; Miki, E.-I. Molecular vibrations and structure of the light-induced metastable state of $[Fe(CN)_5NO]^{2-}$. *Chem. Phys. Lett.* **1998**, *292*, 625–630. [CrossRef]
29. Delley, B.; Schefer, J.; Woike, T. Giant lifetimes of optically excited states and the elusive structure of sodiumnitroprusside. *J. Chem. Phys.* **1997**, *107*, 10067–10074. [CrossRef]
30. Woike, T.; Kirchner, W.; Kim, H.-S.; Haussühl, S.; Rusanov, V.; Angelov, V.; Ormandjiev, S.; Bonchev, T.; Schroeder, A.N.F. Mössbauer parameters of the two long-lived metastable states in $Na_2[Fe(CN)_5NO]\cdot 2H_2O$ single crystals. *Hyperfine Interact.* **1993**, *77*, 265–275. [CrossRef]
31. Boulet, P.; Buchs, M.; Chermette, H.; Daul, C.; Gilardoni, F.; Rogemond, F.; Schläpfer, A.C.W.; Weber, J. DFT Investigation of Metal Complexes Containing a Nitrosyl Ligand. 1. Ground State and Metastable States. *J. Phys. Chem. A* **2001**, *105*, 8991–8998. [CrossRef]
32. Coppens, P.; Novozhilova, I.; Kovalevsky, A. Photoinduced Linkage Isomers of Transition-Metal Nitrosyl Compounds and Related Complexes. *Chem. Rev.* **2002**, *102*, 861–884. [CrossRef] [PubMed]
33. Güdel, H.U. Comment on the nature of the light-induced metastable states in nitroprussides. *Chem. Phys. Lett.* **1990**, *175*, 262–266. [CrossRef]
34. Schefer, J.; Chevrier, G.; Furer, N.; Heger, G.; Schweiss, P.; Vogt, T.; Woike, T. Light-induced structural changes in sodiumnitroprusside $(Na_2(Fe(CN)_5NO)\cdot 2D_2O)$ at 80 K. *Z. Für Phys. B Condens. Matter* **1991**, *83*, 125–130. [CrossRef]
35. Gu, Z.-Z.; Sato, O.; Iyoda, T.; Hashimoto, K.; Fujishima, A. Molecular-Level Design of a Photoinduced Magnetic Spin Coupling System: Nickel Nitroprusside. *J. Phys. Chem.* **1996**, *100*, 18289–18291. [CrossRef]
36. Gu, Z.-Z.; Sato, O.; Iyoda, T.; Hashimoto, K.; Fujishima, A. Spin Switching Effect in Nickel Nitroprusside: Design of a Molecular Spin Device Based on Spin Exchange Interaction. *Chem. Mater.* **1997**, *9*, 1092–1097. [CrossRef]
37. Lynch, M.S.; Cheng, M.; Van Kuiken, B.E.; Khalil, M. Probing the Photoinduced Metal−Nitrosyl Linkage Isomerism of Sodium Nitroprusside in Solution Using Transient Infrared Spectroscopy. *J. Am. Chem. Soc.* **2011**, *133*, 5255–5262. [CrossRef]
38. Kopotkov, V.A.; Sasnovskaya, V.D.; Korchagin, D.V.; Morgunov, R.B.; Aldoshin, S.M.; Simonov, S.V.; Zorina, L.V.; Schaniel, D.; Woike, T.; Yagubskii, E.B. The first photochromic bimetallic assemblies based on Mn(iii) and Mn(ii) Schiff-base (salpn, dapsc) complexes and pentacyanonitrosylferrate. *CrystEngComm* **2015**, *17*, 3866–3876. [CrossRef]
39. Mitra, R.; Jain, D.; Banerjee, A.; Chari, K. Photolysis of sodium nitroprusside and nitroprussic acid. *J. Inorg. Nucl. Chem.* **1963**, *25*, 1263–1266. [CrossRef]
40. Wolfe, S.K.; Swinehart, J.H. Photochemistry of pentacyanonitrosylferrate(2-), nitroprusside. *Inorg. Chem.* **1975**, *14*, 1049–1053. [CrossRef]
41. Stochel, G.; Stasicka, Z. Photoredox chemistry of nitrosylpentacyanoferrate(II) in methanolic medium. *Polyhedron* **1985**, *4*, 1887–1890. [CrossRef]
42. de Oliveira, M.G.; Langley, G.J.; Rest, A.J. Photolysis of the $[Fe(CN)_5(NO)]^{2-}$ ion in water and poly(vinyl alcohol) films: Evidence for cyano radical, cyanide ion and nitric oxide loss and redox pathways. *J. Chem. Soc. Dalton Trans.* **1995**, *12*, 2013–2019. [CrossRef]
43. Stochel, G.; Stasicka, Z. Solvent complexes of the type $[Fe^{III}(CN)_5L]^{n-}$. *Polyhedron* **1985**, *4*, 481–484. [CrossRef]
44. Moggi, L.; Bolletta, F.; Balzani, V.; Scandola, F. Photochemistry of co-ordination compounds—XV: Cyanide complexes. *J. Inorg. Nucl. Chem.* **1966**, *28*, 2589–2597. [CrossRef]
45. Goldstein, S.; Czapski, G. Kinetics of Nitric Oxide Autoxidation in Aqueous Solution in the Absence and Presence of Various Reductants. The Nature of the Oxidizing Intermediates. *J. Am. Chem. Soc.* **1995**, *117*, 12078–12084. [CrossRef]

46. Videla, M.; Braslavsky, S.E. The photorelease of nitrogen monoxide (NO) from pentacyanonitrosyl coordination compounds of group 8 metals. *Photochem. Photobiol. Sci.* **2005**, *4*, 75–82. [CrossRef]
47. Dieckmann, V.; Imlau, M.; Taffa, D.H.; Walder, L.; Lepski, R.; Schaniel, D.; Woike, T. Phototriggered NO and CN release from [Fe(CN)5NO]2− molecules electrostatically attached to TiO2 surfaces. *Phys. Chem. Chem. Phys.* **2010**, *12*, 3283–3288. [CrossRef]
48. Morioka, Y.; Hisamitsu, T.-A.; Inoue, H.; Yoshioka, N.; Tomizawa, H.; Miki, E.-I. Light-Induced Mixed-Valence State of $Fe^{II}[Fe(CN)_5NO]\cdot xH_2O$. *Bull. Chem. Soc. Jpn.* **1998**, *71*, 837–844. [CrossRef]
49. Schaniel, D.; Schefer, J.; Imlau, M.; Woike, T. Light-induced structural changes by excitation of metastable states in $Na_2[Fe(CN)_5NO]2H_2O$ single crystals. *Phys. Rev. B* **2003**, *68*, 104108. [CrossRef]
50. Buchs, M.; Daul, C.A.; Manoharan, P.T.; Schläpfer, C.W. Density functional study of nitroprusside: Mechanism of the photochemical formation and deactivation of the metastable states. *Int. J. Quantum Chem.* **2003**, *91*, 418–431. [CrossRef]
51. Deka, K.; Phukan, P. DFT analysis of the nucleophilicity of substituted pyridines and prediction of new molecules having nucleophilic character stronger than 4-pyrrolidino pyridine. *J. Chem. Sci.* **2016**, *128*, 633–647. [CrossRef]
52. Yoon, H.; Park, S.; Lim, M. Photorelease Dynamics of Nitric Oxide from Cysteine-Bound Roussin's Red Ester. *J. Phys. Chem. Lett.* **2020**, *11*, 3198–3202. [CrossRef] [PubMed]
53. Stepanenko, I.; Zalibera, M.; Schaniel, D.; Telser, J.; Arion, V.B. Ruthenium-nitrosyl complexes as NO-releasing molecules, potential anticancer drugs, and photoswitches based on linkage isomerism. *Dalton Trans.* **2022**, *51*, 5367–5393. [CrossRef] [PubMed]
54. Imlau, M.; Woike, T.; Schaniel, D.; Schefer, J.; Fally, M.; Rupp, R.A. Light-induced extinction originating from holographic scattering. *Opt. Lett.* **2002**, *27*, 2185–2187. [CrossRef] [PubMed]
55. Divó-Matos, Y.; Avila, Y.; Mojica, R.; Reguera, E. Nature of the observed ν(NO) band shift and splitting during the 3D to 2D structural change in transition metal nitroprussides. *Spectrochim. Acta Part A Mol. Biomol. Spectrosc.* **2022**, *276*, 121210. [CrossRef] [PubMed]
56. Ohkoshi, S.-I.; Nakagawa, K.; Imoto, K.; Tokoro, H.; Shibata, Y.; Okamoto, K.; Miyamoto, Y.; Komine, M.; Yoshikiyo, M.; Namai, A. A photoswitchable polar crystal that exhibits superionic conduction. *Nat. Chem.* **2020**, *12*, 338–344. [CrossRef]
57. Komine, M.; Imoto, K.; Namai, A.; Yoshikiyo, M.; Ohkoshi, S.-I. Photoswitchable Nonlinear-Optical Crystal Based on a Dysprosium–Iron Nitrosyl Metal Assembly. *Inorg. Chem.* **2021**, *60*, 2097–2104. [CrossRef]
58. Xu, W.-J.; Romanyuk, K.; Martinho, J.M.G.; Zeng, Y.; Zhang, X.-W.; Ushakov, A.; Shur, V.; Zhang, W.-X.; Chen, X.-M.; Kholkin, A.; et al. Photoresponsive Organic–Inorganic Hybrid Ferroelectric Designed at the Molecular Level. *J. Am. Chem. Soc.* **2020**, *142*, 16990–16998. [CrossRef]
59. Qiu, R.-G.; Chen, X.-M.; Huang, R.-K.; Zhou, D.-D.; Xu, W.-J.; Zhang, W.-X. Nitroprusside as a promising building block to assemble an organic–inorganic hybrid for thermo-responsive switching materials. *Chem. Commun.* **2020**, *56*, 5488–5491. [CrossRef]
60. Avila, Y.; Scanda, K.; Mojica, R.; Rodríguez-Hernández, J.; Cruz-Santiago, L.; González, M.; Reguera, E. Thermally-induced spin transition in $Fe(4,4'-Azopyridine)[Fe(CN)_5NO]$. *J. Solid State Chem.* **2022**, *310*, 123054. [CrossRef]

Article

Infrared Spectrum and UV-Induced Photochemistry of Matrix-Isolated Phenyl 1-Hydroxy-2-Naphthoate

İsa Sıdır [1,2,*], Sándor Góbi [2,3], Yadigar Gülseven Sıdır [1,2] and Rui Fausto [2,*]

1. Department of Physics, Faculty of Science and Arts, Bitlis Eren University, 13000 Bitlis, Turkey; yadigar.gulseven@gmail.com
2. CQC, Department of Chemistry, University of Coimbra, 3004-535 Coimbra, Portugal; sandor.gobi@gmail.com
3. MTA-ELTE Lendület Laboratory Astrochemistry Research Group, Institute of Chemistry, ELTE Eötvös Loránd University, H-1518 Budapest, Hungary
* Correspondence: isidir@beu.edu.tr (İ.S.); rfausto@ci.uc.pt (R.F.)

Abstract: The conformational stability, infrared spectrum, and photochemistry of phenyl 1-hydroxy-2-naphthoate (PHN) were studied by matrix isolation infrared spectroscopy and theoretical computations performed at the DFT(B3LYP)/6-311++G(d,p) level of theory. The main intramolecular interactions determining the relative stability of seven conformers of the molecule were evaluated. According to the calculations, the twofold degenerated O–H···O=C intramolecularly hydrogen-bonded conformer with the phenyl ring ester group ±68.8° out of the plane of the substituted naphtyl moiety is the most stable conformer of the molecule. This conformer is considerably more stable than the second most stable form (by ~15 kJ mol^{-1}), in which a weaker O–H···O–C intramolecular hydrogen bond exists. The compound was isolated in cryogenic argon and N_2 matrices, and the conformational composition in the matrices was investigated by infrared spectroscopy. In agreement with the predicted relative energies of the conformers, the analysis of the spectra indicated that only the most stable conformer of PHN was present in the as-deposited matrices. The matrices were then irradiated at various wavelengths by narrowband tunable UV light within the 331.7–235.0 nm wavelength range. This resulted in the photodecarbonylation of PHN, yielding 2-phenoxynaphthalen-1-ol, together with CO. The extension of the decarbonylation was found to depend on the excitation wavelength.

Keywords: phenyl 1-hydroxy-2-naphthoate; conformational space; infrared spectra in argon and N_2 matrices; photodecarbonylation; 2-phenoxynaphthalen-1-ol

Citation: Sıdır, İ.; Góbi, S.; Gülseven Sıdır, Y.; Fausto, R. Infrared Spectrum and UV-Induced Photochemistry of Matrix-Isolated Phenyl 1-Hydroxy-2-Naphthoate. *Photochem* 2021, 1, 10–25. https://doi.org/10.3390/photochem1010002

Academic Editor: Marcelo I. Guzman

Received: 15 January 2021
Accepted: 21 February 2021
Published: 26 February 2021

Publisher's Note: MDPI stays neutral with regard to jurisdictional claims in published maps and institutional affiliations.

Copyright: © 2021 by the authors. Licensee MDPI, Basel, Switzerland. This article is an open access article distributed under the terms and conditions of the Creative Commons Attribution (CC BY) license (https://creativecommons.org/licenses/by/4.0/).

1. Introduction

Hydroxynaphthoate derivatives show important biological and medicinal properties. For example, methyl 1-hydroxy-2-naphthoate and ethyl 1,6-dihydroxy-2-naphthoate present anti-inflammatory activity [1,2], while 1-hydroxy-2-naphthoic acid itself has a recognized antibacterial action [3]. The 1-hydroxy-2-naphthoate moiety is observed in natural products such as the cytotoxic compounds 3-hydroxymollugin and 3-methoxymollugin [4]. In addition, hydroxynaphthoates have been used as precursors for the synthesis of the anti-carcinogenic compounds taiwanin C [5,6], α- and β-sorigenin methyl ethers [7], and olivin trimethyl ether [8].

Woolfe and Thistlethwaite [9,10] have reported on the photophysics of methyl 3-hydroxy-2-naphthoate and phenyl 1-hydroxy-2-naphthoate. They observed that the first compound gives rise to a dual fluorescence emission, with one component exhibiting a large Stokes shift, in contrast to the second compound, which exhibits normal fluorescence [9,10]. Law and Shoham have also investigated the photophysics of methyl 3-hydroxy-2-naphthoate in different solvent media and at various temperatures, confirming the previous observations regarding its unusual fluorescence pattern, which they interpreted as being the result of a photoinduced excited-state intramolecular proton transfer (ESIPT), leading to the formation of the enol tautomer of the compound that is responsible for the Stokes-shifted emission [11]. The formation of the enol is facilitated by the presence in the ground state of a strong intramolecular hydrogen bond.

Evidence on the existence of intramolecular hydrogen bonding in methyl 2-hydroxy-3-naphthoate has been reported by Bergmann and co-workers [12]. Similarly to methyl 3-hydroxy-2-naphthoate, methyl 2-hydroxy-3-naphthoate exhibits a dual fluorescence emission, which was also explained assuming the occurrence of enolization via ESIPT upon excitation [13]. The ESIPT mechanism in jet-cooled methyl 2-hydroxy-3-naphthoate has been investigated comprehensively, more recently, by McCarthy and Ruth [14]. On the other hand, the lack of ESIPT emission for both methyl 1-hydroxy-2-naphthoate and methyl 2-hydroxy-1-naphthoate has been explained by Catalán et al. [15] in terms of the rapid non-radiative dynamics of their respective lowest-energy keto tautomers. Those authors concluded that the relative position of the intramolecular hydrogen bond in the naphthalene skeleton in this type of compound acts as a switch, and controls the yield of the ESIPT process [15]. These results are also in agreement with the work of Tobita and co-workers on methyl 1-hydroxy-2-naphthoate, in which the fluorescence characteristics of this compound were compared with those exhibited by 1-hydroxy-2-acetonaphthone and 1-hydroxy-2-naphthaldehyde [16]. The occurrence of ESIPT has also been observed for phenyl 1-hydroxy-2-naphthoate (PHN) in a previous study by some of the present authors [17], in which the solvatochromism, intramolecular hydrogen bonding, and ground and excited state dipole moments of the compound were investigated using fluorescence spectroscopy in solutions of different solvents.

In the present study, we focused on the ground-state conformational preferences exhibited by PHN, its infrared spectrum, and UV-induced photochemistry under matrix isolation (Ar and N_2 matrices) conditions. The interpretation of the experimental results received support from structural and spectroscopic data obtained from electronic structure calculations undertaken within the density functional theory (DFT) framework. As described in detail below, intramolecular hydrogen bonding plays a major structural role in PHN. Conformational photo-isomerization, with production of a high-energy conformer of the compound, and photo-induced decarbonylation of PHN were found experimentally to be the major processes taking place upon in situ narrowband UV excitation of the matrix-isolated compound within the 331.7–235.0 nm range, with the extension of the conformational isomerization reaction depending on the excitation wavelength.

2. Experimental Details: Matrix Isolation Infrared and Photochemical Experiments

A commercial sample of PHN was provided by Sigma-Aldrich, St. Louis, MO, USA (powder; 99% purity), and was used without any further purification. In order to prepare the cryogenic matrices, the sample was placed in a mini glass oven attached to the vacuum chamber of the cryostat, and was sublimated using a thermoelectrical heating system, whose main component is a DC Power Supply VITEECOM, model 75-HY5003. The vapor of the compound was co-deposited with a large excess of the host matrix gas (Ar or N_2, obtained from Air Liquide) onto a cold CsI window assembled at the tip of the cryostat, and cooled to 10 K by an APD Cryogenics DE202 closed-cycle refrigerator system. The solute to matrix ratios were kept at ~1:1000 in order to guarantee the adequate isolation of the compound. The temperature of the matrix sample was measured at the sample holder by using a LakeShore Model 331S Temperature Controller, with 0.1 K accuracy.

The infrared spectra were collected in the 4000–400 cm^{-1} range, with 0.5 cm^{-1} resolution, using a Nicolet 6700 FTIR spectrometer (Waltham, MA, USA) equipped with a mercury cadmium telluride (MCT) detector and a KBr beam splitter. The instrument was purged by a stream of dry/CO_2-filtered air in order to avoid interference from atmospheric H_2O and CO_2 vapors.

The photoexcitation of the matrix-isolated PHN was performed through the outer KBr window of the cryostat using tunable narrowband UV light (full-width at half-maximum of 0.2 cm^{-1}) provided by a Spectra Physics MOPO-SL optical parametric oscillator (OPO) pumped by a pulsed (pulse duration 10 ns, repetition rate 10 Hz) Quanta Ray Pro-Series

Nd-YAG laser. The excitations were undertaken within the 331.7–235.0 nm wavelength range, and were performed with different pulse energies and exposure times (see Table S1 in the Supplementary Materials for details).

3. Computational Details

All quantum chemical computations were performed with Gaussian 09 (revision A.02) [18]. The geometries were fully optimized at the DFT level of theory using the B3LYP functional [19–21] and the 6-311++G(d,p) basis set [22]. The harmonic vibrational frequencies and infrared intensities were calculated at the same level of theory, the frequencies being subsequently scaled by a factor of 0.978 in order to approximately correct them for the neglected anharmonicity and method/basis set limitations. The assignment of the vibrational modes was performed with the help of the animation module of GaussView (version 5.0) [23]. In the simulated spectra, the infrared bands were built using Lorentzian functions centered at the calculated (scaled) frequencies, and with a full-width at half-maximum (fwhm) of 2 cm^{-1}.

4. Results and Discussion

4.1. Geometries and Relative Energies of the PHN Conformers

The relative energies of the conformers of PHN, calculated at the B3LYP/6-311++G(d,p) level, are presented in Table 1. The optimized structures of the various conformers are represented in Figure 1, while their Cartesian coordinates are provided in Table S2 (Supporting Information). Table 2 gives the optimized values for the conformationally-relevant dihedral angles of the different conformers of PHN.

Table 1. Calculated relative electronic energies (ΔE), zero-point corrected electronic energies ($\Delta E_{(0)}$), and standard Gibbs energies ($\Delta G°$; 298.15 K; 1 atm) for the different conformers of PHN, dipole moments (μ), and estimated gas phase equilibrium populations at 298.15 K ($p_{(298.15)}$) [a].

	1	2	3	4	5	6	7
ΔE	0.00	14.75	33.13	57.45	55.98	81.46	77.03
$\Delta E_{(0)}$	0.00	14.41	32.35	54.14	53.04	77.36	73.15
$\Delta G°$	0.00	13.02	33.53	49.65	49.48	75.72	73.19
$p_{(298.15)}$	99.5	0.5	0.0	0.0	0.0	0.0	0.0
μ	1.75	1.25	3.80	3.41	3.89	5.55	6.17

[a] Relative energies in kJ mol^{-1}; dipole moments in Debye. Populations in % were estimated using the Boltzmann equation. The calculated absolute values for E, $E_{(0)}$ and $G°$ for conformer **1** are: −880.9763632, −880.730014, and −880.774689 hartrees, respectively.

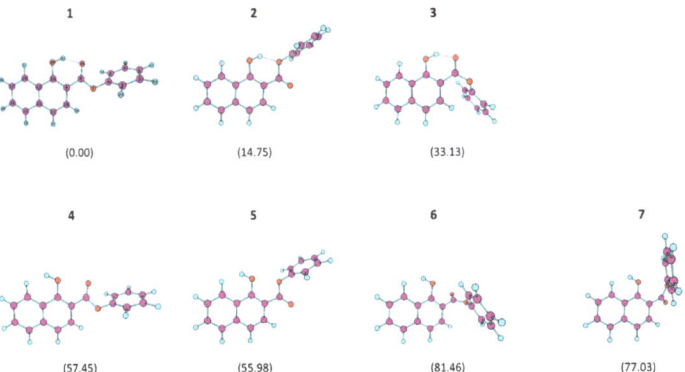

Figure 1. Optimized structures of the PHN conformers. The relative electronic energies are given in parentheses (in kJ mol^{-1}). See Tables 1 and 2 for the details concerning the energies and conformationally-relevant dihedral angles. The full Cartesian coordinates are provided in Table S2.

Table 2. Optimized conformationally-relevant dihedral angles of the different conformers of PHN [a].

Dihedral Angle [b]	1	2	3	4	5	6	7
C–C–O–H	−0.1	0.1	−3.4	−172.5	−171.1	164.3	−170.6
O=C–C–C	−0.2	−179.8	6.6	−10.8	162.5	−59.6	−132.9
C–O–C=O	−0.7	0.7	−118.2	−2.1	−3.3	157.1	−153.1
C–C–O–C	−68.8	76.6	25.7	−63.7	−57.5	−52.3	44.6

[a] Angles in degrees. The indicated values correspond to the form shown in Figure 1; for all of the conformers, there is a symmetry-related minimum where the dihedral angles assume the symmetric values. [b] The carbon atoms in bold correspond to the two ring carbon atoms where the substituents are bound; the C–C–O–C dihedral corresponds to the smaller dihedral made by the phenyl ring of the ester group relative to the plane of the carboxylate moiety.

The molecule of PHN has four a priori conformationally-relevant internal rotation axes, which can be defined as a function of the C–C–O–H, O=C–C–C, C–O–C=O and C–C–O–C dihedrals provided in Table 2. The first and second dihedrals determine the relative orientation of the OH and phenylcarboxy substituents, while the third and fourth dihedrals are related to the conformation assumed by the phenyl ester group. One can group the seven different conformers of PHN into two groups, the first comprising conformers **1–3**, which bear an O–H···O intramolecular hydrogen bond and are of lower energy, and the second including the remaining four conformers (**4–7**), in which no intramolecular hydrogen bond exists, and that are of higher energy. All conformers are doubly-degenerated by symmetry, so that all structures shown in Figure 1—the dihedrals of which are specified in Table 2—have symmetry-related equivalent forms in which all of the dihedral angles have symmetric values. For simplicity, in the discussion below, we will only refer to the structures depicted in the Figure.

In the first group of conformers, the two lower energy forms (**1** and **2**) have the ester group in the intrinsically more stable *cis* conformation (C–O–C=O dihedral equal to ~0°) [24,25], whereas conformer **3** has this group in a distorted *trans* geometry where the C–O–C=O dihedral is −118.2° (the *trans* arrangement corresponds to a C–O–C=O dihedral equal to ~180°). The reasons for the higher stability of the *cis* ester arrangement have been pointed out in detail elsewhere [24]. Essentially, the main factor responsible for the stabilization of a *cis* carboxylic ester arrangement in relation to the *trans* one is electrostatic in nature, and results from the different relative orientation of the dipoles associated with the C=O and O–C (ester) bonds in the two types of geometries. In the *cis* geometry, this alignment is nearly anti-parallel (attractive, stabilizing), while in the *trans* it is approximately parallel (repulsive, destabilizing). In conformer **3**, the distortion relative to the *trans* conformation results from the strong steric repulsion in that geometry between the phenyl and naphtol moieties (see Figure 1). Conformer **1**, in which a strong intramolecular hydrogen bond is established with the carbonyl oxygen atom, is the most stable conformer of PHN, with conformer **2**, where a weaker intramolecular hydrogen bond is established with the ester oxygen atom, being higher in energy than form **1** by 14.75 kJ mol^{-1}. In all of the conformers **1–3**, the phenyl ester group is considerably out of the carboxylic plane in order to minimize repulsions within the carboxylic ester group (for **1** and **2**) and with the naphtol moiety (in the case of **3**). Conformer **3** is higher in energy than the most stable PHN conformer by 33.13 kJ mol^{-1}.

The four higher-energy non-hydrogen bonded conformers (**4–7**) have relative energies from ca. 55 to about 81 kJ mol^{-1} above conformer **1**. In all these conformers, the dominant intramolecular interactions are of repulsive type, and are established between the lone electron pairs of the hydroxylic and the carbonylic or carboxylic ester oxygen atoms. In the case of the highest energy conformers **6** and **7**, to these repulsions one must also add the extra energy resulting from the unfavorable *trans* arrangement of the ester group and the phenyl-naphtol repulsions (in the case of **6**). As for the lower energy conformers, the high-energy conformers of PHN also have the phenyl ester group considerably out of the carboxylic plane in order to minimize repulsions.

It is interesting to examine in more detail the intramolecular hydrogen bonds present in conformers **1–3**. Table 3 presents some structural parameters (bond distances and angles) that are useful to estimate the different strengths of the hydrogen bonds in these conformers. Some of these structural parameters can also be correlated with the different stability of the *cis* and *trans* ester arrangements. As seen in the Table, the strongest hydrogen bond occurs in conformer **3**, as measured by the values of the H···O and O···O distances and, to a less extent, by the O–H bond lengths found in the three conformers under consideration. The hydrogen bond in conformer **1** is slightly less strong than that present in conformer **3**, while that present in conformer **2** (which is established with the ester oxygen atom as acceptor instead of the carbonylic oxygen atom as in **1** and **3**) is much weaker, in particular if one takes into account the smaller value for the O–H···O angle and the much longer H···O distance (see Table 3). If we assume that the energy difference between **1** and **2** is mostly due to the different strengths of the intramolecular hydrogen bonds present in these two forms, we can estimate that the one that is present in the most stable conformer is about 10–15 kJ mol^{-1} stronger than that which is present in conformer **2**. On the other hand, assuming that the hydrogen bonds in **1** and **3** do not differ by more than 3–5 kJ mol^{-1} in energy, the (*trans*-ester)–(*cis*-ester) intrinsic energy difference can be roughly estimated as being of ca. 30 kJ mol^{-1} in PHN, a value that is similar to those found in other carboxylic esters [24,25].

Table 3. Selected optimized geometric parameters (distances and angles) for the intramolecularly hydrogen-bonded conformers **1–3** of PHN [a].

Conformer	C–O (Naphtol)	O–H	H···O	C=O	C–O (Carboxylic)	O···O	<(O–H···O)
1	1.339	0.986	1.715	1.224	1.358	2.593	146.2
2	1.347	0.973	1.751	1.201	1.397	2.597	143.2
3	1.336	0.987	1.700	1.221	1.377	2.579	146.2

[a] Distances in Å; angles in degrees.

The relative values of the C–O (naphtol) and O–H bond lengths and the H···O distances correlate with the relative strength of the intramolecular hydrogen bonds in the three conformers, the first and the last becoming shorter and the second becoming longer when the hydrogen bond becomes stronger, as expected. On the other hand, the relative values of the C=O bond lengths require a somewhat more elaborated explanation, because two factors are relevant to their determination. On the one side, there is the well-known, more important π-electron delocalization in the *cis*-ester arrangement compared to the *trans*-ester geometry, which favors a longer C=O bond in **1** and **2** compared to **3** [24,25], and on the other side, there are the relative strengths of the hydrogen bonds in **1** and **3**, in which the carbonyl oxygen atom acts as the acceptor, and the non-participation of this atom in the hydrogen bond in **2**, which favors a longer C=O bond in the order **3** > **1** > **2**. On the whole, the two factors lead to the observed order for the C=O bond lengths: **1** > **3** >> **2**. The higher degree of π-electron delocalization in the *cis*-ester conformation (as in **1** and **2**), compared to the *trans*-ester geometry (as in **3**) is also in accordance with the shorter C–O (carboxylic) bond length in **1** compared to **3** (form **2** has the longest C–O bond among the three conformers because of its involvement in the intramolecular hydrogen bonding).

In the context of the present work, the relative energies of the different conformers are of particular importance. It is clear from Table 1 that, in the room temperature (298.15 K) gas phase equilibrium, conformer **1** accounts for 99.5 % of the total population. Because the gaseous beam of the compound (and matrix gas: Ar; N$_2$) was at room temperature in the matrix isolation experiments described in the next sections, one can expect that only conformer **1** shall be present in the matrices in a detectable amount. Conformer **2** is predicted by the calculations to account for only 0.5% of the population, which is too low to allow its experimental detection, while the remaining conformers have a negligible population.

4.2. Infrared Spectra of Matrix-Isolated PHN

In the present study, PHN was isolated in low temperature (10 K) Ar and N_2 matrices, as described in Section 2. The recorded experimental mid-IR spectra of the as-deposited matrices in the 1800–1100 and 1100–450 cm^{-1} spectral regions are shown in Figures 2 and 3, in which they are compared with the B3LYP/6-311++G(d,p)-calculated infrared spectrum of conformer **1** (the spectral range 3500–2900 cm^{-1} is shown in Figure S1). The proposed band assignments are provided in Table 4, together with the calculated frequencies and infrared intensities.

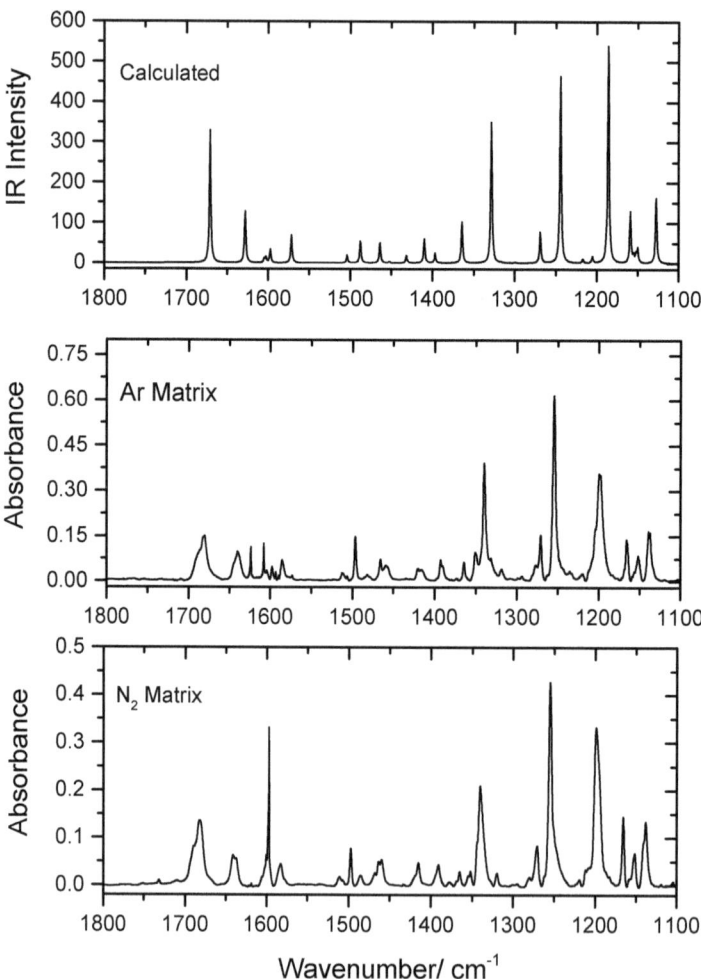

Figure 2. Infrared spectrum of PHN in Ar (**middle**) and N_2 (**bottom**) matrices (10 K), and the B3LYP/6-311++G(d,p) calculated spectra for PHN conformer **1** (**top**) in the 1800–1100 cm^{-1} range. The band at 1597 cm^{-1} in the N_2 matrix spectrum, and those at 1624, 1608, 1593 and 1590 cm^{-1} in the Ar matrix spectrum are due to traces of monomeric water impurity.

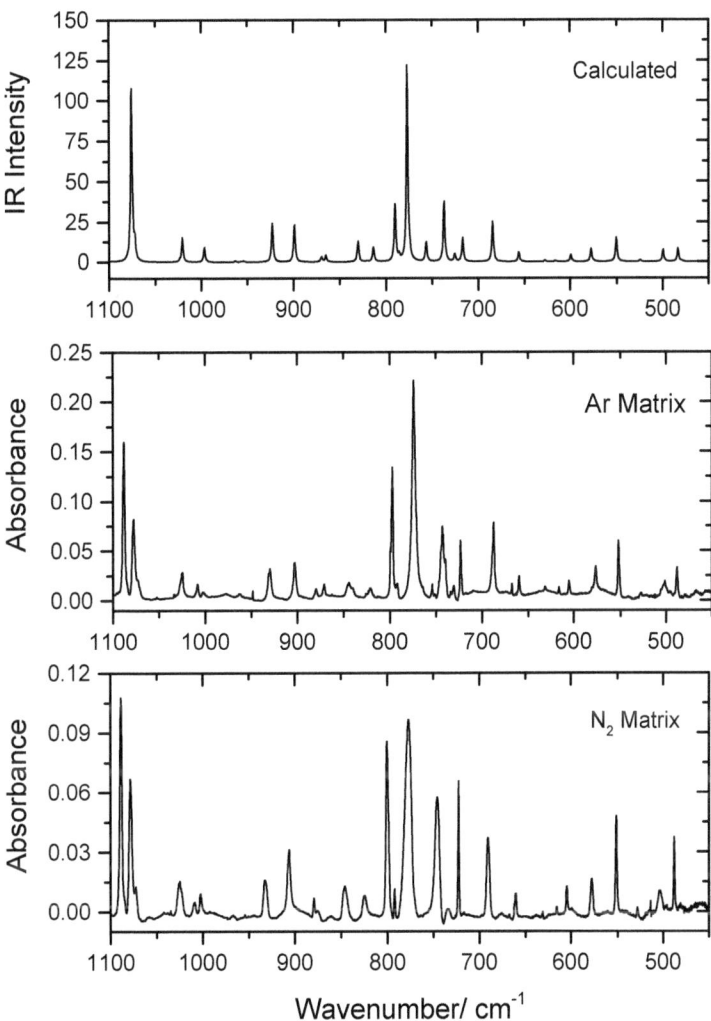

Figure 3. Infrared spectrum of PHN in Ar (**middle**) and N_2 (**bottom**) matrices (10 K), and the B3LYP/6-311++G(d,p) calculated spectra for PHN conformer **1** (**top**) in the 1100–500 cm^{-1} range. The very small bands at ~668 cm^{-1} are due to atmospheric CO_2.

As seen from Figures 2 and 3 and Table 4, the calculated spectrum for conformer **1** reproduces the experimental data very well, demonstrating the sole existence of this conformer in the matrices and making the assignment of the experimental spectra straightforward. The excellent reproduction of the experimental data by the calculations shall be highlighted, because frequently the DFT method does not describe hydrogen bonding precisely, which is an essential intramolecular interaction in the case of the experimentally-relevant conformer of the studied molecule. Nevertheless, the functional/basis set used in the present study has been shown to be reliable for infrared spectra predictions for many other intramolecularly-bonded molecules formed exclusively by atoms of the first and second row of the periodic table [26–28]. In here, only the assignments for the most prominent characteristic vibrations of PHN will be briefly addressed.

Table 4. Assignment of the observed IR spectra of PHN in argon and N_2 matrices (10 K), and the B3LYP/6-311++G(d,p) calculated spectra for conformer 1 [a].

Experimental		Calculated (Form 1)		
Ar Matrix	N_2 Matrix	ν [b]	I^{IR}	Approximate Description [c]
3435/3375	3429/3375	3314	476.4	νO–H
3166	3167	3141	3.3	νC–H ring2
		3138	5.0	νC–H ring3
n.obs.	n.obs.	3133	1.3	νC–H ring1
3124	3131	3128	7.1	νC–H ring1
		3119	22.0	νC–H ring1
3097	3102	3117	21.8	νC–H ring3
3085	3087	3109	7.8	νC–H ring1
		3107	15.8	νC–H ring2
3074	3078	3105	4.0	νC–H ring3
		3100	0.4	νC–H ring1
n.obs.	n.obs.	3093	0.4	νC–H ring3
1691/1686/1682/1680	1693/1689/1682/1681	1671	329.1	νC=O
1644/1639	1641/1638	1628	131.5	ν rings2-3, δO–H
1610	1619	1605	10.4	ν rings2-3
1605	1606	1602	12.8	ν ring1
1598	1600	1597	33.6	ν ring1
1585	1582	1571	69.9	ν rings2-3, δO–H
1512/1506	1511/1507	1504	19.0	ν rings2-3, δC–H rings2-3
1496	1498	1488	55.5	ν ring1, δC–H ring1
1465	1469/1464	1464	51.6	ν rings2-3, δC–H rings2-3, δO–H
1459	1460	1453	3.4	ν ring1, δC–H ring1
1434	1431	1432	18.6	ν rings2-3, δC–H rings2-3, δO–H
1420/1415	1420/1415	1410	62.1	δC–H rings2-3, νC–O(H), δO–H
1393/1390	1395/1391	1397	24.3	δC–H rings2-3, νC–O(H), δO–H
1373/1364	1369/1365	1364	103.0	ν rings2-3, δO–H
1351/1350/1340/1331	1354/1352/1343/1340	1328	354.9	δO–H, νC–C, νC–O, νC–O(H), ν rings2-3
1318	1320	1317	1.4	ν ring1, δC–H ring1
1298/1294	1300/1295	1299	1.4	ν ring1, δC–H ring1
1276/1271	1280/1271	1269	78.2	δO–H, δC–H ring3
1255	1255	1244	464.8	νC–C, νC–O, δC–H rings2-3
1236/1234	1242	1217	10.0	νC–O(H), δ rings2-3, δC–H rings2-3
1223/1219	1223/1219	1205	16.5	δ rings2-3, δC–H rings2-3
1205/1204/1199/1198	1212/1207/1199/1195	1186	549.5	νC–O–C asym.
1166/1165	1165	1159	129.6	δC–H ring1
1159	1158	1155	16.6	δC–H ring1
1156	1152	1152	19.6	δC–H ring3
1151	1151	1150	34.6	δC–H rings2-3
1139/1137	1140/1138	1127	165.6	δC–H rings2-3, δO–H, νC–O
1088	1089	1076	110.6	δ rings2-3, δC–H rings2-3, νC–O(H)
1078/1074	1079/1073	1072	10.3	δC–H ring1
1027	1025	1023	0.8	ν ring3
1025	1022	1020	15.0	δC–H ring1
1008/1002	1009/1002	997	9.3	δ ring1
n.obs.	n.obs.	980	0.0	γC–H ring3
n.obs.	n.obs.	977	0.1	γC–H ring1
963	967	963	0.6	γC–H rings2-3
955	n.obs.	958	0.1	γC–H ring1
953	954	954	0.7	γC–H rings2-3
931/930	933	923	24.5	γC–H ring1
903/902	906	899	24.0	γC–H ring1
880	879	870	3.1	δ ring3

Table 4. Cont.

Experimental		Calculated (Form 1)		
Ar Matrix	N$_2$ Matrix	ν [b]	I^{IR}	Approximate Description [c]
871	875	865	4.0	γC–H rings2-3
845/840	846	830	13.0	γC–H ring1, δC–O–C
826	n.obs.	821	0.5	γC–H ring1
822	825	814	9.2	γC–H rings2-3
800/797	801	790	36.9	γC–H rings2-3
794/792	792	786	3.1	δO–C=O, δ rings1-2-3
774	777	777	122.2	τO–H
754	755	757	12.2	γC=O
745/743/742/740	746	737	38.6	γC–H ring1
734/731	735	726	4.7	γC–H rings2-3, γC=O
723	723	717	14.9	δ rings2-3
687	691	684	25.2	γC–H ring1
660	661	656	5.9	γC–O(H)
632	631	628	0.8	δ rings1-2-3
616	617	616	0.7	δ ring1
605	604	599	4.4	δ rings2-3
576	578	577	8.1	τ rings2-3
552	551	550	15.2	Skeletal deformation
527	528	525	1.1	Skeletal deformation
504/501	503	500	7.5	τ ring1
488	488	483	8.1	δ rings2-3
479	481	479	0.4	τ rings2-3
422	422	423	6.1	τ ring3
n.obs.	n.obs.	412	0.2	τ ring1
n.obs.	n.obs.	408	4.7	δC–O(ph)
		387	1.3	Skeletal deformation
		345	12.4	νH···O
		277	1.1	Skeletal deformation
		268	2.2	Skeletal deformation
		220	2.6	rings2-3 sym. butterfly
		214	0.7	τ ring1
n.i.	n.i.	157	1.5	δC–C=O
		142	0.5	rings2-3 asym. butterfly
		99	0.1	γC–C(ester)
		74	0.2	τC–C
		56	0.7	γC–O(ph)
		28	0.5	τC–O
		15	0.1	τO–C

[a] Frequencies in cm^{-1}; calculated intensities in km mol^{-1}; n.obs., not observed; n.i., not investigated. [b] The calculated frequencies were scaled by 0.978. [c] ν, stretching; δ, bending; γ, rocking; τ, torsion; sym., symmetric; asym., anti-symmetric; ph, phenyl ester group. Rings 1, 2, and 3 correspond to the phenyl ester, phenol, and benzo aromatic moieties, respectively.

4.2.1. 3500–2900 cm^{-1} Region

The stretching mode of the H-bonded O–H group can be expected to give rise to a very broad band, spreading across a wide range of frequencies. Accordingly, in the experimental spectra, a broad feature is observed within the 3460–3350 cm^{-1} range, with maxima at 3435 and 3375 cm^{-1} in the case of the argon matrix spectrum, and at 3429 and 3375 cm^{-1} for the N$_2$ matrix, which are ascribed to the νO–H stretching mode. It is interesting to note that the calculations predict this vibration at 3314 cm^{-1}, i.e., somewhat shifted to lower frequencies when compared to the experimental data. Such a result may indicate that the intramolecular hydrogen bond strength is overestimated to some extent by the DFT calculations, a result that has been found for other molecules bearing strong intramolecular hydrogen bonds, e.g., malonic acid [29]. The fact that the experimentally-observed bands have two major maxima indicates that, in both matrices, PHN molecules have been trapped in two main matrix sites, a conclusion that is reinforced by the observation of other bands exhibiting a doublet-type structure in other spectral regions, as pointed out below.

The bands originating in the aromatic C–H stretching modes give rise to the structured features spreading from 3261 and 2970 cm^{-1}, and from 3229 and 2895 cm^{-1}, in the argon and N$_2$ matrices, respectively. These spectral features result from extensive overlapping bands due to the different νC–H stretching modes (which can also be expected to be site split) and also comprise contributions resulting from Fermi resonance interactions involving the overtones and combination modes of the δC–H bending and νC–C stretching ring modes whose fundamentals are observed in the 1550–1450 cm^{-1} range. This is in accordance with the findings of earlier theoretical studies that found a strong contribution of non-fundamental modes to the C–H stretching region [30].

4.2.2. 1800–1100 cm^{-1} Region

In this spectral range, the IR spectrum of PHN is dominated by the strong bands predicted at 1671 (νC=O), 1328 (a mixed mode with major contributions from δO–H, νC–O(H), νC–O and νC–C), 1244 (corresponding essentially to the anti-symmetric stretching mode involving the C–C and C–O bonds α to the carbonyl) and 1185 (νO–C–O anti-symmetric stretching) cm^{-1}. These bands have experimental counterparts in the experimental spectra obtained in the argon and N$_2$ matricesat 1691/1686/1682/1680, 1351/1350/1340/1331, 1255 and 1205/1204/1199/1198 cm^{-1}, and 1693/1689/1682/1681, 1354/1352/1343/1340, 1255 and 1212/1207/1199/1195 cm^{-1}, respectively (see Table 4).

With the exception of the feature observed at 1255 cm^{-1}, the remaining three modes give rise to four overlapping bands resulting from the simultaneous occurrence of matrix site splitting (two major sites, as seen also for νO–H; see above) and Fermi resonance interactions (tentative interacting modes are respectively the overtone of the mode whose fundamental is observed at 845 cm^{-1}, the combination tone 687 + 660 cm^{-1}, and the overtone of the fundamental at 605 cm^{-1}, in the argon matrix, which in N$_2$ appear at 846, 691+661 and 604 cm^{-1}, respectively). It is also interesting to note that the calculations predict the νC=O stretching at a slightly lower frequency value compared to the observed ones, which supports the above conclusion that the strength of the intramolecular hydrogen bond is somewhat overestimated by the calculations.

A remarkable result that can be extracted from the calculations is the fact that the δO–H bending vibration is mixed extensively with other vibrations, contributing significantly to eight normal modes, whose frequencies span from ca. 1270 up to about 1630 cm^{-1} (see Table 4). As expected, all of these modes thus give rise to bands with a significantly high infrared intensity, which is confirmed by the experimental data (see Table 4 and Figure 2). To a lesser extent, the same can be also stated for both the C–O and C–O(H) stretching vibrations, which are predicted to contribute significantly to four and five normal modes, respectively (in some cases the same to which the δO–H vibration contributes), all of them giving rise to considerably intense experimental infrared bands.

The remaining bands observed in this spectral region in the matrix isolation spectra of PHN are mostly ascribable to vibrations of the aromatic rings and could be assigned straightforwardly in view of the very good agreement with the calculated data (see Table 4).

4.2.3. 1100–450 cm^{-1} Region

In the higher frequency range of this spectral region, from 1100 to 800 cm^{-1}, most of the bands are of low intensity and originate from vibrations of the aromatic moieties, and the experimental spectra are globally very well reproduced by the calculations. Only two vibrations with significant contributions from other molecular fragments are observed in this part of the spectrum. The first corresponds to the intense bands observed in argon and N$_2$ at 1088 and 1089 cm^{-1}, respectively, which are ascribed to a vibration with a relevant contribution from the νC–O(H) stretching (predicted at 1076 cm^{-1}). The second appears at 845/840 and 846 cm^{-1} in the spectra obtained in the argon and N$_2$ matrices, respectively, which—according to the calculations—will also have a non-negligible contribution from the δC–O–C bending (calculated at 830 cm^{-1}; see Table 4). On the other hand, below 800 cm^{-1}, the number of bands ascribable other than the aromatic groups is larger. The most prominent band is due to the τO–H torsion, which is observed at 774 and 777 cm^{-1} in the argon and N$_2$ matrices, respectively, in good agreement with the predicted value (777 cm^{-1}). Besides this, the bands observed at about 800, 755, 735 and 630 cm^{-1} are also observed in this spectral region in both the argon and N$_2$ matrices, and have significant contributions from the δO–C=O, γC=O and γC–O(H) vibrations. Both the frequencies and relative intensities of these bands agree fairly well with the calculated data, as can be seen in Figure 3 and Table 4.

4.3. Narrowband UV-Induced Decarbonylation of PHN

The photochemistry of matrix-isolated PHN (in both argon and N$_2$ matrices) was here investigated for the first time. The matrices were irradiated using narrowband UV-light (see Section 2 for details); the excitation wavelength range used was chosen whilst taking into account the TD-DFT-B3LYP/6-31G(d) calculated UV spectrum of the compound [17]. After each irradiation, an infrared spectrum was collected and analyzed in order to probe the changes taking place in the sample. The irradiation experiments started using UV light with λ = 331.7 nm, which was subsequently decreased stepwise down to λ = 235.0 nm (the sequences of irradiations performed are described in Table S1, which gives the irradiation wavelengths, laser pulse energies and exposure times used). Noticeable spectral changes indicating the transformation of PHN into other molecules were observed for all wavelengths used, the efficiency of the process increasing with the energy of the excitation UV light used.

The spectra of the irradiated matrices reveal the photogeneration of carbon monoxide (CO), which gives rise to the bands at 2138 cm^{-1} (in the Ar matrix) and at 2139 cm^{-1} (in the N$_2$ matrix), which were assigned to the well-isolated monomer, and at 2126/2122/2118/2114 cm^{-1} (Ar) and 2126/2117 cm^{-1} (N$_2$) due to CO molecules interacting with other species in the matrices (Figure 4) [31–35]. These results doubtlessly indicate the occurrence of photodecarbonylation. Together with CO, 2-phenoxynapthalen-1-ol (PNO) shall be produced, which is confirmed by the comparison of the experimental spectra of the irradiated matrices with the calculated spectra of this molecule. This comparison is presented in Figure 5, which shows the 1300–1050 cm^{-1} spectral region of the difference infrared spectra (spectra of the irradiated matrices at the end of the full sequence of irradiations, minus the spectra of the as-deposited matrices), with the simulated difference spectrum built based on the B3LYP/6-311++G(d,p) calculated infrared spectra of *cis* PNO (with positive intensities) and of PHN (with negative intensities). In this figure, the calculated spectrum of the *cis* conformer of PNO is presented, because it was calculated to be more stable than the *trans* form (by 22.9 kJ mol^{-1}) (see Figure 6). The full calculated infrared spectrum of PNO is provided in Figure S2.

The calculations predict that, globally, the infrared spectrum of PNO is less intense than that of PHN. Together with the low efficiency of the decarbonylation reaction (the estimated amount of reactant consumed at the end of the full sequences of irradiation, based on the decrease of intensity of its bands, is less than 5%), this fact makes it possible to detect only the most intense bands of the photoproduct. However, in the 1300–1050 cm^{-1} spectral region (see Figure 5), in which almost all of the most intense bands of PNO are predicted to occur, the emerging bands fit fairly well those calculated for this compound. Specifically, the comparatively intense vibrations predicted at 1264, 1250, 1238, 1205, 1194, 1159, 1146, 1075, and 1074 cm^{-1}, most of them with significant contributions from the δO–H (in most of the cases), νC–O, or/and νC–O(H) coordinates, have experimentally-observed counterparts at 1274/1265, 1251, 1241, 1211, 1195, 1162, 1150, 1087, and 1080 cm^{-1} in the argon matrix, and at 1275/1267, 1248, 1241/1237, 1205, 1195, 1163, 1152, 1090, and 1080 cm^{-1} in the N$_2$ matrix, all being in good agreement with the calculated values (the approximate descriptions of the calculated normal modes for *cis* PNO are given in Table S4).

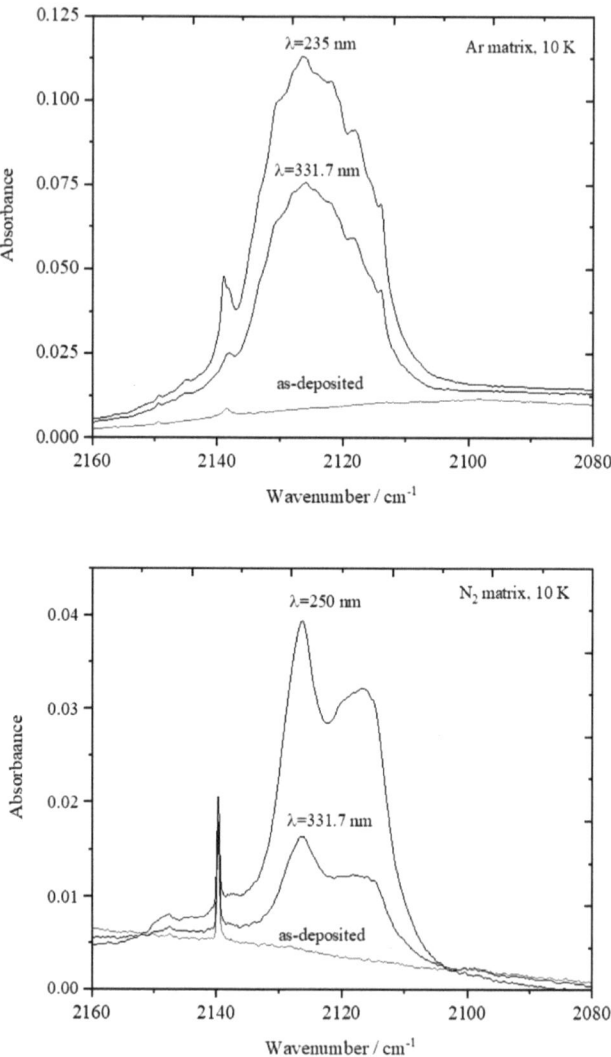

Figure 4. UV-induced generation of carbon monoxide: fragment of the infrared spectra of PHN isolated in the Ar and N_2 matrices recorded after the deposition of the matrix before any irradiation, after irradiation with λ = 331.7 nm and λ = 235.0 nm in an Ar matrix, and after irradiation with λ = 331.7 nm and λ = 250.0 nm in an N_2 matrix.

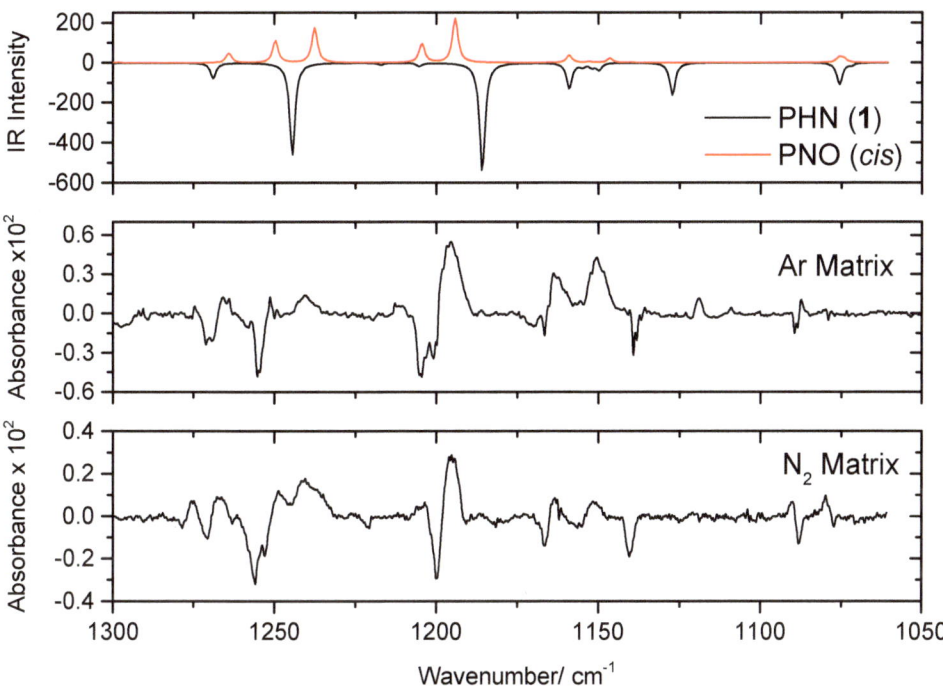

Figure 5. Infrared difference spectra (1300–1050 cm^{-1} region, in which the most intense bands of PNO occur; baseline corrected) showing the results of the full sequence of the performed irradiations (see Table S1) of PHN isolated in Argon (**middle**) and N$_2$ (**bottom**) matrices. The simulated difference spectrum built based on the B3LYP/6-311++G(d,p) calculated IR spectra of PNO (*cis* form) and PHN (conformer **1**) is presented in the (**top**) of the figure. In the calculated spectra, the wavenumbers were scaled by 0.978.

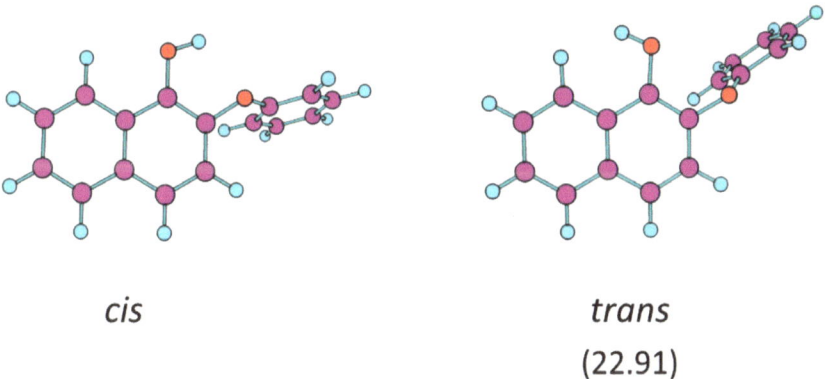

Figure 6. B3LYP/6-311++G(d,p) optimized structures of the *cis* and *trans* conformers of PNO. The ΔE$_{(trans\text{-}cis)}$ value is given in parentheses (in kJ mol^{-1}). The Cartesian coordinates for the optimized structures are given in Table S3.

5. Conclusions

Phenyl 1-hydroxy-2-naphthoate (PHN) was investigated by matrix isolation infrared spectroscopy, complemented by DFT(B3LYP)/6-311++G(d,p) calculations. The investigations were focused on the evaluation of the conformational space of the molecule and reasons for the different stability of its conformers, and on the UV-induced phototransformation of matrix-isolated PHN in argon and N_2 matrices (at 10 K).

The twofold degenerated O–H$\cdots$O=C intramolecularly hydrogen bonded conformer with the phenyl ring ester group $\pm 68.8°$ out of the plane of the substituted naphtyl moiety is the most stable conformer of the molecule, being considerably more stable than the second most stable form (by over 14.75 kJ mol^{-1}), in which a weaker O–H$\cdots$O–C intramolecular hydrogen bond exists. The non-hydrogen bonded conformers were characterized by dominant repulsive intramolecular interactions involving the lone electron pairs of the oxygen atoms as well as the aromatic rings, and were predicted to have much higher energies than the H-bonded forms (over 55 kJ mol^{-1}).

The most stable conformer of PHN was successfully trapped both in argon and nitrogen matrices, and was shown to photodecarbonylate upon UV irradiation within the wavelength range 331.7–235.0 nm, with the decarbonylation process showing a low efficiency which increases for shorter irradiation wavelengths. Together with CO, 2-phenoxynaphthalen-1-ol (PNO) was produced and successfully detected in the photolysed matrices. No evidence of the occurrence of any other photo-induced process was gathered. On the whole, the present study contributes to the better understanding of the structure, intramolecular interactions and photochemistry of PHN, a relevant member of the hydroxynaphtoate family of compounds.

Supplementary Materials: The following are available online https://www.mdpi.com/2673-7256/1/1/1025/s1. Figure S1: Infrared spectrum of PHN in Ar and N_2 matrices (10 K), and B3LYP/6-311++G(d,p) calculated spectra for PHN conformer 1 in the 3500–2900 cm^{-1} range. Figure S2: Graphical representation of the full calculated infrared spectrum of cis PNO. Table S1: Information on the used UV-irradiation sequences. Tables S2 and S3: B3LYP/6-311++G(d,p) optimized Cartesian coordinates for the conformers of PHN and PNO, respectively. Table S4: Calculated infrared data for cis PNO, including the approximate description of the vibrations.

Author Contributions: R.F., conceptualization, formal analysis, supervision, funding acquisition, writing, final version; S.G., methodology, laboratory work, writing—review and editing; İ.S. and Y.G.S., laboratory work, computations, funding acquisition, writing—original draft. All authors have read and agreed to the published version of the manuscript.

Funding: İ.S. and Y.G.S. thank Bitlis Eren University for the computational facilities, the Bitlis Eren University Research Foundation for the financial support (Project BEBAP-2013.04), and the Erasmus Offices of the Bitlis Eren and Coimbra Universities for short term ERASMUS+ Grants at the University of Coimbra. S.G. thanks the LaserLab Coimbra for a researcher position at the Coimbra Chemistry Centre (CQC)–Coimbra. The Coimbra Chemistry Centre (CQC) is supported by the Portuguese Science Foundation (FCT) (Projects UI0313B/QUI/2020 and UI0313P/QUI/2020) and COMPETE-UE. Support from FCT through Project PTDC/QUI-QFI/1880/2020 is also acknowledged.

Data Availability Statement: The data presented in this study are available in Supplementary Materials.

Conflicts of Interest: The authors declare no conflict of interest.

References

1. Zhang, J.-Y.; Jin, H.; Wang, G.-F.; Yu, P.-J.; Wu, S.-Y.; Zhu, Z.-G.; Li, Z.-H.; Tian, Y.-X.; Xu, W.; Zhang, J.-J.; et al. Liposomal Curcumin Targeting Endometrial Cancer Through the NF-κB Pathway. *Inflamm. Res.* **2011**, *60*, 851–859. [CrossRef]
2. Morand, E.F.; Iskander, M.N. Napththalene Derivatives Which Inhibit the Cytokine or Biological Activity of Macrophage Migration Inhibitory Factor (mif). PCT International Application 2003. WO 2003104178A1, 18 December 2003.
3. Baboshin, M.A.; Golovleva, L.A. Increase of 1-Hydroxy-2-naphthoic Acid Concentration as a Cause of Temporary Cessation of Growth for Arthrobacter sp. K$_3$: Kinetic Analysis. *Microbiology* **2009**, *78*, 180–186. [CrossRef]

4. Sastry, M.N.V.; Claessens, S.; Habonimana, P.P.; De Kimpe, N. Synthesis of the Natural Products 3-Hydroxymollugin and 3-Methoxymollugin. *J. Org. Chem.* **2010**, *75*, 2274–2280. [CrossRef]
5. Meyers, A.I.; Avila, W.B. Chemistry of Aryloxazolines. Applications to the Synthesis of Lignan Lactone Derivatives. *J. Org. Chem.* **1981**, *46*, 3881–3886. [CrossRef]
6. Zjawiony, J.; Peterson, J.R. An Improved Synthesis of Naphtho[2,3-d]-1,3-dioxole-5-methoxy-6-carboxylic acid. *Org. Prep. Proced. Int.* **1991**, *23*, 163–172. [CrossRef]
7. Hauser, F.M.; Rhee, R.P. Syntheses of .alpha.- and .beta.-Sorigenin Methyl Ethers. *J. Org. Chem.* **1977**, *42*, 4155–4157. [CrossRef]
8. Franck, R.W.; Bhat, V.; Subramaniam, C.S. The Sstereoselective Total Synthesis of the Natural Enantiomer of Olivin Trimethyl Ether. *J. Am. Chem. Soc.* **1986**, *108*, 2455–2457. [CrossRef]
9. Woolfe, G.J.; Thistlethwaite, P.J. Excited-state Prototropic Reactivity in Salicylamide and Salicylanilide. *J. Am. Chem. Soc.* **1980**, *102*, 6917–6923. [CrossRef]
10. Woolfe, G.J.; Thistlethwaite, P.J. Excited-state Prototropism in Esters of o-Hydroxy-2-naphthoic Acids. *J. Am. Chem. Soc.* **1981**, *103*, 3849–3854. [CrossRef]
11. Law, K.Y.; Shoham, J. Photoinduced Proton Transfers in Methyl Salicylate and Methyl 2-hydroxy-3-naphthoate. *J. Phys. Chem.* **1994**, *98*, 3114–3120. [CrossRef]
12. Bergmann, E.D.; Hirshberg, Y.; Pinchas, S. Ultra-violet Spectrum and Constitution of 3-Hydroxy-2-naphthoic Acid and Related Compounds. *J. Chem. Soc.* **1950**, 2351–2356. [CrossRef]
13. Naboikin, U.V.; Zadorozhnyi, B.A.; Pavlova, E.N. Peculiarities of the Luminescence of Ortho-Disubstituted Aromatic Hydrocarbons: II. *Opt. Spectrosc. (Eng. Transl.)* **1959**, *6*, 312.
14. McCarthy, A.; Ruth, A.A. Fluorescence Excitation and Excited State Intramolecular Relaxation Dynamics of Jet-cooled Methyl-2-hydroxy-3-naphthoate. *Chem. Phys.* **2013**, *425*, 177–184. [CrossRef]
15. Catalán, J.; del Valle, J.C.; Palomar, J.; Díaz, C.; de Paz, J.L.G. The Six-Membered Intramolecular Hydrogen Bond Position as a Switch for Inducing an Excited State Intramolecular Proton Transfer (ESIPT) in Esters of o-Hydroxynaphthoic Acids. *J. Phys. Chem. A* **1999**, *103*, 10921–10934. [CrossRef]
16. Tobita, S.; Yamamoto, M.; Kurahayashi, N.; Tsukagoshi, R.; Nakamura, Y.; Shizuka, H. Effects of Electronic Structures on the Excited-state Intramolecular Proton Transfer of 1-Hydroxy-2-acetonaphthone and Related Compounds. *J. Phys. Chem. A* **1998**, *102*, 5206–5214. [CrossRef]
17. Sıdır, İ.; Gülseven Sıdır, Y. Solvatochromism and Intramolecular Hydrogen-bonding Assisted Dipole Moment of Phenyl 1-Hydroxy-2-naphtoate in the Ground and Excited States. *J. Mol. Liq.* **2016**, *221*, 972–985. [CrossRef]
18. Frisch, M.J.; Trucks, G.W.; Schlegel, H.B.; Scuseria, G.E.; Robb, M.A.; Cheeseman, J.R.; Scalmani, G.; Barone, V.; Petersson, G.A.; Nakatsuji, H.; et al. *Gaussian 09, Revision A.02*; Gaussian, Inc.: Wallingford, CT, USA, 2009.
19. Becke, A.D. Density-functional Exchange-energy Approximation with Correct Asymptotic Behavior. *Phys. Rev. A* **1988**, *38*, 3098–3100. [CrossRef]
20. Lee, C.T.; Yang, W.T.; Parr, R.G. Development of the Colle-Salvetti Correlation-energy Formula into a Functional of the Electron Density. *Phys. Rev. B* **1988**, *37*, 785–789. [CrossRef] [PubMed]
21. Vosko, S.H.; Wilk, L.; Nusair, M. Accurate Spin-dependent Electron Liquid Correlation Energies for Local Spin Density Calculations: A Critical Analysis. *Can. J. Phys.* **1980**, *58*, 1200–1211. [CrossRef]
22. McLean, A.D.; Chandler, G.S. Contracted Gaussian Basis Sets for Molecular Calculations. I. Second Row Atoms, Z=11–18. *J. Chem. Phys.* **1980**, *72*, 5639–5648. [CrossRef]
23. Dennington, R.; Keith, T.; Milam, J. *GaussView (Version 5.0)*; Semichem Inc.: Shawnee Mission, KS, USA, 2009.
24. Fausto, R.; Batista de Carvalho, L.A.E.; Teixeira-Dias, J.J.C. Conformational Analysis of Carbonyl and Thiocarbonyl Ethyl Esters: The HC(=X)Y-CH$_2$CH$_3$ (X, Y = O or S) Internal Rotation. *J. Comput. Chem.* **1992**, *13*, 799–809. [CrossRef]
25. Lopes, S.; Nikitin, T.; Fausto, R. Structural, Spectroscopic, and Photochemical Study of Ethyl Propiolate Isolated in Cryogenic Argon and Nitrogen Matrices. *Spectrochim. Acta A* **2020**, *241*, 118670. [CrossRef] [PubMed]
26. Reva, I.D.; Stepanian, S.; Adamowicz, L.; Fausto, R. Combined FTIR Matrix Isolation and Ab Initio Studies of Pyruvic Acid: Proof for Existence of the Second Conformer. *J. Phys. Chem. A* **2001**, *105*, 4773–4780. [CrossRef]
27. Apóstolo, R.F.G.; Bento, R.F.; Tarczay, G.; Fausto, R. The First Experimental Observation of the Higher-Energy *Trans* Conformer of Trifluoroacetic Acid. *J. Mol. Struct.* **2016**, *1125*, 288–295. [CrossRef]
28. Avadanei, M.; Cozan, V.; Kuş, N.; Fausto, R. Structure and Photochemistry of N-Salicylidene-p-carboxyaniline Isolated in Solid Argon. *J. Phys. Chem. A* **2015**, *119*, 9121–9132. [CrossRef]
29. Maçôas, E.M.S.; Fausto, R.; Lundell, J.; Pettersson, M.; Khriachtchev, L.; Räsänen, M. Conformational Analysis and Near-infrared-induced Rotamerization of Malonic Acid in an Argon Matrix. *J. Phys. Chem. A* **2000**, *104*, 11725–11732. [CrossRef]
30. Beć, K.B.; Grabska, J.; Ozaki, Y.; Hawranek, J.P.; Huck, C.W. Influence of Non-fundamental Modes on Mid-infrared Spectra: Anharmonic DFT Study of Aliphatic Ethers. *J. Phys. Chem. A* **2017**, *121*, 1412–1424. [CrossRef]
31. Wu, L.; Lambo, R.; Tan, Y.; Liu, A.-W.; Hu, S.-M. Infrared Spectroscopy of CO Isolated in Solid Nitrogen Matrix. *Chin. J. Chem. Phys.* **2014**, *27*, 5–8. [CrossRef]

32. Jarmelo, S.; Reva, I.D.; Lapinski, L.; Nowak, M.J.; Fausto, R. Matrix-isolated Diglycolic Anhydride: Vibrational Spectra and Photochemical Reactivity. *J. Phys. Chem. A* **2008**, *112*, 11178–11189. [CrossRef]
33. Giuliano, B.M.; Reva, I.; Fausto, R. Infrared Spectra and Photochemistry of Matrix-isolated Pyrrole-2-carbaldehyde. *J. Phys. Chem. A* **2010**, *114*, 2506–2517. [CrossRef]
34. Breda, S.; Reva, I.; Fausto, R. UV-induced Unimolecular Photochemistry of Diketene Isolated in Cryogenic Inert Matrices. *J. Phys. Chem. A* **2012**, *116*, 2131–2140. [CrossRef] [PubMed]
35. Lapinski, L.; Reva, I.; Gerega, A.; Nowak, M.J.; Fausto, R. UV-induced Transformations of Matrix-isolated 6-Azacytosine. *J. Chem. Phys.* **2018**, *149*, 104301. [CrossRef] [PubMed]

Article

Electronic Absorption, Emission, and Two-Photon Absorption Properties of Some Extended 2,4,6-Triphenyl-1,3,5-Triazines

Alison G. Barnes [1], Nicolas Richy [1], Anissa Amar [2,3], Mireille Blanchard-Desce [4], Abdou Boucekkine [1,*], Olivier Mongin [1] and Frédéric Paul [1,*]

1. Univ. Rennes, CNRS, ISCR (Institut des Sciences Chimiques de Rennes)-UMR 6226, F-35000 Rennes, France; sidereus.nuncius.au@gmail.com (A.G.B.); nicolas.richy@univ-rennes1.fr (N.R.); olivier.mongin@univ-rennes1.fr (O.M.)
2. Laboratoire de Physique et Chimie Quantiques, Faculté des Sciences, Université Mouloud Mammeri de Tizi-Ouzou, Tizi-Ouzou 15000, Algeria; anissa.amar@ummto.dz
3. Faculté de Chimie, Université des Sciences et de la Technologie Houari-Boumediene, Bab-Ezzouar 16111, Algeria
4. Univ. Bordeaux, CNRS, ISM (Institut des Sciences Moléculaires)-UMR 5255, F-33400 Talence, France; mireille.blanchard-desce@u-bordeaux.fr
* Correspondence: abdou.boucekkine@univ-rennes1.fr (A.B.); frederic.paul@univ-rennes1.fr (F.P.); Tel.: +33-02-23-23-59-62 (F.P.)

Abstract: We report herein the linear optical properties of some extended 2,4,6-triphenyl-s-triazines of formula 2,4,6-[(1,4-C$_6$H$_4$)C≡C(4-C$_6$H$_4$X)]$_3$-1,3,5-(C$_3$H$_3$N$_3$) (**3-X**; X = NO$_2$, CN, OMe, NMe$_2$, NPh$_2$) and related analogues **4** and **7-X** (X = H, NPh$_2$), before briefly discussing their two-photon absorption (2PA) cross-sections. Their 2PA performance is discussed in relation to 2PA values previously measured for closely related octupoles such as N,N',N''-triphenylisocyanurates (**1-X**, **5**, and **6-X**) or 1,3,5-triphenylbenzenes (**2-X**). While s-triazines are usually much better two-photon absorbers in the near-IR range than these molecules, especially when functionalised by electron-releasing substituents at their periphery, they present a decreased transparency window in the visible range due to their red-shifted first 1PA peak, in particular when compared with corresponding isocyanurates analogues. In contrast, due to their significantly larger two-photon brilliancy, 2,4,6-triphenyl-s-triazines appear more promising than the latter for two-photon fluorescence bio-imaging purposes. Rationalisation of these unexpected outcomes is proposed based on DFT calculations.

Keywords: 2,4,6-triaryl-s-triazines; octupoles; two-photon absorption; DFT calculations; fluorescence; linear optics

1. Introduction

Planar molecules featuring trigonal symmetry have attracted sustained attention for their second-order nonlinear optical (NLO) properties since the late eighties [1–4]. Initially aroused by the quest for molecules with large second-order NLO properties, these so-called "octupolar" molecules were likely to exhibit sizeable hyperpolarizabilities due to their peculiar symmetry, in reason of the existence of off-diagonal tensorial elements in the electronic coupling matrix between peripheral branches [2]. It was subsequently shown that this symmetry can also be potentially beneficial to third-order NLO properties such as two-photon absorption (2PA). Given the very appealing societal prospects for dyes presenting large 2PA cross-sections [5–8], in particular regarding fluorescence bio-imaging when the dye also fluoresces [6,9], we have started exploring the 2PA properties of various families of molecules, such as extended 1,3,5-triaryltriazinane-2,4,6-triones (**1-X**; Scheme 1) [10,11] or 1,3,5-triphenylbenzene (**2-X**) [12]. Both **1-X** (also known as N,N',N''-triphenylisocyanurates [13]) and **2-X** derivatives proved to be good two-photon absorbers, especially when functionalised by electron-releasing groups at their periphery [11,12]. Actually, in line with these findings as well as with independent reports [14], we observed that the cross-section of the first 2PA peak for these derivatives increased with the polarisation of the peripheral arms, typically when strongly electron-releasing X substituents were present (X = OMe, NMe$_2$, NPh$_2$). However, we also observed that upon progressing from the isocyanurate core (**1-X**) to the slightly less electron-deficient 1,3,5-phenylene core (**2-X**), a slight decrease in the 2PA cross-section occurred for the latter derivatives but only for the most electron-releasing substituents. This suggested that the polarisation induced by the former core was slightly more favourable to 2PA than that induced by the second.

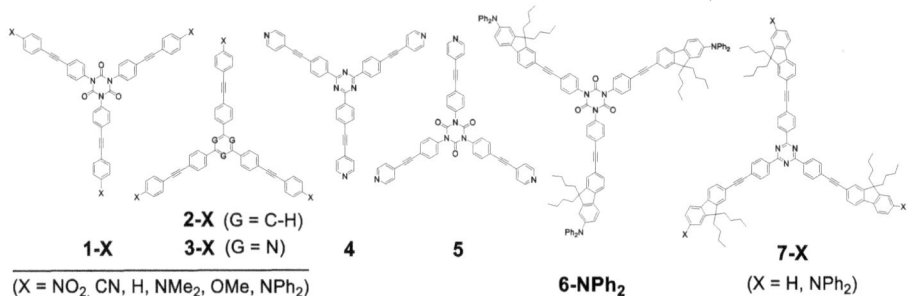

Scheme 1. Molecular structures of known and targeted compounds.

In this respect, surmising that the 1,3,5-triazine core was more electron-attracting than the isocyanurate one, it was now interesting to study the 2PA properties and the fluorescence of related 1,3,5-triazine analogues of **1-X** such as **3-X** (X = NO$_2$, CN, OMe, NMe$_2$, and NPh$_2$). Thus far, emissive triphenyl-s-triazine derivatives and related extended analogues have mainly raised interest in the field of OLEDs or closely related fields [15–18]; however, very few studies were actually focused on the NLO properties of such derivatives. While related derivatives such as extended trialkynyl-s-triazines [19], trialkenyl-s-triazines [20,21], or tris(2-thienyl)-s-triazines [22–25] have given rise to some 2PA investigations, to the best of our knowledge, only one recent theoretical paper deals specifically with the second-order NLO properties of molecules such as **3-X** [26], and only one other single paper addresses the 2PA properties of an s-triazine derivative closely related to **7-H** [27]. Thus, regarding extended triphenyl-s-triazines, essentially styryl-type analogues of **3-X** have been investigated so far for their 2PA properties [28–30]. A general experimental and theoretical study focused on the 2PA properties of octupolar compounds such as **3-X** and **7-X** would therefore be timely. Furthermore, anticipating the well-known propensity of nitro substituents to poison fluorescence in **3-NO$_2$**, we also targeted a derivative such as **4**, which similar to **3-CN**, constitutes another example of a compound featuring electron-withdrawing arms [31]. The optical properties of this new triazine derivative might then be compared with those of its known isocyanurate analogue (**5**) [32]. Finally, given that the extended 4-fluorenyl derivative **6-NPh$_2$** exhibits the most promising optical properties for bio-imaging purposes among all triarylisocyanurates studied so far [33], the study of triazine analogues such as **7-X** (X = H, NPh$_2$) was also required. Accordingly, in the following we will first start with the synthesis of the targeted molecules (**3-X**, **4**, and **7-X**) and study the 2PA properties of their emissive representatives via two-photon excited fluorescence (TPEF). Their optical properties of interest will then be discussed with the help of density functional theory (DFT) and time-dependent DFT (TD-DFT) calculations.

2. Results

2.1. Synthesis and Characterisation

The desired **3-X**, **4**, and **7-X** derivatives are structurally close to known s-triazine derivatives [15–17,27,34–40]. The **3-X** series corresponds to the simplest derivatives in which the 1,3,5-triphenyl-s-triazine core has been extended with a phenyl–alkynyl linker and terminated with X-groups of varying electron-donor/acceptor power. The second series (**7-X**) involves replacing the second phenyl unit with a 2-fluorenyl unit, which is luminescent in its own right. All these derivatives were obtained via Sonogashira coupling reactions (Scheme 2) from the known bromo [17,41–43] or iodo [43] precursors **8-Y**, after reaction with the corresponding aromatic alkynes. Most of these products required chromatographic purification. The use of **8-I** instead of **8-Br** allows generally using smother reaction conditions or leads to higher yields of isolated coupling products under similar conditions. Several **3-X** derivatives have already been reported, such as **3-CN** [17], **3-OMe** [15], and **3-NPh$_2$** [18]. All other compounds were new and were fully characterised by usual techniques. However, synthesis of the nitro derivative **3-NO$_2$**, due solubility issues, was very problematic by such an approach and had to be attempted from the known triyne **8-C≡CH** [16,18], itself obtained in two steps from **8-Br**. Regarding IR-characterization, the in-plane ring stretches of the s-triazine ring resemble those of the phenyl ring and are not so characteristic of triphenyl-s-triazines [44], except perhaps for the fully symmetric stretch (only Raman-active), around 990 cm^{-1} [45,46].

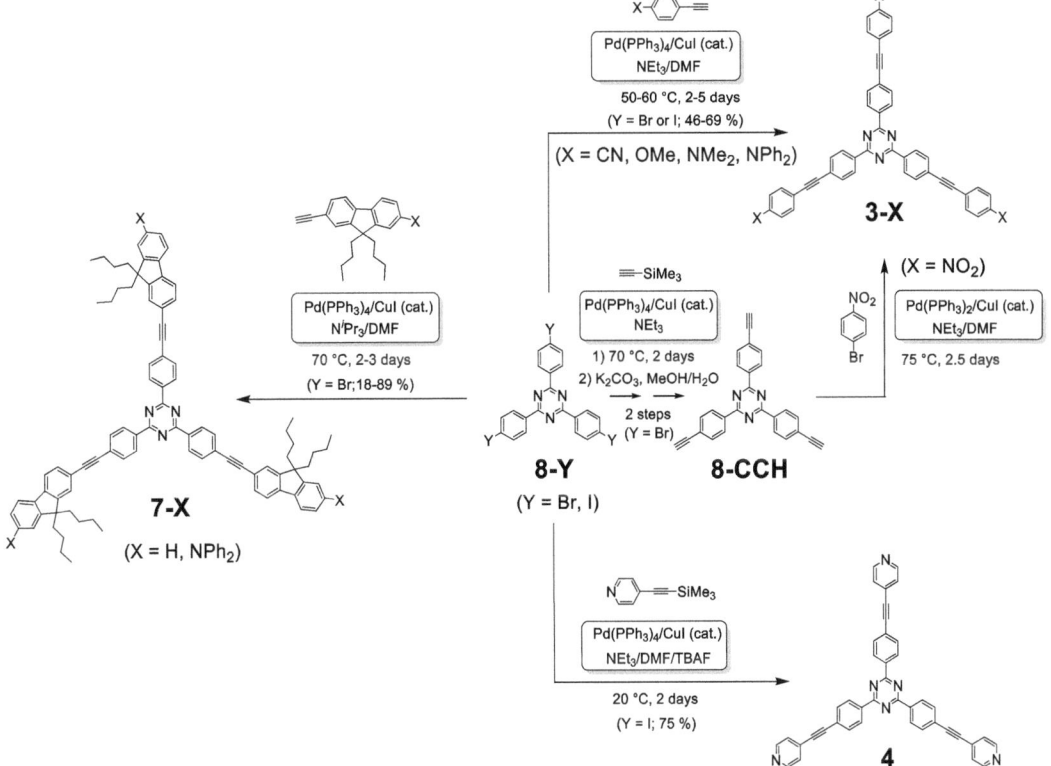

Scheme 2. Synthesis of **3-X**, **4**, and **7-X** derivatives.

2.2. One- and Two-Photon Absorption and Emission Studies

The UV–Vis absorption spectra of the various triazine derivatives were recorded. Except for **3-CN** and **4**, most of the extended compounds (**3-X** and **7-X**) absorbed significantly in the visible range and were strongly coloured (deep yellow in solution), with a lowest-energy absorption entailing significantly above 400 nm (Table 1). Upon progressing from **4** to **3-NPh$_2$** (corresponding to an increase in the electron-releasing nature of the *para*-substituents), a bathochromic shift of the first absorption was clearly observed (Figure 1). A similar trend could also be observed when progressing from **4** to **3-NO$_2$**.

For a given substituent in the **3-X** series, the intensity of the lowest-energy absorption was comparable with that of **1-X** or **2-X** analogues. A higher-energy absorption (at ca. 270–325 nm) was also observed for all these derivatives. The latter became of comparable intensity to the one at the lowest energy for the compounds with the most electron-releasing substituents. Except for **3-NO$_2$**, **3-CN**, and **4**, all extended derivatives were significantly luminescent in CH$_2$Cl$_2$ solutions (Table 1). Luminescence was maximal (Φ_F = 0.80) for the fluorenyl derivative **7-H**. The fluorescence of **3-X** and **4** was comparable to that of their isocyanurate analogues (**1-X** and **5**) but was overall slightly lower than that of their known 1,3,5-triphenylbenzene analogues (**2-X**). In contrast, the fluorescence quantum yield of **7-NPh$_2$** was roughly one-fourth of that previously found for **6-NPh$_2$** (Φ_F = 0.78). In all cases, mirror-symmetry relationships between the first absorption and emission bands and energetic differences between their maxima (see ESI, Figure S7) suggest that the strongly absorbing state at the lowest energy was also the emitting state for all these compounds. Then, as indicated by the corresponding Stokes shifts, larger structural reorganisations and/or solvation energy changes apparently occurred for the compounds featuring the strongest electron-releasing (X = NPh$_2$) or electron-withdrawing (X = NO$_2$) substituents, i.e., 7085 cm^{-1} and 10,685 cm^{-1} in THF, respectively (ESI, Table S1).

In line with similar studies made for **1-X**, **2-X** and related derivatives [11,12,18,27], solvatochromic studies on **3-NPh$_2$** revealed a relatively solvent-insensitive absorption but a very solvent-sensitive emission for such a symmetrical D_{3h} molecule [47]. The shift to lower energy observed on proceeding to the most polar solvent suggests that the excited state was more polarised than the ground state. These results are consistent with the localisation of a (more polar) charge-transfer excited state on one arm of the compound (after relaxation) [14,33,48–50].

The 2PA cross-sections of these derivatives were measured in the near-IR (NIR) range (λ = 700–1000 nm) through an investigation of their two-photon excited fluorescence (TPEF). The excitation was performed with femtosecond pulses from a Ti:sapphire laser (Figure 2 and Table 2). Due to instrumental limitations (1PA below 350 nm for **3-CN**), solubility issues (**3-NO$_2$**), or too weak luminescence (**4**), no TPEF maxima could be detected for several samples. A comparison of these 2PA bands with the 1PA bands for each compound reveals that the 2PA maxima were situated close to twice that of the 1PA maxima detected at the lowest energy in the UV range (see ESI, Figure S9), suggesting that the excited states at the origin of 2PA in the NIR were also active for 1PA or were close in energy to the first allowed 1PA state.

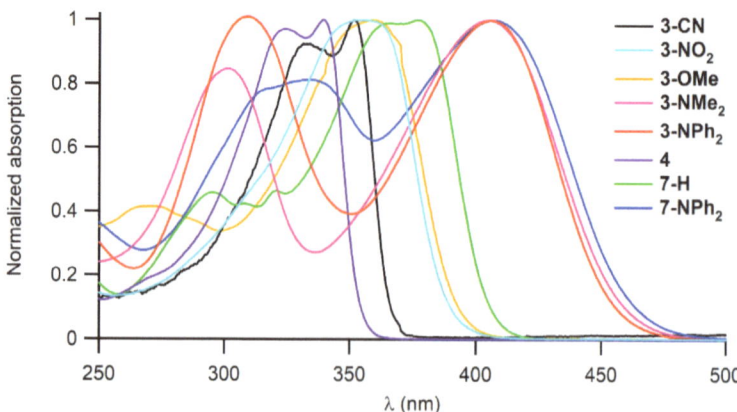

Figure 1. Normalised UV–Vis spectra for **3-X** (X = NO$_2$, CN, OMe, NMe$_2$, NPh$_2$), **4**, and **7-X** (X = H, NPh$_2$) in CH$_2$Cl$_2$ (20 °C).

Table 1. Absorption and emission properties of selected compounds **3-X**, **4**, and **7-X** in CH_2Cl_2 at 25 °C and corresponding TD-DFT computed transitions.

Cmpd	λ_{abs} (nm)	$10^{-3}\,\varepsilon_{max}$ ($M^{-1}\cdot cm^{-1}$)	λ_{em} (nm)	Φ_F [a]	Stokes Shift [b] (cm^{-1})	DFT: λ_{max} (nm) [f] [c] in CH_2Cl_2 CAM-B3LYP/6-31*G [d]	MPW1PW91/6-31*G [d]
3-NO₂ [e]	355	113.0	/	n.e. [f]	/	346 [2.75] 267 [0.04]	392 [2.50] 309 [0.13]
3-CN	352 332 (sh) [e]	154.0 142.0	369	0.02	1308	336 [2.88] 264 [0.02]	374 [2.59] 310 [0.27]
3-OMe	359 270	122.0 51.0	450	0.77	5632	341 [2.64] 258 [0.04]	398 [2.04] 300 [0.57]
3-NMe₂	405 301	105.0 89.0	587	0.27	7656	369 [2.76] 263 [0.10]	456 [1.82] 340 [0.67]
3-NPh₂	406 309	99.0 100.0	583	0.69	7478	371 [3.03] 265 [0.25]	465 [1.83] 340 [0.71]
4	339 270 (sh) [e]	112.0 21.3	405	0.025	4807	322 [2.43] 262 [0.04]	356 [2.15] 289 [0.33]
7-H	377 321 295	173.0 67.0 66.0	455	0.80	4547	352 [3.23] 270 [0.04] /	410 [2.50] 325 [0.62] 296 [0.18]
7-NPh₂	406 334 317	129.0 105.0 101.0	648	0.20	9198	375 [3.73] 277 [0.32] 275 [0.45]	483 [1.71] 388 [1.20] 369 [0.63]

[a] Fluorescence quantum yield in CH_2Cl_2 when excited at λ_{abs} (standard: quinine bisulphate in 0.5 M H_2SO_4).
[b] Stokes shift = $(1/\lambda_{abs} - 1/\lambda_{em})$. [c] f = oscillator strength. [d] Functionals used. [e] Shoulder. [f] Not emissive.

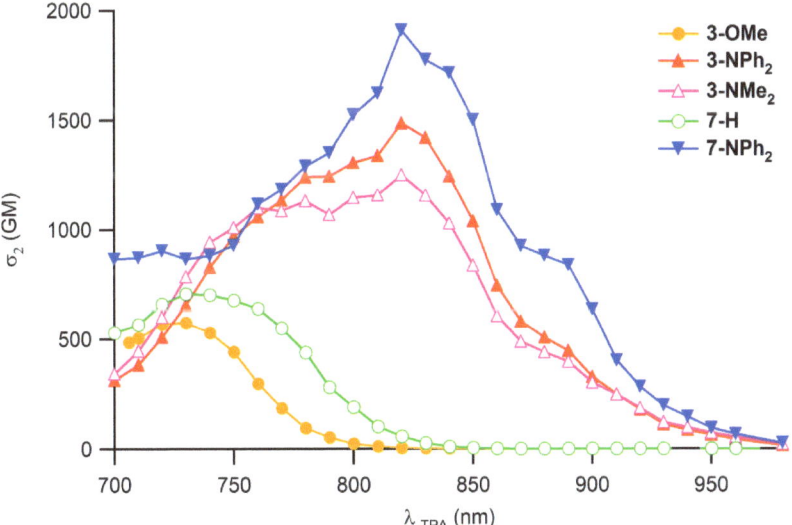

Figure 2. Two-photon absorption spectra for **3-X** (X = OMe, NMe₂, NPh₂) and **7-X** (X = H, NPh₂) in dichloromethane (20 °C).

Table 2. Experimental (TPEF) and calculated 2PA properties of selected derivatives in CH_2Cl_2 at 25 °C.

Cmpd	λ_{1PA} [a] (nm)	λ_{2PA} [b] (nm)	σ_2 [c] (GM)	σ_2/MW [d] (GM/g)	N_{eff} [e]	$\sigma_2/(N_{eff})^2$ (GM)	$\Phi_F \cdot \sigma_2$ [f] (GM)	$\sigma_2 [\lambda_{2PA}]$ [g] (GM [nm])
3-NO_2	355	/	/	/	28.1	/	/	660 [1051]
3-CN	352	/	/	/	31.6	/	/	2735 [918]
3-OMe	359	730	580	0.830	28.1	0.732	447	5391 [954]
3-NMe_2	405	820	1250	1.694	28.1	1.579	338	8701 [1078] [h]
3-NPh_2	407	820	1500	1.351	31.8	1.488	1035	13,875 [1181]
7-H	376	730	710	0.587	38.4	0.481	568	7260 [1033]
7-NPh_2	407	820	1910	1.117	41.1	1.129	382	21,531 [1378]
1-NMe_2 [h]	352	720	360	0.458	24.5	0.600	72	4246 [800] [k]
1-NPh_2 [h]	364	740	410	0.354	28.6	0.502	300	/
2-NMe_2 [i]	357	740	380	0.517	25.0	0.609	262	4322 [800] [k]
2-NPh_2 [i]	369	760	390	0.352	29.0	0.464	285	/
6-NPh_2 [j]	383	770	500	0.284	37.8	0.350	390	/

[a] Wavelength of the one-photon absorption maximum. [b] Wavelength of the two-photon absorption maximum. [c] 2PA cross-section at λ_{2PA}. [d] Specific cross-sections: figure-of-merit relevant for applications in optical limiting or nanofabrication [51]. [e] Effective number of π electrons [52]. [f] Two-photon brightness: figure-of-merit for imaging applications [51]. In these expressions, Φ_F represents the luminescence quantum yield and MW the molecular weight. [g] Computed values using the SAOP functional (see text). [h] For exp. data, see also [11]. [i] For exp. data, see also [12] [j] For exp. data, see also [33]. [k] See: [53].

To understand better the structural dependence of these nonlinear absorptions on the π electrons, we divided their 2PA cross-sections by the square of the effective number of π electrons (N_{eff}^2) [54]. These figures of merit (σ_2/N_{eff}^2) allow normalisation of the cross-sections and should permit a better comparison between them regardless of the different number of π electrons in each compound (see Section 3) [54]. The N_{eff} values were derived according to the method initially proposed by Kuzyk [52], by decomposing the various compounds into a collection of "independently" conjugated π manifolds (ESI, Figure S18).

2.3. Density Functional Theory (DFT) Calculations

DFT calculations (see computational details in Section 4.4) were performed on the **3-X** (X = NO_2, CN, H, OMe, NMe_2, NPh_2), **4** and **7-X** (X = H, NPh_2) derivatives. Similar computations were also performed on the model compounds **9a-b**, **10a-b**, and **11a-b** (Scheme 3) to specifically investigate the impact of the central core on the electronic structure of **1-X**, **2-X**, and **3-X** analogues. The key results of these are disclosed in the next section.

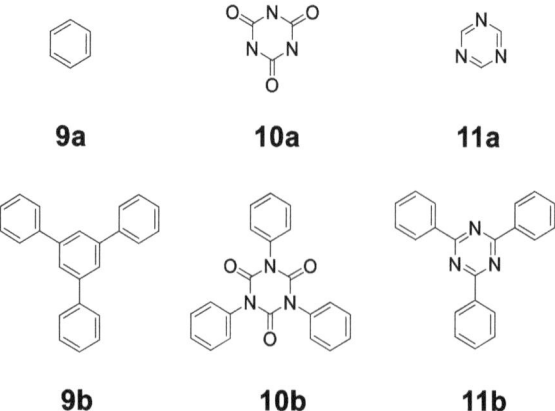

Scheme 3. Model compounds studied for investigating the changes occurring in the FMO when changing the central core.

All derivatives adopted a nearly coplanar conformation after geometrical optimisation relative to the central core upon geometry optimisation in CH_2Cl_2 (ESI, Table S1). As a result, in line with the available crystallographic data for **3-CN** [17] or **3-NPh$_2$**, **3-X** compounds appeared significantly more planar in solution than their **1-X** or **2-X** analogues (55–80°) [18]. Such planar conformations are likely favoured by weak intramolecular hydrogen bonds between *ortho* hydrogen atoms of the first aromatic ring and the triazine nitrogen atoms (with N···H- distances around 2.46 Å in optimised structures) [55,56]. Thus, the permanent dipole moment of **3-X** and **4** compounds with axially symmetric X substituents was nearly zero (ESI, Table S2) [11,33].

The HOMO–LUMO gap in these derivatives (Table 3) was smaller than that in the corresponding isocyanurate (**1-X**) and triphenylbenzene (**2-X**) analogues. Starting from **3-CN**, the calculations revealed that this gap progressively decreased when increasingly electron-releasing substituents were installed at their periphery, mirroring the trend observed experimentally for the lowest-lying intense absorptions of these compounds (Table 1). In line with experimental observations (Figure 1), replacing the peripheral 1,4-phenylene groups with a 2,7-fluorenyl one in **3-X**, when a strong electron-releasing X group such as X = NPh$_2$ was present (Figure 3), led only to a slight bathochromic shift of the first allowed absorption band computed at lowest energy for **7-X** (e.g., 371 nm (f = 3.03) for **3-NPh$_2$** vs. 375 nm (f = 3.73) for **7-NPh$_2$**, using CAM-B3LYP). In contrast, computations predicted that a much more pronounced shift could be expected for a less electron-releasing substituent such as X = H (e.g., 329 nm (f = 2.46) for **3-H** vs. 352 nm (f = 3.23) for **7-H**, using the same functional), reminiscent of observations previously made between **1-X** and **6-X** [33]. Calculations made for **3-X** compounds also indicated that, for X = CN (ESI; Figure S14) and more electron-withdrawing substituents, the direction of the photo-induced charge-transfer process was reversed, leading to a "umpolung" of the polarisation of the peripheral branches in the first allowed excited state, as already observed for the **2-X** analogues [12].

Table 3. Calculated (CAM-B3LYP/6-31G* level in CH_2Cl_2) HOMO–LUMO energy gaps in CH_2Cl_2 for **1-X**, **2-X**, and **3-X**.

X	HOMO–LUMO Gap (eV)		
	2-X	1-X	3-X
NO$_2$	5.62	7.06	5.82
CN	6.09	6.45	5.99
H	6.42	7.47	6.00
OMe	6.23	6.30	5.71
NMe$_2$	5.81	5.86	5.18
NPh$_2$	5.70	5.82	5.11

TD-DFT calculations qualitatively reproduced the energy trends observed for the most intense transitions at the lowest energy for **3-X**, **4**, and **7-X** (Table 1). Transition energies were consistently overestimated when using the CAM-B3LYP functional and underestimated when using the MPW1PW91 functional [57], a better agreement with the experiment being obtained using the CAM-B3LYP functional for the low energy band [58]. However, overall, better results were obtained with MPW1PW91 when the transition moments were also considered (ESI, Figure S13). The nature of the dominant excitations underlying the first absorption band (ESI, Table S4) was the same using either the CAM-B3LYP or the MPW1PW91 functional. It does not change much for all the compounds presently considered. These excitations are a pair of (nearly) degenerate transitions that would correspond to the set of degenerate transitions towards an E-type excited state under strict C_{3v} symmetry (i.e., an E←A transition). In line with previous findings for **1-X** and **2-X** [11,12,33], the first absorption band, therefore, corresponds to a $\pi \rightarrow \pi^*$ symmetric charge transfer (CT) between the peripheral arms and the central core.

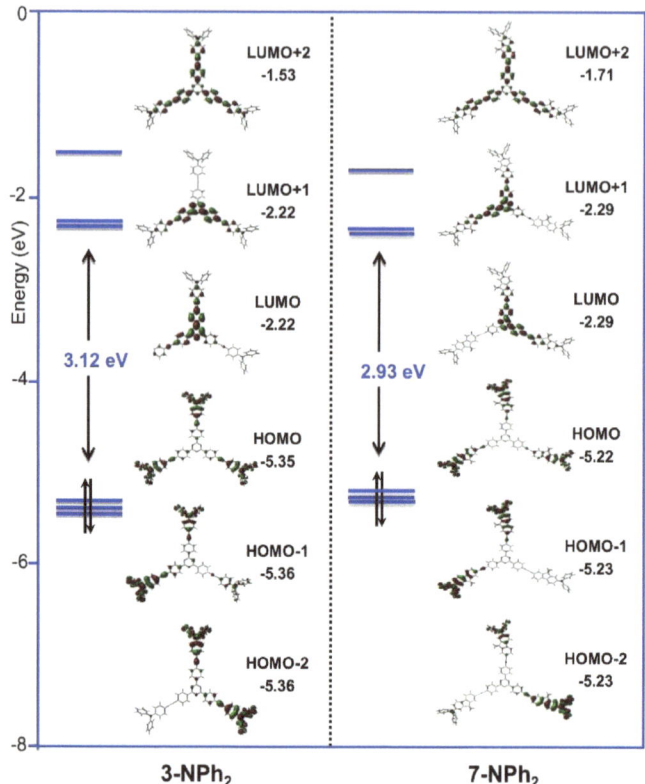

Figure 3. Frontier molecular orbitals involved in the lowest-energy (intense) allowed transition for 3-NPh$_2$ and 7-NPh$_2$ (isocontour 0.02 [e/bohr^3]$^{1/2}$).

Consistent with the acceptor character of the s-triazine ring, the CT occurs usually from the periphery towards the centre for all compounds, when X is electron-releasing [11,12,33]. The next and weaker absorption at higher energy observed for all these derivatives also corresponds to an allowed π–π* transition with a similar CT character but involves deeper-lying occupied MOs. As a result, this transition is more arm-centred than that at the lowest energy. However, it does not correspond to an n–π* transition, as previously proposed [18]. Indeed, according to our calculations, transitions with n–π* character were much weaker ($f < 0.1$) and remained hidden beneath the dominant bands at lowest energy (especially for 7-X derivatives) but also at higher energy.

In line with our previous studies [53,59], the simulations of 2PA spectra were also carried out for selected compounds using the damped cubic response theory of Jensen et al. [60]. The calculations were performed using the SAOP functional, for sake of consistency with previous computations already reported on these compounds (see computational details in Section 4.5) [53]. The simulated 2PA spectra (ESI, Figure S17) revealed a first 2PA band for these compounds at lower energy than that experimentally determined with significantly higher 2PA cross-section values (Table 2). While an overestimation of the real 2PA cross-sections by this method was expected [53], the experimental trends were generally qualitatively fairly well reproduced within a given family of compounds [59]. Based on these SAOP calculations, we speculate that the cross-section of the first 2PA band of **3-CN** or **3-NO$_2$**, which could not be experimentally determined, should be lower than that of the other **3-X** compounds. Interestingly, these computations revealed the existence of another and much larger 2PA peak at about 620 nm for **3-NPh$_2$**. This second 2PA band was hypsochromically shifted for **3-NMe$_2$** and **3-OMe**.

3. Discussion

As for their **1-X** and **2-X** analogues [12,33], the first allowed electronic transition at lowest energy for most of the **3-X** compounds corresponds to a $\pi \rightarrow \pi^*$ transition, with a charge transfer (CT) character directed from the electron-rich peripheral arms towards the central electron-poor core (Figure 3). In line with previous investigations on **1-X** [11,33] and **2-X** [4,14] derivatives, the available solvatochromic studies on some of these molecules [18,27] strongly suggest localisation of the excited state on one branch in the first excited state after vibrational relaxation. As surmised at the start of this study, DFT calculations on compounds **9a-b**, **10a-b**, and **11a-b** (Figure 4), confirm that the s-triazine central unit in **3-X** compounds is more electron-withdrawing than the isocyanurate core in **1-X** analogues. Indeed, regardless of the presence of three peripheral phenyl groups (**11b**) or not (**11a**), for a HOMO of comparable energy, the LUMO is always more stabilised in **11a-b** than in **10a-b**, resulting in lower HOMO–LUMO gaps for the triazine-cored compounds relative to their isocyanurate analogues.

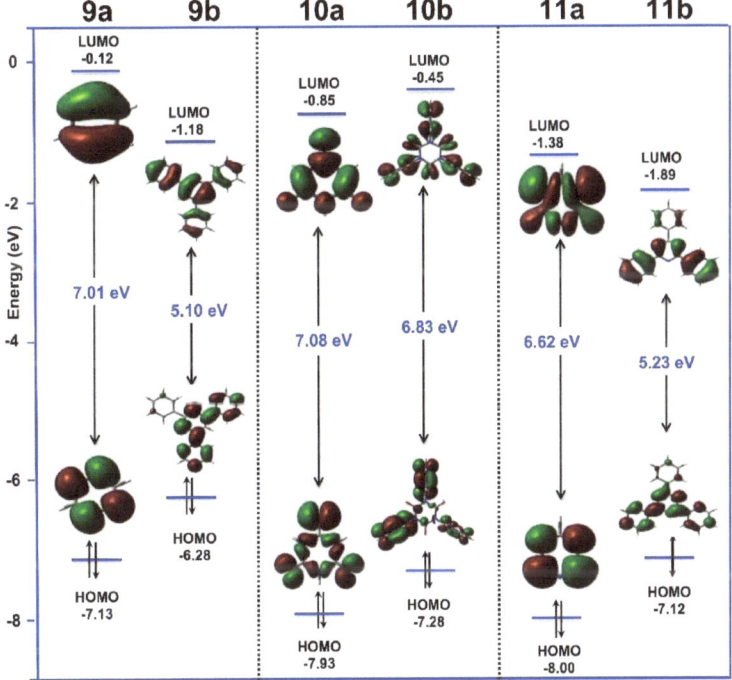

Figure 4. Frontier molecular orbitals of the **9a-11a** and **9b-11b** compounds modelling the central core in **2-X**, **1-X** and **3-X** (isocontour 0.02 [e/bohr^3]$^{1/2}$).

Furthermore, **3-X** compounds were also significantly more fluorescent than their **1-X** counterparts, a trend especially apparent with the weaker electron-releasing substituents (X = OMe, NMe$_2$). As for **2-X** derivatives, this might be related to the stronger transition moments observed for the lowest-energy 1PA absorptions in **3-X** compared with **1-X** analogues. Stronger transition moments should result in larger emission quantum yields, owing to the Strickler–Berg relationship [61], and in the absence of low-lying n–π^* states able to efficiently quench the fluorescence in **3-X** derivatives [11,62].

As previously observed for 1,3,5-triphenylbenzene- or isocyanurate-cored families of analogues, electron-releasing peripheral (X) substituents favour larger 2PA cross-sections (σ_2) [11,12]. Thus, the cross-sections determined by TPEF for the **3-X** derivatives were always significantly higher than those found for **1-X** and **2-X** (Table 2). In accordance with this observation, the σ_2 values calculated for **3-NMe$_2$** and for quite all other **3-X** derivatives (except for X = CN) were also higher than those previously computed for **1-NMe$_2$** or **2-NMe$_2$** (Table 2) [53]. In spite of the fact that SAOP calculations always overestimate 2PA cross-sections [53,59], the present study therefore provides further evidence that the qualitative ordering of the experimental σ_2 values is generally retrieved.

Then, the 2PA cross-section presently measured for **7-H** (710 GM) is significantly lower than the value of 2210 GM independently reported for its dihexyl analogue (**13-H**; Scheme 4) at nearly the same wavelength (730 nm vs. 740 nm) [27,63]. Actually, the value reported for **7-H** is closer to that reported for compound **14** (395 GM at 790 nm), although the latter value was not recorded at the TPA peak maximum (779 nm) and possibly includes some RSA contribution [64].

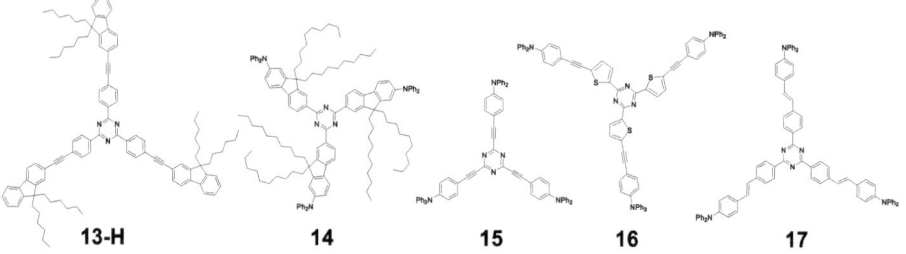

Scheme 4. Related derivative independently reported.

Comparison of the photophysical data presently measured for **3-NPh$_2$** with those reported for structurally related s-triazine derivatives such as **15** [19], **16** [22], and **17** [29] (Scheme 4) revealed larger fluorescence yields in chlorinated organic solvents for **3-NPh$_2$** (69% for the latter compound vs. 52%, 51%, and 27%, respectively). Regarding 2PA, removing the first 1,4-phenyl ring significantly reduced the 2PA cross-section (1500 GM for **3-NPh$_2$** vs. 910 GM reported for **15**), while replacing it with a 2,5-thienyl unit apparently did not impact this figure significantly (1508 GM reported for **16**) apart from slightly red-shifting the 2PA maximum (850 nm vs. 830 nm for **3-NPh$_2$**). Then, changing the alkynyl linkers for alkenyl ones in **3-NPh$_2$** seemed to reduce the 2PA cross-section. However, care should be given here since the value reported for **17** (495 GM) was obtained at a single wavelength (800 nm) which did not exactly match that of the 2PA maximum of this compound.

In terms of NLO activity per volume/mass unit, specific cross-sections [51] (Table 2) revealed that **3-X** derivatives are much more active than **1-X** and **2-X** (Table 2). More precisely, among all triazine derivatives investigated, **3-NPh$_2$** appears to be the best-suited two-photon absorber for elaborating molecular materials. As confirmed by other relevant figures of merit ($\sigma_2/(N_{eff})^2$) [54], these compounds optimise 2PA for a given number of effective electrons (N_{eff}), a feature possibly connected to their enforced planarity [65] but also certainly to their more pronounced (multi)polarity, compared to those of their **1-X** and **2-X** analogues. Both of these features ultimately translate into lower HOMO–LUMO gaps for **3-X** derivatives. Based on a simple perturbational approach, such a reduction in HOMO–LUMO gaps when progressing from **1-X** or **2-X** to **3-X** derivatives is expected to favour 2PA (at least at wavelengths above that of the lowest-energy absorption) [9]. Finally, the near-perfect spectral overlap between the 1PA spectrum plotted at half wavelength and the 2PA spectrum for each of these octupolar compounds (ESI, Figure S9) suggests a near degeneracy of the *A* and *E* excited states, in line with a small electronic/excitonic coupling between the peripheral branches [5,14,66]. As a result, the contribution of the octupolar symmetry to the 2PA improvement in these compounds should not be determining [67,68].

In terms of applied uses of these 2PA absorbers, both the significant bathochromic shift of 1PA and 2PA peaks and the larger intensity of the first absorption band result in a diminished transparency relative to **1-X** analogues, making them less favourable for selected NLO applications [6] such as optical limiting [69] or second harmonic generation in the visible range [70]. However, both the increased two-photon brightness [51] and the bathochromically-shifted 2PA maxima for the various **3-X** compounds (Table 2) points to a larger potential for two-photon fluorescence imaging. Such uses were already reported for structurally different s-triazine derivatives presenting much poorer figures of merit than **3-X** [71], but also for nanoparticles obtained from extended analogues of **17**, and giving rise to aggregation-induced emission (AIE) in water–THF mixtures [30]. Compared with **1-X** or **2-X** derivatives, the best candidate for fluorescence imaging is **3-NPh$_2$**. This compound has indeed a two-photon brightness far above those of its known triphenylbenzene and isocyanurate analogues (**1-NPh$_2$** and **2-NPh$_2$**, respectively) and a 2PA peak at 820 nm, i.e., also significantly bathochromically shifted relative to them (Table 2) [6,9]. In contrast, contrary to the trend previously observed with triarylisocyanurates derivatives [33], the tris(2-fluorenyl)triazine **7-NPh$_2$** presents neither a higher 2PA cross-section value nor a larger two-photon brightness than **3-NPh$_2$**.

4. Materials and Methods

4.1. General

All manipulations were carried out under an inert atmosphere of argon with dried and freshly distilled solvents [72]. Transmittance FTIR spectra were recorded using a Perkin Elmer Spectrum 100 spectrometer (Waltham, MA, USA) equipped with a universal ATR sampling accessory (400–4000 cm^{-1}). Raman spectra of the solid samples were obtained by diffuse scattering on a Bruker IFS 28 spectrometer and recorded in the 400–2500 cm^{-1} range (Stokes emission), with a laser excitation source at 1064 nm or on a Raman LabRAM HR 800 spectrometer with a laser excitation source at 785 nm. Nuclear magnetic resonance spectroscopy was performed using a Bruker AV-300 (300 MHz for ^{1}H, 75 MHz for ^{13}C) a Bruker AV-400 (400 MHz for ^{1}H, 101 MHz for ^{13}C) or a Bruker AV–500 (500 MHz for ^{1}H, 101 MHz for ^{13}C) spectrometers at ambient temperature. ^{1}H and ^{13}C spectra were calibrated using residual solvent peaks [73,74]. MS analyses were performed at the "Centre Regional de Mesures Physiques de l'Ouest" (CRMPO, Université de Rennes) on high resolution Bruker Maxis 4G or Thermo Fisher Q-Extractive Spectrometers. Elemental analyses were also performed at CRMPO. Chromatographic separations (rapid suction filtration (RSF), column chromatography, or flash chromatography) were performed on Merck silica gel (40–63 µm), Aldrich basic Alumina (act 1), or Aldrich neutral Alumina (act 1), with the eluants indicated. Commercial reagents and (pre/co)catalysts were used as received. The trisbromo s-triazine precursor **8-Br** was synthesised using a procedure inspired by the literature [42], which was subsequently extended to the known trisiodo analogue **8-I** (ESI). The triyne **8-C≡CH** was also obtained following modifications of the reported literature procedure (ESI) [16]. The known triazine derivatives **3-CN** [17], **3-OMe** [15], and **3-NPh$_2$** [18] were obtained according to published procedures and fully characterised (ESI), as were the required alkynes [33,75–78].

4.2. Synthesis of the New Triazine Compounds

2,4,6-tris{4'-[(4''-dimethylamino)-2'''-phenylethynyl]phenyl}-1,3,5-triazine (3-NMe$_2$). A dry Schlenk flask was charged with **8-I** (203 mg, 0.295 mmol), 4-ethynyl-*N*,*N*-dimethylaniline (211 mg, 1.453 mmol, 5 eq.), Pd(PPh$_3$)$_4$ (35 mg, 0.030 mmol, 10 mol%), and CuI (11 mg, 0.058 mmol, 20 mol%), and then degassed (4 × vacuum/argon cycles). A degassed mixture of DMF/^{i}Pr$_2$NH (2/1 mixture, 30 mL) was added using a cannula, and the flask was sealed and heated at 50 °C for 2 days. The solvent was removed in vacuo and the residue triturated with Et$_2$O to remove the unreacted starting material. The residue was dissolved in a mixture of pentane/CH$_2$Cl$_2$/NEt$_3$ (200 mL/200 mL/20 mL) and filtered through a short plug (2 cm × 2 cm) of neutral alumina (deactivated with NEt$_3$). The bright orange filtrate was evaporated under reduced pressure, giving the title product as an orange powder (141 mg, 65%). **MP:** > 150 °C (dec.). R_f: 0.38 (petroleum ether/CH$_2$Cl$_2$ [7:3]). **HRMS (ESI, MeOH):** m/z = 739.3542 [M + H]$^+$ (calc. for C$_{51}$H$_{43}$N$_6$: 739.3544). **Anal. Calc. for C$_{51}$H$_{42}$N$_6$•H$_2$O:** C, 80.92, H, 5.86, N, 11.10; found: C, 81.39, H, 6.25, N, 10.24. **^{1}H NMR (500 MHz, CDCl$_3$):** δ 8.74 (d, J = 8.4 Hz, 6H, H_{Ph}); 7.69 (d, J = 8.4 Hz, 6H, H_{Ph}); 7.48 (d, J = 8.8 Hz, 6H, $H_{Ph'}$); 6.71 (d, J = 8.0 Hz, 6H, $H_{Ph'}$); 3.02 (s, 18H, CH$_3$). **^{13}C{^{1}H} NMR (126 MHz, CDCl$_3$):** δ = 171.2 ($C_{triazine}$); 135.1 ($C_{Ph'}$); 133.1 (C_{Ph}H); 131.6 ($C_{Ph'}$H); 130.6 (C_{Ph}); 129.0 (C_{Ph}H); 128.7 (C_{Ph}); 125.7 ($C_{Ph'}$H); 112.0 ($C_{Ph'}$H); 94.1 (C≡C); 87.9 (C≡C); 40.4 (CH$_3$). **IR (KBr, cm^{-1}):** $\bar{v}$ = 2849 (w, C$_{Ar}$-H); 2201 (m, C≡C); 1598 (s, C=C$_{Ar}$); 1566 (m, C=C$_{Ar}$); 1507 (vs, C=N$_{triazine}$). **Raman (neat, cm^{-1}):** $\bar{v}$ = 2205 (s, C≡C); 1600 (vs, C=C$_{Ar}$); 1511 (m, C=N$_{triazine}$); 990 (w, C=N$_{triazine}$).

2,4,6-tris{4'-[(4''-nitro)-2''-phenylethynyl]phenyl}-1,3,5-triazine (3-NO$_2$). The compound was isolated using a workup similar than that described above from **8-C≡CH** (110 mg, 0.289 mmol) and 4-bromonitrobenzene in excess (293 mg, 1.45 mmol, 5 eq.) with Pd(PPh$_3$)$_4$ (10 mol%) and CuI (20 mol%) as catalysts in an DMF/NEt$_3$ mixture at 75 °C for 2.5 days. The volatiles were removed in vacuo and the residue triturated with small portions of CH$_2$Cl$_2$ and Et$_2$O to remove the unreacted 4-bromonitrobenzene and other soluble byproducts that formed during the reaction. The residue (ca. 90 mgs) was then dissolved in THF and filtered through a short plug (2 cm × 2 cm) and recrystallised several times to eventually yield pure fractions of yellow product that were used for characterisations. **MP:** > 250 °C (dec.). R_f: 0.65 (petroleum ether/THF [9:1]). **HRMS (MALDI-TOF, DTCB):** m/z = 745.183 [M + H]$^+$ (calc. for C$_{45}$H$_{25}$N$_6$O$_6$: 745.18301). **^{1}H NMR (300 MHz, THF-d_8):** δ = 8.90 (d, J = 8.7 Hz, AA'XX, 6H, $H_{Ph'}$); 8.31 (d, J = 8.7 Hz, AA'XX, 6H, H_{Ph}); 7.83 (m, 12H, $H_{Ph'}$). **^{13}C{^{1}H} NMR (125 MHz, THF-d_8):** δ = 172.2 ($C_{triazine}$); 148.8 ($C_{Ph'}$); 137.6 (C_{Ph}); 133.6 ($C_{Ph'}$H); 133.1 (C_{Ph}H); 130.4 (C_{Ph}); 130.2 (C_{Ph}H); 127.8 ($C_{Ph'}$); 124.8 ($C_{Ph'}$H); 94.4 (C≡C); 91.3 (C≡C). **IR (KBr, cm^{-1}):** $\bar{v}$ = 2213 (w, C≡C); 1592 (s, C=C$_{Ar}$); 1569 (m, C=C$_{Ar}$); 1510 (vs, NO$_{2-sym}$); 1506 (vs, C=N$_{triazine}$); 1340 (vs, NO$_{2-asym}$). **Raman (neat, cm^{-1}):** $\bar{v}$ = 2217 (vs, C≡C); 1600 (vs, C=C$_{Ar}$); 1510 (m, C=N$_{triaz}$); 1341 (m, NO$_{2-asym}$); 990 (w, C=N$_{triazine}$).

2,4,6-tris{4'-[2'''-(4''-pyridyl)ethynyl]phenyl}-1,3,5-triazine (4). A solution of TBAF (1.0 M in THF, 0.2 mL, 0.2 mmol) was evaporated to dryness in a dry Schlenk flask. The flask was then charged with **8-I** (394 mg, 0.57 mmol), (trimethylsilyl)ethnynyl pyridine (566 mg, 3.23 mmol), Pd(PPh$_3$)$_4$ (78 mg, 0.067 mmol) and CuI (23 mg, 0.121 mmol) and then degassed (4 × vacuum/argon cycles). The flask was wrapped in foil before degassed DMF (20 mL) and NEt$_3$ (9 mL) were added using a cannula. The reaction mixture was then stirred at 25 °C for 2 days. The solvent was removed in vacuo, and the residue was dissolved in CH$_2$Cl$_2$ (100 mL), washed with water (3 × 50 mL), and dried (MgSO$_4$). The solvent was removed under reduced pressure, and the crude material purified using column chromatography (neutral alumina, 5 cm × 6 cm, eluting with a methanol/CH$_2$Cl$_2$ [2:98] mixture). The resulting solid was then precipitated from methanol/CH$_2$Cl$_2$, with the slow evaporation of the CH$_2$Cl$_2$, and collected on a glass frit, washed with methanol, and dried under reduced pressure (high vacuum) for 12 h giving the title product as an off-white coloured powder (261 mg, 75%). **MP:** 310 °C (dec.). R_f: 0.60 (MeOH/CH$_2$Cl$_2$ [2.5:97.5]). **HRMS (ESI, MeOH):** m/z = 613.2134 [M + H]$^+$ (calc. for C$_{42}$H$_{25}$N$_6$: 613.2135). **Anal. Calc. for C$_{42}$H$_{24}$N$_6$•½H$_2$O:** C, 77.91, H, 3.85, N, 12.83; found: C, 78.35, H, 3.80, N, 12.89. **^{1}H NMR (400 MHz, CD$_2$Cl$_2$):** δ = 8.76 (d, J = 8.2 Hz, 6H, H_{Ph}); 8.61 (d, J = 5.6 Hz, 6H, $H_{Ph'}$); 7.76 (d, J = 8.2 Hz, 6H, H_{Ph}); 7.44 (d, J = 5.6 Hz, 6H, $H_{Ph'}$). **^{13}C{^{1}H} NMR (101 MHz, CD$_2$Cl$_2$):** δ = 171.6 ($C_{triazine}$); 150.5 (C_{Py}H); 136.9 (C_{Ph}); 132.7 (C_{Ph}H); 131.4 (C_{Py}); 129.5 (C_{Ph}H); 127.0 (C_{Ph}); 126.1 (C_{Py}H); 93.6 (C≡C); 89.8 (C≡C). **IR (KBr, cm^{-1}):** $\bar{v}$ = 2218 (w, C≡C); 1598 & 1569 (m, C=C$_{Ar}$); 1515 (vs, C=N$_{triazine}$). **Raman (neat, cm^{-1}):** $\bar{v}$ = 2223 (s, C≡C); 1605 (vs, C=C$_{Ar}$); 1520 (w, C=N$_{triazine}$); 992 (w, C=N$_{triazine}$).

2,4,6-tris{4'-[2'''-(9'',9''-dibutyl-2''-fluorenyl)ethynyl]-phenyl}-1,3,5-triazine (7-H). A dry Schlenk flask was charged with **8-Br** (100 mg, 0.183 mmol), 2-ethynyl-9,9-dibutylfluorene (263 mg, 0.869 mmol), Pd(PPh$_3$)$_4$ (24 mg, 0.021 mmol), and CuI (8 mg, 0.042), and then degassed (4 × vacuum/argon cycles). A mixture of dry, degassed DMF/^{i}Pr$_2$NH (1/1 mixture, 30 mL) was added using a cannula. The flask was sealed and the reaction mixture was heated at 70 °C for 3 days. The solvent was removed in vacuo, and the residue was dissolved in CH$_2$Cl$_2$ (100 mL), washed with water (3 × 30 mL) and dried (MgSO$_4$). The solvent was removed under reduced pressure, and the crude material was purified using flash column chromatography (silica gel, 3.5 cm × 15 cm, eluting with an hexanes/CH$_2$Cl$_2$ [9:1] mixture), giving the title product as a pale yellow solid (198 mg, 89%). **MP:** 146 °C. R_f: 0.52 (petroleum ether/CH$_2$Cl$_2$ [85:15]). **HRMS (ESI, MeOH):** m/z = 1210.6970 [M + H]$^+$ (calc. for C$_{90}$H$_{88}$N$_3$: 1210.6973). Anal. Calc. for C$_{90}$H$_{87}$N$_3$: C, 89.29, H, 7.24, N, 3.47; found: C, 89.33, H, 7.54, N, 3.50. **^{1}H NMR (400 MHz, CDCl$_3$):** δ = 8.80 (d, J = 8.4 Hz, 6H, H_{Ph}); 7.79 (d, J = 8.4 Hz, 6H, H_{Ph}); 7.73–7.71 (m, 6H, H_{Flu}); 7.61–7.58 (m, 6H, H_{Flu}); 7.36–7.35 (m, 9H, H_{Flu}); 2.02 (t, J = 8.0 Hz, 12H, CH$_{2\text{-Bu}}$); 1.16–1.07 (m, 12H, CH$_{2\text{-Bu}}$); 0.70 (t, J = 7.4 Hz, 18H, CH$_{3\text{-Bu}}$); 0.66–0.50 (m, 12H, CH$_{2\text{-Bu}}$). **^{13}C{^{1}H} NMR (101 MHz, CDCl$_3$):** δ = 171.3 (C_{triazine}); 151.2; 151.0 (C_{Flu}); 142.0; 140.5; 135.7; 132.0; 131.0; 129.1; 128.0; 127.8; 127.1; 126.2; 123.1; 121.1; 120.2; 119.9; 93.78 (C≡C); 89.5 (C≡C); 55.3 (C_{Flu}); 40.4 (CH$_{2\text{-Bu}}$); 26.1 (CH$_{2\text{-Bu}}$); 23.2 (CH$_{2\text{-Bu}}$); 14.0 (CH$_{3\text{-Bu}}$). **IR (KBr, cm^{-1}):** $\bar{\nu}$ = 2953, 2925, 2854 (m, C_{Ar}-H); 2196 (w, C≡C); 1603 (m, C=C_{Ar}); 1569 (s, C=C_{Ar}); 1507 (vs, C=N_{triazine}). **Raman (neat, cm^{-1}):** $\bar{\nu}$ = 2202 (s, C≡C); 1605 (vs, C=C_{Ar}); 1511 (w, C=N_{triazine}); 991 (vw, C=N_{triazine}).

2,4,6-tris{4'-[2'''-(9'',9''-dibutyl-7''-diphenylamino-2''-fluorenyl)ethynyl]phenyl}-1,3,5-triazine (**7-NPh$_2$**). A dry Schlenk flask was charged with **8-Br** (49 mg, 0.090 mmol), 2-ethynyl-9,9-dibutyl-7-diphenylaminofluorene (198 mg, 0.422 mmol), Pd(PPh$_3$)$_4$ (11 mg, 0.009 mmol), and CuI (4 mg, 0.020 mmol), and then degassed (3 × vacuum/argon cycles). A mixture of dry, degassed DMF/^{i}Pr$_2$NH (2/1 mixture, 20 mL) was added using a cannula. The flask was sealed, and the reaction mixture was heated at 70 °C for 2.5 days. The solvent was removed in vacuo, and the residue was dissolved in CH$_2$Cl$_2$ (70 mL), washed with water (2 × 20 mL), and dried (MgSO$_4$). The solvent was removed under reduced pressure, and the crude material was purified using flash column chromatography (neutral alumina deactivated with NEt$_3$, eluting with mixture of ether/hexanes/NEt$_3$ gradient from 50/450/5 mL up to 245/250/5 mL mixture), giving the title product as a bright yellow waxy solid (27 mg, 18%). **MP:** 140 °C (dec.). R_f: 0.34 (petroleum ether/CH$_2$Cl$_2$ [7:3]). **HRMS (ESI, CHCl$_3$/MeOH [8:2]):** m/z = 855.4542 [M]$^{2+}$ (calc. for C$_{126}$H$_{114}$N$_6$: 855.4547). Anal. Calc. for C$_{126}$H$_{114}$N$_6$•½CH$_2$Cl$_2$: C, 86.58, H, 6.61, N, 4.79; found: C, 86.72, H, 6.89, N, 4.87. **^{1}H NMR (400 MHz, CDCl$_3$):** δ = 8.81 (d, J = 8.6 Hz, AA'XX', 6H, H_{Ph}); 7.80 (d, J = 8.6 Hz, AA'XX', 6H, H_{Ph}); 7.65–7.55 (m, 12H, H_{Flu}); 7.32–7.25 (m, 12H, H_{Ph2N}); 7.19–7.14 (m, 15H, H_{Ph2N} and H_{Ph2N}); 7.08–7.03 (m, 9H, H_{Ph2N} and H_{Ph2N}); 1.92 (m, 12H, CH$_{2\text{-Bu}}$); 1.13 (m, 12H, CH$_{2\text{-Bu}}$); 0.80–0.65 (m, 30H, CH$_{3\text{-Bu}}$ & CH$_{2\text{-Bu}}$). **^{1}H NMR (400 MHz, CDCl$_3$):** δ = 171.3 (C_{triazine}); 152.8; 150.9 (C_{Flu}); 148.0; 147.9; 141.9; 135.6; 135.5; 131.9; 131.1; 129.4; 129.1; 128.1; 126.1; 124.2; 123.4; 122.9; 120.9; 120.3; 119.2; 119.1; 93.8 (C≡C); 89.4 (C≡C); 55.2 (C_{Flu}); 40.1 (CH$_{2\text{-Bu}}$); 26.2 (CH$_{2\text{-Bu}}$); 23.2 (CH$_{2\text{-Bu}}$); 14.0 (CH$_{3\text{-Bu}}$). **IR (KBr, cm^{-1}):** $\bar{\nu}$ = 2194 (w, C≡C); 1593 (s, C=C_{Ar}); 1569 (s, C=C_{Ar}); 1508 (vs, C=N_{triazine}); 1490 (vs, C=N_{triazine}). **Raman (neat, cm^{-1}):** $\bar{\nu}$ = 2199 (w, C≡C); 1601 (vs, C=C_{Ar}); 1569 (w, C=C_{Ar}); 1508 (w, C=N_{triazine}); 989 (w, C=N_{triazine}).

4.3. Fluorescence Measurements

All photophysical measurements were performed with freshly prepared air-equilibrated CH$_2$Cl$_2$ (or THF) solutions (HPLC grade) at room temperature (298 K). UV–Vis absorption spectra were recorded on dilute solutions (ca. 10^{-5} M) by using a Jasco V-570 spectrophotometer (Mary's Court, Easton MD, USA). The samples used to make the solutions were freshly recrystallised or thoroughly washed with cooled ether/pentane prior to the measurements to remove any organic impurity. Steady-state fluorescence studies were performed in diluted air-equilibrated solutions in quartz cells of 1 cm path length (ca. 1 × 10^{-6} M, optical density < 0.1) at room temperature (20 °C), using an Edinburgh Instruments (FLS920) spectrometer (Edinburgh, UK) in photon-counting mode. Fully corrected excitation and emission spectra were obtained, with an optical density at $\lambda_{\text{exc}} \leq 0.1$. The fluorescence quantum yield of each compound was calculated using the integral of the fully corrected emission spectra relative to a standard, quinine bisulphate (QBS, λ_{ex} = 346 nm, Φ_F = 0.546) [79,80]. UV–Vis absorption spectra used for the calculation of the fluorescence quantum yields were recorded using a double-beam Jasco V-570 spectrometer.

4.4. Two-Photon Absorption Experiments

To span the 790–920 nm range, an Nd:YLF-pumped Ti:sapphire oscillator (Chameleon Ultra, Coherent) was used, generating 140 fs pulses at an 80 MHz rate. The excitation power was controlled using neutral density filters of varying optical density mounted in a computer-controlled filter wheel. After a fivefold expansion through two achromatic doublets, the laser beam was focused with a microscope objective (10×, NA 0.25, Olympus, Shinjuku, Tokyo, Japan) into a standard 1 cm absorption cuvette containing the sample. The applied average laser power arriving at the sample was typically between 0.5 and 40 mW, leading to a time-averaged light flux in the focal volume on the order of 0.1–10 mW/mm^2. The fluorescence intensity was measured at several excitation powers in this range owing to the filter wheel. For each sample and each wavelength, the quadratic dependence of the fluorescence intensity (F) on the excitation intensity (P), i.e., the linear dependence of F on P^2 was systematically checked (see ESI, Figure S8). The fluorescence from the sample was collected in epifluorescence mode, using the microscope objective, and reflected by a dichroic mirror (Chroma Technology Corporation, Bellows Falls, VT, USA; "red" filter set: 780dxcrr). This made it possible to avoid the inner filter effects related to the high dye concentrations used (10^{-4} M) by focusing the laser near the cuvette window. Residual excitation light was removed using a barrier filter (Chroma Technology; "red": e750sp–2p). The fluorescence was coupled into a 600 μm multimode fibre with an achromatic doublet. The fibre was connected to a compact CCD-based spectrometer (BTC112-E, B&WTek, Newark DE, USA), which measured the two-photon excited emission spectrum. The emission spectra were corrected for the wavelength dependence of the detection efficiency using correction factors established through the measurement of reference compounds having known fluorescence emission spectra. Briefly, the setup allowed for the recording of corrected fluorescence emission spectra under multiphoton excitation at variable excitation power and wavelengths. Further, 2PA cross-sections (σ_2) were determined from the two-photon excited fluorescence (TPEF) cross-sections ($\sigma_2 \cdot \Phi_F$) and the fluorescence emission quantum yield (Φ_F). TPEF cross-sections of 10^{-4} M dichloromethane solutions were measured relative to fluorescein in 0.01 M aqueous NaOH using the well-established method described by Xu and Webb [81] and the appropriate solvent-related refractive index corrections [82]. To check the absence of aggregation, UV–Vis absorption spectra of the dyes were recorded at this concentration in cells of 1 mm pathlength and compared with those obtained with diluted solutions in cells of 1 cm pathlength.

4.5. DFT Computations

DFT and TD-DFT calculations reported in this study were performed using the Gaussian09 [83] program. The geometries of all compounds were optimised without symmetry constraints using the CAM-B3LYP [84] or the MPW1PW91 [57] functionals and the 6-31G* basis set. The solvent effects were taken into account by means of the polarizable continuum model (PCM) [85]. Calculations of the normal modes of vibration were carried out to confirm the true minima character of the optimised geometries. TD-DFT calculations were performed at the same level of theory using the previously optimised geometries. A value of 0.1 was considered for the damping parameter when simulating the electronic spectra (ESI). Swizard [86] was used to plot the simulated spectra, and GausView [87] was used for the MO plots. Subsequently, the 2PA properties were calculated for selected compounds using the SAOP model potential [88,89] (statistical average of orbital model exchange-correlation potential) using the damped cubic response theory module of Jensen et al. [60] implemented in the ADF program package [90], considering the different molecules in the gas phase with the optimised geometries obtained in solution (CH_2Cl_2). The lifetime of the electronically excited states was included in the theory using a damping parameter of 0.0034 au (~0.1 eV ~ 800 cm^{-1}) value, which was found suitable for 2PA computations [91,92]. The 2PA cross-section σ_2 was obtained from the imaginary part of the third-order hyperpolarizability γ using ad hoc expressions (see ESI) [59]. Cross-section (σ_2) values are usually given in Göppert–Mayer units (1 GM = 10^{-50} cm^4·s·photon^{-1}), so we first evaluated them in atomic units and then multiplied them by $(0.529177 \times 10^{-8}$ cm/a.u.$)^4 \times (2.418884 \times 10^{-17}$ s/a.u.) to obtain values in the conventional units (cm^4·s·photon^{-1}).

5. Conclusions

The two-photon absorption properties of several 2,4,6-triaryl-s-triazines derivatives such as **3-X** (X = NO$_2$, CN, H, OMe, NMe$_2$, NPh$_2$), **4**, and **7-X** (X = H, NPh$_2$) were studied. After characterising the new members of these families (X = **3-NO$_2$**, **3-NMe$_2$**, **4**, **7-H**, and **7-NPh$_2$**), we uncovered evidence that significant 2PA occurs at the near-IR edge (730–820 nm) for most of them. Similar to what was previously shown for 1,3,5-triarylbenzene or N,N',N''-triarylisocyanurate analogues, this nonlinear absorption process apparently populates the π-π* charge-transfer excited state(s) at the lowest energy. As confirmed by DFT calculations, the charge transfer occurring during 2PA corresponds to a shift of electron density from the periphery (arms) toward the central ring, except for strongly electron-withdrawing X substituents (X = NO$_2$ and CN) for which the charge transfer direction reverts. Based on our calculations, compared with **1-X** and **2-X**, the improved 2PA properties of **3-X** derivatives find their origin in the quasi-planar conformation adopted by these molecules in solution and also in the increased polarisation of their π manifold, two features directly resulting from the presence of the central s-triazine unit. The latter favours π-π interactions between peripheral arms and the central core, resulting in a smaller HOMO–LUMO gap for these molecules. We also showed that all **3-X** and **7-X** derivatives featuring neutral to electron-releasing peripheral groups were sufficiently emissive for performing two-photon imaging purposes in solution ($\Phi_F > 0.2$). Actually, in line with extent data for other s-triazines in the literature, they presented larger two-photon brightness than all **1-X**, **2-X**, and **6-X** fluorophores investigated so far. This statement, coupled to the fact that their first 2PA peak is always bathochromically shifted relative to that of their 1,3,5-triarylbenzene or N,N',N''-triaryl isocyanurate analogues, points to a larger potential for **3-X** derivatives, with **3-NMe$_2$** and **3-NPh$_2$** being the most promising molecules in this respect. Provided that such compounds can now be made water-soluble without changing the observed trends, they should give rise to appealing new two-photon dyes for fluorescence bioimaging. Research along these lines is in progress.

Supplementary Materials: The following supporting information can be downloaded at: https://www.mdpi.com/article/10.3390/photochem2020023/s1, Experimental part; NMR, absorption, and emission spectra of **3-X** derivatives; TPEF data for selected **3-X** derivatives; Cartesian coordinates of optimised geometries and selected FMOs for **3-X** and related model derivatives; computed dipole moments and bandgaps; computed 1PA (TD-DFT) and 2PA (-DPZ) spectra; derivation of N_{eff} values.

Author Contributions: Conceptualisation, F.P.; funding acquisition, F.P.; investigation, A.G.B., N.R. and A.A.; methodology, A.G.B. and N.R.; resources, M.B.-D.; supervision, A.B., O.M. and F.P.; writing—original draft preparation, A.G.B. and A.A.; writing—review and editing, A.B., O.M. and F.P. All authors have read and agreed to the published version of the manuscript.

Funding: This project was supported by Région Bretagne (SAD Project *Fotoporf*, N°7205) and ANR (ANR-17-CE07-0033-01 project).

Data Availability Statement: Not applicable.

Acknowledgments: The Region Bretagne (PhD grant for A.G.B.) and ANR (*Isogate* Project) are acknowledged for financial support. We also acknowledge the HPC resources of CINES and of IDRIS under the allocations [2021-x080649] made by GENCI (Grand Equipement National de Calcul Intensif), as well as the SIR platform of ScanMAT at the University of Rennes 1 for technical assistance during the Raman measurements.

Conflicts of Interest: The authors declare no conflict of interest.

Sample Availability: Samples of the compounds are available from the authors.

References and Notes

1. Wolff, J.J.; Wortmann, R. Organic materials for non-linear optics. *J. Prakt. Chem.* **1998**, *340*, 99–111. [CrossRef]
2. Zyss, J.; Ledoux, I. Nonlinear optics in multipolar media: Theory and experiment. *Chem. Rev.* **1994**, *94*, 77–105. [CrossRef]
3. Lee, S.-H.; Park, J.-R.; Jeong, M.-Y.; Kim, H.M.; Li, S.; Song, J.; Ham, S.; Jeon, S.-J.; Cho, B.R. First hyperpolarizability of 1,3,5-Tricyanobenzene Derivatives: Origin of Larger beta Values for the Octupoles than for the Dipoles. *ChemPhysChem* **2006**, *7*, 206–212. [CrossRef] [PubMed]
4. Brunel, J.; Mongin, O.; Jutand, A.; Ledoux, I.; Zyss, J.; Blanchard-Desce, M. Propeller-shaped octupolar molecules derived from triphenylbenzene for nonlinear optics: Synthesis and optical studies. *Chem. Mater.* **2003**, *15*, 4139–4148. [CrossRef]
5. Terenziani, F.; Katan, C.; Badaeva, E.; Tretiak, S.; Blanchard-Desce, M. Enhanced Two-Photon Absorption of Organic Chromophores: Theoretical and Experimental Assessments. *Adv. Mater.* **2008**, *20*, 4641–4678. [CrossRef]
6. He, G.S.; Tan, L.-S.; Zheng, Q.; Prasad, P.N. Multiphoton Absorbing Materials: Molecular Designs, Characterizations, and Applications. *Chem. Rev.* **2008**, *108*, 1245–1330. [CrossRef]
7. Cho, B.R.; Son, K.H.; Lee, S.H.; Song, Y.-S.; Lee, Y.-K.; Jeon, S.-J.; Choi, J.H.; Lee, H.; Cho, M. Two Photon Absorption Properties of 1,3,5-Tricyano-2,4,6-tris(styrl)Benzene Derivatives. *J. Am. Chem. Soc.* **2001**, *123*, 10039–10045. [CrossRef]
8. Lee, W.H.; Lee, S.H.; Kim, J.-A.; Choi, J.H.; Cho, M.; Jeon, S.-J.; Cho, B.R. Two Photon Absorption and Nonlinear Optical Properties of Octupolar Molecules. *J. Am. Chem. Soc.* **2001**, *123*, 10658–10667. [CrossRef]
9. Pawlicki, M.; Collins, H.A.; Denning, R.G.; Anderson, H.L. Two-Photon Absorption and the Design of Two-Photon Dyes. *Angew. Chem. Int. Ed.* **2009**, *48*, 3244–3266. [CrossRef]
10. Argouarch, G.; Veillard, R.; Roisnel, T.; Amar, A.; Boucekkine, A.; Singh, A.; Ledoux, I.; Paul, F. Donor-substituted Triaryl-1,3,5-Triazinane-2,4,6-Triones: Octupolar NLO-phores with a Remarkable Transparency–Nonlinearity Trade-off. *New J. Chem.* **2011**, *35*, 2409–2411. [CrossRef]
11. Argouarch, G.; Veillard, R.; Roisnel, T.; Amar, A.; Meghezzi, H.; Boucekkine, A.; Hugues, V.; Mongin, O.; Blanchard-Desce, M.; Paul, F. Triaryl-1,3,5-Triazinane-2,4,6-Triones (Isocyanurates) Peripherally Functionalized by Donor Groups: Synthesis and Study of their Linear and Nonlinear Optical Properties. *Chem. Eur. J.* **2012**, *18*, 11811–11826. [CrossRef] [PubMed]
12. Streatfield, S.L.; Pradels, C.; Ngo Ndimba, A.; Richy, N.; Amar, A.; Boucekkine, A.; Cifuentes, M.P.; Humphrey, M.G.; Mongin, O.; Paul, F. Electronic Absorption, Emission and Two-Photon Absorption Properties of Some Functional 1,3,5-Triphenylbenzenes. *Chem. Select.* **2017**, *2*, 8080–8085. [CrossRef]
13. Hoffmann, D.K. Model system for a urethane-modified isocyanurate foam. *J. Cell. Plast.* **1984**, *20*, 129–137. [CrossRef]
14. Terenziani, F.; Le Droumaguet, C.; Katan, C.; Mongin, O.; Blanchard-Desce, M. Effect of branching on Two-Photon Absorption in Triphenylbenzene Derivatives. *ChemPhysChem* **2007**, *8*, 723–734. [CrossRef] [PubMed]
15. Lee, C.-H.; Yammamoto, T. Synthesis of Liquid-Crystalline, Highly Luminescent π-Conjugated 1,3,5-Triazine Derivatives by Palladium-Catalyzed Cross-Coupling Reaction. *Mol. Cryst. Liq. Cryst.* **2002**, *378*, 13–21. [CrossRef]
16. Hu, Q.Y.; Lu, W.X.; Tang, H.D.; Sung, H.H.Y.; Wen, T.B.; Williams, I.D.; Wong, G.K.L.; Lin, Z.; Jia, G. Synthesis and photophysical properties of trimetallic acetylide complexes with a 1,3,5-triazine core. *Organometallics* **2005**, *24*, 3966–3973. [CrossRef]
17. Ranganathan, A.; Heisen, B.C.; Dix, I.; Meyer, F. A triazine-based three directional rigid-rod tecton forms a novel 1D channel structure. *Chem. Commun.* **2007**, *43*, 3637–3639. [CrossRef]
18. Das, P.; Kumar, A.; Chowdhurry, A.; Mukherjee, P.S. Aggregation-Induced Emission and White Luminescence from a Combination of π-Conjugated Donor–Acceptor Organic Luminogens. *ACS Omega* **2018**, *3*, 13757–13771. [CrossRef]
19. Zhang, L.; Zou, L.; Xiao, J.; Zhou, P.; Zhong, C.; Chen, X.; Qin, J.; Mariz, I.F.A.; Maçôas, E. Symmetrical and unsymmetrical multibranched D–π–A molecules based on 1,3,5-triazine unit: Synthesis and photophysical properties. *J. Mater. Chem.* **2012**, *22*, 16781–16790. [CrossRef]
20. Cai, Z.-B.; Chen, L.-J.; Li, S.-L.; Ye, Q.; Tian, Y.-P. Synthesis, electronic structure, linear and nonlinear photophysical properties of novel asymmetric branched compounds. *Dye. Pigm.* **2020**, *175*, 108115. [CrossRef]
21. Cui, Y.Z.; Fang, Q.; Xue, G.; Xu, G.B.; Yin, L.; Yu, W.T. Cooperative enhancement of two-photon absorption of multibranched compounds with vinylenes attaching to the s-triazine core. *Chem. Lett.* **2005**, *34*, 644–645. [CrossRef]
22. Zou, L.; Liu, Z.; Yan, X.; Liu, Y.; Fu, Y.; Liu, J.; Huang, Z.; Chen, X.; Qin, J. Star-Shaped D-π-A Molecules Containing a 2,4,6-Tri(thiophen-2-yl)-1,3,5-triazine Unit: Synthesis and Two-Photon Absorption Properties. *Eur. J. Org. Chem.* **2009**, *2009*, 5587–5593. [CrossRef]
23. Mariz, I.F.A.; Maçôas, E.M.S.; Martinho, J.M.G.; Zou, L.; Zhou, P.; Chen, X.; Qin, J. Molecular architecture effects in two-photon absorption: From octupolar molecules to polymers and hybrid polymer nanoparticles based on 1,3,5-triazine. *J. Mater. Chem. B* **2013**, *1*, 2169–2177. [CrossRef] [PubMed]
24. Liu, S.; Lin, K.S.; Churikov, V.M.; Su, Y.Z.; Lin, J.T.S.; Huang, T.-H.; Hsu, C.C. Two-photon absorption properties of star-shaped molecules containing peripheral diarylthienylamines. *Chem. Phys. Lett.* **2004**, *390*, 433–439. [CrossRef]

25. Zou, L.; Liu, Y.; Ma, N.; Macoas, E.; Martinho, J.M.G.; Pettersson, M.; Chen, X.; Qin, J. Synthesis and photophysical properties of hyperbranched polyfluorenes containing 2,4,6-tris(thiophen-2-yl)-1,3,5-triazine as the core. *Phys. Chem. Chem. Phys.* **2011**, *13*, 8838–8846. [CrossRef]
26. Vidya, V.M.; Chetti, P. A DFT probe on the linear and nonlinear optical characteristics of some star-shaped DI-π-DII-A type acetylene-bridged rigid triazines. *J. Phys. Org. Chem.* **2020**, *33*, e4027.
27. Jiménez-Sánchez, A.; Isunza-Manrique, I.; Ramos-Ortiz, G.; Rodríguez-Romero, J.; Farfán, N.; Santillan, R. Strong Dipolar Effects on an Octupolar Luminiscent Chromophore: Implications on their Linear and Nonlinear Optical Properties. *J. Phys. Chem. A* **2016**, *120*, 4314–4324. [CrossRef]
28. Li, B.; Tong, R.; Zhu, R.; Meng, F.; Tian, H.; Qian, S. The Ultrafast Dynamics and Nonlinear Optical Properties of Tribranched Styryl Derivatives Based on 1,3,5-Triazine. *J. Phys. Chem. B* **2005**, *109*, 10705–10710. [CrossRef]
29. Jiang, Y.; Wang, Y.; Wang, J.; He, N.; Qian, S.; Hua, J. Synthesis, Two-Photon Absorption and Optical Limiting Properties of Multi-branched Styryl Derivatives Based on 1,3,5-Triazine. *Chem. Asian J.* **2011**, *6*, 157–165. [CrossRef]
30. Gao, Y.; Qu, Y.; Jiang, T.; Zhang, H.; He, N.; Li, B.; Wu, J.; Hua, J. Alkyl-triphenylamine end-capped triazines with AIE and large two-photon absorption cross-sections for bioimaging. *J. Mat. Chem. C* **2014**, *2*, 6353–6361. [CrossRef]
31. Kukhta, N.A.; Simokaitiene, J.; Volyniuk, D.; Ostrauskaite, J.; Grazulevicius, J.V.; Juska, G.; Jankauskas, V. Effect of linking topology on the properties of star-shaped derivativesof triazine and fluorene. *Synth. Met.* **2014**, *195*, 266–275. [CrossRef]
32. Drouet, S.; Merhi, A.; Argouarch, G.; Paul, F.; Mongin, O.; Blanchard-Desce, M.; Paul-Roth, C.O. Synthesis of Luminescent Supramolecular Assemblies from Fluorenyl Porphyrins and Polypyridyl Isocyanurate-based Spacers. *Tetrahedron* **2012**, *68*, 98–105. [CrossRef]
33. Gautier, Y.; Argouarch, G.; Malvolti, F.; Blondeau, B.; Richy, N.; Amar, A.; Boucekkine, A.; Nawara, K.; Chlebowicz, K.; Orzanowska, G.; et al. Triarylisocyanurate-Based Fluorescent Two-Photon Absorbers. *ChemPlusChem* **2020**, *85*, 411–425. [CrossRef] [PubMed]
34. Lee, C.-H.; Yammamoto, T. Synthesis and characterization of a new class of liquid-crystalline, highly luminescent molecules containing a 2,4,6-triphenyl-1,3,5-triazine unit. *Tet. Lett.* **2001**, *42*, 3993–3996. [CrossRef]
35. Iijima, T.; Lee, C.-H.; Fujiwara, Y.; Shimokawa, M.; Suzuki, H.; Yamane, K.; Yamamoto, T. Photoluminescence of 2,4,6-tris[4-(phenylethynyl)phenyl]-1,3,5-triazines dispersed in polymer films. *Opt. Mater.* **2007**, *29*, 1782–1788. [CrossRef]
36. Jiang, Y.; Wang, Y.; Hua, J.; Tang, J.; Li, B.; Qian, S.; Tian, H. Multibranched triarylamine end-capped triazines with aggregation-induced emission and large two-photon absorption cross-sections. *Chem. Commun.* **2010**, *46*, 4689–4691. [CrossRef]
37. Omer, K.M.; Ku, S.-Y.; Chen, Y.-C.; Wong, K.-T.; Bard, A.J. Electrochemical behavior and electrogenerated chemiluminescence of star-shaped D-A compounds with a 1,3,5-triazine core and substituted fluorene arms. *J. Am. Chem. Soc.* **2010**, *132*, 10944–10952. [CrossRef]
38. Aihara, H.; Tanaka, T.; Satou, M.; Yamakawa, T. Synthesis and Electroluminescence of New Organic Emitters Based on a π-Conjugated 1,3,5-Triazine Core. *Trans. Mat. Res. Soc. Jpn.* **2010**, *35*, 675–680. [CrossRef]
39. Jiao, S.; Men, J.; Ao, C.; Huo, J.; Ma, X.; Gao, G. Synthesis and mesophases of C3h-symmetric 2,4,6-tris(2- hydroxyphenyl)-1,3,5-triazine derivatives with intramolecular hydrogen bonding networks. *Tetrahedron Lett.* **2015**, *56*, 5185–5189. [CrossRef]
40. Data, P.; Zassowski, P.; Lapkowski, M.; Grazulevicius, J.V.; Kukhtad, N.A.; Reghu, R.R. Electrochromic behaviour of triazine based ambipolar compounds. *Electrochim. Acta* **2016**, *192*, 283–295. [CrossRef]
41. Idzik, K.R.; Rapta, P.; Cywinski, P.J.; Beckert, R.; Dunsch, L. Synthesis and electrochemical characterization of new optoelectronic materials based on conjugated donor–acceptor system containing oligo-tri(heteroaryl)-1,3,5-triazines. *Electrochim. Acta* **2010**, *55*, 4858–4864. [CrossRef]
42. Furukawa, H.; Go, Y.B.; Ko, N.; Park, Y.K.; Uribe-Romo, F.J.; Kim, J.; O'Keeffe, M.; Yaghi, O.M. Isoreticular Expansion of Metal–Organic Frameworks with Triangular and Square Building Units and the Lowest Calculated Density for Porous Crystals. *Inorg. Chem.* **2011**, *50*, 9147–9152. [CrossRef] [PubMed]
43. Berger, R.; Hauser, J.; Labat, G.; Weber, E.; Hulliger, J. New symmetrically substituted 1,3,5-triazines as host compounds for channel-type inclusion formation. *Cryst. Eng. Comm.* **2012**, *14*, 768–770. [CrossRef]
44. Reimschuessel, H.K.; McDevitt, N.T. Infrared Spectra of Some 1,3,5-Triazine Derivatives. *J. Am. Chem. Soc.* **1960**, *82*, 3756–3762. [CrossRef]
45. Larkin, P.J.; Makowski, M.P.; Colthup, N.B. The form of the normal modes of s-triazine: Infrared and Raman spectral analysis and ab initio force field calculations. *Spectrochim. Acta A* **1999**, *55*, 1011–1020. [CrossRef]
46. In addition to this stretch, a very strong IR- and Raman-active absorption at ca. 1510 ± 5 cm^{-1} was also always observed as an intense mode in IR for all triazine derivatives. Although the C = N stretches in these vibrational modes are certainly admixed with some C-H stretches, we have still labelled them "C = N" in the experimental part.
47. For **3-NPh$_2$**, a correlation between Φ_F and the solvent polarity was stated (e.g., for acetone: Φ_F[400 nm] = 4.3 %, CH2Cl2: Φ_F[406 nm] = 69.1 %, THF: Φ_F[405 nm] = 73 %, Et2O: Φ_F[408 nm] = 87 % and pentane: Φ_F[392 nm] = 95 %). This solvatofluorochromism was also independently documented in the literature in Das, P. et al. 2018.
48. Stahl, R.; Lambert, C.; Kaiser, C.; Wortmann, R.; Jakober, R. Electrochemistry and photophysics of Donor-Substituted Triarylboranes: Symmetry Breaking in Ground and Excited State. *Chem. Eur. J.* **2006**, *12*, 2358–2370. [CrossRef]

49. Katan, C.; Terenziani, F.; Mongin, O.; Werts, M.H.V.; Porrès, L.; Pons, T.; Mertz, J.; Tretiak, S.; Blanchard-Desce, M. Effects of (Multi)branching of Dipolar Chromophores on Photophysical Properties and Two-Photon Absorption. *J. Phys. Chem. A* **2005**, *109*, 3024–3037. [CrossRef]
50. Bangal, P.R.; Lam, D.M.K.; Peteanu, L.A.; van der Auweraer, M.J. Excited-State localization ina 3-Fold-Symmetric Molecule as Probed by Electroabsorption Spectroscopy. *J. Phys. Chem. B* **2004**, *108*, 16834–16840. [CrossRef]
51. Kim, H.M.; Cho, B.R. Two-photon materials with large two-photon cross sections. Structure-property relationship. *Chem. Commun.* **2009**, *45*, 153–164. [CrossRef]
52. Kuzyk, M.G. Fundamental limits on two-photon absorption cross-sections. *J. Chem. Phys.* **2003**, *119*, 8327–8334. [CrossRef]
53. Amar, A.; Boucekkine, A.; Paul, F.; Mongin, O. DFT study of two-photon absorption of octupolar molecules. *Theo. Chem. Acc.* **2019**, *138*, 1–7. [CrossRef]
54. Kuzyk, M.G. Using fundamental principles to understand and optimize nonlinear-optical materials. *J. Mater. Chem.* **2009**, *19*, 7444–7465. [CrossRef]
55. Webber, A.L.; Yates, J.R.; Zilka, M.; Sturniolo, S.; Uldry, A.-C.; Corlett, E.K.; Pickard, C.J.; Pérez-Torralba, M.; Angeles Garcia, M.; Santa Maria, D.; et al. Weak Intermolecular CH⋯N Hydrogen Bonding: Determination of 13CH–15N Hydrogen-Bond Mediated J Couplings by Solid-State NMR Spectroscopy and First-Principles Calculations. *J. Phys. Chem. A* **2020**, *124*, 560–572. [CrossRef] [PubMed]
56. Desiraju, G.R. Hydrogen bridges in crystal engeneering: Interactions without border. *Acc. Chem. Res.* **2002**, *35*, 565–573. [CrossRef] [PubMed]
57. Adamo, C.; Barone, V. Exchange functionals with improved long-range behavior and adiabatic connection methods without adjustable parameters: The mPW and mPW1PW models. *J. Chem. Phys.* **1998**, *108*, 664–675. [CrossRef]
58. The CAM-B3LYP computed longest wavelengths are within 2100 cm^{-1} of the experimental data for all derivatives (and below 4000 cm^{-1} for MPW1PW91 functional).
59. Amar, A.; El Kechai, A.; Halet, J.-F.; Paul, F.; Boucekkine, A. Two-photon absorption of dipolar and quadrupolar oligothiophene-cored chromophores derivatives containing terminal dimesitylboryl moieties: A theoretical (DFT) structure-property investigation. *New J. Chem.* **2021**, *45*, 15074–15081. [CrossRef]
60. Hu, Z.; Autschbach, J.; Jensen, L. Simulating Third-Order Nonlinear Optical Properties Using Damped Cubic Response Theory within Time-Dependent Density Functional Theory. *J. Chem. Theory Comput.* **2016**, *12*, 1294–1304. [CrossRef]
61. Strickler, S.J.; Berg, R.A. Relationship between absorption intensity and fluorescence lifetime of molecules. *J. Chem. Phys.* **1962**, *37*, 814–822. [CrossRef]
62. Rabouël, I.; Richy, N.; Amar, A.; Boucekkine, A.; Roisnel, T.; Mongin, O.; Humphrey, M.G.; Paul, F. 1,3,5-Triaryl-1,3,5-Triazinane-2,4,6-Trithiones: Synthesis, Electronic Structure and Linear Optical Properties. *Molecules* **2020**, *25*, 5475. [CrossRef]
63. Considering that these compounds have a quite comparable two-photon brightness, part of this discrepancy could originate from underestimation of the quantum yield of **13-H** in THF ($\Phi_F = 0.26$), since we found $\Phi_F = 0.78$ for **7-H** in THF.
64. Kannan, R.; He, G.S.; Lin, T.-C.; Prasad, P.N.; Vaia, R.A.; Tan, L.-S. Toward highly active two-photon absorbing liquids. Synthesis and characterization of 1,3,5-triazine-based octupolar molecules. *Chem. Mater.* **2004**, *16*, 185–194. [CrossRef]
65. In this respect, the larger Neff numbers presently considered for s-triazines derivatives (see ESI) reflect (at least in part) the beneficial effect of having the peripheral units coplanar with the central core. However, even when the 2PA cross-sections are corrected for these larger numbers of effective electrons (N_{eff}) using the ad hoc figures of merit ($\sigma_2/(N_{eff})^2$), the trend previously observed for σ_2 values is still apparent suggesting that another factor is also influential.
66. Beljonne, D.; Wenseleers, W.; Zojer, E.; Shuai, Z.; Vogel, H.; Pond, S.J.K.; Perry, J.W.; Marder, S.R.; Brédas, J.-L. Role of dimensionality on the two-photon absorption response of conjugated molecules: The case of octupolar compounds. *Adv. Funct. Mater.* **2002**, *12*, 631–641. [CrossRef]
67. For instance, neglecting differences in relaxation energies, an estimate of the excitonic coupling of ca. 0.04 eV (350 cm^{-1}) was found for **3-OMe** from the energy difference (3V) between the first degenerate set of allowed transition (E set under strict D_3 symmetry) and the corresponding (forbidden) excitation of A symmetry in MPW1PW91. Likewise, a value of ca. 0.04 eV (320 cm^{-1}) was found for this compound using the SAOP functional. Note that during these computational approaches, any vibronic contribution was not considered, Macak, P. et al. 2000 but the latter should also remain weak according to the apparent good match (degeneracy) between twice the wavelength of the 1PA transition and that of the 2PA transition experimentally stated (ESI, Figure S9) Beljonne, D. et al. 2002.
68. Macak, P.; Luo, Y.; Norman, P.; Ågren, H. Electronic and vibronic contributions to two-photon absorption of molecules with multi-branched structures. *J. Chem. Phys.* **2000**, *113*, 7055–7061. [CrossRef]
69. Spangler, C.W. Recent development in the design of organic materials for optical power limiting. *J. Mater. Chem.* **1999**, *9*, 2013–2020. [CrossRef]
70. Wortmann, R.; Glania, C.; Krämer, P.; Matschiner, R.; Wolff, J.J.; Kraft, S.; Treptow, B.; Barbu, E.; Längle, D.; Görlitz, G. Nondipolar Structures with Threefold symmetry for Nonlinear Optics. *Chem. Eur. J.* **1997**, *3*, 1765–1773. [CrossRef]
71. Zhou, H.; Zheng, Z.; Xu, G.; Yu, Z.; Yang, X.; Cheng, L.; Tian, X.; Kong, L.; Wu, J.; Tian, Y. 1, 3, 5-Triazine-cored derivatives dyes containing triphenylamine based two-photon absorption: Synthesis, optical characterization and bioimaging. *Dye. Pigm.* **2012**, *94*, 570–582. [CrossRef]
72. Shriver, D.F.; Drezdzon, M.A. *The Manipulation of Air-Sensitive Compounds*; Wiley: New York, NY, USA, 1986.

73. Gottlieb, H.E.; Kotlyer, V.; Nudelman, A. NMR Chemical Shifts of Common Laboratory Solvents as Trace Impurities. *J. Org. Chem.* **1977**, *62*, 7512–7515. [CrossRef]
74. Fulmer, G.R.; Miller, A.J.M.; Sherden, N.H.; Gottlieb, H.E.; Nudelman, A.; Stoltz, B.M.; Bercaw, J.E.; Goldberg, K.I. NMR Chemical Shifts of Trace Impurities: Common Laboratory Solvents, Organics, and Gases in Deuterated Solvents Relevant to the Organometallic Chemist. *Organometallics* **2010**, *29*, 2176–2179. [CrossRef]
75. Ziessel, R.; Suffert, J.; Youinou, M.-T. General method for the preparation of alkyne-functionalized oligopyryridine building blocks. *J. Org. Chem.* **1996**, *61*, 6535–6546. [CrossRef]
76. Lavastre, O.; Cabioch, S.; Dixneuf, P.H.; Vohlidal, J. Selective and Efficient Access to *Ortho*, *Meta* and *Para* Ring-substituted Phenylacetylene Derivatives R-[CC-C$_6$H$_4$]$_x$-Y (Y:H, NO$_2$, CN, I, NH$_2$). *Tetrahedron* **1997**, *53*, 7595–7604. [CrossRef]
77. Lim, J.; Albright, T.A.; Martin, B.R.; Miljanić, O.Š. Benzobisoxazole Cruciforms: Heterocyclic Fluorophores with Spatially Separated Frontier Molecular Orbitals. *J. Org. Chem.* **2011**, *76*, 10207–10219. [CrossRef]
78. Malvolti, F.; Rouxel, C.; Triadon, A.; Grelaud, G.; Richy, N.; Mongin, O.; Blanchard-Desce, M.; Toupet, L.; Abdul Razak, F.I.; Stranger, R.; et al. 2,7-Fluorenediyl-bridged Complexes Containing Electroactive "Fe(eta5-C5Me5)(eta2-dppe)C≡C-" Endgroups: Molecular Wires and Remarkable Nonlinear Electrochromes. *Organometallics* **2015**, *34*, 5418–5437. [CrossRef]
79. Demas, N.; Crosby, G.A. Measurement of photoluminescence quantum yields. *J. Phys. Chem.* **1971**, *75*, 991–1024.
80. Eaton, G.R. Reference Materials for Fluorescence Measurement. *Pure Appl. Chem.* **1988**, *60*, 1107–1114. [CrossRef]
81. Xu, C.; Webb, W.W. Measurement of two-photon excitation cross sections of molecular fluorophores with data from 690 to 1050 nm. *J. Opt. Soc. Am. B* **1996**, *13*, 481–491. [CrossRef]
82. Werts, M.H.V.; Nerambourg, N.; Pélégry, D.; Le Grand, Y.; Blanchard-Desce, M. Action cross sections of two-photon excited luminescence of some Eu(III) and Tb(III) complexes. *Photochem. Photobiol. Sci.* **2005**, *4*, 531–538. [CrossRef]
83. Frisch, M.J.; Trucks, G.W.; Schlegel, H.B.; Scuseria, G.E.; Robb, M.A.; Cheeseman, J.R.; Scalmani, G.; Barone, V.; Petersson, G.A.; Nakatsuji, H.; et al. *Gaussian 09*; Revision A.03; Gaussian, Inc.: Wallingford, CT, USA, 2016.
84. Yanai, T.; Tew, D.P.; Handy, N.C. A new hybrid exchange-correlation functional using the Coulomb-attenuating method (CAM-B3LYP). *Chem. Phys. Lett.* **2004**, *393*, 51–57. [CrossRef]
85. Tomasi, J.; Mennucci, B.; Cammi, R. Quantum Mechanical Continuum Solvation Models. *Chem. Rev.* **2005**, *105*, 2999–3093. [CrossRef]
86. Gorelsky, S.I. *SWizard Program*, Revision 4.5; University of Ottawa: Ottawa, ON, Canada, 2013. Available online: http://www.sg-chem.net/ (accessed on 5 February 2022).
87. Dennington, R.; Keith, T.; Millam, J. *GaussView*; Version 5; Semichem Inc.: Shawnee Mission, KS, USA, 2009.
88. Schipper, P.R.T.; Gritsenko, O.V.; van Giesbergen, J.A.; Baerends, E.J. Molecular calculations of excitation energies and (hyper)polarizabilities with a statistical average of orbital model exchange-correlation potentials. *J. Chem. Phys.* **2000**, *112*, 1344–1352. [CrossRef]
89. Jensen, J.F.; van Duijnen, P.T.; Snijders, J.G.J. A discrete solvent reaction field model for calculating frequency-dependent hyperpolarizabilities of molecules in solution. *J. Chem. Phys.* **2003**, *119*, 12998–13005. [CrossRef]
90. *ADF2019*; Vrije Universiteit: Amsterdam, The Netherlands, 2019; Available online: www.scm.com (accessed on 14 May 2022).
91. Jensen, J.F.; Autschbach, J.; Schatz, J. Finite lifetime effects on the polarizability within time-dependent density-functional theory. *J. Chem. Phys.* **2005**, *122*, 224115. [CrossRef]
92. Hu, Z.; Autschbach, J.; Jensen, L. Simulation of resonance hyper-Rayleigh scattering of molecules and metal clusters using a time-dependent density functional theory approach. *J. Chem. Phys.* **2014**, *141*, 124305. [CrossRef] [PubMed]

Article

E–*Z* Photoisomerization in Proton-Modulated Photoswitchable Merocyanine Based on Benzothiazolium and o-Hydroxynaphthalene Platform

Aleksey A. Vasilev [1,2,*], Stanislav Baluschev [3,4], Sonia Ilieva [1] and Diana Cheshmedzhieva [1,*]

[1] Faculty of Chemistry and Pharmacy, Sofia University "Saint Kliment Ohridski", 1 James Bourchier Blvd., 1164 Sofia, Bulgaria; silieva@chem.uni-sofia.bg
[2] Institute of Polymers, Bulgarian Academy of Sciences, Akad. G. Bonchev St., bl 103A, 1113 Sofia, Bulgaria
[3] Faculty of Physics, Sofia University "Saint Kliment Ohridski", 5 James Bourchier Blvd., 1164 Sofia, Bulgaria; balouche@mpip-mainz.mpg.de
[4] Max Planck Institute for Polymer Research, Ackermannweg 10, 55128 Mainz, Germany
* Correspondence: ohtavv@chem.uni-sofia.bg (A.A.V.); dvalentinova@chem.uni-sofia.bg (D.C.); Tel.: +35-92-8161-354 (D.C.)

Abstract: The potential of *E*–*Z* photoisomerization in molecular organic light-to-thermal conversion and storage in an *E*–styryl merocyanine system was studied in a polar acidic medium. A photoswitchable styryl merocyanine dye (*E*)-2-(2-(2-hydroxynaphthalen-1-yl)vinyl)-3,5-dimethylbenzo[*d*]thiazol-3-ium iodide was synthesized for the first time. The reversible *E*–*Z* photoisomerisation of the dye was investigated using UV-Vis spectroscopy and DFT calculations. *E*–*Z* isomerization was induced through the use of visible light irradiation (λ = 450 nm). The obtained experimental and theoretical results confirm the applicability of the *Z* and *E* isomers for proton-triggered light harvesting.

Keywords: *E*–*Z* photoisomerization; spiropyran; organic light-to-thermal systems; merocyanine dyes; DFT

Citation: Vasilev, A.A.; Baluschev, S.; Ilieva, S.; Cheshmedzhieva, D. *E*–*Z* Photoisomerization in Proton-Modulated Photoswitchable Merocyanine Based on Benzothiazolium and o-Hydroxynaphthalene Platform. Photochem 2023, 3, 301–312. https:// doi.org/10.3390/photochem3020018

Academic Editors: Marcelo Guzman, Vincenzo Vaiano and Rui Fausto

Received: 2 May 2023
Revised: 28 May 2023
Accepted: 9 June 2023
Published: 19 June 2023

Copyright: © 2023 by the authors. Licensee MDPI, Basel, Switzerland. This article is an open access article distributed under the terms and conditions of the Creative Commons Attribution (CC BY) license (https:// creativecommons.org/licenses/by/ 4.0/).

1. Introduction

Photoswitchable molecules have been widely investigated due to their potential applications as light-responsive materials. Several classes of organic compounds that can undergo reversible photoisomerization around double bonds (C=C, C=N, N=N) are reported in the literature. The common denominator in these systems is that one of the conformers is thermodynamically stable, and the other is metastable and is formed after irradiation with UV or visible light [1,2]. Examples of photoswitchable molecules include dihydroazulene/vinylheptafulvene (DHA/VHF) [3,4], fulvalene dimetal complexes [5], norbornadiene/quadricyclane (NBD/QC) [6], and azobenzene systems [1,2,7–11]. Qiu et al., provide an overview of recent developments and applications of various types of synthetic photoswitches as molecular solar thermal (MOST) energy storage materials [11]. The application of azobenzenes performing *E*–*Z* photoisomerization upon irradiation is driven by their possible inclusion as light-triggered switches in polymers, surface-modified materials, proteins, and a variety of "molecular machines" [7,12]. Azobenzenes are highly valued photoswitches with applications in biological systems due to the fact that the photoisomerisation process is fast and they have a high photoisomerization quantum yield [2]. Diarylethene molecules, for example, can be converted from a conjugated to a cross-conjugated state upon illumination in the visible region, and this reversible isomerization is the basis of room-temperature conduction switching [13,14]. The class of spiropyrans is among the most important photochromes, with broad applications as smart materials in energy, data storage and photopharmacology [15–18].

The photochromism of spiropyrans is due to the interconversion between the ring-closed spiropyran (SP) structure and ring-open merocyanine (MC) form, in which the

C−O bond is split [17]. The photochemical properties of the two isomers are quite different, making spiropyrans unique as a class of photoswitches. Their applicability to new spiropyran-based dynamic materials is determined by the fact that the molecules undergo a reversible and continuous interconversion in response to variety of stimuli. Spiropyran functionalized materials have been developed due to their potential applications [18]. The fluorescent properties of spiropyran-bound polymers have been studied extensively due to their potential applications in detection and imaging. PULSAR microscopy has been developed with a special application for imaging biological systems and overcoming the significant problem of false positive signals due to cell autofluorescence [19,20]. The functionalization of biomolecules with spiropyran molecular photoswitches contributes to solving various scientific challenges, such as the light-assisted control of natural biological properties [21,22], the photoregulation of enzyme activity [23], and others.

The protonation of the o-hydroxyl group in merocyanines built by 3,3'-dimethylindolenine and phenol end groups has been broadly investigated [24–29]. Acidochromism has been reported to modify the thermally induced conversion to the protonated merocyanine form (MCH+) [24–29]. However, the mechanism and nature of the intermediates formed throughout these reports are not consistent. Fissi et al. [24] and Wojtyk et al. [25], for example, observed that an equilibrium between the non-protonated and the protonated spiropyran is established in the presence of trifluoroacetic acid, lying in favor of the protonated E-merocyanine form. In another report, Rémon et al. [26] proposed that, in an aqueous medium, the protonation only occurs at a pH below 0.5 and that the only protonated species present is the open-ring merocyanine form, which is in thermal equilibrium with the non-protonated closed form. Additionally, Schmidt et al. [27] also noted the formation of the protonated E-merocyanine in ethanol upon the addition of trifluoroacetic acid. Overall, these reports denote that ring opening to the stable protonated E-merocyanine occurs in a strongly acidic environment. In other words, it was postulated that a rapid equilibrium is established between species of the transient spiropyran forms that have a broken Cspiro-O bond and a geometry intermediate to the perpendicular spiro and the planar merocyanine form under sufficiently acidic conditions [24–27]. In the same period, Roxburgh et al., reported the trifluoroacetic acid induced the thermal ring-opening of spiropyrans to their protonated E isomer that was proposed to be via either the unprotonated or protonated Z form [28]. The proposed intermediacy of the protonated Z form was subsequently supported by Shiozaki, who proposed that the protonation of spiropyran in ethanol with sulfuric acid, a stronger acid than trifluoroacetic acid, generated the Z-merocyanine form, which could not only undergo subsequent thermal but also photochemical Z/E isomerization [29]. Shiozaki's interpretation of the changes observed by UV-vis absorption spectroscopy, analogous to the acid-induced ring opening (C-O bond cleavage) observed for the related photochromic spirooxazines [29], which was supported by theoretical results. Kortekaas et al. [30] demonstrated that the extent of acid-induced ring opening is controlled by matching both the concentration and strength of the acid used. The authors show that with strong acids, full ring opening to the Z-merocyanine isomer occurs spontaneously, allowing its characterization via 1H NMR spectroscopy as well as UV/vis spectroscopy. The reversible switching between Z-E isomerization through irradiation with UV and visible light is considered. Under sufficiently acidic conditions, both E- and Z-isomers are thermally stable. The judicious choice of the acid so that its pKa lies between that of the E- and Z-merocyanine forms enables thermally stable switching between spiropyran and E-merocyanine forms, and hence pH-gating between thermally irreversible and reversible photochromic switching. Our group recently demonstrated [31] for the first time the E–Z photoisomerization observed directly by excitation with light, substantially red-shifted compared to the absorption spectrum of the E–Z active moieties via the process of triplet–triplet annihilation upconversion. As photoactive molecular systems, we used a series of rare-earth-free stabilized styryl dye-Ba^{2+} complexes prepared via an improved, easy, reliable two-step synthetic procedure. The goal is to find molecular systems that do not contain metals and are environmentally friendly. Therefore, our efforts have been focused on finding metal-free photoswitchable molecules operating in the visible region (450–600 nm) and testing their ability in terms of proton-triggered E–Z photoisomerization. We are looking for a metastable structure that is the Z-form, which should thermally relax to the E-isomer. Usually, the Z-form is higher in energy and more unstable. The transition from Z- to E-form is expected to result in energy gain.

The aim of the present study is to investigate the E–Z photoisomerization of merocyanine (E)-2-(2-(2-hydroxynaphthalen-1-yl)vinyl)-3,5-dimethylbenzo[d]thiazol-3-ium platform as a molecular organic light-to-thermal conversion system in acidic conditions. We synthesized and investigated the photophysical properties of a promising candidate, which is easy to obtain and that can be subsequently modified via structural changes.

2. Results and Discussion

2.1. Synthesis

The Knövenagel type condensation of the CH-acid 2,3,5-trymethyl-benzothiazolium iodide (**1**) and a slight molar excess of 2-hydroxy-1-napthaldehyde (**2**) in the presence of catalytic amounts of N,N'-diisopropyl-ethylamine (DIPEA) afforded the reaction product **3** in very good yield (71%), Scheme 1. The reaction conditions were modified, thus leading to the complete conversion of reactant **1** and the easy purification of the target product **3**. Only one spot on the TLC was observed after the separation of the reaction product from the ethanol/ethyl acetate solution. Single recrystallization from ethanol/ethyl acetate afforded the target dye **3** in analytical purity.

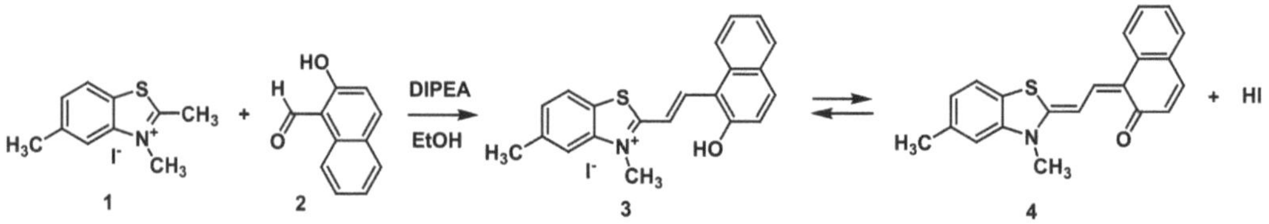

Scheme 1. Synthesis of the photochromic dye **3** and its merocyanine tautomer **4**.

The chemical structure of dye **3** was proved via NMR spectroscopy, ESI-MS spectrometry, melting point and UV-VIS spectroscopy. In the proton NMR spectrum of protonated merocyanine **3** in DMSO-d_6 (Figures S1–S3), all of the characteristic signals of the proposed chemical structure are observed. At 2.52 ppm, a singlet with integral intensity corresponding to three protons is denoted. In our opinion, it corresponds to the methyl group attached to the aromatic core of the benzothiazolium fragment. In a weaker field at 4.06 ppm, a singlet with integral intensity for three protons appears, which corresponds to the methyl protons of the group directly bonded to the quaternary nitrogen atom of the methyl-benzothiazolium fragment. The doublet at 7.01 ppm with a J-constant of 9.2 Hz is characteristic of the methine proton of the styryl group connecting the two aromatic moieties of the conjugated system of the dye. The other similar methine proton signal appears at 7.84 ppm with the same J-constant as the previous one. In the aromatic part of the proton signals, the most characteristic is the singlet appearing at 7.89 ppm, describing the proton between the methyl group of the benzothiazlium fragment and the quaternary nitrogen atom. The number and integral intensity of the remaining signals corresponds to the protons in the aromatic part of dye structure **3**. The presence of only positive signals in the carbon DEPT spectrum of dye **3** (Figure S4) certifies the absence of methylene protons in the structure and confirms the presence of methyl and methine protons, corresponding to the proposed chemical structure. The signals at 21.65 ppm and 56.50 ppm can be assigned to the carbon atoms from the methyl groups.

To the best of our knowledge, dye **3** has not been described in the literature.

2.2. Photophysical Properties of Dye 3

Absorption Spectra

The electronic absorption spectra of dye *3* in solvents with different polarities are demonstrated in Figure 1 and Table 1. The dye displays several different bands between 300 and 600 nm. As expected [18,28], two forms of the dye are present in the solution (Figure 1). The longest wavelength absorption corresponds to the merocyanine form (*MC*) (Scheme 2). The extensive conjugation in the structure of the merocyanine dye leads to strong absorption in the visible region (574–590 nm in the studied solvents). The *trans*-styryl (*TS*) hemicyanine form (Scheme 2) is characterized by absorption maxima from 451 nm to 481 nm, depending on the solvent polarity. It can be concluded from Figure 1 that dye *3* exhibits horizontal and vertical sovatochromism typical for merocyanine and styryl dyes. It is well known that merocyanine dyes are characterized by positive solvatochromism, while styryl hemicyanine dyes are characterized by negative ones.

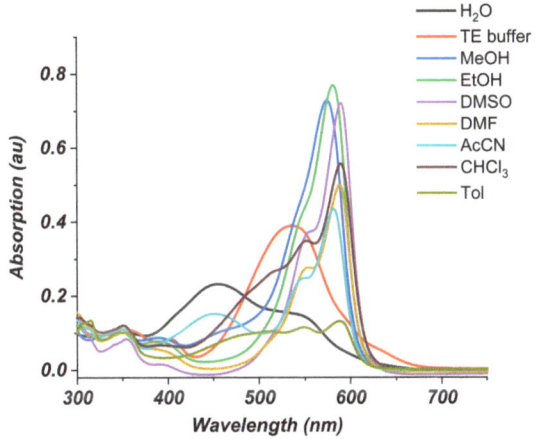

Figure 1. UV-VIS absorption spectra of merocyanine dye *3* in solvents with different polarity.

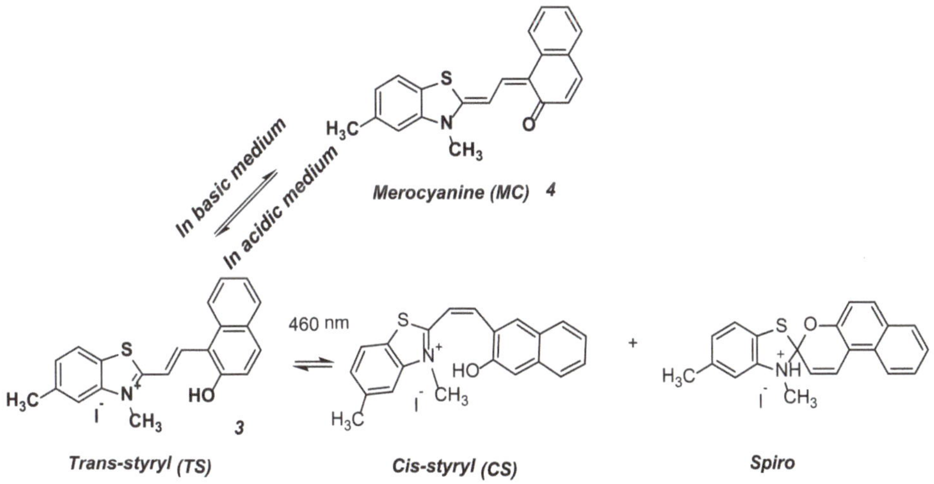

Scheme 2. Isomers of dye *3* in acidic medium.

Table 1. Dependency of the visible absorption of dye 3 from the solvent polarity.

Solvent	Polarity Index	Dielectric Permittivity ε	λ_{max1} (nm) Styryl	ε_1 (L × mol^{-1} × cm^{-1})	λ_{max2} (nm) MC	ε_2 (L × mol^{-1} × cm^{-1})
Toluene	2.4	2.3741	481	9 490	589	13,160
CHCl$_3$	4.1	4.71	-----	-----	590	55,700
MeOH	5.1	32.613	459	10 510	574	72,860
EtOH	5.2	24.85	-----	-----	581	76,980
Acetonitrile (AcCN)	5.8	35.688	451	15 220	582	43,540
DMF	6.4	37.219	-----	-----	588	50,050
DMSO	7.2	46.826	-----	-----	590	72,190
H$_2$O	10.2	78.355	454	23 180	543	14,930
TE buffer pH = 7	-----	-----	407	6 460	535	38,960

As can be seen from Figure 1, the solvatochromism of the merocyanine form is mainly vertical (molar absorptivity from 13,160 L × mol^{-1} × cm^{-1} in toluene till 76,980 L × mol^{-1} × cm^{-1} in ethanol) with a weak red/bathochromic shift of the longest wavelength maximum going from less polar to more polar solvents. This positive solvatochromism is an indication of an increased dipole moment of the excited state relative to the ground state. Meanwhile, for the trans-styryl form, we observed a negative solvatochromism typical for dyes of the styryl cyanine type [22]. The merocyanine form predominates in some solvents (Table 1), while both forms of the dye exist simultaneously in other solvents.

Solvents, such as DMF, DMSO, ethanol, and methanol, stabilizing the merocyanine form, are more suitable for investigating the E-Z photoisomerization of dye 3. The acidochromic properties of dye 3 in DMF were investigated. DMF–an aprotic, basic, polar solvent, can coordinate acidic protons with its lone pair. Dye 3 was titrated with glacial acetic acid in DMF. Figure 2 shows the change in the absorption of dye 3 as a function of the amount of glacial acetic acid added to a solution of the dye in DMF. As the amount of added acid increases, the intensity of the absorption band at 581 nm corresponding to the merocyanine form 4 (MC Scheme 2, Figure 2) decreases, while at the same time the intensity of the absorption band at 451 nm that corresponds to the trans-styryl (or so-called "protonated merocyanine") form 3 (TS Scheme 2, Figure 2) increases. The merocyanine form converts into the protonated species, absorbing at λ_{max} 480 nm upon the addition of acid to the solution. The presence of a distinct isosbestic point is unequivocal proof of the existence of the two discussed tautomeric forms (MC and TS) in DMF–acetic acid solution. The merocyanine form of dye 3 is stable in basic medium. The addition of glacial acetic acid (AcOH) leads to the formation of trans-styryl 3.

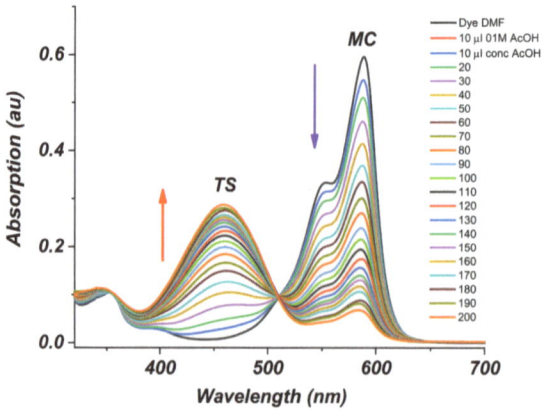

Figure 2. Acetic acid triggered merocyanine to styryl form (3 to 4) isomerization in DMF as a solvent.

The solution was irradiated with a 450 nm laser for different periods of time to induce *cis*-trans isomerization. The irradiation of the solution (dye **3** in DMF and acetic acid) resulted in a substantial decrease in the intensity of the band responsible for the *trans*-styryl form of the dye (Figure 3). As can be seen from Figure 3, the irradiation after the fortieth minute leads to a complete disappearance of the signals for forms **3** and **4**, which is visually registered as a complete discoloration of the solution. This phenomenon is accompanied by an increase in the absorption intensity at 363 nm, which can be associated with the formation of the *cis*-styryl isomer and its eventual incorporation into a spiropyran cycle. The formation of a second isosbestic point at 390 nm (Figure 3) confirms this hypothesis. The color of the solution did not recover, and the system did not return to the original absorbance values at 450 nm, even after six months of relaxation in the dark at room temperature and/or upon heating to 50 °C. The irreversibility of the E–Z isomerization process in DMF–acetic acid makes these conditions unsuitable for the intended application.

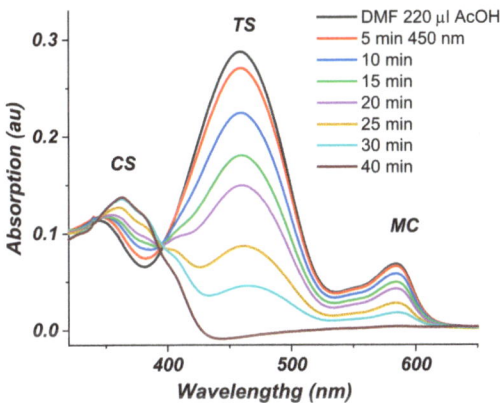

Figure 3. *E-Z* photoisomerization of dye **3** in the presence of acetic acid upon 450 nm laser irradiation.

Our goal was to achieve the transformation of the metastable *cis* form to the most stable *trans*-form upon relaxation of the system. The observations described above forced us to repeat the experiment in a different solvent and with a different acid. Since we are interested in E–Z photoisomerization of merocyanine dye **3**, the process was studied in ethanol as a medium, and the titration was performed with concentrated hydrochloric acid.

In ethanol, the merocyanine form **4** is stable, as can be seen from Figure 4 (black line). Then, 1 M aqueous hydrochloric acid was added to the solution, resulting in the formation of *trans*-styryl form **3** (463 nm, the blue and red lines in Figure 4). In contrast to the experiment in DMF, in ethanol, even small amounts of 1 M aq. HCl acid resulted in a complete conversion of form **4** (*MC*) to form **3** (*TS*). In ethanol–water, the *TS* form of the dye remains stable for at least 24 h, which is a prerequisite for performing the above experiment in the given new medium (ethanol/water). In an acidic environment, the merocyanine form is eliminated and is no longer the most stable form. Even if the spiro form is formed, under these conditions, it will quickly open to the *cis* (metastable) form, which is then stabilized in the *trans* form.

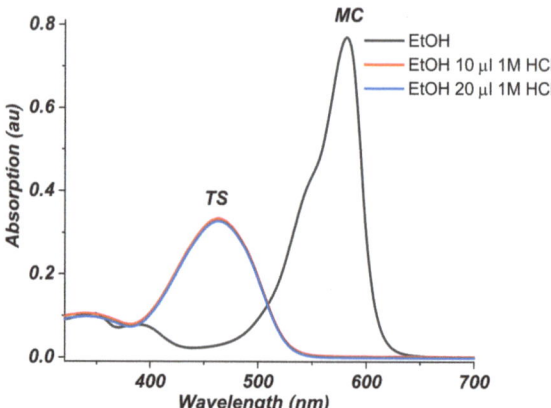

Figure 4. Hydrochloric acid triggered merocyanine to trans-styryl form of dye **3** in ethanol.

Irradiation with 450 nm laser light results in a reduction in the *trans* band (463 nm, brown line, Figure 5) and the formation of the *cis* isomer (321 nm, cyan line, Figure 5). In general, the molar absorptivity of the *cis* and *trans* forms does not commensurate. The *cis* form has very low molar absorptivity. After the irradiation, the solution is allowed to relax. The absorbance of the solution was measured every 5 min. As seen from the spectrum in Figure 5, after 24 h, the *TS* form is fully restored. The process of transition from merocyanine to the *trans*-styryl form will result in the production of energy when irradiated with light.

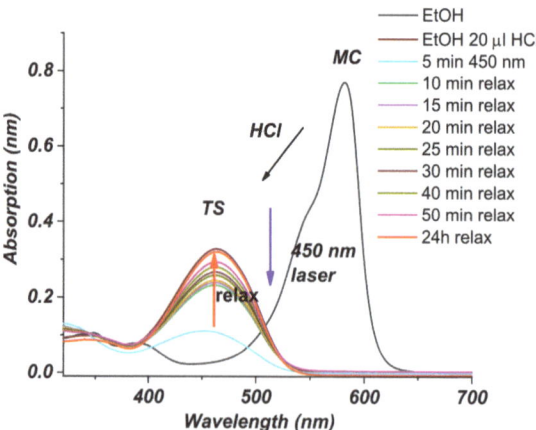

Figure 5. The reversible *cis*-to-*trans* (Z-E) photoisomerization of photoswitch **3** in ethanol–hydrochloric acid system after 5 min 450 nm laser irradiation. Black line: dye **3** in ethanol; brown line: dye **3** with 20 μL hydrochloric acid added; cyan line: the change after 5 min of irradiation with 450 nm laser light.

The experimental studies were supplemented with PCM [32,33] DFT [34] calculations using B3LYP hybrid functional in conjunction with 6-31G(d,p) [35] basis set, and only the iodine SDD basis set and effective core potential were used [36,37]. The quantum chemical computations were performed with the G16 A.03 software package [38]. The optimized ground state structures and computed relative Gibbs free energies for the dye **3** isomers are presented in Figure 6. The optimized structures with iodine counterion are given in Figure S6. The calculated free energies of all the stationary points are reported in kcal/mol relative to the *trans*-styryl form.

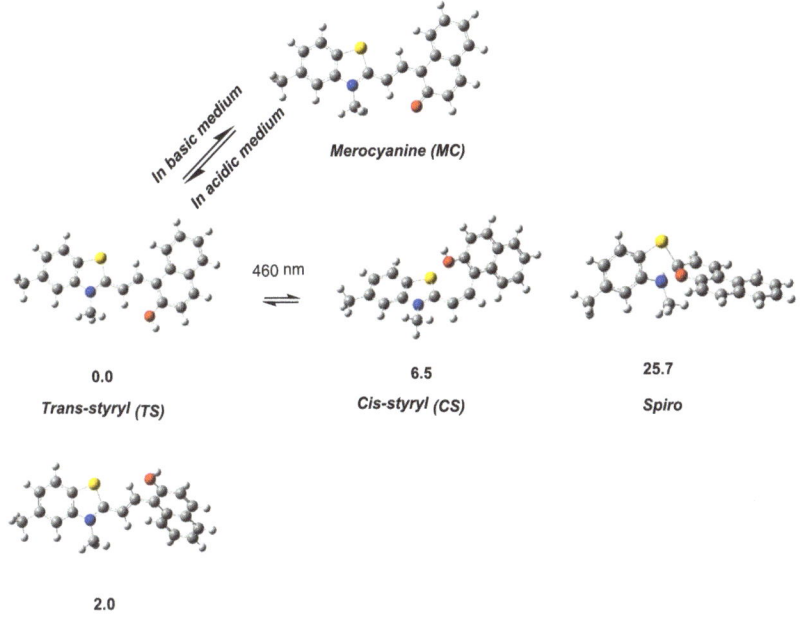

Figure 6. B3LYP/6-31G(d,p) optimized ground state structures (in water medium) and computed relative Gibbs free energies for the dye 3 isomers. Color scheme: gray—C, light gray—H, red—O, blue—N, yellow—S.

The Trans-styryl (TS)-B isomer (Figure 6) is more unstable by 2 kcal/mol because the planar structure is disrupted, which lowers the conjugation in the heterocyclic system. The dihedral angle between the benzothiazole and naphthol fragments is 24.15 degrees. Upon irradiation (450 nm), TS converts to the cis isomer (CS). The spiro form is energetically less favorable than the cis form, but its formation cannot be excluded when the system is energized by light irradiation (450 nm). The free energy profile for the transformation between TS, CS, and spiro forms (in water) is presented in Figure 7.

The transition states structures are characterized via the eigenvector of the imaginary frequency and are proven through intrinsic reaction coordinate (IRC) calculations. The spiro form converts to the cis isomer (CS) through an energy barrier of 30.6 kcal/mol. The trans to cis conversion has a free energy barrier of 33.6 kcal/mol (Figure 7). The theoretical results show that the *trans* form is 6.5 kcal/mol more stable than the *cis* form. Therefore, the transition from *cis* to *trans* results in energy gain.

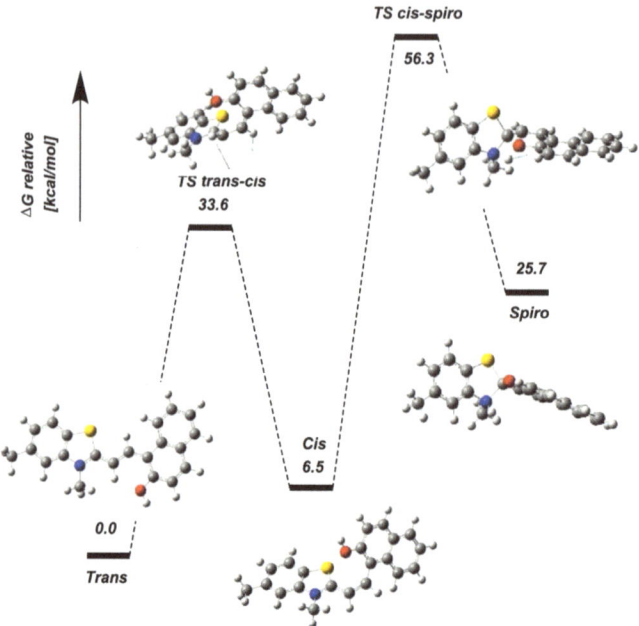

Figure 7. Free energy profile for the transformation between trans (TS), cis (CS), and spiro forms from B3LYP/6-31G(d,p) calculations in water medium. Color scheme: gray—C, light gray—H, red—O, blue—N, yellow—S.

3. Conclusions

The photoswitchable merocyanine dye (*E*)-2-(2-(2-hydroxynaphthalen-1-yl)vinyl)-3,5-dimethylbenzo[*d*]thiazol-3-ium iodide (**3**) was prepared for the first time. The reversible switching between *E* and *Z* isomers and the acidoxchromism of dye **3** in DMF and ethanol was studied through the use of UV/Vis spectroscopy and DFT calculations. The *E*-*Z*-isomerization is induced by irradiation with visible light. In an ethanol acidic medium the studied system is suitable for harvesting visible light. The obtained experimental and theoretical results confirm the applicability of the *TS* and *CS* isomers for proton-triggered light-to-thermal molecular systems. We have a promising candidate, a simple molecule, which is easy to obtain and that can be subsequently modified in order to increase the energy difference between *Z*- and *E*-isomers.

4. Materials and Methods

4.1. General

All of the solvents used in the present work were commercially available (HPLC grade). The synthesis of 2,3,5-trimethylbenzo[*d*]thiazol-3-ium iodide (**1**) was prepared via a previously described procedure [39]. 2-Hydroxy-1-naphthaldehyde (**2**) is commercially available and was used as supplied. The melting points were determined on a Kofler apparatus and were uncorrected. The NMR spectra of the samples in DMSO-d_6 were obtained on a Bruker Avance III 500 DRX 600 MHz spectrometer at the Faculty of Chemistry and Pharmacy, University of Sofia. Mass spectrum acquisitions were conducted at the MPIP, Mainz, Germany on an Advion expression compact mass spectrometer (CMS) with atmospheric pressure chemical ionization (APCI) at high temperature and low fragmentation regime. The MS-spectrum was acquired in the positive ion reflection mode, m/z range from 10 to 1000 m/z, and acquisition speed 10,000 m/z units s^{-1}. The obtained spectrum was analyzed by using Advion CheMS Express software version 5.1.0.2. at the

MPIP, Mainz, Germany. The UV-VIS spectra were measured on a Unicam 530 UV–VIS spectrophotometer in conventional quartz cells of 1 cm path length. The absorption spectra were recorded for solutions with identical total dye concentrations (1×10^{-5} M). A thermal stabilized single-mode diode laser passes through a spatial filter in order to reach nearly TEM_{00} transversal intensity distribution, and the used laser intensity is 1 mW $\times 10^{-2}$ cm, $\lambda = 450$ nm, cw.

4.2. Synthesis of (E)-2-(2-(2-Hydroxynaphthalen-1-yl)vinyl)-3,5-dimethylbenzo[d]thiazol-3-ium Iodide (3)

In a 50 mL round bottom flask equipped with an electromagnetic stirrer and reflux condenser, 0.25 g (0.82 mmol) 2,3,5-trimethylbenzo[d]thiazol-3-ium iodide (1) and 0.17 g (0.98 mmol) 2-hydroxy-1-naphthaldehyde (2) were dissolved in 15 mL ethanol and one drop of DIPEA was added. The reaction mixture was vigorously stirred and refluxed under argon for 2 h. After cooling to room temperature, 30 mL of ethyl acetate was added, and the mixture was stored in a refrigerator for 24 h. The formed precipitate was suction filtered, washed with 20 mL cold ethanol and 30 mL ethyl acetate, and air dried. Yield: 0.27 g (71%). M.p. > 250 °C. ^{1}H-NMR (500 MHz, DMSO-d6, δ (ppm)): 2.52 s (3H, CH$_3$), 4.06 s (3H, CH$_3$), 7.01 (d, 1H, $^3J_{HH}$ = 9.2 Hz), 7.33 (dd, 1H, $^3J_{HH}$ = 7.3 Hz), 7.46 (d, 1H, CH, $^3J_{HH}$ = 8.4 Hz), 7.55 (dd, 1H, $^3J_{HH}$ = 7.6 Hz), 7.75 (d, 1H, $^3J_{HH}$ = 7.8 Hz), 7.84 (d, 1H, CH, $^3J_{HH}$ = 9.2 Hz), 7.89 (s, 1H, CH), 8.08 (d, 1H, CH, $^3J_{HH}$ = 8.2 Hz), 8.18 (d, 1H, CH, $^3J_{HH}$ = 8.4 Hz), 8.26 (d, 1H, CH, $^3J_{HH}$ = 14.6 Hz), 8.46 (d, 1H, CH, $^3J_{HH}$ = 14.7 Hz). ^{13}C NMR DEPT 125 MHz (DMSO-d$_6$, δ(ppm)): 21.65 (CH$_3$), 56.50 (CH$_3$), 114.01 (CH), 115.77 (CH), 121.91 (CH), 123.56 (CH), 123.87 (CH), 128.55 (CH), 128.69 (CH), 129.55 (CH), 136.33 (CH), 139.59 (CH), 141.49 (CH). Calc. for: $C_{21}H_{18}NOS^+$ m/z = 332.4, found: ESI-MS: m/z = 332.2.

Supplementary Materials: The supporting information can be downloaded at: https://www.mdpi.com/article/10.3390/photochem3020018/s1. Figure S1. 1H-NMR spectrum of dye 3 in DMSO-d6. Figure S2. 1H-NMR spectrum of dye 3 in DMSO-d6—methyl groups region. Figure S3. 1H-NMR spectrum of dye 3 in DMSO-d6—aromatic region. Figure S4. 13C-NMR DEPT spectrum of dye 3 in DMSO-d6. Figure S5. ESI-MS (m/z) spectra of dye 3. Figure S6. B3LYP/6-31G(d,p) optimized ground state structures with iodine counterion (in water medium) and computed relative Gibbs free energies for the dye 3 isomers. Cartesian coordinates of the optimized structures from B3LYP/6-31G** calculations in water medium.

Author Contributions: Conceptualization, S.I., A.A.V., S.B., and D.C.; methodology, S.I. and A.A.V.; validation, S.I., A.A.V., S.B., and D.C.; investigation, A.A.V., S.I., and D.C.; resources, S.I.; data curation, S.I. and D.C.; writing—original draft preparation, S.I. and D.C.; writing—review and editing, A.A.V., S.I., and D.C.; visualization, S.I. and D.C.; supervision, A.A.V. and D.C.; project administration, S.B.; funding acquisition, S.B. All authors have read and agreed to the published version of the manuscript.

Funding: This research was funded by the Bulgarian National Science Fund (BNSF), grant number КП-06-Н37/15-06.12.19 SunUp-project.

Data Availability Statement: The data presented in this study are available either in this article itself and Supplementary Materials.

Conflicts of Interest: The authors declare no conflict of interest.

References

1. Volarić, J.; Szymanski, W.; Simeth, N.A.; Feringa, B.L. Molecular photoswitches in aqueous environments. *Chem. Soc. Rev.* **2021**, *50*, 12377–12449. [CrossRef] [PubMed]
2. Olesińska-Mönch, M.; Deo, C. Small-molecule photoswitches for fluorescence bioimaging: Engineering and applications. *Chem. Commun.* **2023**, *59*, 660–669. [CrossRef] [PubMed]
3. Skov, A.B.; Broman, S.L.; Gertsen, A.S.; Elm, J.; Jevric, M.; Cacciarini, M.; Kadziola, A.; Mikkelsen, K.V.; Nielsen, M.B. Aromaticity-Controlled Energy Storage Capacity of the Dihydroazulene-Vinylheptafulvene Photochromic System. *Chem. Eur. J.* **2016**, *22*, 14567. [CrossRef] [PubMed]

4. Mogensen, J.; Christensen, O.; Kilde, M.D.; Abildgaard, M.; Metz, L.; Kadziola, A.; Jevric, M.; Mikkelsen, K.V.; Nielsen, M.B. Molecular Solar Thermal Energy Storage Systems with Long Discharge Times Based on the Dihydroazulene/Vinylheptafulvene Couple. *Eur. J. Org. Chem.* **2019**, *2019*, 1986–1993. [CrossRef]
5. Moth-Poulsen, K.; Ćoso, D.; Börjesson, K.; Vinokurov, N.; Meier, S.K.; Majumdar, A.; Vollhardt, K.P.C.; Segalman, R.A. Molecular solar thermal (MOST) energy storage and release system. *Energy Environ. Sci.* **2012**, *5*, 8534–8537. [CrossRef]
6. Bren', V.A.; Dubonosov, A.D.; Minkin, V.I.; Chernoivanov, V.A. Norbornadiene–quadricyclane—An effective molecular system for the storage of solar energy. *Russ. Chem. Rev.* **1991**, *60*, 451–469. [CrossRef]
7. Bandara, H.M.D.; Burdette, S.C. Photoisomerization in different classes of azobenzene. *Chem. Soc. Rev.* **2012**, *41*, 1809–1825. [CrossRef]
8. Velema, W.A.; Szymanski, W.; Feringa, B.L. Photopharmacology: Beyond Proof of Principle. *J. Am. Chem. Soc.* **2014**, *136*, 2178–2191. [CrossRef]
9. Dong, L.; Feng, Y.; Wang, L.; Feng, W. Azobenzene-based solar thermal fuels: Design, properties, and applications. *Chem. Soc. Rev.* **2018**, *47*, 7339–7368. [CrossRef]
10. Shi, Y.; Gerkman, M.A.; Qiu, Q.; Zhang, S.; Han, G.G.D. Sunlight-activated phase change materials for controlled heat storage and triggered release. *J. Mater. Chem. A* **2021**, *9*, 9798–9808. [CrossRef]
11. Qiu, Q.; Shi, Y.; Han, G.G.D. Solar energy conversion and storage by photoswitchable organic materials in solution, liquid, solid, and changing phases. *J. Mater. Chem. C* **2021**, *9*, 11444–11463. [CrossRef]
12. Volarić, J.; Thallmair, S.; Feringa, B.L.; Szymanski, W. Photoswitchable, Water-Soluble Bisazobenzene Cross-Linkers with Enhanced Properties for Biological Applications. *ChemPhotoChem* **2022**, *6*, e202200170. [CrossRef]
13. Matsuda, K.; Irie, M. Diarylethene as a photoswitching unit. *J. Photochem. Photobiol. C* **2004**, *5*, 169–182. [CrossRef]
14. Kudernac, T.; van der Molen, S.J.; van Wees, B.J.; Feringa, B.L. Uni- and bi-directional light-induced switching of diarylethenes on gold nanoparticles. *Chem. Commun.* **2006**, *34*, 3597–3599. [CrossRef]
15. Broichhagen, J.; Frank, J.A.; Trauner, D. A Roadmap to Success in Photopharmacology. *Acc. Chem. Res.* **2015**, *48*, 1947–1960. [CrossRef]
16. Lerch, M.M.; Hansen, M.J.; van Dam, G.M.; Szymanski, W.; Feringa, B.L. Emerging Targets in Photopharmacology. *Angew. Chem. Int. Ed.* **2016**, *55*, 10978–10999. [CrossRef]
17. Fischer, E.; Hirshberg, Y. Formation of Coloured Forms of Spirans by Low-Temperature Irradiation. *J. Chem. Soc.* **1952**, 4522–4524. [CrossRef]
18. Klajn, R. Spiropyran-based dynamic materials. *Chem. Soc. Rev.* **2014**, *43*, 148–184. [CrossRef]
19. Hu, D.H.; Tian, Z.Y.; Wu, W.W.; Wan, W.; Li, A.D.Q. Single-Molecule Photoswitching Enables High-Resolution Optical Imaging. *Microsc. Microanal.* **2009**, *15*, 840–841. [CrossRef]
20. Tian, Z.Y.; Li, A.D.Q.; Hu, D.H. Super-resolution fluorescence nanoscopy applied to imaging core–shell photoswitching nanoparticles and their self-assemblies. *Chem. Commun.* **2011**, *47*, 1258–1260. [CrossRef]
21. Montagnoli, G.; Pieroni, O.; Suzuki, S. Control of peptide chain conformation by photoisomerising chromophores: Enzymes and model compounds. *Polym. Photochem.* **1983**, *3*, 279–294. [CrossRef]
22. Ciardelli, F.; Fabbri, D.; Pieroni, O.; Fissi, A. Photomodulation of polypeptide conformation by sunlight in spiropyran-containing poly(L-glutamic acid). *J. Am. Chem. Soc.* **1989**, *111*, 3470–3472. [CrossRef]
23. Sakata, T.; Yan, Y.L.; Marriott, G. Optical switching of dipolar interactions on proteins. *Proc. Natl. Acad. Sci. USA* **2005**, *102*, 4759–4764. [CrossRef] [PubMed]
24. Fissi, A.; Pieroni, O.; Angelini, N.; Lenci, F. Photoresponsive Polypeptides. Photochromic and Conformational Behavior of Spiropyran-Containing Poly-l-Glutamate)s under Acid Conditions. *Macromolecules* **1999**, *32*, 7116–7121. [CrossRef]
25. Wojtyk, J.T.C.; Wasey, A.; Xiao, N.N.; Kazmaier, P.M.; Hoz, S.; Yu, C.; Lemieux, R.P.; Buncel, E. Elucidating the Mechanisms of Acidochromic Spiropyran-Merocyanine Interconversion. *J. Phys. Chem. A* **2007**, *111*, 2511–2516. [CrossRef]
26. Remon, P.; Li, S.M.; Grotli, M.; Pischel, U.; Andreasson, J. An Acido- and Photochromic Molecular Device That Mimics Triode Action. *Chem. Commun.* **2016**, *52*, 4659–4662. [CrossRef]
27. Schmidt, S.B.; Kempe, F.; Brügner, O.; Walter, M.; Sommer, M. Alkyl-Substituted Spiropyrans: Electronic Effects, Model Compounds and Synthesis of Aliphatic Main-Chain Copolymers. *Polym. Chem.* **2017**, *8*, 5407–5414. [CrossRef]
28. Roxburgh, C.J.; Sammes, P.G. On the Acid Catalysed Isomerisation of Some Substituted Spirobenzopyrans. *Dyes Pigments* **1995**, *27*, 63–69. [CrossRef]
29. Shiozaki, H. Molecular Orbital Calculations for Acid Induced Ring Opening Reaction of Spiropyran. *Dyes Pigments* **1997**, *33*, 229–237. [CrossRef]
30. Kortekaas, L.; Chen, J.; Jacquemin, D.; Browne, W.R. Proton-Stabilized Photochemically Reversible E/Z Isomerization of Spiropyrans. *J. Phys. Chem. B* **2018**, *122*, 6423–6430. [CrossRef]
31. Vasilev, A.; Dimitrova, R.; Kandinska, M.; Landfester, K. Baluschev. S. Accumulation of the photonic energy of the deep-red part of the terrestrial sun irradiation by rare-earth metal-free E–Z photoisomerization. *J. Mater. Chem. C* **2021**, *9*, 7119–7126. [CrossRef]
32. Cossi, M.; Barone, V.; Cammi, R.; Tomasi, J. Ab initio study of solvated molecules: A new implementation of the polarizable continuum model. *Chem. Phys. Lett.* **1996**, *255*, 327–335. [CrossRef]
33. Tomasi, J.; Mennucci, B.; Cammi, R. Quantum Mechanical Continuum Solvation Models. *Chem. Rev.* **2005**, *105*, 2999–3094. [CrossRef]

34. Labanowski, J.K.; Andzelm, J.W. (Eds.) *Density Functional Methods in Chemistry*; Springer: New York, NY, USA, 1991.
35. Petersson, G.A.; Bennett, A.; Tensfeldt, T.G.; Al-Laham, M.A.; Shirley, W.A.; Mantzaris, J. A complete basis set model chemistry. I. The total energies of closed-shell atoms and hydrides of the first-row elements. *J. Chem. Phys.* **1988**, *89*, 2193–2218. [CrossRef]
36. Leininger, T.; Nicklass, A.; Stoll, H.; Dolg, M.; Schwerdtfeger, P. The accuracy of the pseudopotential approximation. II. A comparison of various core sizes for indium pseudopotentials in calculations for spectroscopic constants of InH, InF, and InCl. *J. Chem. Phys.* **1996**, *105*, 1052–1059. [CrossRef]
37. Haas, J.; Bissmire, S.; Wirth, T. Iodine Monochloride–Amine Complexes: An Experimental and Computational Approach to New Chiral Electrophiles. *Chem. Eur. J.* **2005**, *11*, 5777–5785. [CrossRef]
38. Frisch, M.J.; Trucks, G.W.; Schlegel, H.B.; Scuseria, G.E.; Robb, M.A.; Cheeseman, J.R.; Scalmani, G.; Barone, V.; Petersson, G.A.; Nakatsuji, H.; et al. *Gaussian 16 Revision 16.A.03*; Gaussian, Inc.: Wallingford, CT, USA, 2016.
39. Zonjić, I.; Radić Stojković, M.; Crnolatac, I.; Tomašić Paić, A.; Pšeničnik, S.; Vasilev, A.; Kandinska, M.; Mondeshki, M.; Baluschev, S.; Landfester, K.; et al. Styryl dyes with N-Methylpiperazine and N-Phenylpiperazine Functionality: AT-DNA and G-quadruplex binding ligands and theranostic agents. *Bioorg. Chem.* **2022**, *127*, 105999. [CrossRef]

Disclaimer/Publisher's Note: The statements, opinions and data contained in all publications are solely those of the individual author(s) and contributor(s) and not of MDPI and/or the editor(s). MDPI and/or the editor(s) disclaim responsibility for any injury to people or property resulting from any ideas, methods, instructions or products referred to in the content.

Article

Excited-State Dynamics of Carbazole and *tert*-Butyl-Carbazole in Organic Solvents

Konstantin Moritz Knötig [†], Domenic Gust [†], Thomas Lenzer and Kawon Oum *

Physical Chemistry 2, Department Chemistry and Biology, Faculty IV, School of Science and Technology, University of Siegen, Adolf-Reichwein-Str. 2, 57076 Siegen, Germany; konstantin.knoetig@uni-siegen.de (K.M.K.); domenic.gust@uni-siegen.de (D.G.); lenzer@chemie.uni-siegen.de (T.L.)
* Correspondence: oum@chemie.uni-siegen.de
[†] These authors contributed equally to this work.

Abstract: Carbazole-based molecular units are ubiquitous in organic optoelectronic materials; however, the excited-state relaxation of these compounds is still underexplored. Here, we provide a detailed investigation of carbazole (Cz) and 3,6-di-*tert*-butylcarbazole (*t*-Bu-Cz) in organic solvents using femtosecond and nanosecond UV–Vis–NIR transient absorption spectroscopy, as well as time-resolved fluorescence experiments upon photoexcitation in the deep-UV range. The initially prepared S_x singlet state has a (sub-)picosecond lifetime and decays to the S_1 state by internal conversion (IC). The S_1 state exhibits absorption peaks at 350, 600 and 1100 nm and has a lifetime of 13–15 ns, which is weakly dependent on the solvent. Energy transfer from vibrationally hot S_1 molecules (S_1^*) to the surrounding solvent molecules takes place with a time constant of 8–20 ps. The T_1 triplet state is populated by intersystem crossing (ISC) from S_1 with a typical quantum yield of 51–56% and shows a lifetime which is typically in the few microseconds regime. The S_1 and T_1 states of both carbazole compounds in solution are strongly quenched by O_2. Two-photon excitation leads to the formation of a small amount of the respective radical cation. The influence of the *tert*-butyl substituents on the photophysics is relatively weak and mainly reflects itself in a small increase in the Stokes shift. The results provide important photophysical information for the interpretation of carbazole relaxation in more complex environments.

Keywords: carbazole; excited-state dynamics; deep-UV ultrafast laser spectroscopy; triplet formation

1. Introduction

Carbazole-based compounds represent one of the core components in the field of organic optoelectronic devices. They are used as molecular building blocks, oligomers, dendrimers or polymers and exhibit several distinct advantages, such as low cost of the starting materials, facile access to functionalization at the nitrogen atom and easy linkage through the carbazole backbone [1–5]. In particular, they show beneficial electronic and photophysical properties with respect to organic light-emitting diode (OLED) applications [6], as they feature high-energy S_1 and T_1 states, which are essential for their functions either as host materials or as molecular electron donor moieties in efficient emitters based on donor–acceptor concepts featuring thermally activated delayed fluorescence (TADF) [7–10].

A comprehensive understanding of the function of such optoelectronic materials requires a good knowledge regarding the excited-state dynamics of the parent compound carbazole and its ring-substituted analogs under "isolated" conditions, because in more dense environments, such as thin films, the intramolecular carbazole chromophore relaxation is in competition with processes such as singlet–singlet, singlet–triplet and triplet–triplet annihilation, as well as slow vibrational cooling of the thin film, which lead to a complex spectral and kinetic behavior [11–15]. The majority of the previous studies on Cz have focused on basic photophysical properties, such as absorption and fluorescence spectra,

which have been characterized in considerable detail [6]. For instance, the steady-state absorption, fluorescence and phosphorescence spectra of carbazole have been reported in solution and the gas phase [16–21]. In addition, triplet–triplet absorption spectra have been provided [22,23]. Moreover, the quenching of the Cz T_1 triplet state, intramolecular excimer formation, hydrogen-bonding interactions and hydrogen transfer reactions from carbazole derivatives to bases, such as pyridine, have been studied in detail [24–31].

In contrast, time-resolved studies of the photoinduced dynamics of these compounds have been less frequently reported and focused on the transient fluorescence of the S_1 state. Lifetimes in the range of 7–15 ns have been reported for Cz in different organic solvents [17,26,29,32,33]. Transient broadband absorption measurements mainly involved N-alkylated carbazole derivatives [15,34–36]. Bayda-Smykaj et al. reported Vis–NIR transient absorption spectra for Cz in acetonitrile up to 2.6 ns, but no femtosecond transient absorption (fs-TA) experiments for wavelengths below 450 nm, and also no detailed kinetic analysis of the TA spectra were provided [34]. Hiyoshi et al. employed picosecond and nanosecond transient absorption (ns-TA) spectroscopy covering the spectral range of 380–810 nm mainly to explore the relevance of photodeprotonation and photoionization processes of carbazole and N-ethylcarbazole in organic solvents induced by two-photon excitation. However, they provided only transient spectra at selected time delays and did not perform an in-depth kinetic analysis [37].

Therefore, the goal of the current work is to investigate in detail the excited-state dynamics of carbazole in the organic solvents n-heptane, tetrahydrofuran (THF) and acetonitrile upon photoexcitation in the wavelength range of 260–273 nm. In addition, the closely related 3,6-di-*tert*-butylcarbazole is studied in the same solvents because of the importance of such 3,6-dialkyl-subsituted chromophores as electron donor moieties in donor–acceptor TADF compounds [38,39]. A comprehensive characterization of the photoinduced relaxation mechanism of these carbazole derivatives is achieved by a combination of fs-TA and ns-TA spectroscopy, as well as time-correlated single photon counting (TCSPC).

2. Materials and Methods

2.1. Chemicals Used and Preparation of Solutions

Carbazole (TCI Deutschland, Eschborn, Germany, high purity), 3,6-di-*tert*-butylcarbazole (TCI Deutschland, >98.0%), n-heptane (Merck, Darmstadt, Germany, Uvasol, ≥99.9%), tetrahydrofuran (THF, Merck, Uvasol, ≥99.9%) and acetonitrile (ACN, Merck, Uvasol, ≥99.9%) were used without further purification. For the preparation of the solutions, a defined amount of Cz or *t*-Bu-Cz was freshly dissolved in the organic solvent of interest. The organic solvents were placed in a septum-sealed vial which was pierced by two stainless steel hollow needles, one used for bubbling nitrogen gas (Messer Industriegase GmbH, 4.6, Bad Soden, Germany) through the solution (duration: 90 min) and the other one used as outlet. Cz or *t*-Bu-Cz were dissolved in the nitrogen-saturated solvent of interest and the colorless solution was then passed through a PTFE filter (pore size 0.45 µm). All sample-handling steps were carried out in a glove box under a nitrogen atmosphere in order to minimize unwanted contact with air.

2.2. Steady-State Absorption and Fluorescence

The steady-state absorption measurements were performed by using a double-beam spectrophotometer (Varian Inc., Palo Alto, CA, USA, Cary 5000). Prior to the experiments, a baseline without a sample was recorded and subtracted afterwards. A quartz cuvette (Hellma, Müllheim, Germany) containing the nitrogen-flushed pure solvent served as a reference. The steady-state emission spectra were measured by using a fluorescence spectrophotometer (Agilent, Santa Clara, CA, USA, Cary Eclipse). The slit width was adjusted to 5 nm for excitation and emission, and the emission spectra were recorded with a resolution of 1 nm. The emission spectra were corrected for the wavelength-dependent sensitivity of the detection system. Steady-state experiments were performed in quartz cuvettes with path lengths of 1 or 10 mm (Hellma).

2.3. Time-Correlated Single Photon Counting

The experimental TCSPC arrangement, originally reported by Morgenroth et al. [40], is described below. Briefly, a pulsed (500 ps full width at half maximum (FWHM)) UV-LED with a center wavelength of 273 nm served as the excitation source. The excitation pulses were vertically polarized (0°) by a wire-grid linear polarizer prior to sample excitation. The resulting sample emission was collected by a quartz lens positioned at an angle of 90° with respect to the pulsed LED beam and then passed another wire-grid linear polarizer, which was set at the magic angle of 54.7° to avoid any contributions of orientational relaxation of the chromophores to the emission signal. A part of the emission spectrum was selected by a bandpass filter (center wavelength 370 nm, FWHM 10 nm) and detected by a hybrid-alkali photodetector. Fluorescence decays were recorded at a repetition frequency of 1 MHz. The fluorescence decays were fitted by a single exponential function with a small constant offset using an iterative reconvolution procedure applying an instrument response function, which was obtained from the LED scattering signal of a diluted suspension of colloidal silica nanoparticles in water. Concentrations of Cz and t-Bu-Cz in the TCSPC and steady-state fluorescence experiments were in the range of $6–11 \times 10^{-6}$ mol L^{-1}. The nitrogen-bubbled Cz and t-Bu-Cz solutions were placed under nitrogen in an airtight cuvette.

2.4. Femtosecond and Nanosecond Transient Absorption Spectroscopy

Broadband transient absorption experiments with a time resolution of ca. 80 fs covering delay times up to 1500 ps were carried out at a repetition frequency of 920 Hz on two setups dedicated to measurements in the UV–Vis [41] and NIR ranges [42]. They are both based on the pump-supercontinuum probe (PSCP) method [43]. Excitation was performed at 260 nm by an OPA system with a repetition frequency of 460 Hz and a pulse energy in the range of 300–390 nJ, corresponding to a fluence of 1.1–1.5 mJ cm^{-2}. Measurements up to a delay time of 10 µs were carried out by interfacing the UV–Vis setup with a Q-switched Nd:YAG microlaser, as described previously [44], using its fourth harmonic at 266 nm (FWHM ca. 420 ps) for excitation, with similar fluences as for the femtosecond experiments. Solutions of Cz or t-Bu-Cz (T = 296 K) were placed under nitrogen in an airtight quartz cuvette with a path length of 1 mm, with typical concentrations in the range of $1–5 \times 10^{-4}$ mol L^{-1}. During the measurements, the cuvette was constantly moved in a plane perpendicular to the propagation axis of the probe beam to minimize any photochemical decomposition and also local heating effects of the solutions that could occur by repeated laser excitation of the same sample volume.

3. Results

3.1. Steady-State Absorption and Emission of Carbazole and 3,6-Di-Tert-Butyl-Carbazole

The steady-state absorption and fluorescence spectra of Cz and t-Bu-Cz were investigated in the organic solvents n-heptane, THF and acetonitrile at 296 K (Figure 1). Characteristic parameters are summarized in Table 1. The spectra for Cz (panels a–c) show three main bands: The structured absorption band in the wavelength range of 300–340 nm corresponds to the short-axis polarized $S_0 \rightarrow S_1$ transition. The S_1 state has A_1 symmetry and is a L_a state according to Platt's notation [19,21,28,45,46]. The $S_0 \rightarrow S_2$ transition at about 290 nm is long-axis polarized, and the S_2 state is of B_2 symmetry [28,45]. The absorption bands below 270 nm arise from a combination of several electronic bands with B_2 and A_1 symmetry [45], which we will denote by the collective notation S_x later on. For Cz in acetonitrile, we determined the absolute absorption coefficient for the longest-wavelength peak in the absorption spectrum at 334 nm as 3900 L mol^{-1} cm^{-1}. This value is expected to be only weakly dependent on the solvent.

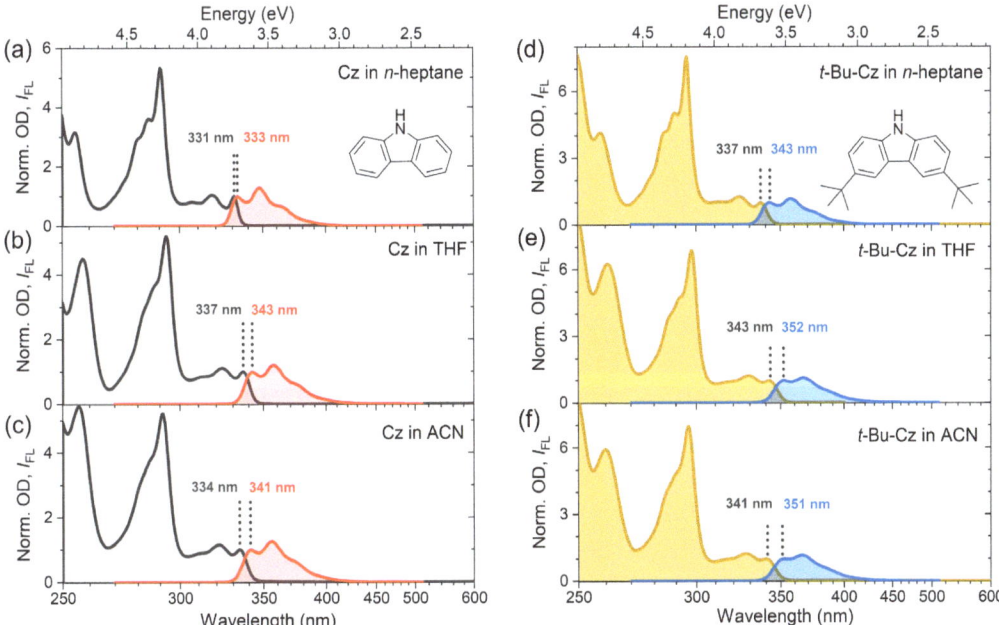

Figure 1. (a–c) Normalized steady-state absorption spectra (black) and normalized fluorescence spectra (red) of carbazole (Cz) in *n*-heptane, THF and acetonitrile, respectively. (d–f) Normalized steady-state absorption spectra (brown) and normalized fluorescence spectra (blue) of 3,6-di-*tert*-butylcarbazole (*t*-Bu-Cz) in the same solvents. The 0–0 transitions of the absorption and emission spectra are indicated by black dotted lines and labeled by the respective wavelength values. Molecular structures of Cz and *t*-Bu-Cz are provided in panels (**a**,**d**), respectively.

For all of the three solvents, the emission bands are observed in the wavelength range of 330–420 nm. They resemble the mirror image of the respective S_1 absorption band and show substantial spectral overlap with the corresponding absorption spectra. The solvatochromic behavior of Cz was analyzed by fitting the structured electronic bands to a sum of Gaussian functions, which provided the positions of the 0–0 transition in absorption and emission as well as the Stokes shift (Table 1). The spectral positions showed only a weak solvent dependence, with a variation of less than 8 nm. The Stokes shifts of Cz were estimated from the difference between the 0–0 transitions of the absorption and emission bands. They are very small and appear to increase slightly with solvent polarity, as identified by the solvent polarity parameter Δf (which is based on the known values for the refractive index n and the dielectric constant ε) [47,48]. They range from 193 cm^{-1} in *n*-heptane to 614 cm^{-1} in acetonitrile, which corresponds to only 2–7 nm in this wavelength range. This suggests that only minor structural changes between the S_0 and S_1 states occur. The finding is supported by previous experiments of Pratt and co-workers using rotationally resolved electronic spectroscopy of isolated Cz molecules in a molecular beam. They found only a slight decrease in the rotational constants of Cz upon excitation from the S_0 to the S_1 state. This was attributed to minor ring expansions, which are typical of π–π^* transitions [21]. Moreover, the planarity of the π-system is conserved in the S_1 state [21]. We note that correlations of the Stokes shift with other parameters, such as the Lorenz–Lorentz function $R(n)$ or the solvent viscosity η, did not provide any systematic trends.

Table 1. Summary of steady-state spectroscopic properties of carbazole (Cz) and 3,6-di-*tert*-butylcarbazole (*t*-Bu-Cz) in organic solvents at 296 K.

Molecule	Solvent	Δf [1]	λ_{abs}^{0-0} (nm)	λ_{fl}^{0-0} (nm)	$\Delta\lambda_{Stokes}^{0-0}$ (nm)	$\Delta\tilde{\nu}_{Stokes}^{0-0}$ (cm^{-1})
Cz	*n*-Heptane	≈0	331	333	2	193
	THF	0.44	337	343	6	525
	Acetonitrile	0.71	334	341	7	614
t-Bu-Cz	*n*-Heptane	≈0	337	343	6	519
	THF	0.44	343	352	9	745
	Acetonitrile	0.71	341	351	10	835

[1] The solvent polarity function was estimated as $\Delta f = R(\varepsilon) - R(n)$, where $R(\varepsilon) = (\varepsilon - 1)/(\varepsilon + 2)$ and $R(n) = (n^2 - 1)/(n^2 + 2)$, with the dielectric constant ε and the refractive index n of the respective solvent and the value for *n*-heptane being essentially identical to that of *n*-hexane [47,48].

The panels d–f of Figure 1 display the corresponding results for *t*-Bu-Cz in the same solvents. The shape of the absorption and emission bands is similar to those of Cz. However, all spectra are systematically shifted to larger wavelengths (by about 6 and 10 nm, for absorption and emission, respectively), and this red shift can be attributed to the larger polarizability of *t*-Bu-Cz. In addition, the resulting Stokes shifts are slightly larger (in the range of 519–835 cm^{-1}, corresponding to 6–10 nm) compared with Cz and also increase with solvent polarity. It therefore appears as if the two *tert*-butyl substituents only have a minor impact on the steady-state spectroscopic properties of the carbazole chromophore in solution.

3.2. Transient Absorption Studies and TCSPC Experiments of Carbazole in Organic Solvents

Femtosecond and nanosecond transient absorption spectroscopy as well as time-correlated single photon counting measurements were performed for Cz in three organic solvents at 296 K. Figures 2–4 summarize the spectral data and the kinetic analysis of the ultrafast excited-state dynamics of Cz in *n*-heptane, THF and acetonitrile, respectively. Figure 5 provides a sketch of the proposed kinetic mechanism, and Table 2 summarizes time constants from the kinetic fits.

The ultrafast excited-state dynamics of carbazole upon photoexcitation at 260 nm were recorded using fs-TA from the sub-picosecond time range up to 1.45 ns over the probe wavelength range of 260–1600 nm. We note that around a zero delay time, there was a pronounced coherent signal from the solvent itself, especially in the deep-UV region. In Figures 2–4, the spectral development is presented as a contour plot in panel a and also for selected time windows in panels b and c. We start with the dynamics of Cz in *n*-heptane (Figure 2). The initially prepared S_x state quickly relaxes to the S_1 state via ultrafast internal conversion with a time constant $\tau_x = k_x^{-1}$ of ca. 0.7 ps. In panels b and c, a ground state bleach feature (GSB, $S_0 \to S_2$) at 290 nm and characteristic $S_1 \to S_n$ ESA bands with peaks at 350, 600 and 1100 nm are clearly visible. In particular, the two S_1 ESA bands at 600 and 1100 nm show a distinct spectral narrowing on the time scale up to 50 ps (panel b), which we assign to collisional energy transfer (CET) from vibrationally hot carbazole molecules (S_1^*) to the surrounding "cold" solvent molecules. The detailed kinetic analysis provides a time constant $\tau_{CET} = k_{CET}^{-1}$ of 19 ps for this spectral evolution, see panel d for the multiexponential fits to the time traces at various probe wavelengths. Such dynamics have been frequently observed for other vibrationally excited molecules in organic solvents, such as, for example, azulene, anthracene derivatives, *trans*-stilbene and carotenoids [49–53].

The kinetic traces of the fs-TA experiments in panel d do not noticeably decay. Therefore, additional experiments using ns-TA spectroscopy and time-resolved emission were performed to elucidate the fate of the S_1 state on longer time scales (panels e–h). As shown in panel h, the fluorescence decay curve of the S_1 state of carbazole in *n*-heptane at 296 K obtained from TCSPC using a bandpass filter with a center wavelength of 370 nm (FWHM of 10 nm) is well described by a monoexponential fit with a small constant offset, resulting in a total S_1 lifetime $\tau_1 = k_1^{-1}$ of 14.7 ns, which was also used as the long component in the fit of the fs-TA kinetics in panel d. The ns-TA spectroscopy experiments using excitation at 266 nm provide further insight into the decay dynamics of the S_1 state; see the contour plot in panel e and the nanosecond transient absorption spectra for selected times in panels f and g. At early times (1–12 ns, panel f), the ESA bands of the S_1 state near 350 nm and 600 nm dominate. At ca. 330 nm, the structured UV ESA band shows clear indications for an overlapping $S_0 \to S_1$ GSB ($0 \to 0'$ transition) and $S_1 \to S_0$ stimulated emission (SE) feature ($0' \to 0$ transition) of the S_1 state, as can be seen by comparison with the inverted steady-state absorption spectrum (magenta) and the steady-state stimulated emission spectrum (cyan). Similar spectral features located at 316 nm ($0 \to 1'$) and 347 nm ($0' \to 1$) can be again assigned to vibronic structure in the GSB and SE bands. The small amplitude of these features underscores that the absorption coefficient of the $S_0 \to S_1$ band (about 3900 L mol^{-1} cm^{-1} at 334 nm, as mentioned above) is much smaller than that of the overlapping, strongly allowed $S_1 \to S_n$ ESA band, which has a value of about 20,000 L mol^{-1} cm^{-1} at 620 nm [54].

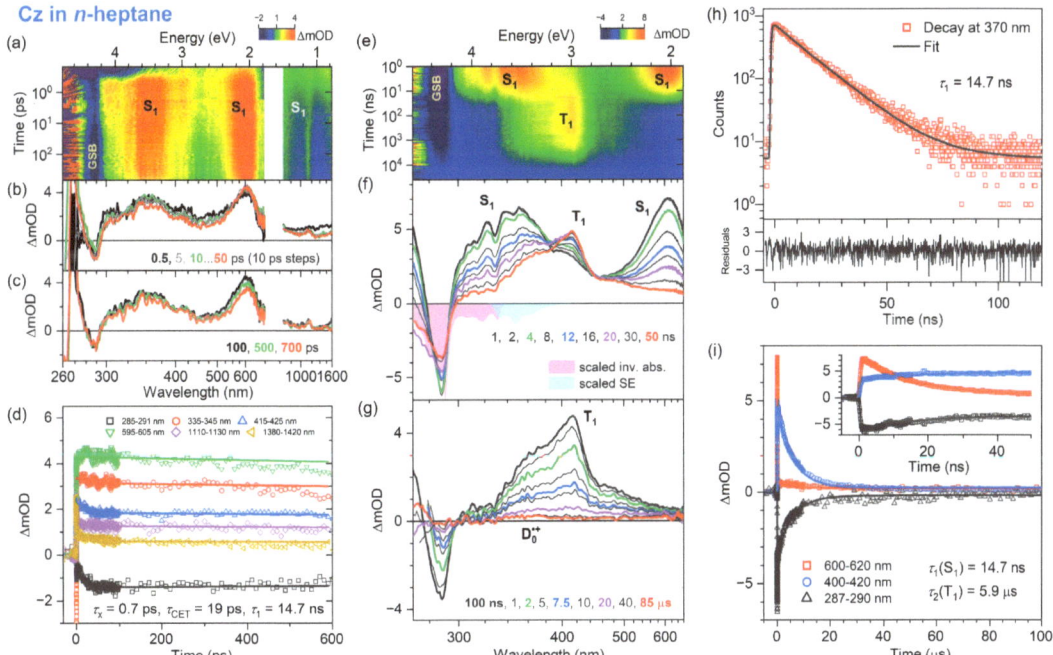

Figure 2. Time-resolved optical spectroscopy of carbazole (Cz) in n-heptane at 296 K ($\lambda_{\text{pump}} = 260$ nm for fs-TA, $\lambda_{\text{pump}} = 266$ nm for ns-TA and $\lambda_{\text{pump}} = 273$ nm for TCSPC). (**a**) fs-TA spectra for the time range up to 700 ps shown as a contour plot. (**b,c**) fs-TA spectra for selected time windows. (**d**) Kinetic traces and fits at selected wavelengths obtained from the fs-TA data. (**e**) Contour plot of ns-TA spectra. (**f,g**) ns-TA spectra at selected times, including the inverted and scaled steady-state absorption spectrum (magenta) and the steady-state stimulated emission spectrum (cyan). (**h**) TCSPC signal (red symbols) obtained using a bandpass filter centered at 370 nm, including a monoexponential fit with a small constant offset (black) and fit residuals (lower panel). (**i**) Kinetic traces and fits of the ns-TA data for the wavelength ranges 600–620 nm, 400–420 nm and 287–290 nm, where S_1, T_1 and ground-state bleaching, respectively, have dominant contributions (cf. panels (**f,g**)). The time constants τ_1 and τ_2 represent the total lifetimes of the S_1 state and the T_1 state, respectively. The constant offset arises from the long-lived carbazole radical cation ($D_0^{\bullet+}$, cf. panel (**g**)).

On even longer time scales (panel g), the S_1 ESA bands have disappeared and a new ESA band near 420 nm has built up, which is assigned to the T_1 state [22,23,37,54]. It decays on the time scale of several microseconds. Kinetic traces extracted from the nanosecond transient absorption data are compared in panel i of Figure 2 for three wavelength ranges: 600–620 nm (red, predominantly decay of S_1), 400–420 nm (blue, mainly formation and decay of T_1) and 286–290 nm (black, mainly GSB and thus recovery of S_0). The kinetic analysis in panel i demonstrates that the time traces in the nanosecond transient absorption signals are well reproduced by biexponential fits, with a time constant of 14.7 ns for the formation of T_1 via intersystem crossing (ISC) from S_1, and a time constant of 5.9 µs for the lifetime of T_1 ($\tau_2 = k_2^{-1}$). Note that the time constant of 14.7 ns obtained from the fit to the ns-TA experiments agrees with the S_1 lifetime found in the TCSPC measurements in panel h. In addition, a long-lived, weakly absorbing species with a broad spectrum was observed (340–660 nm, cf. panel g at 85 µs). We propose that this spectral feature arises from the Cz$^{\bullet+}$ radical cation (doublet ground state, $D_0^{\bullet+}$). First of all, it strongly resembles the flat steady-state absorption spectrum of Cz$^{\bullet+}$ reported by Shida et al., which was obtained by exposing Cz in an organic matrix at 77 K to γ radiation [55]. Secondly, the vertical ionization energy of Cz was previously reported as 7.68 eV (161 nm) [56]. Therefore, resonant two-photon ionization of Cz via the S_x state by our pump laser pulse at 266 nm will be energetically feasible. The radical cation lives longer than the time window covered in the ns-TA experiments, and thus only a lower limit can be given for its lifetime τ_{RCat}.

An accurate determination of the quantum yield for the formation of T_1 from S_1 is not possible; however, an estimate may be provided based on the known absorption coefficients of the S_1 ESA band of carbazole at 620 nm (20,000 L mol^{-1} cm^{-1}), the T_1 ESA band at 420 nm (14,500 L mol^{-1} cm^{-1}) [54] and our ns-TA spectra at 1 ns (predominantly S_1 ESA) and about 50 ns (predominantly T_1 ESA), as shown in panel f of Figure 2, resulting in a triplet quantum yield of about 55% (Table 2). This value is in very good agreement with previous estimates in the range of 45–60% by the groups of Huber and Ware [17,26]. Therefore, the individual contributions to τ_1 in the mechanism of Figure 5 are $\tau_{1,\text{ISC}} = \tau_1/\Phi_{\text{ISC}} \approx 26.7$ ns and $\tau_{1,\text{IC+Fl}} = \tau_1/\Phi_{\text{IC+Fl}} \approx 32.7$ ns where $\tau_1 = (\tau_{1,\text{ISC}}^{-1} + \tau_{1,\text{IC+Fl}}^{-1})^{-1}$ (i.e., $k_1 = k_{1,\text{ISC}} + k_{1,\text{IC+Fl}}$). Regarding the quantum yields for the parallel fluorescence and IC channels from S_1, Ware, Huber and co-workers estimate values of $\Phi_{\text{Fl}} \approx 40\%$ and $\Phi_{\text{IC}} \leq 15\%$, respectively [17,26].

Comparable dynamic behavior is observed for the electronic relaxation of carbazole in mid-polar THF and polar acetonitrile, as demonstrated in Figures 3 and 4, respectively. The fs-TA experiments provide ESA band shapes (panels a–c) and decay kinetics (panel d) which are similar to Cz in *n*-heptane: The lifetime τ_x of the initially prepared S_x state in the two solvents is ultrashort (about 0.2 ps), and the time constant τ_{CET} for collisional energy transfer to the solvent is 19 and 21 ps in THF and acetonitrile, respectively (Table 2). The S_1 lifetimes obtained from TCSPC (panel h) are 13.6 ns in THF and 14.3 ns in acetonitrile (Table 2). The S_1 lifetime of carbazole is therefore only weakly dependent on the solvent. This is understandable because Cz exhibits only a small change in dipole moment upon photoexcitation, from 1.9 D in the ground state to 3.1 D in the S_1 state [57].

The ns-TA spectra (panels e–g) show the decay of the two S_1 ESA bands with peaks at 350 and 600 nm and the subsequent formation of the T_1 state with a band centered at about 420 nm. The decay of the T_1 state occurs with a time constant of 3.4 µs in THF and 10.3 µs in acetonitrile. As for Cz in *n*-heptane, a small-amplitude, long-lived absorption signal arises from the Cz$^{\bullet+}$ radical cation. Therefore, also in these two solvents, the excited-state relaxation processes of Cz are well described by the kinetic scheme provided in Figure 5. The time constants and triplet quantum yields are summarized in Table 2.

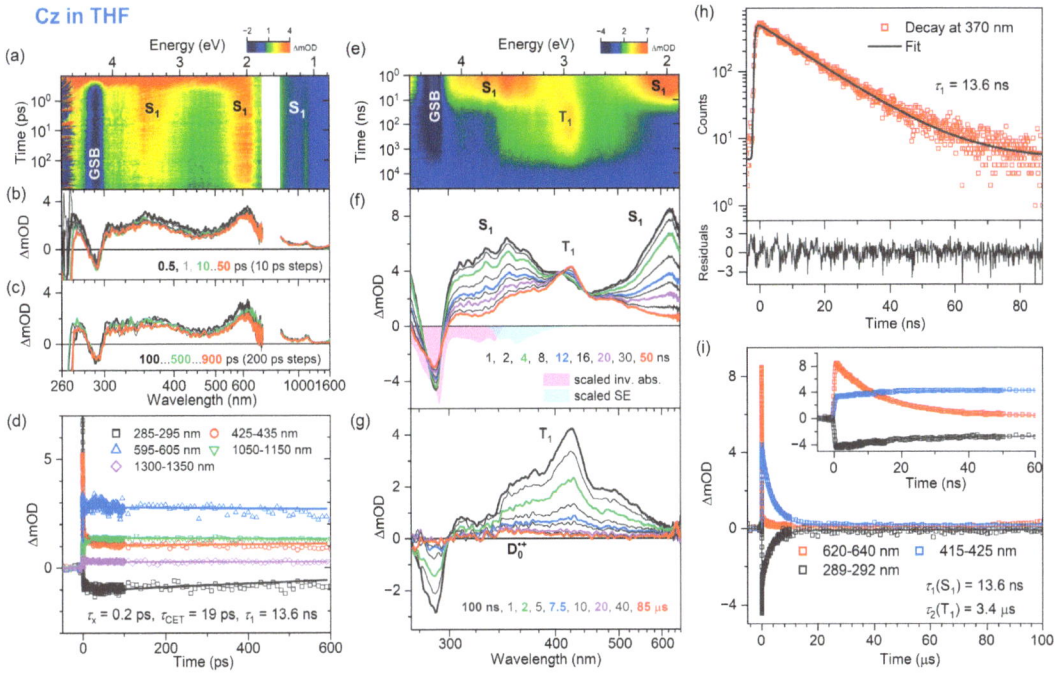

Figure 3. Same as Figure 2 but for carbazole in THF.

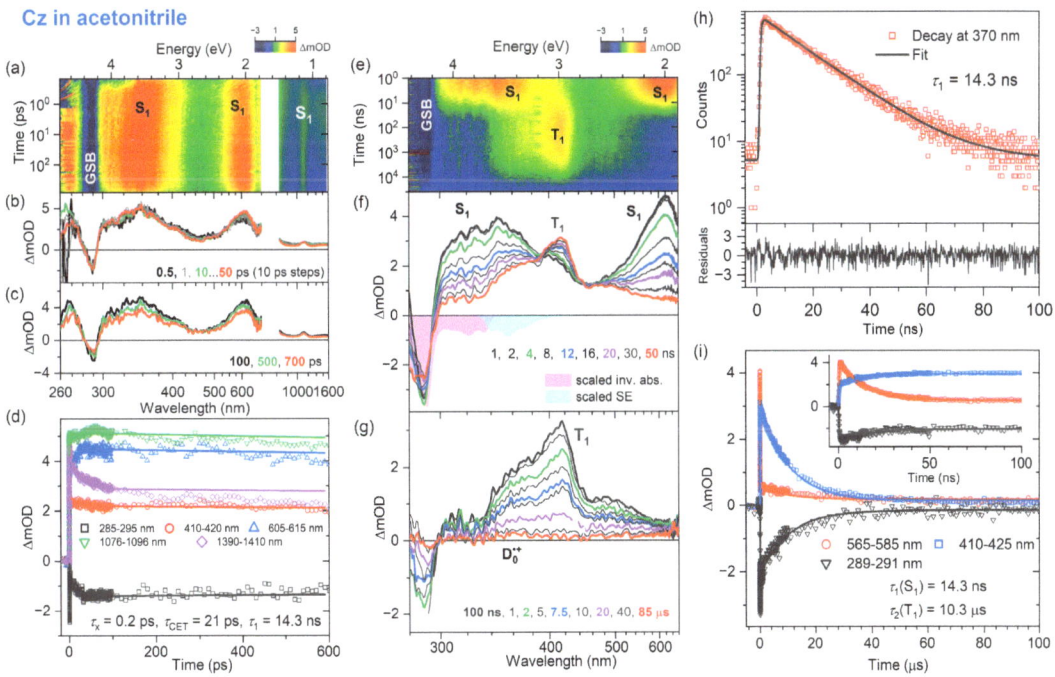

Figure 4. Same as Figure 2 but for carbazole in acetonitrile.

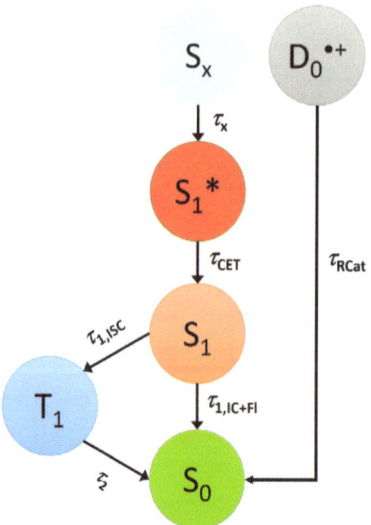

Figure 5. Kinetic scheme describing the different pathways for the relaxation of Cz and *t*-Bu-Cz upon photoexcitation by a laser. The S_x state is prepared by one-photon excitation, whereas the $D_0^{\bullet +}$ radical cation is likely generated by a small fraction of two-photon excitation with subsequent ionization.

Table 2. Summary of time constants of carbazole (Cz) in organic solvents at 296 K obtained from femtosecond and nanosecond transient absorption as well as TCSPC experiments.

Cz	fs-TA		TCSPC		ns-TA	
	τ_x	τ_{CET}	τ_1 (S_1)	τ_2 (T_1)	Φ (T_1)	τ_{RCat}
n-Heptane	0.7 ps	19 ps	14.7 ns	5.9 µs	55%	>50 µs
THF	0.2 ps	19 ps	13.6 ns	3.4 µs	51%	>50 µs
Acetonitrile	0.2 ps	21 ps	14.3 ns	10.3 µs	56%	>50 µs

3.3. Transient Absorption and TCSPC Experiments for 3,6-Di-Tert-Butyl Carbazole in Organic Solvents

The influence of additional alkyl substituents at the aromatic ring system on the excited-state dynamics was explored by investigating the Cz derivative 3,6-di-*tert*-butylcarbazole in *n*-heptane, THF and acetonitrile using fs-TA, ns-TA and TCSPC measurements. The experimental results are summarized in Figures 6–8. Characteristic parameters from the kinetic fitting procedure are compiled in Table 3.

The fs-TA experiments for *t*-Bu-Cz were performed for excitation at 260 nm, and an overview of the dynamics is provided by the contour plots in panel a of each figure. The IC process from the initially populated S_x state to the S_1 state is ultrafast in all cases (τ_x = 1.3, 0.04 and 0.3 ps for *n*-hexane, THF and acetonitrile, respectively). The S_1 state shows the typical ESA bands also observed for Cz, with peaks at 350, 600 and 1100 nm (panels b and c). In addition, a spectral narrowing of the S_1 ESA bands at 600 and 1100 nm is observed, which is assigned to the transfer of vibrational excess energy from vibrationally hot (S_1^*) *t*-Bu-Cz to the solvents, with τ_{CET} = 8.9, 16 and 16 ps for *n*-hexane, THF and acetonitrile, respectively (panel b). On longer time scales (panel c), the kinetics do not decay visibly. This is consistent with the results of the TCSPC experiments (panel h, λ_{pump} = 273 nm), which reveal a slow monoexponential decay of the S_1 state with time constants of 15.0, 12.7 and 14.1 ns in *n*-heptane, THF and acetonitrile, respectively. This behavior is also consistent with our findings for Cz and suggests quite a weak intramolecular charge transfer (ICT) character of the S_1 state.

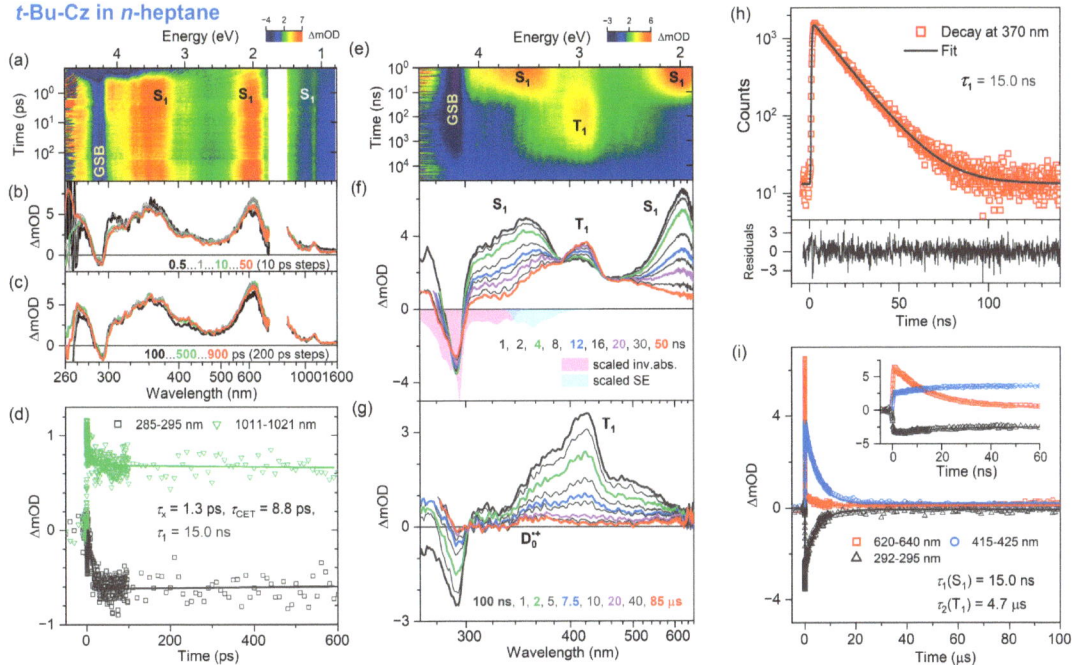

Figure 6. Time-resolved optical spectroscopy of 3,6-di-*tert*-butylcarbazole (*t*-Bu-Cz) in *n*-heptane at 296 K (λ_{pump} = 260 nm for fs-TA, λ_{pump} = 266 nm for ns-TA and λ_{pump} = 273 nm for TCSPC). Arrangement of the panels as in Figure 2.

Table 3. Summary of time constants of 3,6-di-*tert*-butylcarbazole (*t*-Bu-Cz) in organic solvents at 296 K obtained from femtosecond and nanosecond transient absorption as well as TCSPC experiments.

t-Bu-Cz	fs-TA		TCSPC		ns-TA	
	τ_x	τ_{CET}	τ_1 (S_1)	τ_2 (T_1)	Φ (T_1)	τ_{RCat}
n-Heptane	1.3 ps	8.9 ps	15.0 ns	4.7 μs	53%	>50 μs
THF	0.04 ps	16 ps	12.7 ns	4.2 μs	36%	>50 μs
Acetonitrile	0.3 ps	16 ps	14.1 ns	7.5 μs	54%	>50 μs

The excited-state dynamics of *t*-Bu-Cz on longer time scales were investigated using ns-TA experiments over the time range of 1 ns–100 μs (panels e–g, λ_{pump} = 266 nm). The S_1 spectrum with the two distinct peaks at 350 and 600 nm is replaced on longer time scales by the spectrum of the T_1 state, which has a peak at 420 nm and a pronounced shoulder toward 600 nm (panel g). The lifetime of the T_1 state is 4.7, 4.2 and 7.5 μs in *n*-heptane, THF and acetonitrile, respectively (panel i and Table 3). An estimate of the quantum yield for T_1 formation based on the known S_1 and T_1 absorption coefficients of Cz [54] and the transient spectra in panel f of Figures 6–8 provides triplet quantum yields of 55%, 36% and 54%, respectively. The shape of the T_1 spectrum depends only weakly on the solvent. Similar to Cz, we observe the formation of a small fraction of a long-lived broadly absorbing species in all solvents (340–660 nm, see panel g of Figures 6–8 at 85 μs). This spectral feature is assigned to the *t*-Bu-Cz$^{\bullet+}$ radical cation ($D_0^{\bullet+}$), which is generated via resonant two-photon ionization by the 266 nm pump laser pulse, as explained for Cz above. Because of its long lifetime, we can only provide a lower limit for the time constant τ_{RCat} of the *t*-Bu-Cz$^{\bullet+}$ species, as in the case of Cz$^{\bullet+}$.

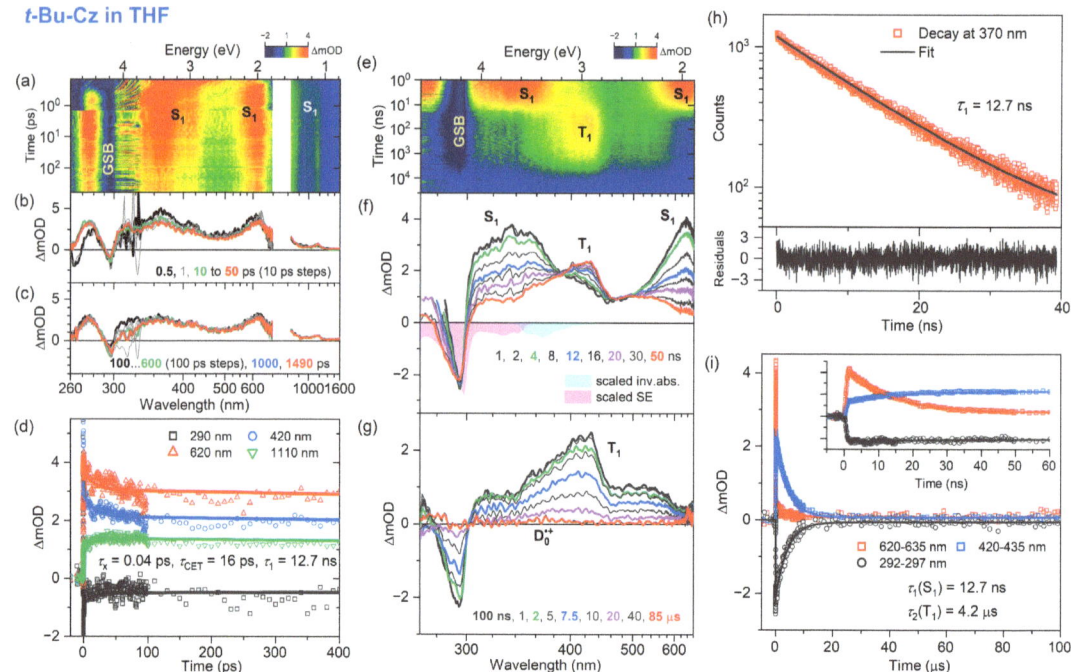

Figure 7. Same as Figure 6 but for the solvent THF.

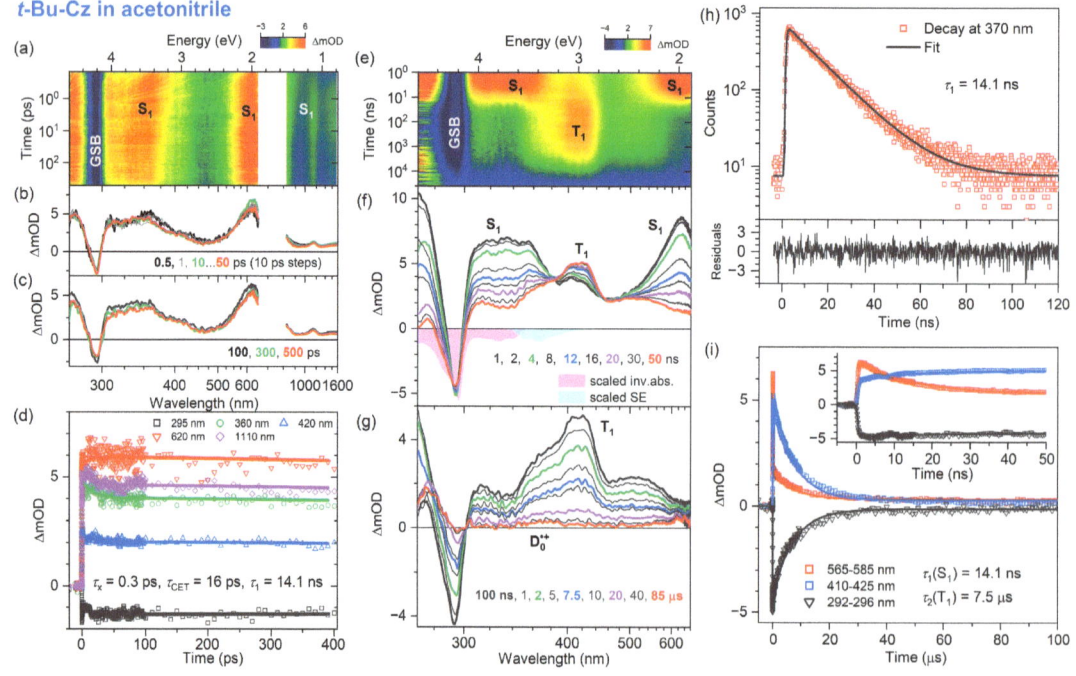

Figure 8. Same as Figure 6 but for the solvent acetonitrile.

4. Discussion

The electronic relaxation of Cz and t-Bu-Cz in organic solvents upon one-photon excitation in the deeper UV range below 300 nm is well described by a kinetic scheme involving S_x, S_1 and T_1 species. The initially populated S_x state has a (sub-)picosecond IC lifetime. The S_1 state shows characteristic ESA bands with peaks at 350, 600 and 1100 nm. The ESA band at 350 nm is reported here for the first time, and the same holds for a pronounced $S_0 \to S_2$ bleach band at 290 nm. The total lifetime of the S_1 state in deaerated solutions is in the range of 13–15 ns and only weakly dependent on the solvent. This result may be compared with previous time-resolved fluorescence experiments for Cz: For isolated jet-cooled carbazole molecules in the gas phase, a value of 29.4 ns for the $S_0 \to S_1$ (0–0) transition [19] located at 324.59 nm (3.8197 eV) [21] was found. Transient fluorescence studies in solution have reported total S_1 lifetimes in the range of 7–15 ns [17,26,29,32,33]. The S_1 lifetimes in the 13–15 ns range obtained for Cz and t-Bu-Cz are confirmed by the nanosecond transient absorption experiments which also provide clean spectral fingerprints of the T_1 state, with a peak absorption at 420 nm and a long absorption tail toward 600 nm. The T_1 state is populated from S_1 by ISC, and its lifetime for both carbazole derivatives is in the range of several microseconds. Based on the relative amplitudes of the S_1 and T_1 absorption in the ns-TA spectra and known absorption coefficients for the two states from the literature [54], we estimate a quantum yield for T_1 triplet formation of typically 51–56% for Cz and t-Bu-Cz in most of the solvents.

The lifetimes of the S_1 and T_1 states of Cz and t-Bu-Cz are very sensitive to the presence of $O_2(^3\Sigma_g^-)$ in the three organic solvents. We note that in another set of TCSPC and ns-TA experiments, we obtained S_1 lifetimes in the range of 6–8 ns in not fully deoxygenated solutions and also observed a drastic reduction of the T_1 lifetime from about 10 µs down to 25–80 ns. Martin and Ware measured S_1 lifetimes of 14.8 and 8.0 ns for Cz in O_2-free and aerated cyclohexane, respectively [26]. Similarly, Bonesi and Erra-Balsells obtained S_1 lifetimes of 14.2/8.70 ns in cyclohexane and 15.1/7.80 ns in acetonitrile under O_2-free/aerated conditions [32].

Different S_1 quenching mechanisms of $O_2(^3\Sigma_g^-)$ need to be considered to understand this effect [58]: Quenching via the pathway $S_1 + O_2(^3\Sigma_g^-) \to T_1 + O_2(^1\Delta_g)$ requires an S_1–T_1 energy gap of at least 0.98 eV [58] to generate $O_2(^1\Delta_g)$ and, therefore, cannot take place in Cz and t-Bu-Cz, because the S_1–T_1 energy gap of Cz is 0.6 eV [17]. Viable pathways are the $O_2(^3\Sigma_g^-)$-enhanced ISC process via $S_1 + O_2(^3\Sigma_g^-) \to T_1 + O_2(^3\Sigma_g^-)$ and also enhancement of the IC channel via $S_1 + O_2(^3\Sigma_g^-) \to S_0 + O_2(^3\Sigma_g^-)$. In addition, the deactivation channel $S_1 + O_2(^3\Sigma_g^-) \to T_2 + O_2(^3\Sigma_g^-)$ (with subsequent fast $T_2 \to T_1$ IC) [58] should be available in Cz and t-Bu-Cz [59].

Next, we consider quenching of the T_1 state by $O_2(^3\Sigma_g^-)$. Because the T_1–S_0 energy gap of the two carbazoles is about 3.1 eV [17] and thus clearly larger than the 1.6 eV required for generating $O_2(^1\Sigma_g^+)$, quenching of the T_1 triplet state by $O_2(^3\Sigma_g^-)$ can occur via the three channels $T_1 + O_2(^3\Sigma_g^-) \to S_0 + O_2(^1\Sigma_g^+)$, $T_1 + O_2(^3\Sigma_g^-) \to S_0 + O_2(^1\Delta_g)$ and $T_1 + O_2(^3\Sigma_g^-) \to S_0 + O_2(^3\Sigma_g^-)$ [58]. For the T_1 state of Cz, Garner and Wilkinson reported a high $O_2(^3\Sigma_g^-)$ quenching rate constant of 5.7×10^9 L mol^{-1} s^{-1} in aerated benzene (for an O_2 concentration of 1.7×10^{-3} mol L^{-1}), which could arise from the influence of Cz charge transfer states [27]. In any case, the impact of $O_2(^3\Sigma_g^-)$ on the S_1 and T_1 lifetimes is substantial and needs to be carefully addressed in the TCSPC and ns-TA experiments. It also explains the spread of 7–15 ns in the S_1 lifetimes reported in the literature [17,26,29,32,33].

Furthermore, we note that we did not detect any steady-state phosphorescence from the T_1 state. In the literature, phosphorescence lifetimes in the range of 7.8–8.0 s have been measured for Cz in solid organic matrices only at a temperature of 77 K [17,22,32], with a phosphorescence quantum yield of 44% [17,32]. However, phosphorescence has so far not been observed at 300 K in organic solvents [60]. Assuming a temperature-independent radiative T_1 lifetime of 18 s [17,32] and taking our total T_1 lifetime of ca. 10 µs from the ns-TA experiments, we estimate a phosphorescence quantum yield of 6×10^{-7} at 296 K, so the phosphorescence will be very hard to detect, and the primary decay channel from T_1 is ISC.

Based on our femtosecond transient absorption data, we did not find clear indications for substantial ICT character of the S_1 state in Cz and t-Bu-Cz. The mirror-image-type structured absorption and emission spectra with small Stokes shifts support this interpretation, suggesting only minor structural and electronic changes in the S_1 state. We also did not find spectral evidence for the formation of carbazyl radicals (Cz$^\bullet$), which were observed earlier by Hiyoshi et al. in their nanosecond transient absorption study [37]. However, in all of the solvents, we found spectral signatures for the formation of a small fraction of Cz$^{\bullet+}$ and t-Bu-Cz$^{\bullet+}$ radical cations, which show a broad absorption band spanning the wavelength range of 340–660 nm. They are generated by two-photon excitation induced by the pump laser.

The current results for "isolated" Cz and t-Bu-Cz molecules in organic solvents provide relevant photophysical and kinetic data for the interpretation of their electronic contributions in more complex environments, such as carbazole-based thin film materials or blue-emitting TADF compounds employing Cz-type derivatives as electron donor group [6,9,61,62], where competing channels, such as, for example, ICT processes, singlet–singlet, singlet–triplet and triplet–triplet annihilation take place simultaneously. For instance, the energetic position of the S_1 state of the Cz-type donor relative to the electronic states in the acceptor part of modern TADF dyes is relevant to the efficient transfer of the initial excitation to the acceptor. Moreover, the high-lying S_1 and T_1 states of carbazole-based host materials are of general importance for OLED operation, as the light-emitting states of the guest molecule are not subject to quenching by triplet energy transfer to the host.

Author Contributions: Conceptualization, K.O. and T.L.; methodology, K.O. and T.L.; investigation, K.M.K., D.G., K.O. and T.L.; writing—original draft preparation, K.O., T.L. and K.M.K.; writing—review and editing, K.O., T.L. and K.M.K.; supervision, K.O. and T.L. All authors have read and agreed to the published version of the manuscript.

Funding: This research received no external funding.

Data Availability Statement: Data are contained within the article.

Acknowledgments: We thank M. Dango for supporting K.M.K. and D.G. in the lab during this project.

Conflicts of Interest: The authors declare no conflicts of interest.

References

1. Blouin, N.; Leclerc, M. Poly(2,7-carbazole)s: Structure-Property Relationships. *Acc. Chem. Res.* **2008**, *41*, 1110–1119. [CrossRef]
2. Beaupré, S.; Leclerc, M. PCDTBT: En route for low cost plastic solar cells. *J. Mater. Chem. A* **2013**, *1*, 11097–11105. [CrossRef]
3. Shizu, K.; Lee, J.; Tanaka, H.; Nomura, H.; Yasuda, T.; Kaji, H.; Adachi, C. Highly efficient electroluminescence from purely organic donor–acceptor systems. *Pure Appl. Chem.* **2015**, *87*, 627–638. [CrossRef]
4. Dumur, F. Carbazole-based polymers as hosts for solution-processed organic light-emitting diodes: Simplicity, efficacy. *Org. Electron.* **2016**, *25*, 345–361. [CrossRef]
5. Ledwon, P. Recent advances of donor-acceptor type carbazole-based molecules for light emitting applications. *Org. Electron.* **2019**, *75*, 105422. [CrossRef]
6. Wex, B.; Kaafarani, B.R. Perspective on carbazole-based organic compounds as emitters and hosts in TADF applications. *J. Mater. Chem. C* **2017**, *5*, 8622–8653. [CrossRef]
7. Kondo, Y.; Yoshiura, K.; Kitera, S.; Nishi, H.; Oda, S.; Gotoh, H.; Sasada, Y.; Yanai, M.; Hatakeyama, T. Narrowband deep-blue organic light-emitting diode featuring an organoboron-based emitter. *Nat. Photonics* **2019**, *13*, 678–682. [CrossRef]
8. Ahn, D.H.; Kim, S.W.; Lee, H.; Ko, I.J.; Karthik, D.; Lee, J.Y.; Kwon, J.H. Highly efficient blue thermally activated delayed fluorescence emitters based on symmetrical and rigid oxygen-bridged boron acceptors. *Nat. Photonics* **2019**, *13*, 540–546. [CrossRef]

9. Hwang, J.; Koh, C.W.; Ha, J.M.; Woo, H.Y.; Park, S.; Cho, M.J.; Choi, D.H. Aryl-Annulated [3,2-a] Carbazole-Based Deep-Blue Soluble Emitters for High-Efficiency Solution-Processed Thermally Activated Delayed Fluorescence Organic Light-Emitting Diodes with $CIE_y < 0.1$. *ACS Appl. Mater. Interfaces* **2021**, *13*, 61454–61462. [PubMed]
10. Lee, H.; Braveenth, R.; Muruganantham, S.; Jeon, C.Y.; Lee, H.S.; Kwon, J.H. Efficient pure blue hyperfluorescence devices utilizing quadrupolar donor-acceptor-donor type of thermally activated delayed fluorescence sensitizers. *Nat. Commun.* **2023**, *14*, 419. [CrossRef]
11. Hofkens, J.; Cotlet, M.; Vosch, T.; Tinnefeld, P.; Weston, K.D.; Ego, C.; Grimsdale, A.; Müllen, K.; Beljonne, D.; Brédas, J.L.; et al. Revealing competitive Förster-type resonance energy-transfer pathways in single bichromophoric molecules. *Proc. Natl. Acad. Sci. USA* **2003**, *100*, 13146–13151. [CrossRef] [PubMed]
12. Ruseckas, A.; Ribierre, J.C.; Shaw, P.E.; Staton, S.V.; Burn, P.L.; Samuel, I.D.W. Singlet energy transfer and singlet-singlet annihilation in light-emitting blends of organic semiconductors. *Appl. Phys. Lett.* **2009**, *95*, 183305. [CrossRef]
13. Hedley, G.J.; Schröder, T.; Steiner, F.; Eder, T.; Hofmann, F.J.; Bange, S.; Laux, D.; Höger, S.; Tinnefeld, P.; Lupton, J.M.; et al. Picosecond time-resolved photon antibunching measures nanoscale exciton motion and the true number of chromophores. *Nat. Commun.* **2021**, *12*, 1327. [CrossRef] [PubMed]
14. Haase, N.; Danos, A.; Pflumm, C.; Stachelek, P.; Brütting, W.; Monkman, A.P. Are the rates of dexter transfer in TADF hyperfluorescence systems optically accessible? *Mater. Horiz.* **2021**, *8*, 1805–1815. [CrossRef] [PubMed]
15. Morgenroth, M.; Lenzer, T.; Oum, K. Understanding Excited-State Relaxation in 1,3-Bis(N-carbazolyl)benzene, a Host Material for Organic Light-Emitting Diodes. *J. Phys. Chem. C* **2023**, *127*, 4582–4593. [CrossRef]
16. Walba, H.; Branch, G.E.K. The Absorption Spectra of Some N-Substituted p-Aminotriphenylmethyl Ions. *J. Am. Chem. Soc.* **1951**, *73*, 3341–3348. [CrossRef]
17. Adams, J.E.; Mantulin, W.W.; Huber, J.R. Effect of Molecular Geometry on Spin-Orbit Coupling of Aromatic Amines in Solution. Diphenylamine, Iminobibenzyl, Acridan, and Carbazole. *J. Am. Chem. Soc.* **1973**, *95*, 5477–5481. [CrossRef]
18. Johnson, G.E. A Spectroscopic Study of Carbazole by Photoselection. *J. Phys. Chem.* **1974**, *78*, 1512–1521. [CrossRef]
19. Auty, A.R.; Jones, A.C.; Phillips, D. Spectroscopy and decay dynamics of jet-cooled carbazole and N-ethylcarbazole and their homocyclic analogues. *Chem. Phys.* **1976**, *103*, 163–182. [CrossRef]
20. Yu, H.; Zain, S.M.; Eigenbrot, I.V.; Phillips, D. Investigation of carbazole derivatives and their van der Waals complexes in the jet by laser-induced fluorescence spectroscopy. *J. Photochem. Photobiol. A* **1994**, *80*, 7–16. [CrossRef]
21. Yi, J.T.; Alvarez-Valtierra, L.; Pratt, D.W. Rotationally resolved $S_1 \leftarrow S_0$ electronic spectra of fluorene, carbazole, and dibenzofuran: Evidence for Herzberg-Teller coupling with the S_2 state. *J. Chem. Phys.* **2006**, *124*, 244302. [CrossRef] [PubMed]
22. Henry, B.R.; Kasha, M. Triplet-Triplet Absorption Studies on Aromatic and Heterocyclic Molecules at 77 °K. *J. Chem. Phys.* **1967**, *47*, 3319–3327. [CrossRef]
23. Fratev, F.; Hermann, H.; Olbrich, G.; Polansky, O.E. Höhere Triplettanregungszustände von Fluoren, Carbazol und Benzologen: CNDO-CI-Berechnungen und Triplett-Triplett-Absorptionsmessungen. *Z. Naturforsch. A Phys. Phys. Chem. Kosmophys.* **1976**, *31*, 84–86. [CrossRef]
24. Johnson, G.E. Intramolecular excimer formation in carbazole double molecules. *J. Chem. Phys.* **1974**, *61*, 3002–3008. [CrossRef]
25. Masuhara, H.; Tohgo, Y.; Mataga, N. Fluorescence quenching processes of carbazole-amine systems as revealed by laser photolysis method. *Chem. Lett.* **1975**, *4*, 59–62. [CrossRef]
26. Martin, M.M.; Ware, W.R. Fluorescence Quenching of Carbazole by Pyridine and Substituted Pyridines. Radiationless Processes in the Carbazole-Amine Hydrogen Bonded Complex. *J. Phys. Chem.* **1978**, *82*, 2770–2776. [CrossRef]
27. Garner, A.; Wilkinson, F. Quenching of triplet states by molecular oxygen and the role of charge-transfer interactions. *Chem. Phys. Lett.* **1977**, *45*, 432–435. [CrossRef]
28. Bigelow, R.W.; Ceasar, G.P. Hydrogen Bonding and N-Alkylation Effects on the Electronic Structure of Carbazole. *J. Phys. Chem.* **1979**, *83*, 1790–1795. [CrossRef]
29. Johnson, G.E. Fluorescence Quenching of Carbazoles. *J. Phys. Chem.* **1980**, *84*, 2940–2946. [CrossRef]
30. Martin, M.M.; Bréhéret, E. Hydrogen Bonding Interaction Effect on Carbazole Triplet State Photophysics. *J. Phys. Chem.* **1982**, *86*, 107–111. [CrossRef]
31. Kikuchi, K.; Yamamoto, S.A.; Kokubun, H. Hydrogen atom transfer reaction from excited carbazole to pyridine. *J. Photochem.* **1984**, *24*, 271–283. [CrossRef]
32. Bonesi, S.M.; Erra-Balsells, R. Electronic spectroscopy of carbazole and N- and C-substituted carbazoles in homogeneous media and in solid matrix. *J. Lumin.* **2001**, *93*, 51–74. [CrossRef]
33. Boo, B.H.; Ryu, S.Y.; Kang, H.S.; Koh, S.G.; Park, C.-J. Time-resolved Fluorescence Studies of Carbazole and Poly(N-vinylcarbazole) for Elucidating Intramolecular Excimer Formation. *Bull. Korean Chem. Soc.* **2010**, *57*, 406–411. [CrossRef]
34. Bayda-Smykaj, M.; Burdzinski, G.; Ludwiczak, M.; Hug, G.L.; Marciniak, B. Early Events in the Photoinduced Electron Transfer between Carbazole and Divinylbenzene in a Silylene-Bridged Donor-Acceptor Compound. *J. Phys. Chem. C* **2020**, *124*, 19522–19529. [CrossRef]

35. Thornton, G.L.; Phelps, R.; Orr-Ewing, A.J. Transient absorption spectroscopy of the electron transfer step in the photochemically activated polymerizations of N-ethylcarbazole and 9-phenylcarbazole. *Phys. Chem. Chem. Phys.* **2021**, *23*, 18378–18392. [CrossRef] [PubMed]
36. Yang, Y.; Jiang, Z.; Liu, Y.; Guan, T.; Zhang, Q.; Qin, C.; Jiang, K.; Liu, Y. Transient Absorption Spectroscopy of a Carbazole-Based Room-Temperature Phosphorescent Molecule: Real-Time Monitoring of Singlet-Triplet Transitions. *J. Phys. Chem. Lett.* **2022**, *13*, 9381–9389. [CrossRef] [PubMed]
37. Hiyoshi, R.; Hiura, H.; Sakamoto, Y.; Mizuno, M.; Sakai, M.; Takahashi, H. Time-resolved absorption and time-resolved Raman spectroscopies of the photochemistry of carbazole and N-ethylcarbazole. *J. Mol. Struct.* **2003**, *661–662*, 481–489. [CrossRef]
38. Notsuka, N.; Nakanotani, H.; Noda, H.; Goushi, K.; Adachi, C. Observation of Nonradiative Deactivation Behavior from Singlet and Triplet States of Thermally Activated Delayed Fluorescence Emitters in Solution. *J. Phys. Chem. Lett.* **2020**, *11*, 562–566. [CrossRef] [PubMed]
39. Kim, H.J.; Kang, H.; Jeong, J.-E.; Park, S.H.; Koh, C.W.; Kim, C.W.; Woo, H.Y.; Cho, M.J.; Park, S.; Choi, D.H. Ultra-Deep-Blue Aggregation-Induced Delayed Fluorescence Emitters: Achieving Nearly 16% EQE in Solution-Processed Nondoped and Doped OLEDs with $CIE_y < 0.1$. *Adv. Funct. Mater.* **2021**, *31*, 2102588.
40. Morgenroth, M.; Scholz, M.; Guy, L.; Oum, K.; Lenzer, T. Spatiotemporal Mapping of Efficient Chiral Induction by Helicene-Type Additives in Copolymer Thin Films. *Angew. Chem. Int. Ed.* **2022**, *61*, e202203075. [CrossRef]
41. Oum, K.; Lenzer, T.; Scholz, M.; Jung, D.Y.; Sul, O.; Cho, B.J.; Lange, J.; Müller, A. Observation of Ultrafast Carrier Dynamics and Phonon Relaxation of Graphene from the Deep-Ultraviolet to the Visible Region. *J. Phys. Chem. C* **2014**, *118*, 6454–6461. [CrossRef]
42. Flender, O.; Scholz, M.; Klein, J.R.; Oum, K.; Lenzer, T. Excited-state relaxation of the solar cell dye D49 in organic solvents and on mesoporous Al_2O_3 and TiO_2 thin films. *Phys. Chem. Chem. Phys.* **2016**, *18*, 26010–26019. [CrossRef]
43. Dobryakov, A.L.; Kovalenko, S.A.; Weigel, A.; Pérez Lustres, J.L.; Lange, J.; Müller, A.; Ernsting, N.P. Femtosecond pump/supercontinuum-probe spectroscopy: Optimized setup and signal analysis for single-shot spectral referencing. *Rev. Sci. Instrum.* **2010**, *81*, 113106. [CrossRef] [PubMed]
44. Merker, A.; Scholz, M.; Morgenroth, M.; Lenzer, T.; Oum, K. Photoinduced Dynamics of $(CH_3NH_3)_4Cu_2Br_6$ Thin Films Indicating Efficient Triplet Photoluminescence. *J. Phys. Chem. Lett.* **2021**, *12*, 2736–2741. [CrossRef] [PubMed]
45. Ljubić, I.; Sabljić, A. CASSCF/CASPT2 and TD-DFT Study of Valence and Rydberg Electronic Transitions in Fluorene, Carbazole, Dibenzofuran, and Dibenzothiophene. *J. Phys. Chem. A* **2011**, *115*, 4840–4850. [CrossRef] [PubMed]
46. Platt, J.R. Classification of Spectra of Cata-Condensed Hydrocarbons. *J. Chem. Phys.* **1949**, *17*, 484–495. [CrossRef]
47. Lohse, P.W.; Bürsing, R.; Lenzer, T.; Oum, K. Exploring 12′-Apo-β-carotenoic-12′-acid as an Ultrafast Polarity Probe for Ionic Liquids. *J. Phys. Chem. B* **2008**, *112*, 3048–3057. [CrossRef]
48. Lide, D.R. (Ed.) *Handbook of Chemistry and Physics*, 85th ed.; CRC Press: Boca Raton, FL, USA, 2004.
49. Schwarzer, D.; Troe, J.; Votsmeier, M.; Zerezke, M. Collisional deactivation of vibrationally highly excited azulene in compressed liquids and supercritical fluids. *J. Chem. Phys.* **1996**, *105*, 3121–3131. [CrossRef]
50. Schwarzer, D.; Hanisch, C.; Kutne, P.; Troe, J. Vibrational Energy Transfer in Highly Excited Bridged Azulene-Aryl Compounds: Direct Observation of Energy Flow through Aliphatic Chains and into the Solvent. *J. Phys. Chem. A* **2002**, *106*, 8019–8028. [CrossRef]
51. Kovalenko, S.A.; Schanz, R.; Hennig, H.; Ernsting, N.P. Cooling dynamics of an optically excited molecular probe in solution from femtosecond broadband absorption spectroscopy. *J. Chem. Phys.* **2001**, *115*, 3256–3273. [CrossRef]
52. Andersson, P.O.; Gillbro, T. Photophysics and dynamics of the lowest excited singlet state in long substituted polyenes with implications to the very long-chain limit. *J. Chem. Phys.* **1995**, *103*, 2509–2519. [CrossRef]
53. Lenzer, T.; Ehlers, F.; Scholz, M.; Oswald, R.; Oum, K. Assignment of carotene S* state features to the vibrationally hot ground electronic state. *Phys. Chem. Chem. Phys.* **2010**, *12*, 8832–8839. [CrossRef] [PubMed]
54. Martin, M.; Bréhéret, E.; Tfibel, F.; Lacourbas, B. Two-Photon Stepwise Dissociation of Carbazole in Solution. *J. Phys. Chem.* **1980**, *84*, 70–72. [CrossRef]
55. Shida, T.; Nosaka, Y.; Kato, T. Electronic Absorption Spectra of Some Cation Radicals as Compared with Ultraviolet Photoelectron Spectra. *J. Phys. Chem.* **1978**, *82*, 695–698. [CrossRef]
56. Haink, H.J.; Adams, J.E.; Huber, J.R. The Electronic Structure of Aromatic Amines: Photoelectron Spectroscopy of Diphenylamine, Iminobibenzyl, Acridan and Carbazole. *Ber. Bunsenges. Phys. Chem.* **1974**, *78*, 436–440. [CrossRef]
57. Liptay, W. Electrochromism and Solvatochromism. *Angew. Chem. Int. Ed.* **1967**, *8*, 177–188. [CrossRef]
58. Schweitzer, C.; Schmidt, R. Physical Mechanisms of Generation and Deactivation of Singlet Oxygen. *Chem. Rev.* **2003**, *103*, 1685–1757. [CrossRef] [PubMed]
59. Hernández, F.J.; Crespo-Otero, R. Excited state mechanisms in crystalline carbazole: The role of aggregation and isomeric defects. *J. Mater. Chem. C* **2021**, *9*, 11882–11892. [CrossRef]
60. Chen, C.; Chong, K.C.; Pan, Y.; Qi, G.; Xu, S.; Liu, B. Revisiting Carbazole: Origin, Impurity, and Properties. *ACS Mater. Lett.* **2021**, *3*, 1081–1087. [CrossRef]

61. Hosokai, T.; Matsuzaki, H.; Nakanotani, H.; Tokumaru, K.; Tsutsui, T.; Furube, A.; Nasu, K.; Nomura, H.; Yahiro, M.; Adachi, C. Evidence and mechanism of efficient thermally activated delayed fluorescence promoted by delocalized excited states. *Sci. Adv.* **2017**, *3*, e1603282. [CrossRef]
62. Godumala, M.; Choi, S.; Cho, M.J.; Choi, D.H. Recent breakthroughs in thermally activated delayed fluorescence organic light emitting diodes containing non-doped emitting layers. *J. Mater. Chem. C* **2019**, *7*, 2172–2198. [CrossRef]

Disclaimer/Publisher's Note: The statements, opinions and data contained in all publications are solely those of the individual author(s) and contributor(s) and not of MDPI and/or the editor(s). MDPI and/or the editor(s) disclaim responsibility for any injury to people or property resulting from any ideas, methods, instructions or products referred to in the content.

Article

Excited-State Dynamics of Carbazole and *tert*-Butyl-Carbazole in Thin Films

Konstantin Moritz Knötig, Domenic Gust, Kawon Oum and Thomas Lenzer *

Physical Chemistry 2, Department Chemistry and Biology, Faculty IV: School of Science and Technology, University of Siegen, Adolf-Reichwein-Str. 2, 57076 Siegen, Germany; konstantin.knoetig@uni-siegen.de (K.M.K.); domenic.gust@uni-siegen.de (D.G.); oum@chemie.uni-siegen.de (K.O.)
* Correspondence: lenzer@chemie.uni-siegen.de

Abstract: Thin films of carbazole (Cz) derivatives are frequently used in organic electronics, such as organic light-emitting diodes (OLEDs). Because of the proximity of the Cz units, the excited-state relaxation in such films is complicated, as intermolecular pathways, such as singlet–singlet annihilation (SSA), kinetically compete with the emission. Here, we provide an investigation of two benchmark systems employing neat carbazole and 3,6-di-*tert*-butylcarbazole (*t*-Bu-Cz) films and also their thin film blends with poly(methyl methacrylate) (PMMA). These are investigated by a combination of atomic force microscopy (AFM), femtosecond and nanosecond transient absorption spectroscopy (fs-TA and ns-TA) and time-resolved fluorescence. Excitonic J-aggregate-type features are observed in the steady-state absorption and emission spectra of the neat films. The S_1 state shows a broad excited-state absorption (ESA) spanning the entire UV–Vis–NIR range. At high S_1 exciton number densities of about 4×10^{18} cm^{-3}, bimolecular diffusive S_1–S_1 annihilation is found to be the dominant SSA process in the neat films with a rate constant in the range of 1–2×10^{-8} cm^3 s^{-1}. SSA produces highly vibrationally excited molecules in the electronic ground state (S_0^*), which cool down slowly by heat transfer to the quartz substrate. The results provide relevant photophysical insight for a better microscopic understanding of carbazole relaxation in thin-film environments.

Keywords: carbazole; thin films; ultrafast laser spectroscopy; singlet–singlet annihilation; picosecond ultrasonics

Citation: Knötig, K.M.; Gust, D.; Oum, K.; Lenzer, T. Excited-State Dynamics of Carbazole and *tert*-Butyl-Carbazole in Thin Films. *Photochem* **2024**, *4*, 179–197. https://doi.org/10.3390/photochem4020011

Academic Editors: Marcelo Guzman, Vincenzo Vaiano and Rui Fausto

Received: 31 January 2024
Revised: 31 March 2024
Accepted: 2 April 2024
Published: 9 April 2024

Copyright: © 2024 by the authors. Licensee MDPI, Basel, Switzerland. This article is an open access article distributed under the terms and conditions of the Creative Commons Attribution (CC BY) license (https:// creativecommons.org/licenses/by/ 4.0/).

1. Introduction

Thin films composed of carbazole-based compounds are ubiquitous materials in organic and hybrid organic–inorganic electronics, where they serve as light harvesters for organic solar cells [1], host materials [2], hole transport layers [3] and efficient emitters of visible light [4]. In these applications, thin films of carbazole-containing materials are used, which are either produced by physical vapor deposition (PVD) or wet-chemical methods such as spin-coating [5]. Under typical operating conditions involving high exciton number densities, e.g., in OLEDs, the emission process of carbazole-containing compounds is in kinetic competition with intermolecular channels, such as singlet–singlet annihilation, singlet–triplet annihilation (STA), singlet–heat annihilation (SHA) or triplet–triplet annihilation (TTA) [6,7]. Therefore, it is important to understand the dynamics of these processes and extract relevant kinetic parameters for the annihilation steps, which can be used in the modeling of OLED operation.

In the current contribution, we focus on thin films of the parent compound Cz and its substituted analog *t*-Bu-Cz. Basic optical properties of monocrystalline Cz thin films have been explored previously [8–12], but time-resolved optical studies of Cz and related film compounds combining ultrafast transient absorption and emission techniques have been performed rarely [13]. Yet, such techniques can provide profound insight into the singlet exciton dynamics in these films. For other systems, time-resolved studies of SSA processes have been performed largely by fluorescence-based techniques [7,14–23]. Such measurements detect the decay of the S_1 exciton emission and are sensitive even at low exciton number densities [24]. However, they might have some drawbacks in terms of interference by fluorescing impurities and limited time resolution (especially for studies at high exciton number densities). Transient absorption studies tracking the decay of the $S_1 \rightarrow S_n$ excited-state absorption bands have been performed more rarely [25–29]. While not being as sensitive as fluorescence detection, such TA methods with femtosecond time resolution can detect SSA processes even at very high exciton number densities and are at the same time much less prone to the interference of impurities, in contrast to emission studies, where a trace impurity or an impurity-induced crystal defect with high radiative quantum yield can overwhelm the signal of interest [24,30,31].

Here, we provide a detailed investigation of spin-coated Cz and *t*-Bu-Cz thin films and their blends with PMMA using a combination of atomic force microscopy, steady-state absorption and emission, transient fluorescence and ultrafast broadband transient absorption spectroscopy to extract key kinetic parameters, such as the S_1 lifetime at low exciton number densities, the rate constant for the SSA processes at high exciton number densities and the time scale for the cooling of the highly vibrationally excited ground state species produced by the SSA process.

2. Materials and Methods

2.1. Preparation of Thin Films

Solutions of carbazole (TCI Deutschland, Eschborn, Germany, high purity), 3,6-di-*tert*-butyl-carbazole (TCI Deutschland, >98.0%) and PMMA (Alfa Aesar, Ward Hill, MA, USA) were prepared in N_2-saturated tetrahydrofuran (Merck, Darmstadt, Germany, Uvasol, $\geq$99.9%) to obtain solutions corresponding to a mass concentration of 5 mg mL^{-1}. These solutions were passed through a PTFE filter (pore size 0.45 µm) to remove any residual particles and subsequently employed to produce solutions of PMMA:Cz and PMMA:*t*-Bu-Cz with nominal weight percent (wt%) ratios of 0:100, 70:30, 90:10 and 98:2. Quartz substrates (Tempotec Optics Co., Ltd., Fuzhou, China, JGS1) were pre-cleaned with acetonitrile (Merck, Uvasol, $\geq$99.9%) and irradiated by UV light (Dinies, Villingendorf, Germany, 2 UVC lamps with 253.7 nm emission, 11 W each) for 60 min to remove organic contaminants. Spin-coating was performed under a nitrogen atmosphere (Messer, Bad Soden im Taunus, Germany, 5.0) in a glovebox by depositing ca. 200 µL of a solution on the quartz substrate, followed by a loading time of 30 s and a spinning period of 1 min at 500 rpm.

2.2. Atomic Force Microscopy and X-ray Diffraction Experiments

Topography images of the Cz and *t*-Bu-Cz thin films were obtained using an atomic force microscope (PSIA XE-100, Park Systems Corp., Suwon, Republic of Korea) with a silicon tip cantilever in non-contact mode. Images were taken with a resolution of 128 times 128 pixels for an area of 30 × 30 µm^2 and a scan rate of 0.1 Hz to provide an overview of the surfaces. Furthermore, to estimate the thickness of the corresponding thin film, the surface was intentionally scratched. Alongside this scratch, AFM images were recorded with a resolution of 4096 times 64 pixels for an area of 60 × 2 µm^2 and a scan rate of 0.1 Hz. All images were evaluated using XEI software (version 4.3.4, Park Systems Corp.), and visualization was performed by using OriginPro 2023b (OriginLab Corp., Northampton, MA, USA). The forward and backward scans were averaged, and a background correction was carried out for all AFM images.

Thin-film X-ray diffraction experiments were carried out using a PANalytical X'Pert MPD PRO diffractometer employing Cu radiation (Kα_1 = 1.54060 Å, Kα_2 = 1.54443 Å). The diffractograms were simulated using Rietveld refinement, as implemented in the program MAUD, including the fitting of the baseline and also considering texture effects [32].

2.3. Steady-State Absorption and Photoluminescence

Steady-state absorption measurements of the thin films were carried out in transmission at normal incidence using a double-beam spectrophotometer (Varian Inc., Palo Alto,

CA, USA, Cary 5000, spectral bandwidth: 0.5 nm). Steady-state emission and photoluminescence excitation spectra were recorded on a fluorescence spectrophotometer (Agilent, Santa Clara, CA, USA, Cary Eclipse, spectral bandwidth for excitation and emission: 5 nm). The photoluminescence spectra were corrected by a calibration function that included the wavelength-dependent sensitivity of the emission spectrograph and the detector.

2.4. Time-Correlated Single-Photon Counting (TCSPC)

Pulses of a UV-LED (Becker & Hickl, Berlin, Germany, UVL-FB-270, wavelength: 273 nm, pulse width: 500 ps) at a repetition frequency of 1 MHz were sent through a wire-grid linear polarizer (Thorlabs, Newton, NJ, USA, WP25M-UB) set at vertical polarization (0°) to excite the respective thin-film sample at an angle of 45° relative to the surface normal, with the thin-film layer facing the beam. The photoluminescence emitted at an angle of 90° was collimated by a quartz lens and then passed through another wire-grid linear polarizer, with its plane of polarization oriented at 54.7° (magic angle). A well-defined spectral window of the PL emission was selected by appropriate bandpass filters (Thorlabs) and imaged onto a hybrid-alkali photodetector (Becker & Hickl, HPM-100-07). Its signal was fed into a TCSPC module operated in reverse start–stop configuration (Becker & Hickl, SPC-130IN). The kinetic traces were analyzed by the FAST program (Edinburgh Instruments, Livingston, UK, version 3.5.2) using a sum of exponential functions and an iterative reconvolution procedure that involved an instrument response function (IRF), which was obtained either from the LED scattering signal of a diluted suspension of colloidal SiO_2 nanoparticles (Merck, Ludox AS-40) in water or a quartz diffuser (Thorlabs DGUV10-120 or DGUV10-1500).

2.5. Broadband Transient Absorption Spectroscopy

Ultrafast broadband transient absorption experiments were carried out using the pump–supercontinuum probe (PSCP) method [33] on two setups dedicated to measurements in the UV–Vis [34] and NIR regions [35]. They are both based on a regeneratively amplified titanium:sapphire laser system (Coherent, Santa Clara, CA, USA, Libra USP-HE) operating at a repetition frequency of 920 Hz. The thin films were excited at a repetition frequency of 460 Hz by an OPA system running at a wavelength of 260 nm (Coherent, OPerA Solo). The overall time resolution was about 80 fs, and time delays up to 1500 ps were accessible. In order to perform measurements covering longer delay times up to 10 µs, the fourth harmonic (266 nm) of a Q-switched Nd:YAG microlaser (Standa, Vilnius, Lithuania, Standa-Q1TH, pulse length 420 ps) was employed [36]. The laser was electronically triggered at 460 Hz and synchronized with the probe beam of the ultrafast broadband transient absorption system by means of a delay generator (Stanford Research Systems, Sunnyvale, CA, USA, DG535). The thin films were mounted inside a home-built N_2-flushed aluminum cell and constantly translated by two piezo stages in the x–y plane within a sample area of 2×2 mm^2 during the experiments to minimize photochemical decomposition and local heating of the thin films, which could occur because of the exposure of the same sample area to repeated laser excitation.

3. Results

3.1. AFM Images and Picosecond Ultrasonics of Cz and t-Bu-Cz Thin Films on Quartz

Neat Cz and *t*-Bu-Cz thin films on quartz substrates were investigated by atomic force microscopy to determine their surface morphology. Figure 1a shows AFM images of a 30×30 µm^2 area of two neat Cz thin films of different average thicknesses. They consist of microcrystallites with dimensions on the order of 10 µm. The film thickness was obtained by generating a histogram (height h vs. its frequency of occurrence in the AFM image), as shown in Figure 1b. The sharp spike near $h = 0$ arises from uncoated areas on the quartz substrate (yellow ranges in panel a). The broad distribution at larger values of h was fitted by a Gaussian function, leading to an average thickness d of 109 nm and 153 nm for the two films. Additional AFM images were recorded for two *t*-Bu-Cz thin films of different thicknesses, as shown in Figure 1c. Here, the size of the crystallites was much smaller, around 1 µm and below. Analysis of the histograms in panel d provided the average thickness of these two thin films as 138 nm and 358 nm, respectively.

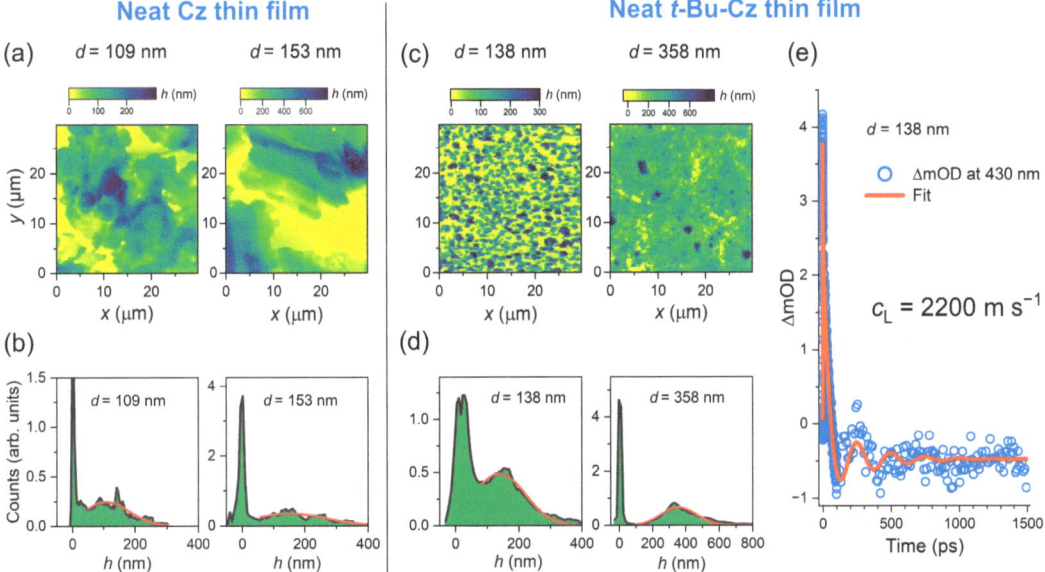

Figure 1. (a) AFM images of two neat Cz thin films with an average thickness d of 109 and 153 nm, respectively. (b) The distribution of the thickness of the two thin films shown in panel a is plotted as counts (i.e., the frequency of occurrence) versus height h. In each case, the red solid line is a fit to a Gaussian function, where the peak of the distribution corresponds to the average thickness. (c,d) Same as in panels a and b, but for two t-Bu-Cz thin films of different thicknesses (138 nm and 358 nm). (e) Transient absorption signal at 430 nm for the 138 nm thick t-Bu-Cz film showing coherent acoustic phonon oscillations (blue circles). A fit using a sum of two exponentials and a damped cosine function (red solid line) provides an oscillation period of 251 ps, from which the longitudinal sound velocity c_L of t-Bu-Cz was determined as 2200 m s^{-1}.

In addition, we applied laser-based picosecond ultrasonics [37] to the thin films. These are transient absorption experiments in which an ultrafast pump laser pulse (here, 260 nm) generates a coherent acoustic phonon in the film. This longitudinal sound wave propagates back and forth between the two film interfaces (thin film/quartz and thin film/nitrogen gas) and induces a periodic modulation of the transient optical signal [38–42]. Damping of the longitudinal sound wave occurs only at the thin film/quartz interface [42]. A representative kinetics for the neat t-Bu-Cz thin film with a thickness of 138 nm is depicted in Figure 1e and shows a damped oscillatory decay. Such measurements provide "contact-free noninvasive" access to the longitudinal sound velocity of the thin film material via the expression $c_L = 4d/\tau_a$ [38], where d is the film thickness and τ_a is the oscillation period of the coherent acoustic phonon. Fitting the kinetics in Figure 1e to the sum of two exponential decay functions and a damped cosine function provides a value of (251 ± 6) ps for τ_a, corresponding to a frequency of (3.99 ± 0.09) GHz. Using the known value of 138 nm for the average film thickness d, one obtains a value of (2200 ± 60) m s^{-1} for the longitudinal sound velocity c_L. This value is in very good agreement with results from a previous study on the closely related compound N-isopropyl-carbazole, which provided values of 2150 m s^{-1} and 2250 m s^{-1} from a piezoelectric resonance–antiresonance measurement and an ultrasonic pulse-echo experiment [43]. It also compares well with previous studies of other organic layers, such as polyfluorene-based copolymers [42], where values of about 2500 m s^{-1} were obtained. Knowing this longitudinal sound velocity makes it possible to optically determine the thickness of other t-Bu-Cz thin films via the relation $d = 0.25 \cdot c_L \cdot \tau_a$ in an all-optical fashion.

The value of c_L for Cz thin films should be very similar; however, we note that we were not able to obtain clean oscillatory kinetics exceeding the noise level of the transient absorption experiments for Cz thin films. It could be that the interface of the large crystallites (cf. Figure 1a) is very rough, so the reflection of the sound wave at the thin film/nitrogen gas interface occurs in several directions. This could result in a very weak oscillatory response. In contrast, the situation is much less critical for smooth copolymer films with homogeneous thickness, as demonstrated in previous measurements using picosecond ultrasonics [42].

3.2. Steady-State Absorption, Photoluminescence and Transient Emission of the Thin Films

The panels a and b in Figure 2 contain the absolute and normalized steady-state absorption spectra (recorded in transmission at normal incidence) of neat Cz and t-Bu-Cz thin films as well as PMMA:Cz and PMMA:t-Bu-Cz thin film blends with weight percent ratios of 98:2, 90:10 and 70:30. The structured absorption band rising below 350 nm corresponds to the $S_0(A_1) \rightarrow S_1(A_1)$ transition, which is short-axis polarized [10,44–46]. In contrast, the $S_0(A_1) \rightarrow S_2(B_2)$ transition located at about 290 nm is long-axis polarized [10,45,47]. The absorption features at wavelengths below 270 nm represent a superposition of several electronic bands of A_1 and B_2 symmetry [47]. For a low content of Cz/t-Bu-Cz in the film (98:2 (green), 90:10 (blue)), the 0–1 transition of the $S_0 \rightarrow S_1$ band is higher than the 0–0 transition, and also the amplitude of the $S_0 \rightarrow S_2$ band is much higher than that of the $S_0 \rightarrow S_1$ band. Both findings are in line with the typical behavior of Cz in organic solvents [45,48,49].

Drastic changes are observed for the neat Cz film (black) and the 70:30 wt% PMMA:Cz blend (red). For the neat Cz thin film, the $S_0(A_1) \rightarrow S_2(B_2)$ transition is not observable, and the absorbance below 270 nm is considerably reduced. This suggests that absorption contributions from long-axis polarized $A_1 \rightarrow B_2$ transitions are "missing". Based on the known crystal structure of carbazole [50,51], one can conclude that there must be a preferential growth of the microcrystallites (cf. Figure 1a), with the crystallographic a–c plane of the unit cell oriented parallel to the substrate's surface. We confirmed this in separate X-ray diffraction measurements (see Figures S1 and S2, Supporting Information). Therefore, the transition dipole moment of the $A_1 \rightarrow B_2$ transitions lies parallel to the light propagation direction, and, as a result, these states cannot be photoexcited. Thus, only short-axis polarized $A_1 \rightarrow A_1$ transitions can be observed in the spectrum of the neat thin films. For the 70:30 blend, the $A_1 \rightarrow B_2$ transitions are still visible, but already their spectral amplitude is substantially reduced. Therefore, there already appear to be Cz domains that also exhibit preferential growth.

Another interesting effect in the spectra of the neat Cz film and the 70:30 wt% PMMA:Cz blend is the emergence of a pronounced absorption peak at 340 nm, which is red-shifted by about 50 meV (5 nm) compared with the $S_0(A_1) \rightarrow S_1(A_1)$ transition of the thin films having low Cz content. Based on the comparison with findings for thin films of aromatic molecules, such as tetracene [24], or polymer compounds [52], the spectral signature is compatible with the formation of J aggregates. The crystal structure of Cz [50,51] features a herringbone-type packing [53]. As shown by previous plane-wave DFT calculations [53], nearest-neighbor Cz–Cz dimer-type interactions in this crystal are predominantly of the J-aggregate-type, with a Cz–Cz spacing of 4.8–5.7 Å, whereas H-type dimer-type arrangements in the same structure show a much larger Cz–Cz spacing (9.9–11.5 Å). Therefore, it is understandable that contributions from J-type aggregation should have a more substantial impact on the absorption spectrum of these films. The fairly small aggregation-induced red-shift of the J-aggregate absorption band observed in our absorption spectrum of the Cz film is also compatible with the results of calculations for J-type dimers in the Cz crystal [53]. As a result, the experimental absorption spectrum can be loosely described as a superposition of monomer-like absorption (from the only weakly coupling H-aggregate-type arrangements) plus contributions of more pronounced J-type aggregate absorption. This spectral appearance of the lowest-energy absorption band is still present even in the 70:30 PMMA-Cz blend, which suggests that there still must be Cz domains giving rise to aggregation. We note that a neat PMMA thin film only shows weak absorption starting below 250 nm (see Figure S3, Supporting Information), so contributions of PMMA to the absorption spectra reported in Figure 2 can be neglected.

Figure 2c contains the photoluminescence excitation spectra (recorded at the emission wavelength 370 nm), and panels d and e show the absolute and normalized PL spectra. The photoluminescence spectra of the PMMA:Cz blends with low Cz content (98:2 and 90:10) represent almost a mirror image of the respective absorption spectra and are similar to those reported previously in organic solvents [45,48,49]. For the 90:10 blend, the long-wavelength tail of the emission band is already slightly enhanced (panel e, e.g., above 450 nm), which may be attributed to a contribution of excimer emission, similar to what was previously observed for related carbazole-based systems, such as mCP [13]. In addition, photoexcitation at shorter wavelengths leads to a relative increase in the amplitude of the excimer band compared with the main peak (panel f), similar to the effect previously observed for mCP [13]. This could be due to the fact that at higher energy, the initial excitation might migrate over a larger distance, leading to an enhanced probability of finding an excimer site for emission [13].

More pronounced changes are observed for the thin films of neat Cz and the PMMA:Cz blend with a weight percent ratio of 70:30, where the intensity of the most intense emission peak (the $1'$–0 vibrational transition of the $S_1 \rightarrow S_0$ band) is strongly enhanced (panel d). Similar to the observation for the absorption spectra, we trace this observation back to excitonic effects (as a result of the overlapping emission of mainly J-aggregate-type dimer species) and also reabsorption effects (especially for the films with the highest Cz and t-Bu-Cz content) because of the small Stokes shift (Table 1). The enhanced peak in the center of the emission spectrum could then be an indication of superradiance, i.e., an enhanced radiative decay rate of the lowest-lying exciton state [24,54,55]. Characteristic parameters of the steady-state spectra are summarized in Table 1.

Table 1. Characteristic photophysical parameters of the steady-state absorption and photoluminescence spectra of the thin films of Cz and t-Bu-Cz and their blends with PMMA.

Thin Film Composition	Mixing Ratio (wt%:wt%)	λ_{abs}^{0-0} (nm)	λ_{PL}^{0-0} (nm)	λ_{max}^{PL} (nm)	$\Delta\lambda_{Stokes}$ (nm) [1]	$\Delta\tilde{\nu}_{Stokes}$ (meV) [1]	$\Delta\tilde{\nu}_{Stokes}$ (cm^{-1}) [1]
PMMA:Cz	0:100	341 [2]	344	356	3	32	260
	70:30	340 [2]	344	356	4	42	340
	90:10	336	342	358	6	64	520
	98:2	336	341	357	5	55	440
PMMA:t-Bu-Cz	0:100	344 [2]	342	349	5	52	420
	70:30	342	341	351	9	93	750
	90:10	342	341	348	6	62	500
	98:2	342	341	349	7	73	590

[1] The Stokes shift was determined as the difference in 0–0 vibrational bands of the lowest electronic states, which were determined by fitting a sum of Gaussian functions to the absorption and PL bands of the thin films. [2] For the neat Cz and t-Bu-Cz thin films and the PMMA:Cz thin film blend with a weight percent ratio of 70:30, this is the position of the excitonic absorption peak.

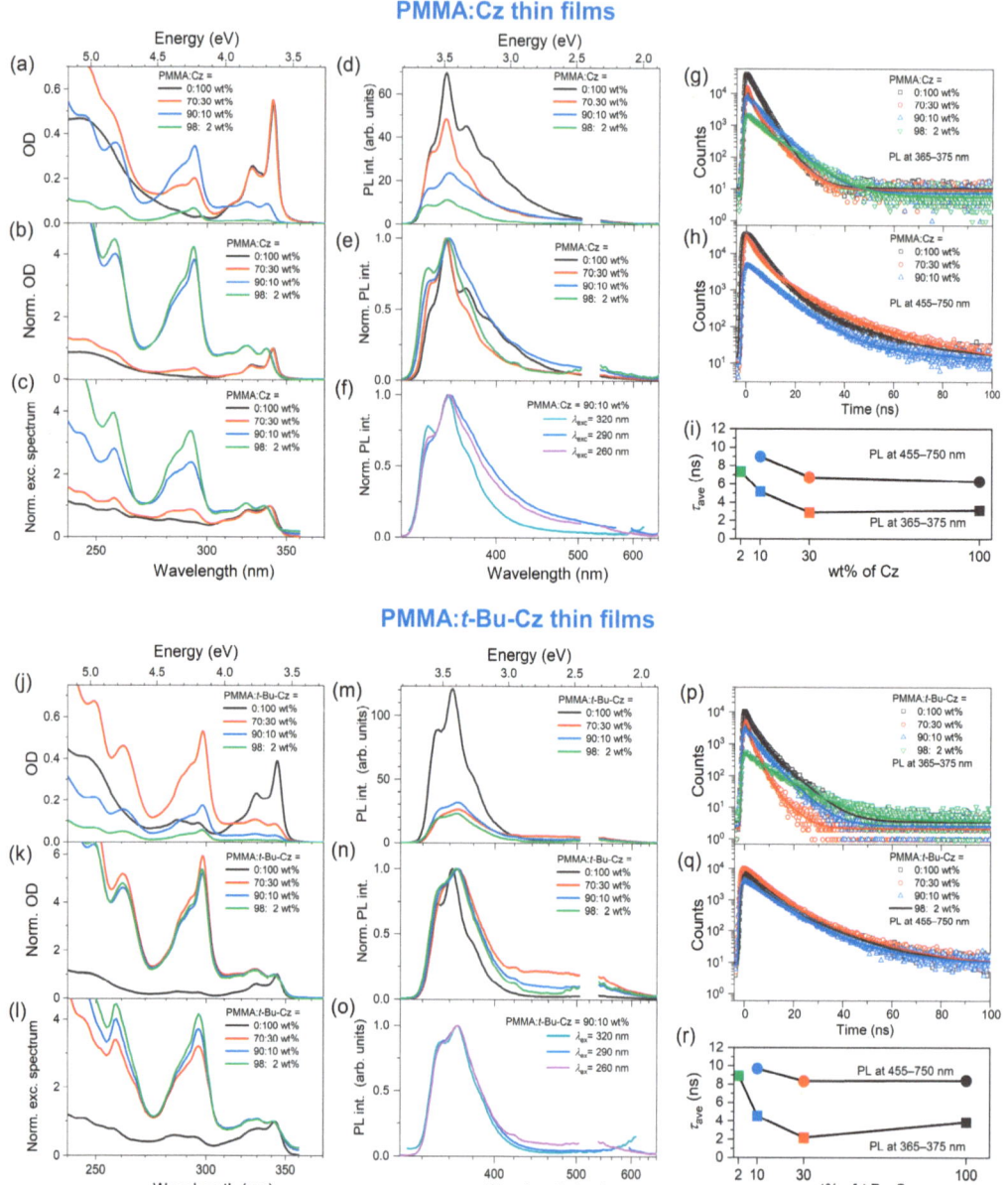

Figure 2. Summary of steady-state absorption and photoluminescence (PL) spectra of PMMA:Cz and PMMA:*t*-Bu-Cz blends with weight percent ratios of 100:0 (i.e., neat Cz and *t*-Bu-Cz thin films, black), 70:30 (red), 90:10 (blue) and 98:2 (green). (**a**) OD of thin films. (**b**) Normalized OD. (**c**) PL excitation spectra recorded at 370 nm. (**d**) PL intensity of the thin films excited at 260 nm. (**e**) Same PL spectra but normalized to their peak. (**f**) Normalized PL spectra of thin films of the PMMA:Cz 90:10 blends, excited at 260, 290 or 320 nm. (**g**,**h**) Transient PL decays of the thin films using two different bandpass filters of 365–375 nm ("monomer" or "exciton" region) and 455–750 nm ("excimer" region), respectively. (**i**) Average lifetimes of the thin film blends for the PL ranges of 365–375 nm (squares) and 455–750 nm (circles), respectively. (**j**–**r**) Same as panels (**a**–**i**), but for the PMMA:*t*-Bu-Cz thin films.

Time-resolved fluorescence decays obtained from TCSPC measurements for the neat Cz thin film and the different PMMA:Cz blends are displayed in panel g ("monomer" or "exciton" emission, averaged over the emission wavelength range of 365–375 nm) and panel h ("excimer" emission, averaged over the emission wavelength range of 455–750 nm) of Figure 2. The initial exciton number density in these experiments was about 6×10^{12} cm^{-3}. The decays were fitted to a sum of up to four exponentials. The resulting time constants τ_i and associated amplitudes A_i are summarized in Table 2, including the average lifetime $\langle \tau \rangle$ ($= \tau_{ave}$), calculated as $\langle \tau \rangle = (A_1\tau_1^2 + A_2\tau_2^2 + A_3\tau_3^2)/(A_1\tau_1 + A_2\tau_2 + A_3\tau_3)$ for the example of a triexponential decay [13]. As shown in panel i, the average lifetime decreases with increasing Cz content in the films. For the "monomer" band of the 98:2 PMMA:Cz blend, the longest lifetime of 7.4 ns is observed, which decreases to about 3 ns for the 70:30 blend and the neat Cz film. For comparison, the S_1 lifetime of "isolated" Cz in O_2-free organic solvents has been reported to be in the range of 14–15 ns [48,49,56,57].

Table 2. Summary of TCSPC data of the thin films of Cz and *t*-Bu-Cz and their blends with PMMA at 296 K (λ_{pump} = 273 nm).

Thin Film Composition	Mixing Ratio (wt%:wt%)	λ_{probe} (nm)	τ_1 (ns) [1]	τ_2 (ns) [1]	τ_3 (ns) [1]	τ_4 (ns) [1]	$\langle \tau \rangle$ (ns) [1]
PMMA:Cz	0:100	365–375	3.0 (97%) [1]	5.7 (3%)	–	–	3.1
		455–750	1.1 (46%)	2.9 (23%)	4.9 (30%)	16.66 (1%)	4.5
	70:30	365–375	0.6 (64%)	2.2 (26%)	5.2 (10%)	–	2.9
		455–750	2.1 (67%)	6.5 (30%)	18.8 (3%)	–	6.7
	90:10	365–375	3.7 (62%)	6.6 (38%)	–	–	5.2
		455–750	1.0 (33%)	7.2 (65%)	20.5 (2%)	–	8.1
	98:2	365–375	3.9 (37%)	8.3 (63%)	–	–	7.4
		455–750 [2]	–	–	–	–	–
PMMA:*t*-Bu-Cz	0:100	365–375	1.0 (47%)	3.1 (40%)	6.5 (13%)	–	3.9
		455–750	0.3 (40%)	5.4 (51%)	14.2 (9%)	–	8.0
	70:30	365–375	1.1 (58%)	2.5 (39%)	5.3 (3%)	–	2.2
		455–750	0.8 (38%)	6.3 (59%)	17.2 (3%)	–	7.4
	90:10	365–375	1.3 (24%)	3.6 (46%)	6.0 (30%)	–	4.5
		455–750	1.2 (26%)	7.9 (71%)	22.3 (3%)	–	8.9
	98:2	365–375	1.2 (14%)	3.8 (22%)	9.8 (64%)	–	8.9
		455–750 [2]	–	–	–	–	–

[1] The value in parentheses is the relative amplitude for the respective time constant. [2] For the 98:2 blends, the TCSPC signals recorded in the wavelength range of 455–750 nm (exciton band) were too weak, so the concentration of excimers in these thin films with low Cz and *t*-Bu-Cz content was below our detection limit.

This shortening of the S_1 lifetime may be explained as follows: The $S_0 \to S_1$ absorption spectrum and the $S_1 \to S_0$ emission spectrum of Cz show considerable overlap, see, e.g., panels a and d of Figure 2, and this is also underlined by the small Stokes shifts summarized in Table 1. Therefore, already starting for blends with low Cz content (PMMA:Cz 90:10) and much more strongly for even higher Cz content, Cz(S_1)–Cz(S_0) Förster resonance energy transfer (FRET) based on a dipole–dipole coupling mechanism becomes possible [58–60]. Because of this homo-FRET process (Cz(S_1) → Cz(S_0) accompanied by Cz(S_0) → Cz(S_1)), the excitation can efficiently hop through the thin film and eventually reach an excimer site, which acts as a trap because the emission of the excimer species does not overlap with the Cz absorption band. Therefore, further migration of the excitation through the film is not possible. The average lifetimes of the excimers (as measured using a bandpass filter covering the wavelength range of 455–750 nm) are in general longer (Figure 2i), varying between 8.1 ns (PMMA:Cz 90:10) and 4.5 ns (neat Cz). The weak emission intensity of the excimer suggests that its radiative rate constant is small.

Coming to the measurements for the different PMMA:*t*-Bu-Cz thin films (Figure 2j–r, Tables 1 and 2), we observe trends that are similar to the corresponding Cz-based films. Notable differences are the slightly larger Stokes shift than for the Cz-based systems (Table 1) and the less pronounced exciton peak in the absorption and emission spectra of the neat *t*-Bu-Cz film (panels j, k, m and n). The bulky *tert*-butyl substituents give rise to a slightly larger spacing between the *t*-Bu-Cz chromophores [61] and thus, the J-aggregate interactions compared with those in Cz-based films are weaker. Also, breaking up the *t*-Bu-Cz–*t*-Bu-Cz interactions by PMMA appears to be easier than for Cz–Cz. This is underlined by the fact that the absorption and emission of the PMMA:*t*-Bu-Cz blend with the weight percent ratio of 70:30 already behaves similarly to the films with low *t*-Bu-Cz content (90:10 and 98:2), whereas the corresponding spectra for the PMMA:Cz blend with the weight percent ratio of 70:30 are much more similar to those of the neat Cz thin film.

3.3. Femtosecond and Nanosecond UV-Vis-NIR Transient Absorption Spectra

Figure 3 summarizes the transient broadband absorption spectra and kinetics obtained from the fs-TA experiments (λ_{pump} = 260 nm) and ns-TA experiments (λ_{pump} = 266 nm) for a neat Cz thin film and two different PMMA:Cz blends with weight percent ratios of 70:30 and 90:10. Relevant parameters from the kinetic analysis are summarized in Table 3. We start with the results for the neat Cz thin film in panels a–c. The contour plots in panel a provide an overview of the dynamics for the subpicosecond to several hundred nanosecond range. The transient spectra around zero delay time show the formation of a ground state bleach (GSB) band below 350 nm and broad excited-state absorption above. The ESA is assigned to the S_1 state, which is populated by ultrafast (sub-100 fs) internal conversion (IC) from the initially photoexcited higher singlet state S_x. The ESA band of the S_1 band has almost completely disappeared by 100 ps. This is at first sight quite unusual because the S_1 lifetime of carbazole in O_2-free organic solvents is about 14–15 ns [48,49,56,57]. However, efficient S_1–S_1 annihilation takes place in the neat Cz film at the initial exciton number densities of about 4×10^{18} cm^{-3} in this experiment (cf. Section 3.4). The SSA process produces an excited S_n state (which quickly relaxes back to S_1 by IC) and vibrationally hot S_0 molecules, denoted as S_0^*. In fact, the sharp "hot band" absorption at 345 nm (close to the red edge of the inverted steady-state absorption spectrum, magenta) and the sharp negative signals at 340 nm and 325 nm are characteristic spectral features of vibrationally hot S_0^* molecules formed during this process [13]. The cooling of the S_0^* molecules is slow and extends into the several hundred nanosecond range. The kinetic analysis in panels b and c provides a time constant of 12 ps (fs-TA), which is related to the singlet–singlet annihilation process. The time constants of 640 ps (fs-TA), 3 ns, 41 ns and 790 ns (ns-TA) describe the slow cooling process, which involves heat transfer to the quartz substrate and encompasses the picosecond to the several hundred nanosecond range.

Next, we analyze the corresponding dynamics in panels d–f for the PMMA:Cz thin-film blend with a weight percent ratio of 70:30. The spectral evolution looks quite similar to that of the neat Cz film, with a broad S_1 ESA band and the formation of vibrationally hot S_0^* molecules. However, the S_1 ESA band decays on a considerably slower time scale, because the SSA process is less efficient, as the average spacing between the S_1 molecules becomes larger because of the presence of PMMA. SSA will be shown to be a diffusive second-order process in Section 3.4. Thus, it takes longer for two S_1 excitons to come into a critical contact distance for annihilation. The kinetic analysis reflects this change in the dynamics: Besides a weak ultrafast component (0.1 ps, likely IC from the S_x state), the fs-TA data reveal time constants of 7 and 107 ps, which effectively describe the slowed-down diffusive SSA. The time constants of 580 ps (fs-TA) and 13 ns, 113 ns and 51 µs (ns-TA) are again assigned to heat transfer from hot S_0^* Cz molecules to the quartz substrate.

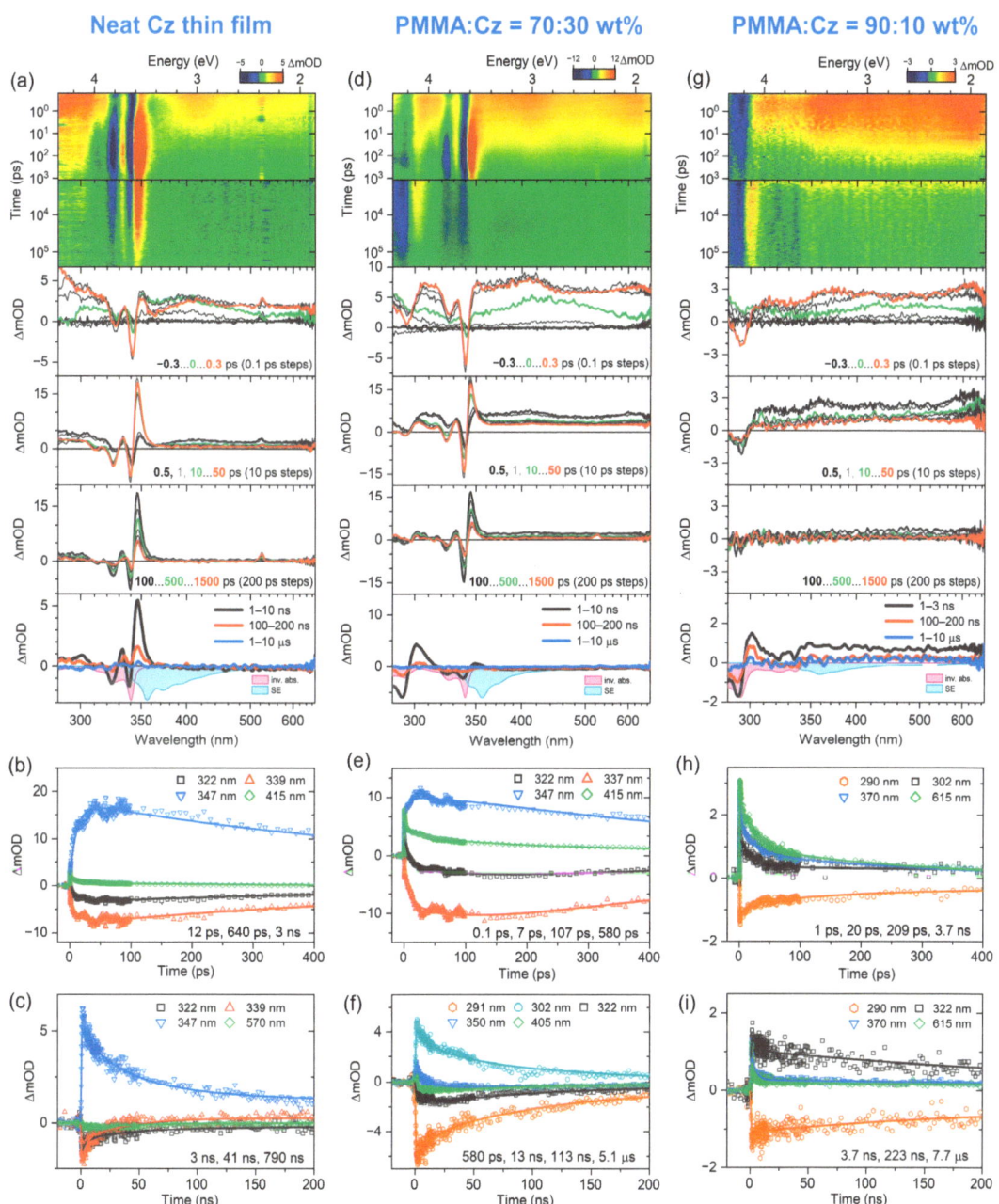

Figure 3. Results from TA experiments for carbazole-based thin films. (**a–c**) For a neat Cz thin film from top to bottom: fs-TA and ns-TA spectra as contour plots, TA spectra at selected delay times (indicated by different colors) including the inverted steady-state absorption spectrum (magenta) and the stimulated emission spectrum (cyan), selected fs-TA and ns-TA kinetics including fits by a sum of exponentials (including time constants). (**d–f**) Same as in panels (**a–c**), but for the 70:30 PMMA:Cz blend. (**g–i**) Same as in panels (**a–c**), but for the 90:10 PMMA:Cz blend.

Table 3. Summary of time constants and relative amplitudes in the GSB range for thin films of neat Cz and different PMMA:Cz blends obtained from the kinetic analysis by using a sum of exponential functions.

Expt.	λ_{pump} (nm)	PMMA:Cz (wt%:wt%)	λ_{probe} (nm)	τ_1 (ns)	τ_2 (ns)	τ_3 (ns)	τ_4 (ns)
fs-TA [1]	260	0:100	global	0.012 (40%)	640 (59%)	3 [2] (1%)	–
		70:30	global	0.0001 (9%)	0.007 (18%)	0.107 (24%)	0.580 (49%)
		90:10	global	0.001 (15%)	0.020 (25%)	0.209 (35%)	3.7 [2] (25%)
ns-TA [1]	260	0:100	global	3 (47%)	41 (37%)	790 (16%)	–
		70:30	global	0.580 (24%)	13 (20%)	113 (49%)	5100 (7%)
		90:10	global	3.7 (20%)	223 (50%)	7700 (30%)	–

[1] The kinetic traces at the different probe wavelengths were fitted simultaneously with a sum of three or four exponentials. As an example, the relative amplitudes in the GSB range are provided in parentheses. These are in the case of the compositions 0:100, 70:30 and 90:10, respectively: 339 nm, 337 nm and 290 nm (for fs-TA); 322 nm, 291 nm and 290 nm (for ns-TA). [2] Fixed value based on the fit results obtained from the ns-TA experiments. Note that the maximum pump–probe delay time in the fs-TA experiments was 1.5 ns.

Finally, in panels g–i, we deal with the dynamics for the PMMA:Cz thin-film blend with a weight percent ratio of 90:10. Here, the Cz–Cz spacing is even larger, which leads to a further slowing-down of the S_1–S_1 annihilation process. Still, this decay is much faster than the aforementioned time constant of 14–15 ns for the S_1 lifetime of Cz in organic solvents. Because of the much higher "dilution" of Cz in PMMA, the spectral features of the hot S_0^* molecules are also much weaker and less pronounced: While in a neat Cz film, the excess energy would be only distributed among the Cz molecules, here, the excess energy is predominantly transferred from "hot" Cz to the "cold" PMMA matrix. As PMMA has no absorption bands in the spectral range investigated here, the heating-up of PMMA by this process is not visible in the TA spectra [13]. Consequently, the spectral features in the GSB range of 280–320 nm in panel g (1–3 ns) and also panel d (1–10 ns) look considerably different, as they are not so strongly influenced by the spectral signature of the S_0^* molecules in panel a (1–10 ns). The kinetic analysis provides time constants of 1 ps, 20 ps and 209 ps for the SSA process (fs-TA) as well as 3.7 ns, 223 ns and 7.7 μs (ns-TA), which we assign to the cooling of the S_0^* molecules. The 3.7 ns component will also contain some contributions of the radiative decay of Cz molecules in S_1 (cf. $\langle\tau\rangle$ = 3.1 ns from TCSPC, Table 2) because intramolecular radiative relaxation becomes more competitive for the wider Cz–Cz distances in the 90:10 thin film blend.

3.4. Kinetic Analysis of Singlet–Singlet Annihilation Processes in Cz and t-Bu-Cz Films

The fs-TA kinetics for the neat Cz (and also *t*-Bu-Cz) thin films show that at the S_1 exciton number densities of ca. 4×10^{18} cm^{-3} employed in these experiments, the lifetime of S_1 is drastically reduced compared with the S_1 lifetime determined from the time-resolved emission experiments, which were recorded at much lower exciton number densities N_1 of ca. 6×10^{12} cm^{-3} (Figure 2 and Table 2). Whereas the total lifetimes of S_1 at the excitation conditions of the TCSPC experiments are 3.14 ns and 3.85 ns for Cz and *t*-Bu-Cz, respectively, most of the S_1 decay of the thin films in the transient absorption experiments already occurred by 100 ps. This is demonstrated in Figure 4a by the measured transient absorption kinetics in the NIR range averaged over the wavelength range of 1100–1300 nm, which monitor the decay of an ESA band exclusively assigned to S_1 excitons. To understand this drastic shortening of the S_1 lifetime with increasing number density N_1, the NIR transients were subjected to a kinetic modeling procedure, which is described below.

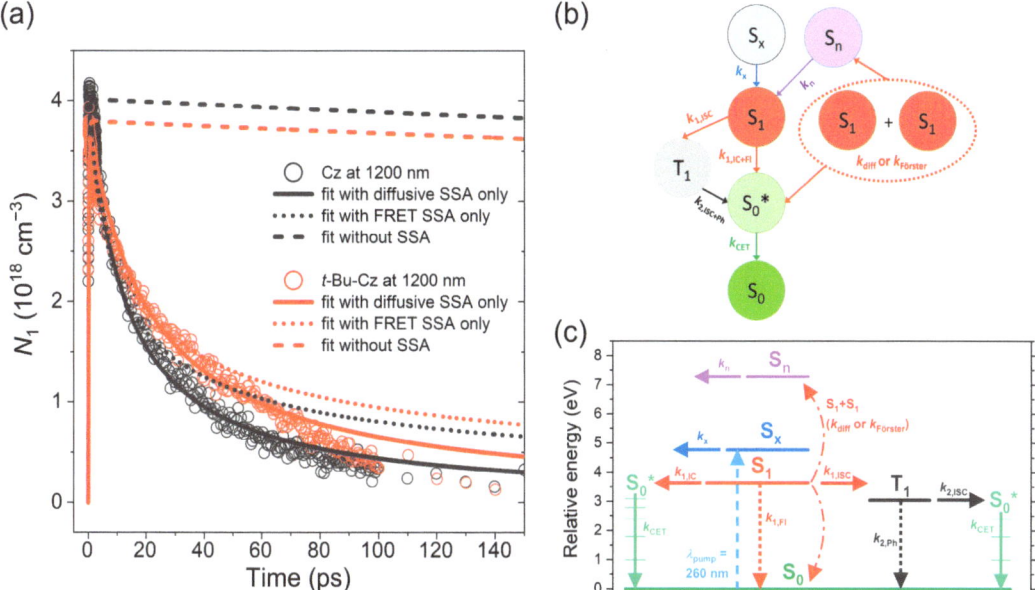

Figure 4. (a) Kinetic traces of the S_1 number densities for a Cz thin film (black circles) and a *t*-Bu-Cz thin film (red circles) averaged over the wavelength range of 1100–1300 nm. The solid lines are corresponding fits based on the kinetic modeling procedure using diffusive singlet–singlet annihilation only. The dotted lines are modeling results using singlet–singlet annihilation based only on S_1–S_1 FRET. The dashed lines are the simulation results without considering any SSA processes. (b) Kinetic scheme of the relaxation pathways of Cz and *t*-Bu-Cz in thin films. (c) Perrin–Jablonski diagram highlighting the energetics of the different states and the photophysical processes involved including the associated rate constants (solid arrows: nonradiative processes of a single electronic state, dotted downward arrows: radiative processes, red curved dash-dotted double arrow: singlet–singlet annihilation). S_0^* denotes the vibrationally hot ground electronic state, where the vibrational excited levels are schematically indicated by thin green lines. Vibrational levels of the S_1, S_x, S_n and T_1 states are omitted for the sake of clarity. Blue dashed upward arrow: laser excitation by a femtosecond pump pulse at 260 nm.

The kinetic scheme for modeling the relaxation of the neat Cz and *t*-Bu-Cz thin films is shown in Figure 4b, and a schematic Perrin–Jablonski diagram based on the data from Table 1 and available information regarding the energy of the triplet state [56] is depicted in Figure 4c. Relevant kinetic parameters are summarized in Table 4. The mechanism can be divided into contributions from intramolecular relaxation, singlet–singlet annihilation and collisional relaxation. We begin with steps 1–4, which summarize the intramolecular relaxation pathways of Cz and *t*-Bu-Cz molecules (left side of the scheme in Figure 4b):

$$S_x \xrightarrow{k_x} S_1 \qquad (1)$$

$$S_1 \xrightarrow{k_{IC+Fl,1}} S_0^* \qquad (2)$$

$$S_1 \xrightarrow{k_{ISC,1}} T_1 \qquad (3)$$

$$T_1 \xrightarrow{k_{ISC+Ph,2}} S_0^* \tag{4}$$

Table 4. Summary of the kinetic rate constants and time constants, thin film and laser parameters used in the kinetic modeling procedure for the neat Cz and *t*-Bu-Cz thin films at 296 K.

Physical Quantity	Cz	*t*-Bu-Cz	Physical Quantity	Cz	*t*-Bu-Cz
N_x (cm^{-3})	4.0×10^{18}	3.8×10^{18}	F (mJ cm^{-2}) [1]	1.03	0.92
k_x (s^{-1})	5.0×10^{12}	5.0×10^{12}	τ_x (fs)	200	200
$k_{1,IC+Fl}$ (s^{-1})	1.6×10^{8}	1.3×10^{8}	$\tau_{1,IC+Fl}$ (ns)	6.3	7.7
$k_{1,ISC}$ (s^{-1})	1.6×10^{8}	1.3×10^{8}	$\tau_{1,ISC}$ (ns)	6.3	7.7
$k_{2,ISC+Ph}$ (s^{-1})	1.25×10^{6}	1.25×10^{6}	$\tau_{2,ISC+Ph}$ (μs)	0.8	0.8
k_{CET} (s^{-1})	1.57×10^{8}	1.57×10^{8}	τ_{CET} (ns)	0.64	0.64
k_{diff} (cm^3 s^{-1})	2.00×10^{-8}	1.25×10^{-8}	$k_{F,1}$ (cm^6 s^{-1}) [2]	2.40×10^{-27}	1.80×10^{-27}
k_n (s^{-1})	5.0×10^{13}	5.0×10^{13}	τ_n (fs)	20	20
d (nm) [3]	153	138	A [4]	0.397	0.371

[1] Fluence F of the pump laser beam at 260 nm. [2] These are the best values to fit the initial S_1 population decays of Cz and *t*-Bu-Cz by using S_1–S_1 FRET as the only SSA process. However, they do not describe the kinetics at long times correctly (see Figure 4a) because of the apparent third-order behavior (see text). [3] Average thickness d of the thin films determined by AFM. [4] Absorbance A of the thin films at 260 nm.

The S_x exciton state, which is prepared by photoexcitation at 260 nm, decays via IC to S_1 (step 1). Because its lifetime is very short ($\tau_x = k_x^{-1} \approx 200$ fs), it does not show clear spectral fingerprints in the transient absorption spectra at early times (Figure 3). S_1 can decay either by internal conversion (IC) or fluorescence to a vibrationally excited ground state (S_0^*) with the rate constant $k_{IC+Fl,1} = \tau_{IC+Fl,1}^{-1}$ (step 2) or by intersystem crossing (ISC) to the T_1 triplet state with the rate constant $k_{ISC,1} = \tau_{ISC,1}^{-1}$ (step 3). The sum of these two rate constants represents the total rate constant for depopulation of the S_1 exciton state ($k_{total,1} = k_{IC+Fl,1} + k_{ISC,1} = \tau_{total,1}^{-1}$). The total lifetime $\tau_{total,1}$ of S_1 was determined in our TCSPC experiments as 3.14 ns and 3.85 ns for Cz and *t*-Bu-Cz, respectively (Figure 1). For our modeling here, we assumed a quantum yield of 50% for ISC from S_1, which is compatible with findings from previous experiments for the relaxation of Cz in organic solvents [49,56,57]. This provides time constants $\tau_{IC+Fl,1} = \tau_{ISC,1}$ of 6.3 ns and 7.7 ns for Cz and *t*-Bu-Cz, respectively (Table 4). The spin-forbidden relaxation of T_1 by ISC and phosphorescence (step 4) is much slower, and based on the results from our ns-TA experiments (Figure 3c), we use a value of $\tau_{ISC+Ph,2} = k_{ISC+Ph,2}^{-1}$ of 0.8 μs (Table 4).

The vibrationally hot Cz and *t*-Bu-Cz molecules in the electronic ground state (S_0^*) formed by steps 2 and 4 decay by collisional energy transfer (CET), where the excess vibrational energy is transferred first to the surrounding room-temperature carbazole molecules and finally to the underlying quartz substrate (step 5):

$$S_0^* \xrightarrow{k_{CET}} S_0 \tag{5}$$

For modeling the decay of the fs-TA kinetics, we use the time constant $\tau_{CET} = k_{CET}^{-1}$ of 3.0 ns extracted from Figure 3b. Modeling the decay of the S_1 exciton population for Cz and *t*-Bu-Cz in Figure 4a with this set of kinetic parameters results in the black dashed and red dashed fit lines, respectively. Both of them decay too slowly, in agreement with the aforementioned total S_1 lifetimes of 3.14 ns (Cz) and 3.85 ns (*t*-Bu-Cz).

Therefore, additional S_1 relaxation steps need to be considered to understand the fast decay of the fs-TA kinetics in Figure 4a. With increasing initial number densities of S_1 excitons in the thin film, singlet–singlet annihilation becomes increasingly important. One possible mechanism is the diffusive singlet–singlet annihilation of two S_1 states:

$$S_1 + S_1 \xrightarrow{k_{\text{diff}}} S_0^* + S_n \tag{6}$$

This bimolecular encounter results in the formation of an S_0^* ground state species and a higher excited S_n singlet state (located at an energy of (maximum) two times the energy of the S_1 state) with the second-order rate constant k_{diff}. The rate-determining step of this diffusion process is repeated hopping of the S_1 excitation based on Förster resonance energy transfer [58–60], which is possible because of the strong overlap between the $S_0 \to S_1$ absorption and the $S_1 \to S_0$ emission bands of Cz and t-Bu-Cz arising from the very small Stokes shift (see e.g., Figures 2 and 3a). The net equation for this homo-FRET process between identical carbazole molecules is:

$$S_1 + S_0 \xrightarrow{k_{\text{Förster},0}(R)} S_0 + S_1 \tag{7}$$

The excitation transfer is based on nonradiative dipole–dipole coupling, i.e., the two coupled first-order steps 8a and 8b comprising de-excitation of an S_1 species (energy donor) accompanied by excitation of an S_0 species (energy acceptor):

$$S_1 \xrightarrow{k_{\text{Förster},0}(R)} S_0 \tag{8a}$$

$$S_0 \xrightarrow{k_{\text{Förster},0}(R)} S_1 \tag{8b}$$

The first-order rate constant $k_{\text{Förster},0}$ is given by [58,59]:

$$k_{\text{Förster},0} = \tau_1^{-1}(R_0/R)^6 \tag{9}$$

where τ_1 is the total S_1 lifetime of the donor in the absence of energy transfer, R is the donor–acceptor distance, and the Förster radius R_0 corresponds to the distance, where the rate constants for FRET and intramolecular decay of S_1 are equal. Significant FRET contributions are expected up to donor–acceptor distances of about 10 nm, with typical Förster radii for singlet exciton systems in the range of 1.4–7 nm [18,60,62–65]. As there are always several carbazole molecules in S_0 available close to an S_1 molecule (i.e., R is small), the FRET process leads to efficient migration of the S_1 excitation through the thin film. Such a random walk of the excitation eventually brings two S_1 molecules close to each other, so that efficient homo-FRET-based singlet–singlet annihilation between the two S_1 carbazole molecules takes place with the net equation:

$$S_1 + S_1 \xrightarrow{k_{\text{Förster},1}(R)} S_0^* + S_n \tag{10}$$

where one S_1 exciton is destroyed (forming carbazole ground state molecules with some excess energy, S_0^*), and at the same time, one S_1 molecule is promoted to a higher S_n state:

$$S_1 \xrightarrow{k_{\text{Förster},1}(R)} S_0^* \tag{11a}$$

$$S_1 \xrightarrow{k_{\text{Förster},1}(R)} S_n \tag{11b}$$

Such an S_1–S_1 FRET process is feasible because the emission spectrum of the S_1 state strongly overlaps with broad S_1 excited-state absorption (see Figure 3a). Summarizing the diffusive SSA mechanism, step 6 used in our kinetic modeling procedure is of second-order because the random walk of the two S_1 molecules is rate-determining.

Alternatively, we also consider the direct homo-FRET SSA process of steps 11a and 11b to occur without prior diffusion. A technical difficulty to implement these two steps in a kinetic mechanism is that the rate constant $k_{\text{Förster},1}$ is R-dependent and thus depends on the spatial distribution of S_1 molecules, which changes with time. However, others [26] and we [66] demonstrated that in the case of randomly distributed S_1 molecules in thin films, the alternative kinetic formulation based on steps 12a and 12b can be used:

$$3\,S_1 \xrightarrow{k_{F,1}} S_0^* + 2\,S_n \qquad (12a)$$

$$3\,S_1 \xrightarrow{k_{F,1}} 2\,S_0^* + S_n \qquad (12b)$$

Note that these two steps also lead to the same net equation as in step 10. Importantly, the S_1 population decay due to S_1–S_1 FRET results in an apparent third-order behavior, with the rate constant $k_{FRET,1} = 6\,k_{F,1}$ [66].

The SSA processes in steps 6 and 10 both lead to the formation of a high-energy S_n singlet exciton species. Such states typically show ultrafast IC on a time scale of sub-50 femtoseconds, repopulating S_1 directly or via an ultrafast cascade of IC steps:

$$S_n \xrightarrow{k_n} S_1 \qquad (13)$$

Here, we assume a value $\tau_n = k_n^{-1}$ of 20 fs for its lifetime. The S_n exciton state is located at a high excitation energy (roughly twice the S_1 energy, which corresponds to about 7.3 eV or 170 nm for the two carbazoles, see Figure 2). At this high energy, S_n states may also form an electron–hole pair, where the electron and hole are located on individual carbazole molecules as an $S_0^{\bullet-}$ radical anion and an $S_0^{\bullet+}$ radical cation, respectively [65–67]. However, as we do not see any long-lived absorption bands assignable to such species, we conclude that this channel is below the detection limit of the fs-TA and ns-TA experiments. Singlet fission ($S_0 + S_1 \to T_1 + T_1$) [68] is not possible either, because the 0–0 energy of the T_1 state of Cz ($E(T_1) \approx 3.061$ eV) is located at more than half of the energy of the S_1 state ($E(S_1) \approx 3.658$ eV) [56]. Finally, triplet–triplet annihilation does not need to be considered either because the population of the T_1 state accumulated on the sub-200 ps time scale considered in Figure 4a is negligible.

The kinetic scheme of Figure 4b containing steps 1–6, 12a, 12b and 13 was implemented using the program Tenua [69]. The initial S_x number density was determined by using the known fluence of the pump laser beam at 260 nm, the thickness of the thin films (measured by AFM) and the absorbance of the thin film samples at 260 nm (Table 4). Because the lifetime of the S_x state is ultrashort, the peak value of the kinetics in Figure 4a essentially corresponds to the initial number density of S_1 molecules. The NIR transient absorption kinetics therefore describe the time-dependent decay of the S_1 population.

To learn about the dominant singlet–singlet annihilation mechanism, we consider the following two limiting cases: Using steps 1–6 and 13 (only diffusive SSA, but no S_1–S_1 homo-FRET, i.e., leaving out steps 12a and 12b), the best fits for the decays of Cz (black solid line) and t-Bu-Cz (red solid line) are obtained using the rate constants $k_{diff} = 2.00 \times 10^{-8}$ cm^3 s^{-1} and 1.25×10^{-8} cm^3 s^{-1}, respectively. Using steps 1–5, 12a, 12b and 13 (only S_1–S_1 homo-FRET, but no diffusive SSA, i.e., leaving out step 6) one obtains the dotted lines when using the rate constants $k_{F,1} = 2.40 \times 10^{-27}$ cm^6 s^{-1} (black) and 1.80×10^{-27} cm^6 s^{-1} (red), respectively.

The diffusive SSA model provides a very good fit to the Cz and t-Bu-Cz data. In contrast, the simulation based on S_1–S_1 homo-FRET SSA can only describe the earliest decay satisfactorily, but it decays too slowly afterward. This is a direct consequence of its apparent third-order behavior, whereas the diffusive SSA is second-order. In addition, the Förster radii, which can be calculated from $k_{F,1}$ as described previously [66], would be 18.3 nm and 17.5 nm for Cz and t-Bu-Cz, respectively, which is much larger than the typically observed 1.4–7 nm [18,60,62–65]. Our modeling results therefore suggest that the SSA mechanism operative in these carbazole thin films is of the diffusive type. A summary of the kinetic fit parameters is provided in Table 4.

4. Discussion

Our combined approach using steady-state absorption, time-resolved photoluminescence and broadband transient absorption experiments has provided a comprehensive overview of the complicated excited-state dynamics in Cz and t-Bu-Cz thin films and their blends with PMMA. In the neat films, excitonic effects in the steady-state absorption and emission spectra are clearly visible in terms of a pronounced excitonic peak, which likely arises from J-type aggregate structures. Yet, these excitonic contributions appear to be less pronounced than in films of "classical" excitonic materials, such as tetracene or different types of polymer thin films [24].

Time-resolved emission experiments using TCSPC at low S_1 number densities of about 6×10^{12} cm^{-3} reveal efficient Cz(S_1)–Cz(S_0) homo-FRET with the ultimate population of excimer sites. The excimer contributions appear to be less important than in other carbazole-type materials, such as 1,3-bis(N-carbazolyl)benzene (mCP). Also, no additional structured red-shifted aggregate emission is observed, in contrast to thin films of mCP [13]. At a higher S_1 exciton number density of about 4×10^{18} cm^{-3}, the S_1 lifetime is considerably reduced because of diffusive S_1–S_1 annihilation with the rate constants k_{diff} of 2.00×10^{-8} cm^3 s^{-1} (Cz) and 1.25×10^{-8} cm^3 s^{-1} (t-Bu-Cz) obtained from ns-TA experiments. These values may be compared with previous results for other carbazole derivatives: For thin films of 4′-bis(9-carbazolyl)-2,2′-biphenyl (CBP), Ruseckas et al. [17] obtained values for the annihilation rate constant γ of 6×10^{-9} cm^3 s^{-1}, respectively, using luminescence-based techniques. This rate constant is slightly smaller but still in reasonable agreement with our values from the fs-TA experiments. Later on, Morgenroth et al. obtained a value of 1.40×10^{-8} cm^3 s^{-1} for k_{diff} of mCP [13], in very good agreement with our present results. The current measurements on Cz and t-Bu-Cz thin films provide valuable information regarding this important kinetic parameter, which is of particular relevance for the efficiency roll-off behavior of OLEDs at high current densities. The fs-TA experiments also show that "direct" S_1–S_1 FRET without prior S_1–S_0 homo-FRET diffusion does not play any role in the singlet–singlet annihilation process. The SSA process generates highly vibrationally excited molecules in the ground state (S_0^*), which can be clearly identified in our fs-TA and ns-TA experiments in terms of a sharp hot-band absorption on the red edge of the $S_0 \rightarrow S_1$ absorption band and sharp bleach features in the GSB region. The dissipation of this excess energy to the quartz substrate spans a wide time scale from several hundred picoseconds to microseconds. As valuable side information, picosecond ultrasonics experiments employing the fs-TA setup provide a value of 2200 m s^{-1} for the longitudinal sound velocity of Cz. This value can be conveniently used in future studies to measure the thickness of Cz and other Cz-based films with a thickness in the range of about 50–500 nm in an all-optical fashion.

There remain several open questions for further research on these films. For instance, we focused on spin-coated films, and their microcrystalline structure is not ideal for applications. It would therefore be of interest to carry out similar studies with Cz and t-Bu-Cz films produced by physical vapor deposition and study the impact of the film preparation method on the steady-state and time-resolved optical properties. From the dynamics side, the current approach using optical excitation initially populates only singlet states. In situations where the initial exciton number density is high, triplet formation cannot compete with the singlet exciton channels, such as bimolecular diffusive SSA. It would be therefore valuable to study such films under pulsed electrical excitation, as this will initially populate the excited states in a statistical T_1:S_1 ratio of 3:1. This way, the kinetics of channels involving the T_1 triplet state, such as triplet–triplet annihilation and singlet–triplet annihilation, can be accessed. These topics will be addressed in future work.

Supplementary Materials: The following supporting information can be downloaded at: https://www.mdpi.com/article/10.3390/photochem4020011/s1, Figures S1 and S2: X-ray diffraction patterns of neat Cz and t-Bu-Cz thin films, respectively, on quartz including simulations based on a Rietveld refinement procedure considering texture effects and an amorphous background signal; Figure S3: Absorption spectrum of a neat PMMA thin film produced by spin-coating on quartz.

Author Contributions: Conceptualization, K.O. and T.L.; methodology, K.O. and T.L.; investigation, K.M.K., D.G., K.O. and T.L.; writing—original draft preparation, K.O., T.L. and K.M.K.; writing—review and editing, K.O., T.L. and K.M.K.; supervision, K.O. and T.L. All authors have read and agreed to the published version of the manuscript.

Funding: This research received no external funding.

Data Availability Statement: The data are contained within this article.

Acknowledgments: We thank T. Staedler and T. Guo (Institute of Materials Engineering, University of Siegen) for kindly providing the setup for the AFM measurements. We also thank M. Killian and T. Kowald (Chemistry and Structure of Novel Materials, University of Siegen) for recording the thin-film X-ray diffraction data.

Conflicts of Interest: The authors declare no conflicts of interest.

References

1. Li, J.; Grimsdale, A.C. Carbazole-based polymers for organic photovoltaic devices. *Chem. Soc. Rev.* **2010**, *39*, 2399–2410. [CrossRef] [PubMed]
2. Oner, S.; Bryce, M.R. A review of fused-ring carbazole derivatives as emitter and/or host materials in organic light emitting diode (OLED) applications. *Mater. Chem. Front.* **2023**, *7*, 4304–4338. [CrossRef]
3. Radhakrishna, K.; Manjunath, S.B.; Devadiga, D.; Chetri, R.; Nagaraja, A.T. Review on Carbazole-Based Hole Transporting Materials for Perovskite Solar Cell. *ACS Appl. Electron. Mater.* **2023**, *6*, 3635–3664. [CrossRef]
4. Ledwon, P. Recent advances of donor-acceptor type carbazole-based molecules for light emitting applications. *Org. Electron.* **2019**, *75*, 105422. [CrossRef]
5. Godumala, M.; Choi, S.; Cho, M.J.; Choi, D.H. Recent breakthroughs in thermally activated delayed fluorescence organic light emitting diodes containing non-doped emitting layers. *J. Mater. Chem. C* **2019**, *7*, 2172–2198. [CrossRef]
6. Nakanotani, H.; Sasabe, H.; Adachi, C. Singlet-singlet and singlet-heat annihilations in fluorescence-based organic light-emitting diodes under steady-state high current density. *Appl. Phys. Lett.* **2005**, *86*, 213506. [CrossRef]
7. Hasan, M.; Shukla, A.; Ahmad, V.; Sobus, J.; Bencheikh, F.; McGregor, S.K.M.; Mamada, M.; Adachi, C.; Lo, S.-C.; Namdas, E.B. Exciton–Exciton Annihilation in Thermally Activated Delayed Fluorescence Emitter. *Adv. Funct. Mater.* **2020**, *30*, 2000580. [CrossRef]
8. Bree, A.; Zwarich, R. Vibrational Assignment of Carbazole from Infrared, Raman, and Fluorescence Spectra. *J. Chem. Phys.* **1968**, *49*, 3344–3355. [CrossRef]
9. Chakravorty, S.C.; Ganguly, S.C. Polarized Absorption Spectra of Carbazole Single Crystal. *J. Chem. Phys.* **1970**, *52*, 2760–2762. [CrossRef]
10. Tanaka, M. Electronic States of Fluorene, Carbazole and Dibenzofuran. *Bull. Chem. Soc. Jpn.* **1976**, *49*, 3382–3388. [CrossRef]
11. Nakhimovsky, L.A.; Fuchs, R.; Martin, D.; Small, G.J. Optical Properties of Carbazole Thin Monocrystalline Films. *Mol. Cryst. Liq. Cryst.* **1988**, *154*, 89–105. [CrossRef]
12. Nguyen, D.D.; Trunk, J.; Nakhimovsky, L.; Spanget-Larsen, J. Electronic transitions of fluorene, dibenzofuran, carbazole, and dibenzothiophene: From the onset of absorption to the ionization threshold. *J. Mol. Spectrosc.* **2010**, *264*, 19–25. [CrossRef]
13. Morgenroth, M.; Lenzer, T.; Oum, K. Understanding Excited-State Relaxation in 1,3-Bis(N-carbazolyl)benzene, a Host Material for Organic Light-Emitting Diodes. *J. Phys. Chem. C* **2023**, *127*, 4582–4593. [CrossRef]
14. Scully, S.R.; McGehee, M.D. Effects of optical interference and energy transfer on exciton diffusion length measurements in organic semiconductors. *J. Appl. Phys.* **2006**, *100*, 034907. [CrossRef]
15. Lewis, A.J.; Ruseckas, A.; Gaudin, O.P.M.; Webster, G.R.; Burn, P.L.; Samuel, I.D.W. Singlet exciton diffusion in MEH-PPV films studied by exciton–exciton annihilation. *Org. Electron.* **2006**, *7*, 452–456. [CrossRef]
16. Shaw, P.E.; Ruseckas, A.; Samuel, I.D.W. Exciton Diffusion Measurements in Poly(3-hexylthiophene). *Adv. Mater.* **2008**, *20*, 3516–3520. [CrossRef]
17. Ruseckas, A.; Ribierre, J.C.; Shaw, P.E.; Staton, S.V.; Burn, P.L.; Samuel, I.D.W. Singlet energy transfer and singlet-singlet annihilation in light-emitting blends of organic semiconductors. *Appl. Phys. Lett.* **2009**, *95*, 183305. [CrossRef]
18. Lunt, R.R.; Giebink, N.C.; Belak, A.A.; Benziger, J.B.; Forrest, S.R. Exciton diffusion lengths of organic semiconductor thin films measured by spectrally resolved photoluminescence quenching. *J. Appl. Phys.* **2009**, *105*, 053711. [CrossRef]
19. Cook, S.; Furube, A.; Katoh, R.; Han, L. Estimate of singlet diffusion lengths in PCBM films by time-resolved emission studies. *Chem. Phys. Lett.* **2009**, *478*, 33–36. [CrossRef]
20. Cook, S.; Liyuan, H.; Furube, A.; Katoh, R. Singlet Annihilation in Films of Regioregular Poly(3-hexylthiophene): Estimates for Singlet Diffusion Lengths and the Correlation between Singlet Annihilation Rates and Spectral Relaxation. *J. Phys. Chem. C* **2010**, *114*, 10962–10968. [CrossRef]
21. Shaw, P.E.; Ruseckas, A.; Peet, J.; Bazan, G.C.; Samuel, I.D.W. Exciton-Exciton Annihilation in Mixed-Phase Polyfluorene Films. *Adv. Funct. Mater.* **2010**, *20*, 155–161. [CrossRef]

22. Masri, Z.; Ruseckas, A.; Emelianova, E.V.; Wang, L.; Bansal, A.K.; Matheson, A.; Lemke, H.T.; Nielsen, M.N.; Nguyen, H.; Coulembier, O.; et al. Molecular Weight Dependence of Exciton Diffusion in Poly(3-hexylthiophene). *Adv. Energy Mater.* **2013**, *3*, 1445–1453. [CrossRef]
23. Lin, J.D.A.; Mikhnenko, O.V.; Chen, J.; Masri, Z.; Ruseckas, A.; Mikhailovsky, A.; Raab, R.P.; Liu, J.; Blom, P.W.M.; Loi, M.A.; et al. Systematic study of exciton diffusion length in organic semiconductors by six experimental methods. *Mater. Horiz.* **2014**, *1*, 280–285. [CrossRef]
24. Bardeen, C.J. The Structure and Dynamics of Molecular Excitons. *Annu. Rev. Phys. Chem.* **2014**, *65*, 127–148. [CrossRef] [PubMed]
25. Gulbinas, V.; Valkunas, L.; Kuciauskas, D.; Katilius, E.; Liuolia, V.; Zhou, W.; Blankenship, R.E. Singlet-Singlet Annihilation and Local Heating in FMO Complexes. *J. Phys. Chem.* **1996**, *100*, 17950–17956. [CrossRef]
26. Stevens, M.A.; Silva, C.; Russell, D.M.; Friend, R.H. Exciton dissociation mechanisms in the polymeric semiconductors poly(9,9-dioctylfluorene) and poly(9,9-dioctylfluorene-co-benzothiadazole). *Phys. Rev. B* **2001**, *63*, 165213. [CrossRef]
27. King, S.M.; Dai, D.; Rothe, C.; Monkman, A.P. Exciton annihilation in a polyfluorene: Low threshold for singlet-singlet annihilation and the absence of singlet-triplet annihilation. *Phys. Rev. B* **2007**, *76*, 085204. [CrossRef]
28. Zaushitsyn, Y.; Jespersen, K.G.; Valkunas, L.; Sundström, V.; Yartsev, A. Ultrafast dynamics of singlet-singlet and singlet-triplet exciton annihilation in poly(3-2′-methoxy-5′-octylphenyl)thiophene films. *Phys. Rev. B* **2007**, *75*, 195201. [CrossRef]
29. Völker, S.F.; Schmiedel, A.; Holzapfel, M.; Renziehausen, K.; Engel, V.; Lambert, C. Singlet–Singlet Exciton Annihilation in an Exciton-Coupled Squaraine-Squaraine Copolymer: A Model toward Hetero-J-Aggregates. *J. Phys. Chem. C* **2014**, *118*, 17467–17482. [CrossRef]
30. Chen, C.; Chi, Z.; Chong, K.C.; Batsanov, A.S.; Yang, Z.; Mao, Z.; Yang, Z.; Liu, B. Carbazole isomers induce ultralong organic phosphorescence. *Nat. Mater.* **2021**, *20*, 175–180. [CrossRef] [PubMed]
31. Chen, C.; Chong, K.C.; Pan, Y.; Qi, G.; Xu, S.; Liu, B. Revisiting Carbazole: Origin, Impurity, and Properties. *ACS Mater. Lett.* **2021**, *3*, 1081–1087. [CrossRef]
32. Lutterotti, L. Total pattern fitting for the combined size-strain-stress-texture determination in thin film diffraction. *Nucl. Instrum. Methods Phys. Res. Sect. B* **2010**, *268*, 334–340. [CrossRef]
33. Dobryakov, A.L.; Kovalenko, S.A.; Weigel, A.; Pérez Lustres, J.L.; Lange, J.; Müller, A.; Ernsting, N.P. Femtosecond pump/supercontinuum-probe spectroscopy: Optimized setup and signal analysis for single-shot spectral referencing. *Rev. Sci. Instrum.* **2010**, *81*, 113106. [CrossRef] [PubMed]
34. Oum, K.; Lenzer, T.; Scholz, M.; Jung, D.Y.; Sul, O.; Cho, B.J.; Lange, J.; Müller, A. Observation of Ultrafast Carrier Dynamics and Phonon Relaxation of Graphene from the Deep-Ultraviolet to the Visible Region. *J. Phys. Chem. C* **2014**, *118*, 6454–6461. [CrossRef]
35. Flender, O.; Scholz, M.; Klein, J.R.; Oum, K.; Lenzer, T. Excited-state relaxation of the solar cell dye D49 in organic solvents and on mesoporous Al_2O_3 and TiO_2 thin films. *Phys. Chem. Chem. Phys.* **2016**, *18*, 26010–26019. [CrossRef] [PubMed]
36. Merker, A.; Scholz, M.; Morgenroth, M.; Lenzer, T.; Oum, K. Photoinduced Dynamics of $(CH_3NH_3)_4Cu_2Br_6$ Thin Films Indicating Efficient Triplet Photoluminescence. *J. Phys. Chem. Lett.* **2021**, *12*, 2736–2741. [CrossRef]
37. Ruello, P.; Gusev, V.E. Physical mechanisms of coherent acoustic phonons generation by ultrafast laser action. *Ultrasonics* **2015**, *56*, 21–35. [CrossRef] [PubMed]
38. Thomsen, C.; Strait, J.; Vardeny, Z.; Maris, H.J.; Tauc, J.; Hauser, J.J. Coherent Phonon Generation and Detection by Picosecond Light Pulses. *Phys. Rev. Lett.* **1984**, *53*, 989–992. [CrossRef]
39. Grahn, H.T.; Maris, H.J.; Tauc, J. Picosecond Ultrasonics. *IEEE J. Quantum Electron.* **1989**, *25*, 2562–2569. [CrossRef]
40. Kanner, G.S.; Vardeny, Z.V.; Hess, B.C. Picosecond acoustics in polythiophene thin films. *Phys. Rev. B* **1990**, *42*, 5403–5406. [CrossRef] [PubMed]
41. Kaake, L.G.; Welch, G.C.; Moses, D.; Bazan, G.C.; Heeger, A.J. Influence of Processing Additives on Charge-Transfer Time Scales and Sound Velocity in Organic Bulk Heterojunction Films. *J. Phys. Chem. Lett.* **2012**, *3*, 1253–1257. [CrossRef] [PubMed]
42. Scholz, M.; Morgenroth, M.; Cho, M.J.; Choi, D.H.; Lenzer, T.; Oum, K. Coherent acoustic phonon dynamics in chiral copolymers. *Struct. Dyn.* **2019**, *6*, 064502. [CrossRef] [PubMed]
43. Nowak, R.; Bernstein, E.R. On the phase transition in N-isopropylcarbazole. *J. Chem. Phys.* **1986**, *85*, 6858–6866. [CrossRef]
44. Auty, A.R.; Jones, A.C.; Phillips, D. Spectroscopy and decay dynamics of jet-cooled carbazole and N-ethylcarbazole and their homocyclic analogues. *Chem. Phys.* **1976**, *103*, 163–182. [CrossRef]
45. Bigelow, R.W.; Ceasar, G.P. Hydrogen Bonding and N-Alkylation Effects on the Electronic Structure of Carbazole. *J. Phys. Chem.* **1979**, *83*, 1790–1795. [CrossRef]
46. Yi, J.T.; Alvarez-Valtierra, L.; Pratt, D.W. Rotationally resolved $S_1 \leftarrow S_0$ electronic spectra of fluorene, carbazole, and dibenzofuran: Evidence for Herzberg-Teller coupling with the S_2 state. *J. Chem. Phys.* **2006**, *124*, 244302. [CrossRef] [PubMed]
47. Ljubić, I.; Sabljić, A. CASSCF/CASPT2 and TD-DFT Study of Valence and Rydberg Electronic Transitions in Fluorene, Carbazole, Dibenzofuran, and Dibenzothiophene. *J. Phys. Chem. A* **2011**, *115*, 4840–4850. [CrossRef]
48. Bonesi, S.M.; Erra-Balsells, R. Electronic spectroscopy of carbazole and N- and C-substituted carbazoles in homogeneous media and in solid matrix. *J. Lumin.* **2001**, *93*, 51–74. [CrossRef]
49. Knötig, K.M.; Gust, D.; Lenzer, T.; Oum, K. Excited-State Dynamics of Carbazole and *tert*-Butyl-Carbazole in Organic Solvents. *Photochem* **2024**, *4*, 163–178. [CrossRef]

50. Gajda, K.; Zarychta, B.; Kopka, K.; Daszkiewicz, Z.; Ejsmont, K. Substituent effects in nitro derivatives of carbazoles investigated by comparison of low-temperature crystallographic studies with density functional theory (DFT) calculations. *Acta Crystallogr. Sect. C Struct. Chem.* **2014**, *70*, 987–991. [CrossRef]
51. Yan, Q.; Gin, E.; Wasinska-Kalwa, M.; Banwell, M.G.; Carr, P.D. A Palladium-Catalyzed Ullmann Cross-Coupling/Reductive Cyclization Route to the Carbazole Natural Products 3-Methyl-9*H*-carbazole, Glycoborine, Glycozoline, Clauszoline K, Mukonine, and Karapinchamine A. *J. Org. Chem.* **2017**, *82*, 4148–4159. [CrossRef] [PubMed]
52. Spano, F.C. The Spectral Signatures of Frenkel Polarons in H- and J-Aggregates. *Acc. Chem. Res.* **2010**, *43*, 429–439. [CrossRef] [PubMed]
53. Hernández, F.J.; Crespo-Otero, R. Excited state mechanisms in crystalline carbazole: The role of aggregation and isomeric defects. *J. Mater. Chem. C* **2021**, *9*, 11882–11892. [CrossRef]
54. Dicke, R.H. Coherence in Spontaneous Radiation Processes. *Phys. Rev.* **1954**, *93*, 99–110. [CrossRef]
55. Blach, D.D.; Lumsargis, V.A.; Clark, D.E.; Chuang, C.; Wang, K.; Dou, L.; Schaller, R.D.; Cao, J.; Li, C.W.; Huang, L. Superradiance and Exciton Delocalization in Perovskite Quantum Dot Superlattices. *Nano Lett.* **2022**, *22*, 7811–7818. [CrossRef] [PubMed]
56. Adams, J.E.; Mantulin, W.W.; Huber, J.R. Effect of Molecular Geometry on Spin-Orbit Coupling of Aromatic Amines in Solution. Diphenylamine, Iminobibenzyl, Acridan, and Carbazole. *J. Am. Chem. Soc.* **1973**, *95*, 5477–5481. [CrossRef]
57. Martin, M.M.; Ware, W.R. Fluorescence Quenching of Carbazole by Pyridine and Substituted Pyridines. Radiationless Processes in the Carbazole-Amine Hydrogen Bonded Complex. *J. Phys. Chem.* **1978**, *82*, 2770–2776. [CrossRef]
58. Förster, T. Zwischenmolekulare Energiewanderung und Fluoreszenz. *Ann. Phys.* **1948**, *437*, 55–75. [CrossRef]
59. Förster, T. Transfer Mechanisms of Electronic Excitation. *Discuss. Faraday Soc.* **1959**, *27*, 7–17. [CrossRef]
60. Mikhnenko, O.V.; Blom, P.W.M.; Nguyen, T.-Q. Exciton diffusion in organic semiconductors. *Energy Environ. Sci.* **2015**, *8*, 1867–1888. [CrossRef]
61. Stalindurai, K.; Krishnan, K.G.; Nagarajan, E.R.; Ramalingan, C. Experimental and theoretical studies on new 7-(3,6-di-*tert*-butyl-9*H*-carbazol-9-yl)-10-alkyl-10*H*-phenothiazine-3-carbaldehydes. *J. Mol. Struct.* **2017**, *1130*, 633–643. [CrossRef]
62. Hofkens, J.; Cotlet, M.; Vosch, T.; Tinnefeld, P.; Weston, K.D.; Ego, C.; Grimsdale, A.; Müllen, K.; Beljonne, D.; Brédas, J.L.; et al. Revealing competitive Förster-type resonance energy-transfer pathways in single bichromophoric molecules. *Proc. Natl. Acad. Sci. USA* **2003**, *100*, 13146–13151. [CrossRef] [PubMed]
63. Lam, A.; St-Pierre, F.; Gong, Y.; Marshall, J.D.; Cranfill, P.J.; Baird, M.A.; McKeown, M.R.; Wiedenmann, J.; Davidson, M.W.; Schnitzer, M.J.; et al. Improving FRET dynamic range with bright green and red fluorescent proteins. *Nat. Methods* **2012**, *9*, 1005–1012. [CrossRef] [PubMed]
64. Müller, S.M.; Galliardt, H.; Schneider, J.; Barisas, B.G.; Seidel, T. Quantification of Förster resonance energy transfer by monitoring sensitized emission in living plant cells. *Front. Plant Sci.* **2013**, *4*, 413. [CrossRef] [PubMed]
65. Murawski, C.; Leo, K.; Gather, M.C. Efficiency Roll-Off in Organic Light-Emitting Diodes. *Adv. Mater.* **2013**, *25*, 6801–6827. [CrossRef] [PubMed]
66. Morgenroth, M.; Scholz, M.; Cho, M.J.; Choi, D.H.; Oum, K.; Lenzer, T. Mapping the broadband circular dichroism of copolymer films with supramolecular chirality in time and space. *Nat. Commun.* **2022**, *13*, 210. [CrossRef] [PubMed]
67. Morgenroth, M.; Scholz, M.; Guy, L.; Oum, K.; Lenzer, T. Spatiotemporal Mapping of Efficient Chiral Induction by Helicene-Type Additives in Copolymer Thin Films. *Angew. Chem. Int. Ed.* **2022**, *61*, e202203075. [CrossRef] [PubMed]
68. Smith, M.B.; Michl, J. Singlet Fission. *Chem. Rev.* **2010**, *110*, 6891–6936. [CrossRef] [PubMed]
69. Wachsstock, D. Tenua 2.1—The Kinetics Simulator for Java. 2007. Available online: http://bililite.com/tenua/ (accessed on 5 January 2024).

Disclaimer/Publisher's Note: The statements, opinions and data contained in all publications are solely those of the individual author(s) and contributor(s) and not of MDPI and/or the editor(s). MDPI and/or the editor(s) disclaim responsibility for any injury to people or property resulting from any ideas, methods, instructions or products referred to in the content.

Article

Inverse Problems in Pump–Probe Spectroscopy

Denis S. Tikhonov [1], Diksha Garg [1] and Melanie Schnell [1,2,*]

[1] Deutsches Elektronen-Synchrotron DESY, Notkestr. 85, 22607 Hamburg, Germany; denis.tikhonov@desy.de (D.S.T.); diksha.garg@desy.de (D.G.)
[2] Institute of Physical Chemistry, Christian-Albrechts-Universität zu Kiel, 24118 Kiel, Germany
* Correspondence: melanie.schnell@desy.de

Abstract: Ultrafast pump–probe spectroscopic studies allow for deep insights into the mechanisms and timescales of photophysical and photochemical processes. Extracting valuable information from these studies, such as reactive intermediates' lifetimes and coherent oscillation frequencies, is an example of the inverse problems of chemical kinetics. This article describes a consistent approach for solving this inverse problem that avoids the common obstacles of simple least-squares fitting that can lead to unreliable results. The presented approach is based on the regularized Markov Chain Monte-Carlo sampling for the strongly nonlinear parameters, allowing for a straightforward solution of the ill-posed nonlinear inverse problem. The software to implement the described fitting routine is introduced and the numerical examples of its application are given. We will also touch on critical experimental parameters, such as the temporal overlap of pulses and cross-correlation time and their connection to the minimal reachable time resolution.

Keywords: pump–probe spectroscopy; inverse problems; Monte-Carlo sampling

1. Introduction

Pump–probe spectroscopy is a powerful tool in the investigation of ultrafast kinetics [1]. The basic scheme of pump–probe studies is the following [1]. First, we initiate the system's dynamics with a pump laser pulse, then wait some time for the excited molecular system to evolve, and then with a second probe laser pulse, we turn otherwise unobservable (due to their short lifetimes) metastable species into something new and observable by the detectors. By varying the delay time between pump and probe pulses (so-called pump–probe delay), we can measure the real-time photochemical dynamics of the molecules. Under the umbrella term "pump–probe spectroscopy", multiple experimental methods hide that differ by what kind of observable is being monitored in the experiment [1–4]. The measured experimental parameters can be the absorption of the probe photons [5], mass spectra [6–8], photoelectron spectra [6,9], ion velocity map imaging [6,7,10], fluorescence [11], and so on [2,3]. We can extract viable information about the rates of various molecular processes from these pump–probe experimental results. To do that, we need to perform the experimental data analysis, which is at the center of this article.

The current manuscript presents a novel approach and its software implementation for performing such data analysis. It is based on mixed linear and nonlinear optimization [12], Monte-Carlo sampling [13,14], and regularization of fitting parameters [15–17], solving many issues of the ill-posed inverse problem of pump–probe experimental data analysis. First, we will formally examine the basic experimental data analysis and the least-squares fitting. Then, we will discuss the model of the pump–probe spectroscopy. This will be followed by a detailed description of the proposed data analysis algorithm and a short description of the implementation of such a method. In the end, a few numerical examples will be given. In the electronic supporting information (ESI), we also provide a user manual, an up-to-date software version, and the numerical examples described in the last section of this article.

order Taylor expansion, then we get $\Phi(\mathbf{P}) \approx \Phi_{\min} + \frac{1}{2}\Delta \mathbf{P}^T \Phi^{(2)} \Delta \mathbf{P}$, where $\Phi_{\min} = \Phi(\mathbf{P}_{\exp})$, $\Delta \mathbf{P} = \mathbf{P} - \mathbf{P}_{\exp}$, and $\Phi^{(2)}$ is the matrix of second derivatives of the Φ over parameters $\mathbf{P}$ at a set of parameters $\mathbf{P}_{\exp}$. The linear term in this expansion is zero because we expand for deviations from the minimum. Upon substitution of this expansion into Equation (5), we get [12,20]:

$$p(\mathbf{P}) \propto \exp\left(-\frac{1}{2}\Delta \mathbf{P}^T \Phi^{(2)} \Delta \mathbf{P}\right). \tag{10}$$

This equation is a normal distribution for parameters $\mathbf{P}$ with $\mathbf{P}_{\exp}$ being the mean values and $\Phi^{(2)}$ being the inverse variance matrix. The diagonal elements of $(\Phi^{(2)})^{-1}$ are the squared standard deviations of the parameters $\mathbf{P}$, and the off-diagonal elements correspond to correlations between the parameters.

Using this methodology, we can solve virtually any inverse problem using the following general iterative procedure [12]:

1. Initialize minimization procedure with a set of initial guess parameters $\mathbf{P}_{\text{ini}}$. Then, we find the corresponding set of observables $\mathcal{O}_{\text{ini}}$ by solving a direct problem (Equation (2)) and compute the LSQ function (Equation (6)) value $\Phi(\mathbf{P}_{\text{ini}})$.

2. Using one of the minimization algorithms, such as the Conjugate Gradient Method [21] or Powell's algorithm [22], we first get iteration trial parameter values $\mathbf{P}_{\text{trial}}^{(1)}$. Depending on the chosen minimization algorithm, we may require either calculation of the LSQ function's gradient ($\nabla \Phi(\mathbf{P}_{\text{ini}})$), Hessian ($\nabla^2 \Phi(\mathbf{P}_{\text{ini}})$), or evaluate the LSQ function's values in a few neighboring points around $\mathbf{P}_{\text{ini}}$.

3. Then, we again compute the $\Phi(\mathbf{P}_{\text{trial}}^{(1)})$ and find second ($\mathbf{P}_{\text{trial}}^{(2)}$), third ($\mathbf{P}_{\text{trial}}^{(3)}$), fourth ($\mathbf{P}_{\text{trial}}^{(4)}$), and so on, values of the parameters, trying to minimize the value of the LSQ function (Equation (6)).

4. We halt this iterative procedure when we reach a pre-defined convergence criterion. For instance, if the change of the parameter value from iteration to iteration is smaller than some small value (δ), e.g., as $\sqrt{\left(\mathbf{P}_{\text{trial}}^{(n+1)} - \mathbf{P}_{\text{trial}}^{(n)}\right)^2} \leq \delta$, then the convergence criterion can be said to satisfy. In this case, we take the last value obtained in the procedure to be our solution, and then we estimate the uncertainties of the parameters and correlations between them using an approximate normal distribution computed from the second derivatives of the LSQ function (see Equation (10)).

However, this procedure can still be an ill-posed problem. Because of the strong nonlinearity of the inverse problem in pump–probe spectroscopy, there might be several local minima in the LSQ function, which means nonuniqueness of the inverse problem solution [13,14,16]. A schematic illustration of such a generic case is given in Figure 1. In this figure, we can see multiple solutions of the inverse problem, each of those will be obtained dependent on the choice of the initial guess $\mathbf{P}_{\text{ini}}$. Then, upon obtaining one of the solutions with the previously described iterative algorithm, we will get an estimation of the parameter uncertainty based on Equation (10), which in the multivariate case will be much smaller than the actual width of the whole multivariate distribution (Equation (5)). Therefore, each of the possible solutions $\mathbf{P}_k \pm \sigma_k$ ($k = 1, 2, 3$) will be a poor estimate of the actual distribution for the system [13,14].

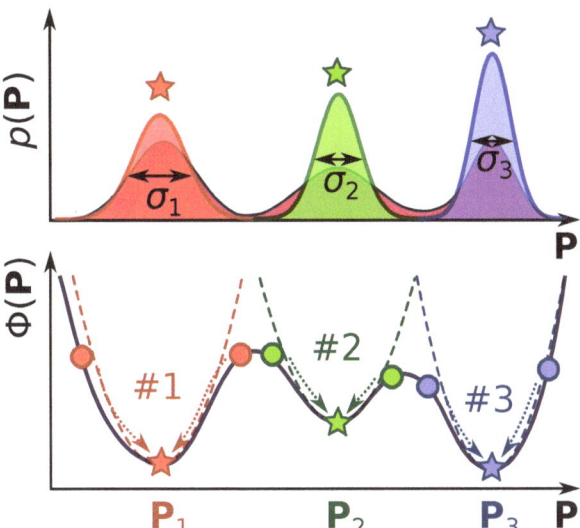

Figure 1. A schematic illustration of the ill-posed LSQ problem. The bottom plot illustrates a LSQ function $\Phi(\mathbf{P})$ (solid curve) with multiple local minima (#1, #2, and #3). The circles illustrate different initial sets of parameters ($\mathbf{P}_{\text{ini}}$) that converge to different solutions ($\mathbf{P}_1$, $\mathbf{P}_2$, $\mathbf{P}_3$), shown with stars. The dashed parabolas in each of the minima represent a local approximation of the nonlinear LSQ function (Equation (6)) with a Taylor expansion up to the second order (see Equation (10)). The top plot shows the corresponding probability distributions $p(\mathbf{P}) \propto \exp(-\Phi(\mathbf{P}))$. The three Gaussian functions represent the approximations of the localized solutions (Equation (10)) with standard deviations σ_1, σ_2, and σ_3. The solid curve in the back represents an actual multivariate probability distribution according to Equation (5).

2.3. Regularization of the Least-Squares Inverse Problem

The main method of how to deal with ill-posed inverse problems is the so-called regularization [14–16]. The idea of this method is simple: since our experimental data do not provide a single solution to the problem, and we cannot decide which of the solutions is the correct one, we need to get some *a priori* information to decide on this. In other words, we take some external information on the system and modify the inverse problem to account for this constraining information. We can have pre-known estimate values for some subset of the parameters $\mathbf{p} \in \mathbf{P}$, or even for all of the parameters ($\mathbf{p} = \mathbf{P}$). Let us denote our known parameters as $\mathbf{p}_{\text{reg}}$. To impose soft constraints from these parameters to the LSQ minimization problem (Equation (8)), we can add a penalty function $\Phi_{\text{reg}}(\mathbf{P})$ modifying the inverse problem to:

$$\Phi(\mathbf{P}) + \Phi_{\text{reg}}(\mathbf{P}) \to \min. \quad (11)$$

The two most common ways to define the regularization term are based on the L_1 or L_2 norms in parameter space. The first one is the so-called L_1-regularization (or lasso regression) with the penalty function defined as the scaled absolute deviation of the minimized subset of parameters $\mathbf{p}$ from the pre-known regularization parameters $\mathbf{p}_{\text{reg}}$ (i.e., $\Phi_{\text{reg}}(\mathbf{P}) = \lambda \cdot |\mathbf{p} - \mathbf{p}_{\text{reg}}|$) [23,24]. The second term is the so-called L_2-regularization (or ridge regression) [15,17]. In this case, the penalty function is defined as a squared deviation of the minimized parameters from the regularization values:

$$\Phi_{\text{reg}}(\mathbf{P}) = \lambda \cdot \left(\mathbf{p} - \mathbf{p}_{\text{reg}}\right)^2, \quad (12)$$

where $\lambda \geq 0$ here (and also in the L_1 case) is a regularization parameter. This free parameter determines how strongly the inverse problem in Equation (11) is penalized, i.e., it measures the "softness" of constraints. The effect of the regularization parameter choice is visualized in Figure 2. Let us assume that the LSQ inverse problem has two equivalent solutions. Adding the penalty term skews the results towards one of the two solutions in the regularized inverse problem (Equation (11)). If the regularization parameter is too small, we still may end up with two solutions, and this case is the so-called underregularized inverse problem. If the parameter λ is too large, the penalty term becomes larger than the initial experimental LSQ function ($\Phi(\mathbf{P})$), which results in the final solution of the inverse problem to be close to the regularization parameters (i.e., $\mathbf{p}_{exp} \approx \mathbf{p}_{reg}$). This case is called overregularized inverse problem. The sweet spot between underregularized and overregularized cases results in the best possible solution. However, this requires a proper choice of the regularization parameter λ. The recipes for the choice of λ are called regularization criteria, and they are widely discussed in the scientific literature [16,25].

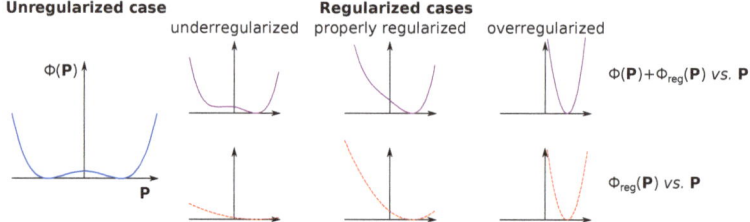

Figure 2. A schematic illustration of the effect of regularization on the ill-posed LSQ problem. The unregularized case on the left shows a pure unregularized LSQ function ($\Phi(\mathbf{P})$) with two equivalent solutions. The three regularized cases on the right show plots of the regularization function $\Phi_{reg}(\mathbf{P})$ (bottom plots) and total function, as a sum of the $\Phi(\mathbf{P}) + \Phi_{reg}(\mathbf{P})$ (top), as functions of parameter values ($\mathbf{P}$).

An alternative to regularization in the ill-posed nonlinear inverse problems is fixing some of the parameters in $\mathbf{P}$ to the pre-computed or pre-known values, e.g., by forcefully setting $\mathbf{p} = \mathbf{p}_{reg}$. This reduction of parameter space, in some cases, can make the ill-posed problem to be well-posed. Depending on the field, this approach may have different names [26], but we will call it rigid constraints. This can also be considered an extreme case of regularization, where the regularization parameter is infinite ($\lambda \to \infty$). The less widely discussed problem of the regularization and rigid constraints is if the constraints' values ($\mathbf{p}_{reg}$) do not come from the experiment but from theoretical calculations or even as arbitrary assumptions. In this case, the nonexperimental assumptions will strongly influence both the resulting parameters' values and their uncertainty estimations, as given by Equation (10), making the inverse problem solution only partially experiment-based [20,27]. In the worst case scenarios, the approach of fixing parameters can even lead to nonphysical solutions [28].

2.4. Monte-Carlo Importance Sampling of the Parameter Space

A more direct approach to obtaining a reliable estimation of the parameters than the LSQ fitting is the Monte-Carlo (MC) sampling of the associated probability distribution, describing the discrepancy between the experimental observations and the model prediction (Equation (5)) [13,14]. In this case, we try to obtain a valid representation of the probability distribution rather than using a local normal distribution (Equation (10)) to approximate the uncertainties of and correlations between the model parameters (see Figure 1).

An effective approach of MC, compared to the naive direct sampling of the whole parameters' space, is the Metropolis algorithm [29], which is a kind of Markov chain MC [30] method developed for usage in computational physics [31]. In general, the algorithm works as follows [12].

1. We start from the initial set of parameters $\mathbf{P}_{\text{ini}}$, which we assign to be the current state of the simulation $\mathbf{P}_{\text{current}}$ (i.e., we set $\mathbf{P}_{\text{current}} = \mathbf{P}_{\text{ini}}$). Generally, we can use any set of parameters for the initial guess, but a faster simulation convergence is reached if we provide initial values in the desired solution region, e.g., as the solution of the LSQ fitting problem (Equation (9)).
2. From the current state, we generate a new trial set of parameters $\mathbf{P}_{\text{trial}}$, and then we compute the transition probability $p(\text{current} \to \text{trial})$. This value describes a chance of changing our current state $\mathbf{P}_{\text{current}}$ to a new state $\mathbf{P}_{\text{trial}}$ (i.e., reassigning $\mathbf{P}_{\text{current}} = \mathbf{P}_{\text{trial}}$). The probability $p(\text{current} \to \text{trial})$ should be related to the probabilities given in Equation (5), and we will discuss it in detail further in the text (see Equation (21)).
3. Then, we draw a random value $\tilde{p} \in [0; 1)$ from a uniform distribution between 0 and 1, and compare the $\tilde{p}$ with $p(\text{current} \to \text{trial})$.
   - If $\tilde{p} \leq p(\text{current} \to \text{trial})$, then $\mathbf{P}_{\text{trial}}$ becomes the new state of the system, i.e., we reassign $\mathbf{P}_{\text{current}} = \mathbf{P}_{\text{trial}}$. This state we will call an accepted step.
   - If $\tilde{p} > p(\text{current} \to \text{trial})$, this means that the transition does not happen (we disregard the $\mathbf{P}_{\text{trial}}$). The new state of the system becomes the same old value $\mathbf{P}_{\text{current}}$. We will call this state a declined step.
4. By repeating steps #2 and #3 for a sufficient amount of iterations (N), we generate a trajectory of states $\mathbf{P}^{(n)}$, where index $n = 1, 2 \ldots, N$ denotes the state $\mathbf{P}_{\text{current}}$ at the n-th iteration of the algorithm. Naturally, some sets of parameters will be repeated multiple times throughout the trajectory. Furthermore, this trajectory $\{\mathbf{P}^{(n)}\}_{n=1}^{N}$ will encode inside the desired distribution given in Equation (5). In practice; the initial part of the trajectory (e.g., first 10% of steps) is disregarded as an equilibration phase. The acceptance ratio refers to the accepted steps in algorithm step #3 (N_{acc}) to the total number of steps ($N_{\text{tot}} = N_{\text{acc}} + N_{\text{rej}}$, where N_{rej} is the number of rejected steps). A general requirement for the simulation to be reasonably good is that this ratio should not be too big or too small. A simple rule of thumb can be that the acceptance rate $R_{\text{acc}} = 100\% \cdot N_{\text{acc}} / N_{\text{tot}}$ should be in the range $10\% \leq R_{\text{acc}} \leq 50\%$.
5. From the obtained trajectory $\{\mathbf{P}^{(n)}\}_{n=1}^{N}$, we can compute all the required parameters. For instance, the mean value of parameter P_k (from the set of parameters $\mathbf{P}$) can be computed as:

$$\langle P_k \rangle = \frac{1}{N} \sum_{n=1}^{N} P_k^{(n)}, \tag{13}$$

where $P_k^{(n)}$ is the value of P_k in the parameter set $\mathbf{P}^{(n)}$. Similarly, we can compute the standard deviation σ_k of parameter P_k from the mean value as:

$$\sigma_k^2 = \langle P_k^2 \rangle - \langle P_k \rangle^2 = \frac{1}{N} \sum_{n=1}^{N} \left(P_k^{(n)} \right)^2 - \left(\frac{1}{N} \sum_{n=1}^{N} P_k^{(n)} \right)^2. \tag{14}$$

The covariance between the parameters P_k and P_l will be given similarly, as:

$$\text{cov}(P_k, P_l) = \langle P_k P_l \rangle - \langle P_k \rangle \cdot \langle P_l \rangle =$$
$$= \frac{1}{N} \sum_{n=1}^{N} \left(P_k^{(n)} \cdot P_l^{(n)} \right) - \left(\frac{1}{N} \sum_{n=1}^{N} P_k^{(n)} \right) \cdot \left(\frac{1}{N} \sum_{n=1}^{N} P_l^{(n)} \right). \tag{15}$$

From standard deviations (Equation (14)) and covariances (Equation (15)), we can also calculate the Pearson's correlation coefficients between parameters P_k and P_l as:

$$\rho(P_k, P_l) = \frac{\text{cov}(P_k, P_l)}{\sqrt{\sigma_k \cdot \sigma_l}}. \tag{16}$$

Using this algorithm, we can effectively sample the possible solutions and provide a more general estimation of parameter values and their uncertainties.

The only aspect missing in the algorithm is the actual expression for the transition probability $p(\text{current} \to \text{trial})$. For that, we need to consider a condition of the detailed balance. In order for Equations (13)–(15) to work, the probability from Equation (5) should be inherently contained within the trajectory $\{\mathbf{P}^{(n)}\}_{n=1}^{N}$ of N steps. It means that for a given state $\mathbf{P}$, it should be contained within the trajectory $N(\mathbf{P})$ times, which should be given through the probability of this state:

$$p(\mathbf{P}) = \mathcal{N} \cdot \exp(-\Phi(\mathbf{P})) \approx \frac{N(\mathbf{P})}{N}, \tag{17}$$

according to Equation (5). Note that, in general, we do not know the normalization factor $\mathcal{N}$. Since we have a random procedure, if we take two possible parameter sets, say $\mathbf{P}$ and $\mathbf{P}'$, and consider that we have a possibility to go from one to another, and *vice versa*, these jumps between $\mathbf{P}$ and $\mathbf{P}'$ should not spoil the inherent probability in the trajectory (Equation (17)). This can be fulfilled if we balance the number of jumps from $\mathbf{P}$ to $\mathbf{P}'$, and back, to be equal. The number of transitions from $\mathbf{P}$ to $\mathbf{P}'$ ($N(\mathbf{P} \to \mathbf{P}')$) is simply the total number of points with state $\mathbf{P}$ times the transition probability $p(\mathbf{P} \to \mathbf{P}')$, that is:

$$N(\mathbf{P} \to \mathbf{P}') = N(\mathbf{P}) \cdot p(\mathbf{P} \to \mathbf{P}') = N \cdot \mathcal{N} \cdot \exp(-\Phi(\mathbf{P})) \cdot p(\mathbf{P} \to \mathbf{P}'). \tag{18}$$

A similar expression can be derived for the number of transitions from $\mathbf{P}'$ to $\mathbf{P}$:

$$N(\mathbf{P}' \to \mathbf{P}) = N(\mathbf{P}') \cdot p(\mathbf{P}' \to \mathbf{P}) = N \cdot \mathcal{N} \cdot \exp(-\Phi(\mathbf{P}')) \cdot p(\mathbf{P}' \to \mathbf{P}). \tag{19}$$

By setting $N(\mathbf{P} \to \mathbf{P}') = N(\mathbf{P}' \to \mathbf{P})$, we can assure that the MC procedure will not change the proper probability distribution. This condition is called the detailed balance. If the MC procedure follows this principle, it produces the desired trajectory with the probability encoded inside the distribution of points within the trajectory. Substitution of Equations (18) and (19) to the detailed balance results in:

$$\exp(-\Phi(\mathbf{P})) \cdot p(\mathbf{P} \to \mathbf{P}') = \exp(-\Phi(\mathbf{P}')) \cdot p(\mathbf{P}' \to \mathbf{P}). \tag{20}$$

Any transition probability $p(\mathbf{P} \to \mathbf{P}')$ fulfilling the detailed balance given in Equation (20) is suitable for the MC simulations [12]. The simplest choice for the transition probability is the following [29]:

$$p(\mathbf{P} \to \mathbf{P}') = S \cdot \min\{\exp(\Phi(\mathbf{P}) - \Phi(\mathbf{P}')), 1\} =$$
$$= S \cdot \begin{cases} 1, & \text{if } (\Phi(\mathbf{P}) - \Phi(\mathbf{P}')) \geq 0, \\ \exp(\Phi(\mathbf{P}) - \Phi(\mathbf{P}')), & \text{if } (\Phi(\mathbf{P}) - \Phi(\mathbf{P}')) < 0 \end{cases}. \tag{21}$$

Let us comment on Equation (21). The parameter $S > 0$ is just an arbitrary scaling factor that could be used to change the acceptance rate R_{acc} in the MC simulation. We will consider it for now to be $S = 1$. If $\Phi(\mathbf{P}) > \Phi(\mathbf{P}')$, i.e., the LSQ function (Equation (6)) for the new set of parameters $\mathbf{P}'$ is smaller than for the old set $\mathbf{P}$, we get $\exp(\Phi(\mathbf{P}) - \Phi(\mathbf{P}')) > 1$, which will give us the probability $p(\mathbf{P} \to \mathbf{P}') = 1$. In other words, if a trial set of parameters is better than the previous one, we definitely accept the new parameters. If the new set of parameters is worse than the old one ($\Phi(\mathbf{P}) < \Phi(\mathbf{P}')$), the probability of accepting the new configuration will be $\exp(\Phi(\mathbf{P}) - \Phi(\mathbf{P}')) < 1$, and the worse the new parameters are compared to the old ones, the less probable their acceptance will be.

As was discussed above, the optimal acceptance rate of the MC sampling (R_{acc}) should be in the range of $10\% \leq R_{\text{acc}} \leq 50\%$. This rate is being controlled by two factors: (1) generation procedure of the trial parameters $\mathbf{P}_{\text{trial}}$ and (2) transition probability (Equation (21)). Therefore, we have two ways of controlling the R_{acc} in the MC sampling procedure. First, we can adjust the trial parameters generation, e.g., reducing/increasing the size of the steps from the current configuration to increase/decrease R_{acc}, respectively. Secondly, we can decrease/increase the scaling factor S value in Equation (21) to decrease/increase R_{acc}, respectively. However, the second approach is less delicate than the first one.

3. Fitting Model of Pump–Probe Spectroscopy

3.1. General Considerations

As explained in the previous Section 2, we require a model of the experiment to solve both the direct (Equation (1)) and inverse problems (Equation (3)). The pump–probe experiment can be imagined as shown in Figure 3. It consists of multiple experiments that only differ in the delay time between pump and probe pulses (pump–probe delay, t_{pp}). With the pump pulse, we initiate the reactions, and with the probe pulse, we change the stable and metastable intermediate products of interaction into the observable results [1,2,4,18]. In all the further discussion here, we will use a convention shown in Figure 3, that is:

- If both the pump and the probe pulses act on the molecule simultaneously, then $t_{pp} = 0$ (this temporal overlap of the pump and probe pulses is also called t_0);
- If the probe pulse interacts with the molecule before the pump pulse, then $t_{pp} < 0$;
- If the probe pulse interacts with the molecule after the pump pulse, then $t_{pp} > 0$.

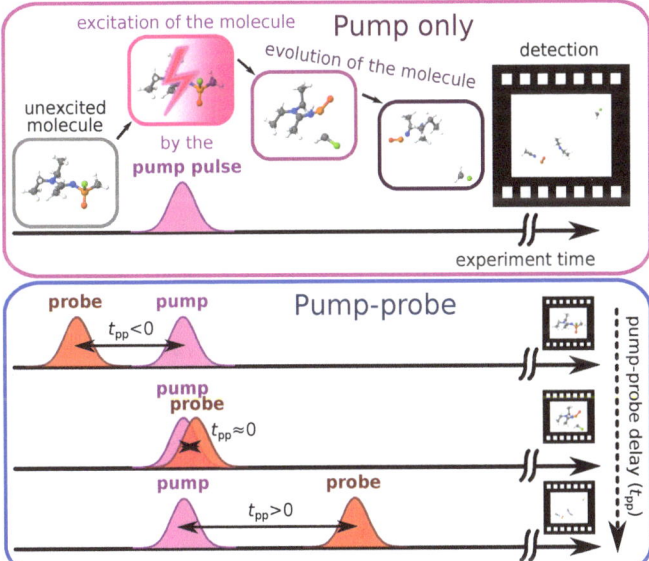

Figure 3. A schematic illustration of the idea of the pump–probe spectroscopic experiment. The (**top**) part of the image illustrates what happens in time in the experiment if only the pump pulse is given. First, we introduce the molecular system in the apparatus, then we excite our molecular system with the pump pulse, and then the photochemically induced changes happen in the system on various timescales. In the end, we collect the signal produced by the molecular system at an infinitely distant time. In the pump–probe case (**bottom**), we perform multiple experiments of such sort, but introducing the second pulse, the probe, with some delay with respect to the pump pulse (pump–probe delay, t_{pp}). The changes in the observed signal as a function of t_{pp} form the pump–probe signal, potentially carrying information of the intermediate species.

We can model all the photochemical processes happening in the pump–probe experiment with chemical kinetic equations, where we define the cross-sections of various processes, such as the interaction with light, branching ratios, and rate constants for various relaxation processes, and then integrate the resulting set of first-order differential equations in time to get our observables [32,33]. In the cases with quantum beating between different states, we also can switch from simple chemical kinetics equations to the Maxwell–Bloch equations formulated in terms of the density matrix of the molecular system [4,34]. The general shape of the equations is very similar, but in Maxwell–Bloch equations, the evolution of the off-diagonal density matrix elements is added. Both explicit dynamics approaches successfully disentangled complex dynamics in various pump–probe experiments [5,10,35,36].

However, modeling the rate dynamics explicitly is rather complicated and time-consuming. First, the cost of solving the differential equations is higher than many other numerical problems. This can be a significant drawback, especially for inverse problems where we are trying to evaluate multiple parameters. In the case of standard LSQ minimization (see Section 2.2), where we need a lot of calculations with various parameters at each minimization iteration, a slow direct model can drastically limit the number of varying parameters. It is even worse in the case of MC sampling (see Section 2.4), where we need as many steps as possible to get the best possible sampling of the parameter space. Second, solving the differential equations can be nontrivial. For instance, the rates of the intramolecular relaxation processes can be on the order of a few tens of femtoseconds, while the isomerization and formation of the fragments that are observed with the mass spectrometer can take multi-picosecond timescales [10]. Such a large variety of timescales requires a careful choice of integration techniques, especially for large pump–probe delays. Another complication for explicit kinetic modeling is the existence of parasitic processes, in which the molecular system's dynamics is initiated not by the pump but by the probe pulse and then probed by the pump pulse [6,37]. Such processes require additional integration of the kinetic equations in the "backward" time direction.

Therefore, in the cases of many observables and parameters, the model-based approaches that describe observables in terms of given functions are more advantageous for solving the inverse problem [38]. Here, we will, thus, only consider such a model approach. We will generally follow the classical work of Pedersen and Zewail [18], but try to show these well-known results from a simpler perspective and also extend the classical equations to the case of coherent oscillations in pump–probe observables [8,10].

3.2. Delta-Shaped Pump–Probe Model

3.2.1. Assumptions of the Model

First, we will consider an ideal case of the pump–probe experiment in which pump and probe pulses have zero temporal width. Both pump and probe pulses instantly convert the molecular system into something else upon interaction. We will also consider the pump–probe delay t_{pp} to be exactly known without any imperfections. In this case, we can try to devise what the results of our pump–probe experiment will be that we will observe. We will use photochemical reaction schemes and chemical kinetics equations to obtain numerical expressions for the pump–probe delay-dependent signals [39,40]. In all cases, we will assume that all the chemical reactions induced are monomolecular first-order reactions. Such an assumption is certainly true for all gas-phase experiments and usually also holds for ultrafast pump–probe processes in the medium [33,38].

3.2.2. Step Function Dynamics

First, let us consider this simple pump–probe reaction scheme:

$$\begin{cases} M + \text{pump} \to A \\ A + \text{probe} \to B \end{cases}, \tag{22}$$

where we have our initial molecule (M), which interacts with photons of the pump pulse to produce the initially unexistent stable chemical A. Then, this newly formed product A can interact with the photons of the probe pulse to produce compound B. Such a pump–probe reaction scheme is quite ubiquitous in real-life experiments. For instance, in the case of polycyclic aromatic hydrocarbons (PAHs), upon interaction with high-energy extreme ultraviolet (XUV) photons, stable mono- or dications can be formed, which then can fragment into smaller ions upon interaction with the probe pulse (see, e.g., [6,7]).

Let us now discuss what would be observed in the reaction scheme (22) for species A and B, and the number of unabsorbed probe photons. At $t_{pp} < 0$ (i.e., when the probe acts before the pump), we would see a constant amount of the product A, no molecules B, and a fully unabsorbed probe pulse. Then, for $t_{pp} > 0$, we would also see a constant amount of molecules A, but smaller than it was for $t_{pp} < 0$, since now some of the A would be lost by the second reaction in the reaction scheme (Equation (22)). A similar behavior would be observed for the probe pulse: its intensity would be constant but depleted since some of the probe photons would be lost in interaction with A. For B, we would see some constant amount of signal. At $t_{pp} = 0$, we would have an instant switch of behavior for all of the observables. Therefore, we can formalize the signal observed for all species (A, B) and the probe pulse intensity as follows:

$$f(t_{pp}) = f_0 + \begin{cases} 0, t_{pp} < 0 \\ f_1, t_{pp} \geq 0 \end{cases} = f_0 + f_1 \cdot \theta(t_{pp}), \tag{23}$$

where f_0 and f_1 are constants, f_0 indicates the initial/final amount of the given species, and f_1 characterizes the cross-section for the interaction between A and probe photons. Function $\theta(t)$ is the Heaviside step function, defined as:

$$\theta(t) = \begin{cases} 0, t < 0 \\ 1, t \geq 0 \end{cases}. \tag{24}$$

For the amount of A and intensity of the probe pulse, we will get $f_0 > 0$, $f_1 < 0$, and for B, it will be $f_0 = 0$ and $f_1 > 0$. Although, theoretically, the magnitude of f_1 for A and B in the reaction scheme (22) should be equal, in practice, the signal magnitude can be different due to the various factors, such as parasitic reactions, different detector efficiency, or even differences in the integration windows for the signal (e.g., in the mass spectra). Nevertheless, the general shape of the signals will be similar.

3.2.3. Instant Dynamics

Another simplistic pump–probe behavior is the instant dynamics, in which the observable product A can be produced from the initial molecule M only if pump and probe pulses simultaneously act on M. This type of photochemical reaction can be formally written as:

$$M + \text{pump} + \text{probe} \rightarrow A. \tag{25}$$

Processes of this type can also be found in real-life experiments. For instance, the dynamics of the formation of photoelectron side bands follows this kinetic scheme (see Refs. [6,41]). Such instant dynamics processes are useful in ultrafast experiments with X-ray free electron lasers (FELs) because they allow for precise determination of the temporal overlap between pump and probe pulses (t_0 or $t_{pp} = 0$).

Since both the pump and probe pulses in our model have zero duration, the only possible way to formally write the yield of A as a function of pump–probe delay t_{pp} is as follows:

$$f(t_{pp}) = f_0 \cdot \delta(t_{pp}), \tag{26}$$

where f_0 is a constant, characterizing the cross-section of reaction (25), and $\delta(t)$ is the Dirac delta function (i.e., $\delta(0) \rightarrow \infty$ and $\delta(t) = 0$ for $t \neq 0$).

3.2.4. Transient Pump–Probe Signatures of Metastable Species

The pump–probe studies' desirable products are the metastable species' signatures. The basic model of their dynamic behavior can be described with the following reaction scheme:

$$\begin{cases} M + \text{pump} \to A^* \\ A^* \xrightarrow{k_r} C \\ A^* + \text{probe} \to B \end{cases}, \quad (27)$$

where we again denote the initial molecule as M and the final observable species as C and B, but now we have a metastable species A^* that spontaneously converts into C with a first-order reaction rate k_r. We assume this decay rate to be too fast for A^* itself to be detectable with the experimental setup. However, while we still have A^* present in the system, we can convert it into the observable species B.

For the reaction scheme (27), we now need to consider the evolution of A^* in real experimental time, which we will denote as $t \geq 0$. At $t = 0$, we obtain $[A^*]_0$ as the initial concentration of A^*, which is created by the pump pulse. Then, at $t > 0$, A^* decays into C with the rate constant k_r. The rate equation for the concentration of A^* molecules ($[A^*] = [A^*](t)$) from the scheme (27) is written as [39,40]:

$$\frac{d[A^*]}{dt} = -k_r[A^*], \quad (28)$$

which results in the following solution (see Appendix A.1):

$$[A^*](t) = [A^*]_0 \cdot \exp(-k_r t) = [A^*]_0 \cdot 2^{-\frac{t}{\tau_{1/2}}} = [A^*]_0 \cdot \exp(-t/\tau_r), \quad (29)$$

where we provide the three most common ways to write the result, using the rate constant k_r, half-life time $\tau_{1/2} = \ln(2)/k_r$, which gives the time at which the concentration $[A^*]$ becomes half of which it was initially, and decay time:

$$\tau_r = \frac{1}{k_r} = \frac{\tau_{1/2}}{\ln(2)}, \quad (30)$$

which indicates the time at which the concentration $[A^*]$ becomes $e \approx 2.72$ times smaller than initially. In further discussions, we will only use the decay time τ_r.

Knowing the decay of A^*, we can also find the evolution of C in the experimental real time. The rate equation for the concentration of C ($[C]$) from the reaction scheme (27) is [39,40]:

$$\frac{d[C]}{dt} = +k_r[A^*]. \quad (31)$$

At the initial time, we do not have any C, and thus, $[C](0) = 0$. Integration of Equation (31) with $[A^*]$ given by Equation (29) (see Appendix A.1) is as follows:

$$[C](t) = [A^*]_0 \cdot (1 - \exp(-t/\tau_r)). \quad (32)$$

Knowing the dynamics of A^* and C (Equations (29) and (32)), we can describe how the change in yields of C and B would behave as a function of the pump–probe delay t_{pp}. At $t_{pp} < 0$, A^* is produced after the probe pulse has already passed the system. Thus, the total amount of B is zero. As for C, we assume that the time of detection by the experimental instrument t_{instr} is infinite compared to the internal dynamics time τ_r ($t_{instr} \gg \tau_r$). This means that the amount of registered C molecules at t_{instr} from Equation (32) is $[C](t_{instr}) \approx [A^*]_0$, i.e., all A^* intermediate is fully converted into C before the detection. At $t_{pp} \geq 0$, the probe pulse will instantly convert part of A^* into B according to the last equation in the reaction scheme (27). If we denote the conversion efficiency of the probe pulse interaction as $0 \leq p \leq 1$, then the resulting concentration of B is $[B] = p \cdot [A^*](t_{pp}) = p \cdot [A^*]_0 \cdot \exp(-t_{pp}/\tau_r)$. The amount of C at $t_{pp} \geq 0$, that we will register at t_{instr}, will be the difference between the full conversion result and the part of A^* lost upon conversion into B by the probe, i.e., $[C](t_{instr}, t_{pp}) = [A^*]_0 - [A^*](t_{pp}) = [A^*]_0 \cdot (1 - p \cdot \exp(-t_{pp}/\tau_r))$.

We can combine the described yields of B and C as functions of pump–probe delay t_{pp} using the Heaviside function (Equation (24)) as:

$$\begin{cases} [B](t_{pp}) = [A^*]_0 \cdot p \cdot \theta(t_{pp}) \exp(-t_{pp}/\tau_r), \\ [C](t_{pp}) = [A^*]_0 \cdot (1 - p \cdot \theta(t_{pp}) \exp(-t_{pp}/\tau_r)) \end{cases} \quad (33)$$

The yield of C resembles the formation kinetics of C in real experimental time (Equation (32)), while for B it resembles the real experimental time dynamics of the unstable intermediate A^* (Equation (29)). Nevertheless, we can summarize both of these dependencies with a general expression of the form:

$$f(t_{pp}) = f_0 + f_1 \cdot \theta(t_{pp}) \exp(-t_{pp}/\tau_r), \quad (34)$$

where f_0 and f_1 are again constants proportional to the cross-section of the pump/probe photons interacting with the molecular species.

Equation (34) has an intriguing similarity with Equation (23), describing the pump–probe dynamics according to reaction scheme (22). This similarity is not a coincidence, since if A^* is stable (i.e., $k_r = 0$, or $\tau_r \to \infty$), the reaction scheme (27) collapses into the reaction scheme (22). Equation (34) will be transferred into Equation (23), if $|t_{pp}| \ll \tau_r$, since in this case $\exp(-t_{pp}/\tau_r) \approx 1$. Similarly, the reaction scheme transforms into reaction scheme (25) if A^* is too unstable (i.e., $k_r \to \infty$, or $\tau_r \to 0$). In this case, the decay exponent becomes localized near $t_{pp} = 0$, which can be approximated by Equation (26).

3.2.5. Coherent Oscillations without Decay

We worked with standard chemical kinetics equations in the previous model (Equation (27)). However, if we want to discuss periodic oscillation features that are being observed in some pump–probe experiments [8,10,42], such semiclassical description turns out to be insufficient, and a quantum-mechanical model has to be used. Here, we will provide the simplest model for such behavior based on a two-level quantum system. First, we will discuss a basic model without the decay dynamics, and then we will modify our model to include the decay effects [43,44].

Let us imagine that our molecular system is described with a Hamiltonian $\hat{H}$, which has eigenstates $|0\rangle$ and $|1\rangle$, that are solutions to the stationary Schrödinger equation $\hat{H}|k\rangle = E_k|k\rangle$ ($k = 0, 1$). These states will be considered to be orthonormal, i.e., $\langle k|l\rangle = \delta_{kl}$ for $k, l = 0, 1$. We will take the energy E_0 of the ground state $|0\rangle$ as a reference, i.e., as zero ($E_0 = 0$). The energy of the excited state $|1\rangle$ will be denoted as $E_1 = \hbar\omega$, where $\hbar$ is the reduced Planck constant and ω is the angular frequency of the photon that provides excitation from the ground to the excited state ($|0\rangle \to |1\rangle$). Note that ω is related to the normal frequency ν as $\omega = 2\pi\nu$. Now, we will consider the evolution of this system in real-time t with explicit inclusion of the effects of the pump and probe pulses.

Suppose now that the system was initially in the ground state, i.e., before the pump pulse acted on the system, the wavefunction of the system was as follows:

$$|\psi_{ini}\rangle = |0\rangle. \quad (35)$$

After instant action, the pump pulse at real-time $t = 0$ has created a superposition state, transferring some population to the excited state. The new state of the system right after the pump pulse at $t = 0$ is described as:

$$|\psi(0)\rangle = c_0|0\rangle + c_1 \exp(-i\varphi)|1\rangle, \quad (36)$$

where c_0 and c_1 are the coefficients, showing the amount of the system in the ground and excited states as $|c_0|^2$ and $|c_1|^2$, respectively, with a condition of $|c_0|^2 + |c_1|^2 = 1$, and φ is the phase of the excited state related to the ground state, which is imposed, e.g., by the phase of the excitation pump pulse [45,46].

To propagate the dynamics of the state after interaction with the pump pulse (Equation (36)), let us switch to the density matrix formalism. The density matrix $\hat{\rho}$ for our two-level system in the basis of states $|0\rangle$ and $|1\rangle$ is written as:

$$\hat{\rho} = \rho_{00}|0\rangle\langle 0| + \rho_{01}|0\rangle\langle 1| + \rho_{10}|1\rangle\langle 0| + \rho_{11}|1\rangle\langle 1|, \tag{37}$$

where the coefficients ρ_{kl} ($k,l = 0,1$) describe the state of the system. This form can be rewritten in the form of an actual matrix by placing the coefficients accordingly as:

$$\varrho = \begin{pmatrix} \rho_{00} & \rho_{01} \\ \rho_{10} & \rho_{11} \end{pmatrix}. \tag{38}$$

The elements of this matrix should fulfill two requirements. First, the trace of this matrix should be equal to one ($\mathrm{tr}(\varrho) = \rho_{00} + \rho_{11} = 1$). Second, the matrix ϱ should be Hermitian, which means $\rho_{01} = \rho_{10}^*$.

The density matrix $\hat{\rho}$ for our system at time $t = 0$ can be obtained from the initial state $|\psi(0)\rangle$ (Equation (36)) as:

$$\hat{\rho}(0) = |\psi(0)\rangle\langle\psi(0)| = \\ = \underbrace{|c_0|^2}_{\rho_{00}(0)}|0\rangle\langle 0| + \underbrace{c_0 c_1^* \exp(+i\varphi)}_{\rho_{01}(0)}|0\rangle\langle 1| + \underbrace{c_0^* c_1 \exp(-i\varphi)}_{\rho_{10}(0)}|1\rangle\langle 0| + \underbrace{|c_1|^2}_{\rho_{11}(0)}|1\rangle\langle 1|. \tag{39}$$

Let us consider c_0 and c_1 as real values, concealing all the complex behavior into the phase φ. We can also denote $c_1^2 = p$ and $c_0^2 = (1-p)$, where $0 \leq p \leq 1$ is the efficiency of the pump pulse excitation $|0\rangle \to |1\rangle$. In this case, the initial parameters of the matrix from Equation (38) will be:

$$\begin{cases} \rho_{00}(0) = 1 - p, \\ \rho_{01}(0) = \sqrt{p \cdot (1-p)} \exp(+i\varphi), \\ \rho_{10}(0) = \sqrt{p \cdot (1-p)} \exp(-i\varphi), \\ \rho_{11}(0) = p. \end{cases} \tag{40}$$

As can be seen, the diagonal elements (ρ_{00} and ρ_{11}) encode the population in a given state, and the off-diagonal elements (ρ_{01} and ρ_{10}) encode coherence between the levels.

The density matrix ϱ evolves according to the von Neumann equation [44]:

$$i\hbar \frac{d\varrho}{dt} = [\mathcal{H}, \varrho], \tag{41}$$

where $[a,b] = ab - ba$ is the commutator and $\mathcal{H}$ is the Hamiltonian matrix composed of the elements $\langle k|\hat{H}|l\rangle$, which in our basis of orthonormal eigenstates looks as follows:

$$\mathcal{H} = \begin{pmatrix} \langle 0|\hat{H}|0\rangle & \langle 0|\hat{H}|1\rangle \\ \langle 1|\hat{H}|0\rangle & \langle 1|\hat{H}|1\rangle \end{pmatrix} = \begin{pmatrix} 0 & 0 \\ 0 & \hbar\omega \end{pmatrix}. \tag{42}$$

Substituting the density matrix (Equation (38)) and the Hamiltonian matrix (Equation (42)) into the von Neumann equation (Equation (41)), we find the following equations for the evolution of each of the density matrix elements (details in Appendix A.2):

$$\begin{cases} \frac{d\rho_{00}}{dt} = 0, \\ \frac{d\rho_{11}}{dt} = 0, \\ \frac{d\rho_{01}}{dt} = +i\omega\rho_{01}, \\ \frac{d\rho_{10}}{dt} = -i\omega\rho_{10}. \end{cases} \quad (43)$$

Solutions of these equations (see in Appendix A.2) with initial conditions given in Equation (40) are:

$$\begin{cases} \rho_{00}(t) = 1 - p, \\ \rho_{11}(t) = p, \\ \rho_{01}(t) = \sqrt{p \cdot (1-p)} \cdot \exp(+i\varphi + i\omega t), \\ \rho_{10}(t) = \sqrt{p \cdot (1-p)} \cdot \exp(-i\varphi - i\omega t). \end{cases} \quad (44)$$

These results describe the state of the free-evolving molecular system at experiment times $t \geq 0$ after the pump excitation. Based on this solution, we want to numerically quantify the observables that we will get upon interaction with the probe pulse.

Let us consider that observables $\mathcal{O}$ in quantum mechanics are given in the form of operators $\hat{O}$. Therefore, we can assume that the result of measurement by the probe will be given by an observable operator $\hat{O}$. In the case of our two-level system, we can represent this operator in the matrix form similar to the Hamiltonian (Equation (42)):

$$\mathcal{O} = \begin{pmatrix} \mathcal{O}_{00} & \mathcal{O}_{01} \\ \mathcal{O}_{10} & \mathcal{O}_{11} \end{pmatrix}, \quad (45)$$

where the matrix elements are the following integrals:

$$\mathcal{O}_{kl} = \langle k|\hat{O}|l\rangle, \quad k, l = 0, 1. \quad (46)$$

We will assume that all these integrals are real, and thus, $\mathcal{O}_{01} = \mathcal{O}_{10}$.

The mean result of the observable measurement $\langle \mathcal{O} \rangle$ of the system described by the density matrix ϱ (Equation (38)) is given as a trace of the product between matrices $\mathcal{O}$ and ϱ:

$$\langle \mathcal{O} \rangle = \mathrm{tr}(\mathcal{O}\varrho) = \mathcal{O}_{00}\rho_{00} + \mathcal{O}_{01}\rho_{10} + \mathcal{O}_{10}\rho_{01} + \mathcal{O}_{11}\rho_{11}. \quad (47)$$

By measuring the state described by the density matrix with elements from Equation (44) in real time equal to the pump–probe delay ($t = t_{pp}$), and taking into account that $\mathcal{O}_{01} = \mathcal{O}_{10}$ as well as Euler's formula ($\exp(ix) = \cos(x) + i\sin(x)$), we obtain the pump–probe signal at $t_{pp} \geq 0$ to be:

$$\langle \mathcal{O} \rangle(t_{pp}) = \mathcal{O}_{00} \cdot \overbrace{\rho_{00}(t_{pp})}^{(1-p)} + \mathcal{O}_{01} \cdot \overbrace{\rho_{10}(t_{pp})}^{\sqrt{p \cdot (1-p)} \cdot \exp(-i\varphi - i\omega t_{pp})} +$$
$$+ \overbrace{\mathcal{O}_{10}}^{\mathcal{O}_{01}} \cdot \overbrace{\rho_{01}(t_{pp})}^{\sqrt{p \cdot (1-p)} \cdot \exp(+i\varphi + i\omega t_{pp})} + \mathcal{O}_{11} \cdot \overbrace{\rho_{11}(t_{pp})}^{p} =$$
$$= \underbrace{\mathcal{O}_{00} \cdot (1-p) + \mathcal{O}_{11} \cdot p}_{f_1} + \underbrace{2\mathcal{O}_{01}\sqrt{p \cdot (1-p)}}_{f_2} \cdot \cos(\omega t_{pp} + \varphi) =$$
$$= f_1 + f_2 \cdot \cos(\omega t_{pp} + \varphi) \quad (48)$$

At pump–probe delay times $t_{pp} < 0$ (i.e., when the probe acts before the pump), the probe pulse will observe the initial state of the system (Equation (35)), which can be represented (similarly to that in Equation (39)) by a density matrix:

$$\varrho_{ini} = \begin{pmatrix} 1 & 0 \\ 0 & 0 \end{pmatrix}, \quad (49)$$

which upon substitution to Equation (47) will provide us with probe results at delays $t_{pp} < 0$:

$$\langle \mathcal{O} \rangle (t_{pp}) = \text{tr}(\mathcal{O} \varrho_{\text{ini}}) = \mathcal{O}_{00} = f_0 \,. \tag{50}$$

Now, we can merge the results of the pump–probe experiment result $(f(t_{pp}) = \langle \mathcal{O} \rangle (t_{pp}))$ for our two-level molecular system for pump–probe delays $t_{pp} < 0$ (Equation (50)) and $t_{pp} \geq 0$ (Equation (48)) to obtain:

$$f(t_{pp}) = f_0 + f_1 \cdot \theta(t_{pp}) + f_2 \cdot \theta(t_{pp}) \cdot \cos(\omega t + \varphi) \,. \tag{51}$$

In this equation, the coefficients f_k ($k = 0, 1, 2$) are again proportional to the cross-sections of pump/probe interaction with the molecules, the oscillation frequency ω encodes the energy difference between two coherent states $|0\rangle$ and $|1\rangle$ through the Planck relation $E_1 - E_0 = \hbar \omega$, and initial phase of the oscillation φ is an imprint of the pump pulse's phase. In the incoherent regime of the system's excitation, the oscillating term disappears, and Equation (51) converts into the classical Equation (23) (see Appendix A.3 for details). Such correspondence between quantum and classical cases is not a coincidence: both cases refer to the same reaction scheme (22), wherein the quantum case, M is $|0\rangle$, A is $|1\rangle$, and instead of providing a concrete yield of observable, we use a more generic treatment of the probe with the operator $\hat{O}$.

3.2.6. Coherent Oscillations with Decay

Now, we will discuss the dynamics of the two-level system in which the excited state can decay back into the ground state after the excitation. The derivation procedure will be the same as in the previous Section 3.2.5. Therefore, we only highlight the changes that will lead to a new result.

We start with the same system described generally by a 2×2 density matrix (Equation (38)). Before the pump, the molecular system is described by a wavefunction from Equation (35) and right after the pumping, by a wavefunction from Equation (36). Equivalently, they are given by a density matrix from Equations (40) and (49), respectively. Therefore, we again need to propagate the pumped state.

To do the propagation, we will replace the von Neumann Equation (41) with its modified version, the Lindblad equation, which considers the decay between states. In our case, we can represent it as follows [44,47]:

$$i\hbar \frac{d\varrho}{dt} = [\mathcal{H}, \varrho] + i\hbar \gamma \left(\sigma^- \varrho \sigma^+ - \frac{1}{2} \{\sigma^+ \sigma^-, \varrho\} \right), \tag{52}$$

where the first part of the equation is the same as in Equation (41), and in the added decay term, γ is the rate of the decay, $\{a, b\} = ab + ba$ is the anticommutator, and the $\sigma^\pm$ matrices are the excitation/deexcitation operators of the following form [44,47]:

$$\sigma^+ = \begin{pmatrix} 0 & 0 \\ 1 & 0 \end{pmatrix} \quad \text{and} \quad \sigma^- = \begin{pmatrix} 0 & 1 \\ 0 & 0 \end{pmatrix}. \tag{53}$$

We can rewrite Equation (52) for our system (similarly to Equation (43)) as follows (see Appendix A.4):

$$\begin{cases} \frac{d\rho_{00}}{dt} = +\gamma \rho_{11}, \\ \frac{d\rho_{01}}{dt} = +i\omega \rho_{01} - \frac{1}{2}\gamma \rho_{01}, \\ \frac{d\rho_{10}}{dt} = -i\omega \rho_{10} - \frac{1}{2}\gamma \rho_{10}. \\ \frac{d\rho_{11}}{dt} = -\gamma \rho_{11}. \end{cases} \tag{54}$$

Solving these equations with initial conditions given in Equation (40) results in the following results:

$$\begin{cases} \rho_{00}(t) = 1 - p \cdot \exp(-\gamma t), \\ \rho_{11}(t) = p \cdot \exp(-\gamma t), \\ \rho_{01}(t) = \sqrt{p \cdot (1-p)} \cdot \exp(+i\varphi + i\omega t) \cdot \exp\left(-\frac{\gamma t}{2}\right), \\ \rho_{10}(t) = \sqrt{p \cdot (1-p)} \cdot \exp(-i\varphi - i\omega t) \cdot \exp\left(-\frac{\gamma t}{2}\right), \end{cases} \quad (55)$$

where we obtain decay dynamics with the rate $k_r = \gamma$. Therefore, to be in line with the previous notation, we will replace γ with the decay time τ_r according to Equation (30).

Now, substituting the density matrix elements from Equation (55) into Equation (47), describing the observable at $t_{pp} \geq 0$, we obtain a following analog of Equation (48):

$$\langle \mathcal{O} \rangle(t_{pp}) = \overbrace{\mathcal{O}_{00}}^{f_1} + \overbrace{p \cdot (\mathcal{O}_{00} + \mathcal{O}_{11})}^{f_2} \cdot \exp\left(-\frac{t_{pp}}{\tau_r}\right) +$$

$$+ \underbrace{2\mathcal{O}_{01}\sqrt{p \cdot (1-p)}}_{f_3} \cdot \cos(\omega t_{pp} + \varphi) \cdot \exp\left(-\frac{t_{pp}}{2\tau_r}\right) =$$

$$= f_1 + f_2 \cdot \exp\left(-\frac{t_{pp}}{\tau_r}\right) + f_3 \cdot \cos(\omega t_{pp} + \varphi) \cdot \exp\left(-\frac{t_{pp}}{2\tau_r}\right). \quad (56)$$

The behavior of this system at delays $t_{pp} < 0$ stays the same as before (Equation (50)). By combining Equations (50) and (56), we get the coherent decay dynamics observables in the pump–probe domain in the following form:

$$f(t_{pp}) = f_0 + f_1 \cdot \theta(t_{pp}) + f_2 \cdot \theta(t_{pp}) \cdot \exp\left(-\frac{t_{pp}}{\tau_r}\right) +$$

$$+ f_3 \cdot \theta(t_{pp}) \cdot \cos(\omega t + \varphi) \cdot \exp\left(-\frac{t_{pp}}{2\tau_r}\right). \quad (57)$$

This equation closely reminds us of Equation (51), which can be obtained again in the limit of $\tau_r \to \infty$ ($\gamma = 0$). At the same time, coherent oscillation dynamics resemble a pump–probe yield from Equation (34), which is restored in the classical limit (see Appendix A.3). This is again no coincidence since the currently discussed pump–probe system is a modification of the reaction scheme (27) with M being $|0\rangle$ and A^* being $|1\rangle$. The difference between our two-level quantum model and rate scheme (27) is again in a generic view on the probing, but also in the decay of A^* back to M ($A^* \to M$ instead of $A^* \to C$ in scheme (27)). We could also include the third state emulating C. This would require extending the two-level system to three levels. However, in the three-level system, the pump–probe observables will still be described with Equation (57).

3.2.7. More Complicated Dynamics Models

Now, let us take a look at more complicated reaction models for the pump–probe dynamics than we have looked at before (Equations (22), (25) and (27)). The three extensions can be seen in Figure 4, where scheme (a) shows a possibility of branching reactions when a single metastable intermediate A^* can result in multiple products, scheme (b) demonstrates the possibility of having multiple interconverting metastable intermediates, and scheme (c) shows how multiple pathways can produce the same product. However, if we evaluate the observables from these schemes in the pump–probe domain, we see that all of them can be described with the following expression (see Appendixes A.5–A.7):

$$f(t_{pp}) = f_0 + f_1 \cdot \theta(t_{pp}) + f_2 \cdot \theta(t_{pp}) \cdot \exp\left(-\frac{t}{\tilde{\tau}_2}\right) + f_3 \cdot \theta(t_{pp}) \cdot \exp\left(-\frac{t}{\tilde{\tau}_3}\right), \quad (58)$$

where $\tilde{\tau}_i$ ($i = 2,3$) are effective rate constants computed from original elementary rate constants $\tau_{r,j}$ and f_i ($0 \leq i \leq 3$) are effective parameters, dependent on the pump/probe interaction cross-sections and on the reaction scheme.

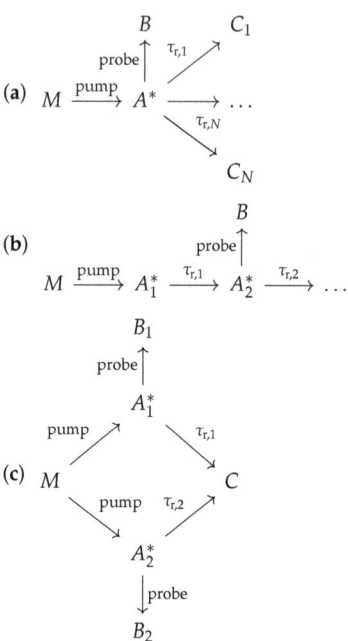

Figure 4. Three examples of pump–probe dynamics reaction schemes, extending the basic ones given in Equations (22), (25) and (27): (**a**) reaction scheme with multiple outcomes from the same intermediate metastable state; (**b**) pump–probe scheme with sequential decay of the intermediate states; (**c**) example of processes with multiple independent intermediate states that lead to the formation of independent and same observables. Solutions for schemes (**a**)–(**c**) are given in Appendixes A.5–A.7.

The apparent simplicity of Equation (58) can be considered a lucky coincidence for pre-designed schemes, but it is not. In fact, it can be explicitly shown (see Appendix A.8) that if the reaction scheme for the pump-induced dynamics consists only of first-order reactions (i.e., of the type $A \to C_1 + C_2 + \ldots$) and the probing dynamics is given only as instant interconversion between species (i.e., of the type $A + \text{pump} \to B$), then the pump–probe yield for any of the products is given as:

$$f(t_{pp}) = f_0 \cdot 1 + f_1 \cdot \theta(t_{pp}) + \sum_{i \geq 2} f_i \cdot \theta(t_{pp}) \cdot \exp\left(-\frac{t}{\tau_{r,i}}\right), \tag{59}$$

where $\tau_{r,i}$ ($i = 2, \ldots$) are some effective rate constants composed of the rate constants for individual reactions and the sum over $i = 2, \ldots$ covers all the effective decay pathways.

We can generalize Equation (59) in the following form:

$$f(t_{pp}) = \sum_{i=1}^{N} f_i \cdot b_i(t_{pp}), \tag{60}$$

which is just a linear combination of N linearly independent basis functions $b_i(t_{pp})$ with coefficients f_i. Equation (59) has the following three types of basis functions:

- The first type is simply the constant ("c") defined as:

$$b_c(t_{pp}) = 1.\quad (61)$$

This function (with coefficient f_0 in Equation (59)) has no parameters and describes the background of the pump–probe experiment.
- The second type is the step or "switch" function ("s"):

$$b_s(t_{pp}) = \theta(t_{pp}).\quad (62)$$

This basis function (with coefficient f_1 in Equation (59)) describes the switching of the background between $t_{pp} < 0$ and $t_{pp} \geq 0$ regimes.
- The third type is the transient ("t") function:

$$b_t(t_{pp}) = \theta(t_{pp}) \cdot \exp\left(-\frac{t}{\tau_r}\right).\quad (63)$$

This type of basis function (with coefficients f_i, $i \geq 2$ in Equation (59)) describes the pump-induced decay dynamics, and it depends on a parameter τ_r, which is an effective decay time.

We can augment the three types of basis functions (Equations (61)–(63)) with three additional functions.
- First, it is the instant ("i") dynamics (Equation (25)) found in Equation (26):

$$b_i(t_{pp}) = \delta(t_{pp}).\quad (64)$$

This type of dynamics describes unresolvably fast relaxation dynamics.
- The second additional function, describing nondecaying coherent oscillation ("o"), can be taken from Equation (51):

$$b_o(t_{pp}) = \theta(t_{pp}) \cdot \cos(\omega t + \varphi).\quad (65)$$

This basis function has two parameters: the oscillation frequency ω and the initial phase φ.
- The last additional function, describing a transient coherent oscillation ("to"), can be taken from Equation (57):

$$b_{to}(t_{pp}) = \theta(t_{pp}) \cdot \exp\left(-\frac{t}{2\tau_r}\right) \cdot \cos(\omega t + \varphi).\quad (66)$$

This basis function has three parameters: the oscillation frequency ω, the initial phase φ, and the decay time τ_r.

Using these six basis functions, b_c, b_s, b_t, b_i, b_o, and b_{to} given with Equations (61)–(66), we can describe any pump–probe observable using expression (60). In the previous discussion, we assumed that the pump pulse only initialized the dynamics and that the probe pulse only changed the species produced with the pump. However, this is not always the case: sometimes the probe pulse can initiate some processes, and the pump can probe it, i.e., the probe acts similar to the pump, and *vice versa* [6,37]. These cases can be easily described with the same dynamic equations by simply inverting the t_{pp}, e.g., by using $b_t(-t_{pp})$ instead of $b_t(t_{pp})$. Therefore, the proper basis set functions in Equation (60) are given as $b_x^\pm$ with:

$$b_i(t_{pp}) = \begin{cases} b_x^+(t_{pp}) = b_x(t_{pp}), \\ b_x^-(t_{pp}) = b_x(-t_{pp}), \end{cases} x = c, s, t, i, o, to.\quad (67)$$

In other words, each basis function requires two identifiers: index "x", which selects one of the basis function types from Equations (61)–(66), and index $\pm$, which sorts between the cases of pump acting as a pump (and probe acting as a probe, denoted with "+") and probe acting as a pump (and pump acting as a probe, denoted with "−"). However, we must note that the index "±" is essentially useless for the basis types of b_c and b_i (Equations (61) and (64)), since these functions are symmetric with respect to the replacement of t_{pp} with $-t_{pp}$.

3.3. Accounting for Finite Duration of the Pulses and Experiment Jitters

After sorting out the basic model for the pump–probe dynamics given by Equation (60) with basis functions described in Equation (67), we are ready to account for deviations of the dynamics from delta-shaped pump–probe measurements. There are two main reasons why real experimental observations do not exactly follow the trends described in Section 3.2 [41,48–50]:

- The real pulses are not delta-shaped but have a finite duration.
- Real experimental setups have fluctuations (jitters) of the pump–probe delay, arising from different physical processes.

The first reason (pulse durations) is inherent to all pump–probe experiments and can be further separated into the pump and probe pulses' durations. The second reason (experiment jitter) can have multiple sources and strongly depends on the experimental setup. Such jitters are especially important for the experiments at the FELs, where optical lasers are used for pumping/probing, because it is technically challenging to keep two separate setups of tens to thousands of meters in size synchronized [49].

To account for these fluctuations, the following approach can be used. Suppose we have a pump–probe observable in a perfect experiment with dynamics given by function $f(t_{pp})$. However, due to various fluctuations, the processes can be initiated imperfectly at times $t_{pp} - t_i$ with a probability p_i, where t_i is the offset from the ideal pump–probe delay times. Let us assume that we have N fixed possible offsets, and the probability is normalized as $\sum_{i=1}^{N} p_i = 1$. Therefore, the actual observed measurement result ($F(t_{pp})$) of an experimental system at a given pump–probe delay t_{pp} will be given as:

$$F(t_{pp}) = \sum_{i=1}^{N} f(t_{pp} - t_i) \cdot p_i . \qquad (68)$$

We can replace the discrete distribution $\{p_i\}_{i=1}^{N}$ with N outcomes by a continuous distribution $p(t)$ ($\int_{-\infty}^{+\infty} p(t)dt = 1$). Then, by replacing the sum in Equation (68) over offset times t, we get the corrected observable to be given as the convolution $f \circledast p$ of the observable f with the pump–probe delay fluctuation distribution:

$$F(t_{pp}) = \int_{-\infty}^{+\infty} f(t_{pp} - t) \cdot p(t)dt = f \circledast p . \qquad (69)$$

When we have more than one (say, N) contributing factors for the offset described with independent distributions $p_i(t)$ ($i = 1, \ldots, N$), Equation (68) can be extended to multiple convolutions:

$$F(t_{pp}) = f \circledast p_1 \circledast \ldots \circledast p_N = f \circledast_{i=1}^{N} p_i =$$

$$= \int_{-\infty}^{+\infty} \ldots \int_{-\infty}^{+\infty} f\left(t_{pp} - \sum_{i=1}^{N} t_i\right) \cdot \left(\prod_{i=1}^{N} p_i(t_i)\right) dt_1 \ldots dt_N . \qquad (70)$$

In general, Equation (70) requires a specific evaluation for each given type of distribution p_i. Therefore, to produce a workable analytical expression, we assume that all our pump–probe fluctuation distributions p_i are simply normal distributions of the form:

$$p(t) = \frac{1}{\sqrt{2\pi}\sigma} \exp\left(-\frac{t^2}{2\sigma^2}\right) = \frac{2\sqrt{\ln(2)}}{\sqrt{\pi} \cdot \text{FWHM}} \exp\left(-\frac{4\ln(2) \cdot t^2}{\text{FWHM}^2}\right) =$$
$$= \frac{1}{\sqrt{\pi}\tau} \exp\left(-\frac{t^2}{\tau^2}\right). \quad (71)$$

We choose the expectation value for each of the distributions p_i to be zero. This means that if there is a systematic shift of the pump–probe delay t_{pp}, we account for it before performing the convolution, shifting the position of the temporal overlap between pulses t_0 into a new value. There are three equivalent alternative forms of normal distribution in Equation (71), which are defined through the standard deviation $\sigma = \text{FWHM}/(2\sqrt{2\ln(2)}) = \tau/\sqrt{2}$, full width at half-maximum ($\text{FWHM} = 2\sqrt{2\ln(2)} \cdot \sigma = 2\sqrt{\ln(2)} \cdot \tau$), and effective width $\tau = \sqrt{2} \cdot \sigma = \text{FWHM}/(2\sqrt{\ln(2)})$. Note that one should always pay attention to which of these forms is used to define the pump/probe pulse width and jitter parameters. Further in the text, we will use the latter form, defined with τ, i.e., each distribution p_i is characterized by its width τ_i.

By choosing all the distributions p_i in Equation (70) to be normal distributions, defined by Equation (71), evaluation of the multiple convolutions become simple and results in a final expression of the form (see Appendix B.1 for proof):

$$F(t_{pp}) = f \circledast \prod_{i=1}^{N} p_i = f \circledast p = \frac{1}{\sqrt{\pi}\tau_{cc}} \int_{-\infty}^{+\infty} f(t_{pp} - t) \cdot \exp\left(-\frac{t^2}{\tau_{cc}^2}\right) dt, \quad (72)$$

where $p(t)$ is an effective normal distribution (Equation (71)) with width τ_{cc} defined as:

$$\tau_{cc}^2 = \sum_{i=1}^{N} \tau_i^2. \quad (73)$$

This effective width of the pump–probe delay fluctuation is called cross-correlation time or instrument response function [10,38], and it is an effective measure of the pump and probe pulses' duration and the instrumental jitter.

To calculate the cross-correlation time given in Equation (73), we require three basic components:

- Pump pulse duration τ_{pump}.
- Probe pulse duration τ_{probe}.
- Instrument jitter magnitude. τ_{jitter}.

The last component of cross-correlation time (τ_{jitter}) can itself be a composite value by a similar expression to Equation (73).

To combine these three values (τ_{pump}, τ_{probe}, and τ_{jitter}) into the cross-correlation time τ_{cc} (Equation (73)), we need to consider a previously ignored issue, namely the number of pump and probe photons used to form the observable. Let us assume that our pump and probe stages (similar to reaction schemes (4), (22), (25), and (27)) are given by the three equations:

$$\begin{cases} M + n_{pump} \times \hbar\omega_{pump} \to A_1^* \\ A_1^* \to \ldots \to A_i^* \to \ldots \\ A_i^* + n_{probe} \times \hbar\omega_{probe} \to B \end{cases}, \quad (74)$$

where ω_{pump} and ω_{probe} denote the angular frequencies of the pump and probe photons, and n_{pump} and n_{probe} denote the numbers of pump and probe photons used in the pump/probe photochemical reactions.

The probability of forming the product is proportional to the light intensity to the power of the number of photons [51]. Therefore, in the case of pump and probe laser pulses, we need to convolute the pump–probe response with $p_{pump}^{n_{pump}}$ and $p_{probe}^{n_{probe}}$, respectively.

These functions are also normal distributions (Equation (71)) but with the effective widths $\tau_{\text{pump, eff}} = \tau_{\text{pump}}/\sqrt{n_{\text{pump}}}$ and $\tau_{\text{probe, eff}} = \tau_{\text{probe}}/\sqrt{n_{\text{probe}}}$. Therefore, upon combination of factors of fluctuating pump–probe delay t_{pp} we will get the cross-correlation time (Equation (73)) in the following form:

$$\tau_{\text{cc}}^2 = \frac{\tau_{\text{pump}}^2}{n_{\text{pump}}} + \frac{\tau_{\text{probe}}^2}{n_{\text{probe}}} + \tau_{\text{jitter}}^2. \tag{75}$$

Now, we can apply the transformation from Equation (72) to the observables for pump–probe experiments with ideal instant excitations (Equation (60)) to produce the adequate model for the real-life experimental observables:

$$F(t_{\text{pp}}) = f \circledast p = \sum_{i=1}^{N} f_i \cdot \overbrace{(b_i \circledast p)}^{B_i(t_{\text{pp}})} = \sum_{i=1}^{N} f_i \cdot B_i(t_{\text{pp}}). \tag{76}$$

The resulting equation is similar to the initial (Equation (60)), in which the basis functions $b_i(t_{\text{pp}})$, given in Equation (67), are replaced with their counterparts $B_i(t_{\text{pp}})$, which are convolutions of $b_i(t_{\text{pp}})$ with the effective distribution $p(t)$ (Equation (72)) with effective width given as cross-correlation time (Equation (73)).

Since the effective Gaussian distribution is symmetric with respect to time inversion ($t \to -t$), the resulting basis functions are given similar to Equation (67):

$$B_i(t_{\text{pp}}) = \begin{cases} B_x^+(t_{\text{pp}}) = (b_x \circledast p)(t_{\text{pp}}) & = B_x(t_{\text{pp}}), \\ B_x^-(t_{\text{pp}}) = (b_x \circledast p)(-t_{\text{pp}}) & = B_x(-t_{\text{pp}}), \end{cases} \quad x = c, s, t, i, o, to. \tag{77}$$

Therefore, to apply Equation (76) for real experimental data fitting, we need to find expressions for the six types of the basis functions: B_c, B_s, B_t, B_i, B_o, and B_{to}, which are convoluted analogs of the six basis functions b_c, b_s, b_t, b_i, b_o, and b_{to} given with Equations (61)–(66).

The actual evaluation of all six basis functions is provided in Appendix B.2. Here, we will only give their final expressions and their visualization (Figure 5). The physical meanings of these expressions are the same as for their idealized counterparts (see comments for Equations (61)–(66)).

- The first function is the constant ("c") function:

$$B_c(t_{\text{pp}}) = 1. \tag{78}$$

- The second type is the step function ("s"):

$$B_s(t_{\text{pp}}) = \frac{1}{2} \cdot \left(1 + \text{erf}\left(\frac{t_{\text{pp}}}{\tau_{\text{cc}}}\right)\right), \tag{79}$$

where $\text{erf}(x) = (2/\sqrt{\pi}) \cdot \int_0^x \exp(-q^2) dq$ is the error function.
- The third type is the transient ("t") function:

$$B_t(t_{\text{pp}}) = \frac{1}{2} \exp\left(\frac{\tau_{\text{cc}}^2}{4\tau_r^2}\right) \cdot \exp\left(-\frac{t_{\text{pp}}}{\tau_r}\right) \cdot \left(1 + \text{erf}\left(\frac{t_{\text{pp}}}{\tau_{\text{cc}}} - \frac{\tau_{\text{cc}}}{2\tau_r}\right)\right). \tag{80}$$

- The fourth type is the instant ("i") dynamics function:

$$B_i(t_{\text{pp}}) = \exp\left(-\frac{t_{\text{pp}}^2}{\tau_{\text{cc}}^2}\right). \tag{81}$$

- The fifth type is the nondecaying coherent oscillation ("o") function:

$$B_o(t_{pp}) = \frac{1}{2}\cos(\omega t_{pp} + \varphi) \cdot \left(1 + \text{erf}\left(\frac{t_{pp}}{\tau_{cc}}\right)\right). \quad (82)$$

- Furthermore, the sixth type is the decaying (transient) coherent oscillation ("to") function:

$$B_{to}(t_{pp}) = \frac{1}{2}\exp\left(\frac{\tau_{cc}^2}{16\tau_r^2}\right) \cdot \exp\left(-\frac{t_{pp}}{2\tau_r}\right) \cdot \cos(\omega t_{pp} + \varphi) \cdot \left(1 + \text{erf}\left(\frac{t_{pp}}{\tau_{cc}} - \frac{\tau_{cc}}{4\tau_r}\right)\right). \quad (83)$$

With these basis set functions, we can fit virtually any pump–probe dependent observables according to Equations (76) and (77).

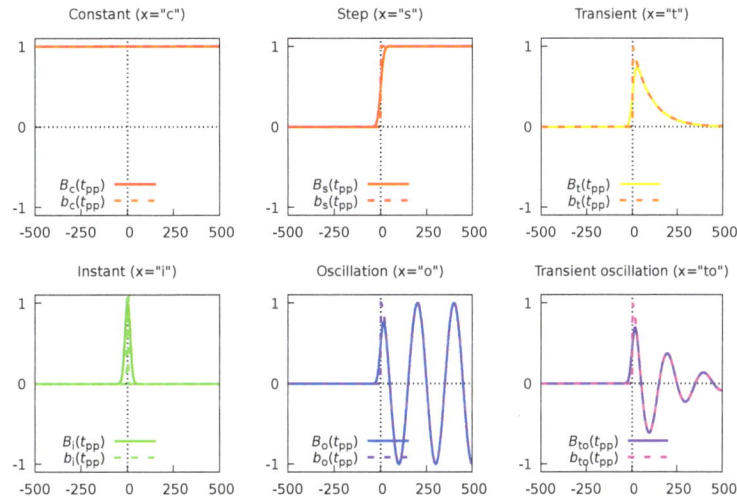

Figure 5. Basis functions for fitting the instant and fluctuating t_{pp} pump–probe kinetics with Equations (60) and (76), respectively. The expressions for the basis functions are given in Equations (61) and (78) for constant behavior, (62) and (79) for step function, (63) and (80) for transient feature, (64) and (81) for instant feature, (65) and (82) for oscillation, and (66) and (83) for transient oscillation. Parameters for plotting of the functions are $\tau_r = 100$ fs, $\tau_{cc} = 20$ fs, vibrational period $\tau_v = 2\pi/\omega = 200$ fs.

4. Estimation Procedure for the Parameters and Their Uncertainties

4.1. Single Dataset Case

Now, we will combine the general ideas of solving inverse problems described in Section 2 with the pump–probe observables model, derived in Section 3, to get a consistent procedure for obtaining reliable estimations of molecular response parameters from the experimental data. Let us assume that we have a dataset of pump–probe data $\{Y_m + \sigma_m\}_{m=1}^M$ consisting of M measured points, where $Y_m = \mathcal{O}(t_{pp,m})$ is the value of the observable $\mathcal{O}$ at the pump–probe delay $t_{pp,m}$, and σ_m is the uncertainty (standard deviation or standard error) of the corresponding value. For each of these points, we can provide a model value $F_m = F(t_{pp,m})$ computed with Equation (76), and consisting of N basis set functions.

Our pump–probe model actually has two types of parameters.

1. The first ones are the linear coefficients $\{f_i\}_{i=1}^N$ before basis functions. These parameters depict effective cross-sections and quantum yields for a given dynamics. We will represent these parameters as an N-dimensional vector $\mathbf{f} = (f_1, f_2, \ldots, f_N)$.

2. The second set of parameters defines each basis function $B_i(t_{pp})$. There are several types of actual parameters.
    - The first type is t_0, representing the temporal overlap of the pump and the probe pulses on the molecular sample. This parameter is not always known in advance from the experimental setup (e.g., in the cases of experiments at the FELs with conventional lasers [6]); it might be needed to be fit. In this case, the parameter is provided to a given basis function $B_i(t_{pp})$ by replacing it with $B_i(t_{pp} - t_0)$. In most cases, t_0 is a shared parameter for all the basis functions and datasets. However, in some cases, some of the basis functions can have a different t_0 parameter to account for Wigner time delay in photoionization [52].
    - The second type of parameter is the cross-correlation time τ_{cc}. This parameter might differ for various basis functions since some processes can require different numbers of photons to be pumped/probed.
    - The third type is the decay time τ_r. Various decay processes usually have different parameters.
    - The fourth type is the coherent oscillation frequency ω.
    - Furthermore, the last, fifth, parameter type is the oscillation phase φ.

These values are required to fully describe the model of the observable. We will denote all of these parameters with a vector $\mathbf{p}$.

With that, we can denote points F_m of the model as $F_m(\mathbf{f}, \mathbf{p})$, i.e., as functions of two sets of parameters, $\mathbf{f}$ and $\mathbf{p}$.

To solve the inverse problem, we need to construct the LSQ function (Equation (6)) given as [12]:

$$\Phi(\mathbf{f}, \mathbf{p}) = \sum_{m=1}^{M} \frac{1}{2\sigma_m^2} (F_m(\mathbf{f}, \mathbf{p}) - Y_m)^2 = \frac{1}{2}(\mathcal{B}_\mathbf{p}\mathbf{f} - \mathbf{Y})^T \mathcal{W}(\mathcal{B}_\mathbf{p}\mathbf{f} - \mathbf{Y}), \tag{84}$$

where in the vector compressed form on the right, the vector $\mathbf{Y} = (Y_1, Y_2, \ldots, Y_M)$ is the M-dimensional vector of the experimental points, the $M \times M$ diagonal matrix of weights $\mathcal{W} = \mathrm{diag}(\sigma_1^{-2}, \sigma_2^{-2}, \ldots, \sigma_M^{-2})$, and the nonlinear parameters dependent matrix $\mathcal{B}_\mathbf{p}$ of size $M \times N$ is composed of the elements $B_{mi} = B_i(t_{pp,m})$.

Let us fix the nonlinear parameters $\mathbf{p}$, and notice that the LSQ inverse problem (Equation (8)) for linear parameters $\mathbf{f}$ has a single explicit solution [12,39]. For this, we need to solve equation $\partial_\mathbf{f} \Phi(\mathbf{f}, \mathbf{p}) = \mathcal{B}_\mathbf{p}^T \mathcal{W} \mathcal{B}_\mathbf{p} \mathbf{f} - \mathcal{B}_\mathbf{p}^T \mathcal{W} \mathbf{Y} = 0$. This is a system of linear equations $\mathcal{A}\mathbf{f} = \mathbf{y}$ with an $N \times N$ matrix $\mathcal{A} = \mathcal{B}_\mathbf{p}^T \mathcal{W} \mathcal{B}_\mathbf{p}$ made of elements $A_{ij} = \sum_{m=1}^{M} B_i(t_{pp,m}) \cdot B_j(t_{pp,m}) / \sigma_m^2$ and N-dimensional right-side vector $\mathbf{y} = \mathcal{B}_\mathbf{p}^T \mathcal{W} \mathbf{Y}$ made of elements $y_i = \sum_{m=1}^{M} Y_m \cdot B_i(t_{pp,m}) / \sigma_m^2$. The solution of this system of equations is:

$$\mathbf{f}_{\min}(\mathbf{p}) = \arg\min_\mathbf{f} \Phi(\mathbf{f}, \mathbf{p}) = \left(\mathcal{B}_\mathbf{p}^T \mathcal{W} \mathcal{B}_\mathbf{p}\right)^{-1} \mathcal{B}_\mathbf{p}^T \mathcal{W} \mathbf{Y}. \tag{85}$$

The requirement for this solution to exist is the invertibility of the matrix $\mathcal{A} = \mathcal{B}_\mathbf{p}^T \mathcal{W} \mathcal{B}_\mathbf{p}$. Substituting this solution (Equation (85)) to the initial LSQ function (Equation (84)), we obtain an effective LSQ function that depends only on the nonlinear parameters $\mathbf{p}$:

$$\Phi_{\min}(\mathbf{p}) = \frac{1}{2}(\mathcal{B}_\mathbf{p} \mathbf{f}_{\min}(\mathbf{p}) - \mathbf{Y})^T \mathcal{W} (\mathcal{B}_\mathbf{p} \mathbf{f}_{\min}(\mathbf{p}) - \mathbf{Y}). \tag{86}$$

Since this effective function depends only on nonlinear parameters $\mathbf{p}$, which means we greatly reduced the problem's dimensionality. By performing the local or global nonlinear minimization of this effective function, we can get an initial optimal solution for both nonlinear and linear parameters as:

$$\begin{cases} \mathbf{p}_{\text{opt}} = \arg\min_{\mathbf{p}} \Phi_{\min}(\mathbf{p}), \\ \mathbf{f}_{\text{opt}} = \mathbf{f}_{\min}(\mathbf{p}_{\text{opt}}). \end{cases} \qquad (87)$$

By using this set of parameters as a starting point of a MC procedure, described in Section 2.4, with probability $p(\mathbf{p}) \propto \exp(-\Phi_{\min}(\mathbf{p}))$, we can get a proper estimation of parameters $\mathbf{p}$ and $\mathbf{f}$.

4.2. Multiple Dataset Case

We can extend this case of a single dataset to a case of multiple $K \geq 1$ datasets, which we will indicate with an upper index $k = 1, 2, \ldots, K$. Let us assume that the k-th dataset has $M^{(k)}$ data points $\{Y_m^{(k)} \pm \sigma_m^{(k)}\}_{m=1}^{M^{(k)}}$ and the k-th model has $N^{(k)}$ basis set functions with linear parameters $\mathbf{f}^{(k)}$ and nonlinear parameters $\mathbf{p}^{(k)}$. We can define the function $\Phi^{(k)}(\mathbf{f}^{(k)}, \mathbf{p}^{(k)})$ in the same way as in Equation (84). The linear parameters $\mathbf{f}^{(k)}$ will be unique for each of the datasets; therefore, in each k-th case, we can find an optimal solution $\mathbf{f}_{\min}^{(k)}(\mathbf{p}^{(k)})$ using Equation (85) and define function $\Phi_{\min}^{(k)}(\mathbf{p}^{(k)})$ through Equation (86). However, unlike for linear parameters, the nonlinear parameters $\mathbf{p}^{(k)}$ might be shared between different datasets, e.g., t_0 position, cross-correlation times, or decay times of the same processes with various observables, etc. Thus, we can form a unique set of N_{nl} nonlinear variables, describing a nonredundant set of variables needed to describe all datasets. We will represent these parameters with an N_{nl}-dimensional vector $\mathbf{P}$ ($\mathbf{p}^{(k)} \in \mathbf{P}$ for all k). In this case, we can formally write that $\mathbf{f}_{\min}^{(k)}(\mathbf{p}^{(k)}) = \mathbf{f}_{\min}^{(k)}(\mathbf{P})$ and $\Phi_{\min}^{(k)}(\mathbf{p}^{(k)}) = \Phi_{\min}^{(k)}(\mathbf{P})$. Therefore, we can define a general experimental LSQ function:

$$\Phi_{\min}^{(\text{exp})}(\mathbf{P}) = \sum_{k=1}^{K} \Phi_{\min}^{(k)}(\mathbf{P}). \qquad (88)$$

We can now, similarly to Equation (87), find the optimal parameters $\mathbf{P}_{\text{opt}} = \arg\min_{\mathbf{P}} \Phi_{\min}^{(\text{exp})}(\mathbf{P})$ and $\{\mathbf{f}_{\text{opt}}^{(k)} = \mathbf{f}_{\min}^{(k)}(\mathbf{P}_{\text{opt}})\}_{k=1}^{K}$, which can be used as starting points for MC sampling of parameters $\mathbf{P}$ and $\{\mathbf{f}^{(k)}\}_{k=1}^{K}$ (see Section 2.4).

4.3. Inverse Problem Regularization

We can also include the regularization of the nonlinear parameters (see Section 2.3) in this procedure by adding a penalty function $\Phi_{\text{reg}}(\mathbf{P})$ to the experimental LSQ function $\Phi_{\min}^{(\text{exp})}(\mathbf{P})$, such as we work with an effective function:

$$\Phi_{\text{eff}}(\mathbf{P}) = \Phi_{\min}^{(\text{exp})}(\mathbf{P}) + \Phi_{\text{reg}}(\mathbf{P}). \qquad (89)$$

In this case, we do exactly the same minimization/MC sampling as for the pure experimental LSQ function (Equation (88)).

The first type of regularization that we will consider is when we have independent estimates for some ($1 \leq N_{\text{reg}} \leq N_{\text{nl}}$) of the nonlinear parameters $\mathbf{p} \in \mathbf{P}$ (dim($\mathbf{p}$) = N_{reg}), e.g., an independent measurement of t_0 or estimates for τ_{cc}. Suppose we have N_{reg} values of $p_{\text{reg},l}$ parameters with their corresponding uncertainties ς_l ($l = 1, 2, \ldots, N_{\text{reg}}$). In this case, we can define a penalty function $\Phi_{\text{reg}}(\mathbf{P})$, which provides enforcing of these *a priori* assumptions for parameters $\mathbf{p}$:

$$\Phi_{\text{reg}}^{(\text{I})}(\mathbf{P}) = \frac{1}{2}(\mathbf{p} - \mathbf{p}_{\text{reg}})^{\text{T}} \mathcal{W}_{\text{reg}} (\mathbf{p} - \mathbf{p}_{\text{reg}}), \qquad (90)$$

using the N_{reg}-dimensional vector $\mathbf{p}_{\text{reg}} = (p_{\text{reg},1}, p_{\text{reg},2}, \ldots, p_{\text{reg},N_{\text{reg}}})$ and weight matrix $\mathcal{W}_{\text{reg}} = \text{diag}(\varsigma_1^{-2}, \varsigma_2^{-2}, \ldots, \varsigma_{N_{\text{reg}}}^{-2})$ of size $N_{\text{reg}} \times N_{\text{reg}}$. This equation has a form of L_2-regularization (Equation (12)), with the regularization parameter for each of the variables p_l being $1/(2\varsigma_l^{-2})$ ($l = 1, 2, \ldots, N_{\text{reg}}$). A statistical meaning of the probability

$p(\mathbf{P}) \propto \exp(-\Phi_{\text{reg}}(\mathbf{P}))$ is that we enforce the normal distribution for parameters $\mathbf{p}$ with variances ς_l [14].

The second type of regularization does not have any conceptual justification but is rather a numerical necessity. If one of the observables has more than one step, decay, or instant increase basis function, they can become linearly dependent upon their parameters approaching each other. In this case, we can account for problems in finding the solution for linear parameters (Equation (85)). Even worse, this can affect the MC sampled parameters, since the two distinct dynamical channels can become interchanged, leading to mixed distributions for different variables. To prevent that, we can add an artificial repelling term $\Phi_{ij}(\mathbf{P})$ for two entangled variables p_i and p_j in $\mathbf{P}$, such as $\Phi_{ij}(\mathbf{P}) \to 0$ if these values are far apart ($|p_i - p_j|$ is large) and $\Phi_{ij}(\mathbf{P}) \to \infty$ if they approach each other ($|p_i - p_j| \to 0$). The simplest choice is the Coulomb-like expression:

$$\Phi_{ij}(\mathbf{P}) = \frac{\alpha_{ij}}{|p_i - p_j|}, \qquad (91)$$

where $\alpha_{ij} \geq 0$ is an arbitrary regularization factor, determining the strength of the repulsion. If $\alpha_{ij} = 0$, no repelling regularization for parameters p_i and p_j is applied. Combining all the possible pairs of parameters, we can define a general penalty function:

$$\Phi_{\text{reg}}^{(\text{II})}(\mathbf{P}) = \sum_{i=1}^{N} \sum_{j=1}^{j<i} \Phi_{ij}(\mathbf{P}). \qquad (92)$$

This general expression, that includes summing over all nonlinear parameters, allows simultaneous treatment of both the unregularized pairs of values (for which $\alpha_{ij} = 0$) and those pairs of parameters, that are artificially constrained from being too close to each other ($\alpha_{ij} > 0$).

4.4. Inverse Problem Solution Algorithm

Now, we can summarize the proposed algorithm for solving the inverse problems in pump–probe spectroscopy.

1. Obtain $K \geq 1$ datasets of pump–probe observables and construct a model for each of them. This means defining a unique set of nonlinear parameters $\mathbf{P}$, a basis set for each dataset, which provides linear parameters $\mathbf{f}_{\min}^{(k)}(\mathbf{P})$ for each of the $1 \leq k \leq K$ datasets (Equation (85)), and the experimental LSQ function $\Phi_{\exp}(\mathbf{P})$ (88).
2. Construct a regularization functional for parameters $\mathbf{P}$. Two types are available.
    (a) If there are some *a priori* expectations on some of the parameters, they can be included through the penalty function $\Phi_{\text{reg}}^{(\text{I})}(\mathbf{P})$ (Equation (90)).
    (b) If, for some observables, there are multiple basis functions of the same type, they can be protected from linear dependency using $\Phi_{\text{reg}}^{(\text{II})}(\mathbf{P})$ (Equation (92)).

    The total regularization function $\Phi_{\text{reg}}(\mathbf{P})$ can be either:
    - $\Phi_{\text{reg}}(\mathbf{P}) = \Phi_{\text{reg}}^{(\text{I})}(\mathbf{P}) + \Phi_{\text{reg}}^{(\text{II})}(\mathbf{P})$, if both regularization cases are applicable;
    - $\Phi_{\text{reg}}(\mathbf{P}) = \Phi_{\text{reg}}^{(\text{I})}(\mathbf{P})$ or $\Phi_{\text{reg}}(\mathbf{P}) = \Phi_{\text{reg}}^{(\text{II})}(\mathbf{P})$, if only one regularization case in demand;
    - $\Phi_{\text{reg}}(\mathbf{P}) = 0$, if no regularization is required.
3. Define an effective function $\Phi_{\text{eff}}(\mathbf{P})$ (Equation (89)) as a sum of experimental and regularization functions.
4. Find a solution of the LSQ problem as $\mathbf{P}_{\text{opt}} = \arg\min_{\mathbf{P}} \Phi_{\text{eff}}(\mathbf{P})$ using local or global fitting.
5. Start a MC sampling procedure (see Section 2.4) with probability $p(\mathbf{P}) \propto \exp(-\Phi_{\text{eff}}(\mathbf{P}))$ to sample nonlinear ($\mathbf{P}$) and linear ($\mathbf{f}_{\min}^{(k)}(\mathbf{P})$) parameters.

6. The final values for parameters will be the mean values from MC ($\langle \mathbf{P} \rangle$ and $\langle \mathbf{f}_{\min}^{(k)}(\mathbf{P}) \rangle$). The uncertainties of the mean values can be given as their respective standard deviations ($\sqrt{\langle \mathbf{P}^2 \rangle - \langle \mathbf{P} \rangle^2}$ and $\sqrt{\langle \left(\mathbf{f}_{\min}^{(k)}(\mathbf{P}) \right)^2 \rangle - \langle \mathbf{f}_{\min}^{(k)}(\mathbf{P}) \rangle^2}$), see Equations (13) and (14).
7. In addition to the values and uncertainties, Pearson correlation coefficients (Equation (16)), histograms of parameter distributions, and higher distribution moments can also be calculated from the MC trajectory.

The first application of this algorithm was in Ref. [6] with generic Python scripts, but the development of general software for such fitting followed soon after. In the next section, we will discuss this software.

5. PP(MC)³Fitting Software

The PP(MC)³Fitting (pump–probe multichannel Markov chain Monte-Carlo fitting) is software that implements the algorithm from Section 4.4 for solving inverse problems of pump–probe spectroscopy. It is written in Python language and is composed of an application programming interface library libMCMCMCFitting.py and an actual script ppmc3fitting.py that provides communication with the user by a command line interface and a set of required and optional input files. The software also features a set of unit tests and basic examples of applications.

The general scheme of working with the PP(MC)³Fitting software is given in Figure 6. There are three required input files for the software to work. The first one, the dataset definition file, provides the names of the files with the data that need to be fitted and the basis functions, which should represent the observables via Equation (76). The second input file, the channels' definition file, provides the definition of the basis functions, i.e., which types of functions are there and which nonlinear variables they depend on. The last compulsory input file (variables' definition file) initializes the nonlinear variables (t_0, τ_{cc}, and τ_r): their minimal and maximal values, and optionally the initial value and the maximal step for the MC sampling routine. A fourth additional file is the regularization definition file, where additional constraints of the fitting can be defined. The data files are simple text files with three or four columns, where the first column provides the pump–probe delay, the second gives the yield of the observable, and the last column provides the yield uncertainty. The units of the first column of all data files define the time units of all corresponding nonlinear parameters.

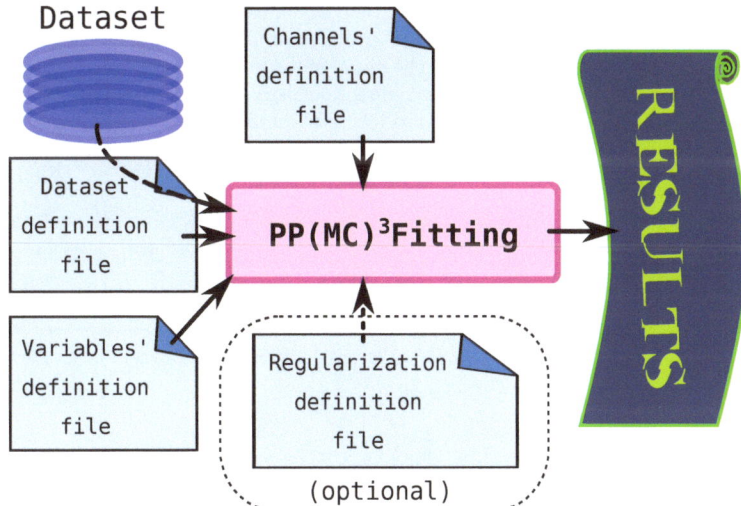

Figure 6. A schematic representation of PP(MC)³Fitting workflow.

The fitting procedure follows the algorithm described in Section 4.4. First, the local or global optimization is run on the data to find the minimum of the effective LSQ function (Equation (89)). For this, the minimization routines from the SciPy library are used [53]. The second step is MC sampling to obtain statistical estimates for the parameters. In the end, the program prints out both optimized and sampled values for nonlinear variables, Pearson correlation coefficients matrix, histograms for the variables, and also the numerical representation of the fitted function and the basis sets to plot alongside the fitted data. A more detailed application manual, the code itself, and examples of the fitting are provided in the electronic supporting information (ESI).

The software provides the following types of basis functions to fit pump–probe dynamics: "constant" (B_c, Equation (78)), "switch" (B_s, Equation (79)), "transient" (B_t, Equation (80)), and "Gaussian" (B_i, Equation (81)). The coherent oscillations functions (Equations (82) and (83)) were not implemented because these dynamics are not always present in the pump–probe data but require more parameters to be included, such as the oscillation frequency and the phase. Instead, the software also prints out the residuals of the fit, which will contain the oscillatory dynamics, if present. Therefore, the coherent oscillations can be fitted a posteriori from the fit, using Equations (82) and (83), similar to Ref. [8].

The `PP(MC)`^{3}`Fitting` software was successfully applied to disentangle complex dynamics of PAHs and published in Refs. [7,54,55]. For instance, in Ref. [7], a total of 31 decay times were extracted from the rich fragmentation dynamics of fluorene ($C_{13}H_{10}$), which corresponded to the lifetimes of the excited mono- and dications of this PAH molecule. Here, we will not show any complicated analysis, but rather provide a few numerical demonstrations of the capabilities of the `PP(MC)`^{3}`Fitting` software and of the approaches and concepts that could be used to work with the real-life experimental data.

6. Numerical Examples

6.1. Multiple Datasets with Shared Parameters

As a first example of the application of the `PP(MC)`^{3}`Fitting` software, we will consider the case of multiple datasets with shared parameters. For this, photoelectron sidebands will be used as an example [56]. We will base the discussion on the actual experimental pump–probe photoelectron spectra collected at the CAMP end-station [57,58] of the soft X-ray free-electron laser FLASH (DESY, Hamburg) [59,60] during the beamtime F-20191568. In this experiment, helium (He) atoms were pumped with XUV photons with an energy of $h\nu_{XUV} = 40.8$ eV (wavelength $\lambda = 30.3$ nm), produced by FLASH, and then probed by infrared (IR) photons with an energy of $h\nu_{IR} = 1.5$ eV ($\lambda = 810$ nm). More details on the experiment can be found in Ref. [6].

He atoms resonantly absorb XUV photons with the energy of 40.8 eV, which is the He II line [61], producing an ionization event according to the reaction:

$$He + XUV \rightarrow He^+ + e^-(KE_0), \tag{93}$$

where He^+ is the He monocation, $e^-(KE_0)$ is the photoelectron with kinetic energy of $KE_0 = h\nu_{XUV} - IP(He) = 16.2$ eV, and $IP(He) = 24.6$ eV is the ionization potential of He [62].

However, in the presence of the IR strong field, the absorption or induced emission of photons by the ionized He can occur, leading to the gain or loss of the photoelectron energy, respectively, [56]. We can describe this process through the following reaction:

$$He + XUV \pm n \cdot IR \rightarrow He^+ + e^-(KE_n), \tag{94}$$

where $KE_n = KE_0 \pm nh\nu_{IR}$ is the new kinetic energy of the photoelectron, and $n = \ldots, -1, 0, +1, \ldots$ is the number of the absorbed photons. If $n = 0$, Equation (94) becomes Equation (93), producing photoelectrons with energy KE_0, and this photoelectron line is also called a main line. However, if any IR photons are absorbed/emitted ($|n| \geq 1$), then the photoelectron energy $KE_n \neq KE_0$, and the resulting signal is called the sideband of the n-th order [56].

We can note that the sideband formation reaction (Equation (94)) is the same as the instant probing scheme (Equation (25)). Therefore, the temporal behavior of these photoelectron signals will be described with a basis function $B_i(t_{pp})$ (Equation (81)). Near the temporal overlap of both pulses (t_0), the main line will be depleted, and the sidebands will be formed. According to the cross-correlation time expression (Equation (75)), we expect that the higher the sideband order $|n|$ is, the smaller is the τ_{cc} of this line. Furthermore, the sidebands should be symmetric, i.e., the width of the line at kinetic energy KE_n should be the same as for KE_{-n}.

The experimental data (Figure 7) show exactly the behavior we described. We can extract the pump–probe photoelectron yields at given values of the photoelectrons' kinetic energies KE_n, corresponding to the intensities of the main line and the sidebands. From the data, in addition to the main line, we see sidebands of orders $n = \pm 1, \pm 2, \pm 3, +4$. All of these datasets can be fitted simultaneously with the PP(MC)³Fitting software with a routine described in Section 4.4. The model function (Equation (76)) in all of the cases is given by:

$$F(t_{pp} - t_0) = f_c \cdot B_c(t_{pp} - t_0) + f_i \cdot B_i(t_{pp} - t_0), \qquad (95)$$

that is, it is a sum of the constant function $B_c = 1$ (Equation (78)) describing the baseline and instant dynamic B_i (Equation (81)). All eight datasets for the main line and seven sidebands share the parameter t_0. The fit for all data should also have five cross-correlation time parameters $\tau_{cc}^{(n)}$, with $n = 0, 1, 2, 3, 4$ denoting the main line ($n = 0$) and sideband order ($n \geq 1$) according to Equation (94). Since the $\pm n$ sidebands are produced with the same amount of IR photons, they should have the same cross-correlation time.

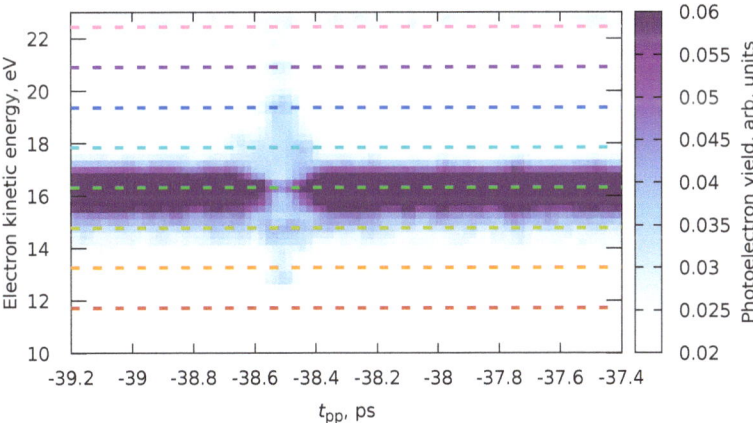

Figure 7. Experimental XUV-IR pump–probe photoelectron spectrum of helium, obtained in beamtime F-20191568. The horizontal dashed lines show the expected position of the photoelectron main line (at 16.2 eV) and sidebands. The experimental photoelectron maxima for higher-order sidebands are offset due to imperfection of the radius-to-energy conversion. Details are provided in the text.

We performed two types of fits: a global fit for all eight datasets and separate fits for the main line (Fit #0) and sidebands of each order (Fits #1 to #4). The results are shown in Figure 8 and Table 1. As one can see, generally, the parameters obtained from both the global fit and the separate fits agree. However, the global fit allows us to reduce the uncertainty for some parameters and avoid inconsistencies between the datasets (e.g., Fit #1 or Fit #4). This, in particular, leads to an unambiguous and precise definition of $t_0 = -38.522 \pm 0.001$ ps that can be used in further analysis, similar to Ref. [6].

Table 1. Final values of the nonlinear parameters for fitting the He photoelectron main line and sidebands. Fitted curves are shown in Figure 8.

Parameter	Value, fs					
	Global	Fit #0	Fit #1	Fit #2	Fit #3	Fit #4
$t_0 + 38.5$ ps	-22 ± 1	-22 ± 3	-35 ± 4	-23 ± 2	-20 ± 2	-15 ± 2
$\tau_{cc}^{(0)}$	97 ± 3	97 ± 3	—	—	—	—
$\tau_{cc}^{(1)}$	138 ± 5	—	143 ± 6	—	—	—
$\tau_{cc}^{(2)}$	88 ± 2	—	—	88 ± 3	—	—
$\tau_{cc}^{(3)}$	63 ± 3	—	—	—	62 ± 3	—
$\tau_{cc}^{(4)}$	62 ± 7	—	—	—	—	61 ± 7

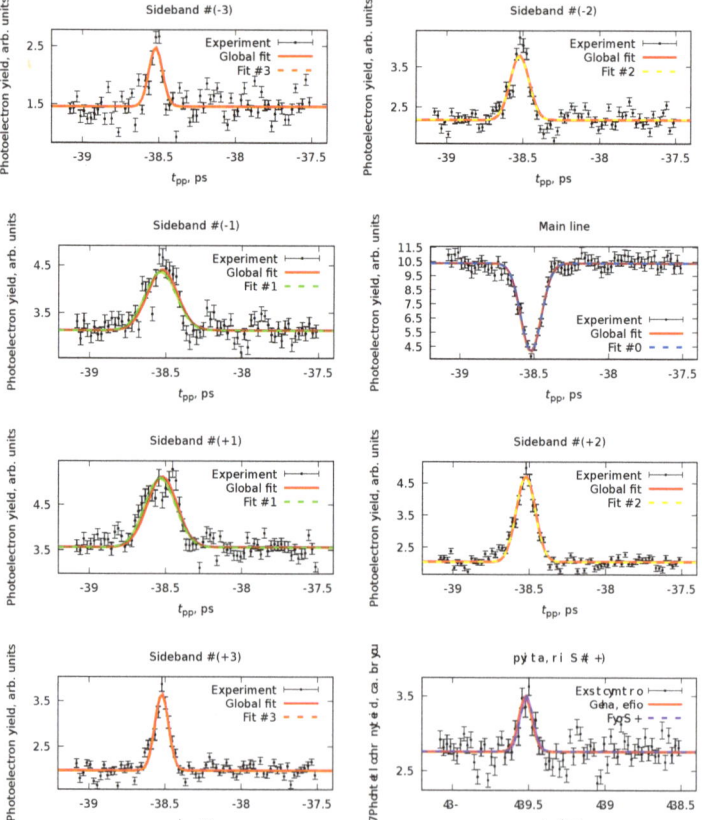

Figure 8. Photoelectron yields at the sidebands of various order and of the main line, corresponding to helium ionization (Reaction (94)), and their fits with the model from Equation (95). The points shown here were obtained as horizontal slices from Figure 7.

6.2. Forward-Backward Channel Dataset

Here, we will consider a case with two pump–probe channels, where the pump acts as a pump and the probe as a probe, and an inverted case, where the probe pulse acts as a pump, and the pump pulse acts as a probe. We will take the formation of the fluorene dication from Ref. [6] as an example. In that work, the neutral three-ring PAH fluorene

($C_{13}H_{10}$) was investigated in an XUV-IR pump–probe experiment with the same parameters as given in the previous Section 6.1. Upon a first look at the experimental ion yield of the fluorene dication $C_{13}H_{10}{}^{2+}$ (Figure 9), one would think that there is a single transient peak and a switch function. However, the temporal overlap $t_0 = 12.650 \pm 0.005$ ps determined with the help of helium sidebands [6], similar to that in Section 6.1, does not allow to fit the resulting behavior using a single transient. The simplest model to explain this anomaly is that the peak is composed of two transients: one with the XUV pulse acting as a pump and the second with the IR pulse acting as a pump.

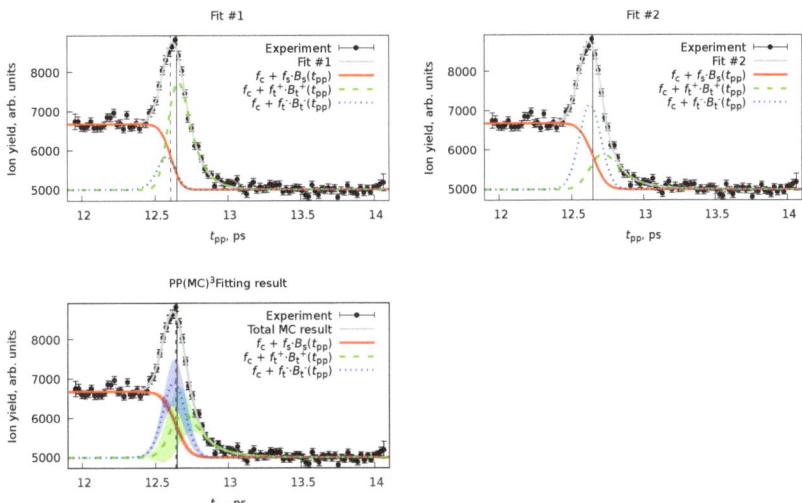

Figure 9. Experimental ion yield of the fluorene dication $C_{13}H_{10}{}^{2+}$ in the XUV-IR pump–probe experiment. Three plots show three independent fits of the same experimental data by the same model, given in Equation (96). Fits #1 and #2 are the results of standard fitting using Gnuplot [63], while the PP(MC)³Fitting result was obtained using the software procedure from Section 4.4 using PP(MC)³Fitting software. The solid vertical line indicates the t_0 from He sidebands, while the dotted vertical lines are the results of the corresponding fit result.

The model (Equation (76)) proposed for the fitting of such a signal is the following:

$$F(t_{pp} - t_0) = f_c \cdot B_c(t_{pp} - t_0) + \\ + f_s^- \cdot B_s^-(t_{pp} - t_0) + f_t^+ \cdot B_t^+(t_{pp} - t_0) + f_t^- \cdot B_t^-(t_{pp} - t_0), \quad (96)$$

where "switch" and transient functions use a single cross-correlation time τ_{cc}, and two transients $B_t^{\pm}$ are described with rate constants $\tau_r^{\pm}$, where τ_r^+ describes lifetime of an excited fluorene monocation $(C_{13}H_{10}{}^+)^*$ and τ_r^- describes the lifetime of an excited neutral fluorene $(C_{13}H_{10})^*$ [6].

Simple fitting of Equation (96) to the experimental data given in Figure 9 is a good example of the ill-posed problem. To illustrate that, we performed two fits with two sets of initial values of nonlinear parameters t_0, τ_{cc}, τ_r^+, and τ_r^- using the Marquardt–Levenberg algorithm [64,65] with the Gnuplot software [63]. The initial t_0 in both cases was taken as a value from the He sidebands, and the initial τ_{cc} was taken from the MC result in Ref. [6]. The only difference is that the initial decay times were taken as 50 fs in the first fit, whereas in the second, they were 150 fs.

Figure 9 and Table 2 show results for both fits. The overall description of the data looks equivalent in both cases, but the actual values of the nonlinear parameters are quite different. While some of the parameters are equivalent to each other due to huge uncertainties, the

τ_r^+ have significantly different values, judging from the standard deviations computed according to Equation (10). The LSQ function values are equivalent in both cases, making these solutions indistinguishable, using, e.g., χ^2-statistics [12] or various criteria, such as the Akaike information criterion [66]. However, even by looking at the components of the fit (Figure 9), we can be sure that these are two alternative solutions.

Table 2. Nonlinear parameters of the model for the fluorene dication ion yield (Equation (96)) obtained with different fit models. "Ini." and "Fin." denote the initial and final values according to the Marquardt–Levenberg algorithm [64,65]. All values are given in fs.

Parameter	Fit #1		Fit #2		PP(MC)^{3}Fitting
	Ini.	Fin.	Ini.	Fin.	
$t_0 - 12.650$ ps	0	-41 ± 3441	0	-2 ± 33	-11 ± 20
τ_{cc}	97	80 ± 1018	97	104 ± 21	101 ± 10
τ_r^+	50	88 ± 7	150	130 ± 30	133 ± 37
τ_r^-	50	36 ± 58	150	21 ± 40	22 ± 20

Applying the algorithm from Section 4.4 is especially advantageous in such cases, since we can automatically sample through all various equivalent solutions, providing an adequate statistical representation of the result. The results of the application of the PP(MC)^{3}Fitting software are also given in Figure 9 and Table 2. Although the total fit looks exactly the same as Fit #1 and Fit #2, the uncertainties of the individual transient components ($f_t^+ \cdot B_t^+(t_{pp} - t_0)$ and $f_t^- \cdot B_t^-(t_{pp} - t_0)$) are significant. By examining the distributions for individual nonlinear variables (Figure 10), we realize that the MC procedure sampled through many equivalent solutions, including Fit #1 and Fit #2. A more accurate and realistic solution can be obtained by applying regularization of t_0 with the value from He sidebands, similar to Ref. [6].

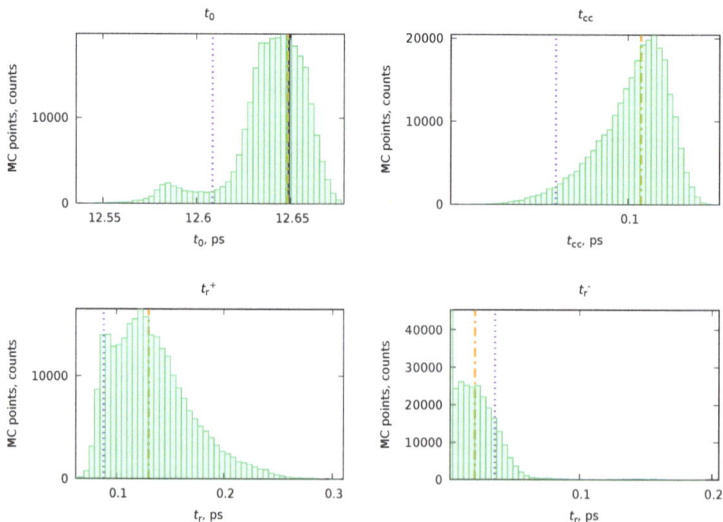

Figure 10. Distributions of the nonlinear parameters of the model for the fluorene dication ion yield (Equation (96)) from the MC sampling procedure. The dashed vertical lines illustrate the values from Fit #1 and Fit #2 (violet and red, respectively, see Table 2). The vertical solid line for t_0 shows the result from the He sidebands measurements.

6.3. Treatment of the Data with Coherent Oscillations

As was said in Section 5, the PP(MC)³Fitting software does not implement the basis functions for the coherent oscillations (Equations (82) and (83)). Nevertheless, this software can still be used to fit such datasets as well. As an example, we will use the ion yield of the indole-water ($C_8H_7N \cdot H_2O$) monocation from Ref. [10], where oscillatory dynamics were observed. In the original publication, a five-state model given by the Maxwell–Bloch equations was used, which produced three rate constants $\tau_{r,1} = 0.45 \pm 0.07$ ps, $\tau_{r,2} = 13 \pm 2$ ps, and $\tau_{r,3} = 96 \pm 10$ ps, and also an oscillation frequency $\omega = 3.77$ THz with phase $\varphi = 1.77$.

To fit the same dataset for the indole-water monocation, we formulated an effective model with three effective decay times:

$$F(t_{pp} - t_0) = f_c \cdot B_c(t_{pp} - t_0) + f_s^- \cdot B_s^-(t_{pp} - t_0) + \sum_{i=1}^{3} f_t^+ \cdot B_{t,i}^+(t_{pp} - t_0), \quad (97)$$

where the t_0 was regularized with a value of $t_{0,\text{reg}} = 0 \pm 1$ fs, while all the rest of the nonlinear parameters τ_{cc} and $\tau_{r,i}$ ($i = 1, 2, 3$), each from function $B_{t,i}$, were left unregularized. The values of $\tau_{r,i}$ were expected to be sufficiently different. Therefore, no repulsion regularization (Equation (92)) was applied in this case. Instead, the nonoverlapping regions for decay times were set: $0.01 \leq \tau_{r,1} \leq 2$ ps, $5 \leq \tau_{r,2} \leq 40$ ps, and $40 \leq \tau_{r,3} \leq 300$ ps. The resulting parameters obtained with PP(MC)³Fitting were $\tau_{cc} = 0.34 \pm 0.02$ ps, $\tau_{r,1} = 0.9 \pm 0.1$ ps, $\tau_{r,2} = 26 \pm 5$ ps, and $\tau_{r,3} = 234 \pm 34$ ps. The obtained decay times are reasonably close to the ones obtained in Ref. [10] (see above), as well as to the experimentally estimated $\tau_{cc} = 0.381$ ps. An exact match was not expected since, in the original paper, a microscopic model was used, which produces the elementary rate constants, while our approach utilizes the effective rate constants.

The obtained fit (Figure 11) by design (Equation (97)) does not contain any of the oscillations, and this means that we have fitted only the noncoherent part of the signal (see Equations (51) and (57)). We can consider the fit's residual to find the signal's coherent part. The fit residual is defined as $\mathbf{y} = \mathbf{Y} - \mathcal{B}_{p_{opt}} \mathbf{f}_{opt}$ (see Equations (86) and (87)) with the same uncertainties as of the original values. Since the desired oscillation signal is proportional to $\cos(\omega t + \varphi)$ (Equations (51) and (57)), we can perform the Fourier transform (FT) of the residual part of the spectrum, to find the initial guess for the oscillation frequency ω and phase φ [12]. Since we know the residuals' uncertainties and the dataset does not contain uniformly spaced data, instead of the fast FT algorithm [12], we can use a least-squares spectral analysis (LSSA) [67,67]. In particular, here, we used the regularized weighted LSSA (rwLSSA) introduced in Ref. [68] (see details in Appendix C). The rwLSSA spectrum of delay range of $-0.7 \leq t_{pp} < 30$ ps for a frequency range of $0 < \omega \leq 7$ THz is shown in Figure 12. The maximal intensity is observed for a point of $\omega = 3.82$ THz with phase $\varphi = 1.4$ at this point, which is already reasonably close to the parameters obtained in Ref. [10] ($\omega = 3.77$ THz and $\varphi = 1.77$).

We then performed a fit of the signal in the range $-0.7 \leq t_{pp} < 10$ ps with the highest density of points. We represented a signal as an oscillation function (Equation (82)):

$$f(t_{pp}) = f_o \cdot B_o(t_{pp}) = \frac{f_o}{2} \cos(\omega t_{pp} + \varphi) \cdot \left(1 + \text{erf}\left(\frac{t_{pp}}{\tau_{cc}}\right)\right) \quad (98)$$

with $\tau_{cc} = 0.34$ ps taken from the PP(MC)³Fitting and rwLSSA peak parameters as initial conditions. By simple LSQ fitting, we get $\omega = 3.72 \pm 0.05$ THz and $\varphi = -1.2 \pm 0.3$. This result is shown in Figure 13. With that, we produced a full description of the pump–probe dataset for both the incoherent decay part of the signal and the coherent oscillations.

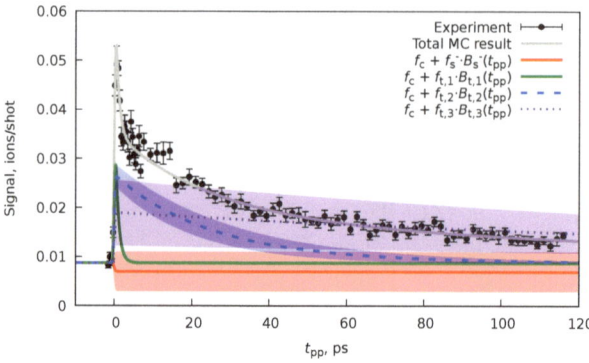

Figure 11. Experimental and fitted ion yield of the indole-water monocation. Experimental data were taken from Ref. [10]. Colored areas indicate the MC uncertainty of a corresponding component of the fit. The model is given in Equation (97).

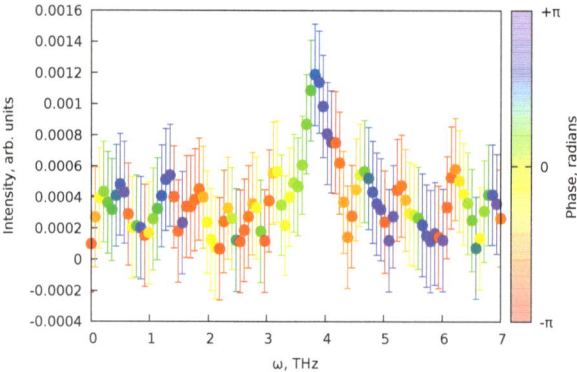

Figure 12. Spectrum of the indole-water monocation ion yield residuals from the fit given in Figure 11. The "intensity" is the absolute value of the rwLSSA spectral intensities obtained from the fit residuals, and the corresponding phase is shown with the point's color.

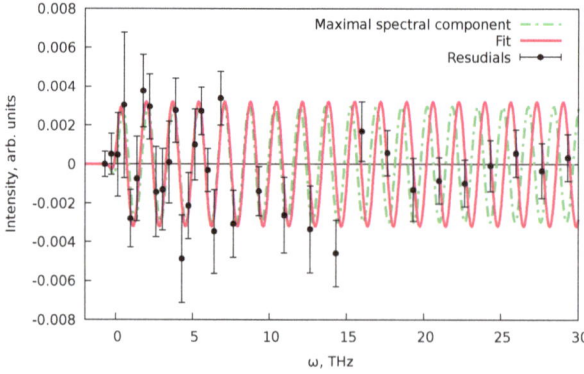

Figure 13. Indole-water monocation ion yield fit residuals, the maximal intensity component from the rwLSSA spectrum (Figure 12) and the fit result of the residual according to Equation (98).

Now, we can summarize a universal analysis procedure that can be applied to the pump–probe datasets with the possible presence of coherent oscillations.

1. First, with the algorithm from Section 4.4 implemented in PP(MC)³Fitting, we fit the nonoscillating part of the dynamics.
2. Then, we perform an rwLSSA analysis [68] (see Appendix C) for the residuals of the fit that are also printed by the PP(MC)³Fitting. This analysis will allow us to check whether the signal contains any systematic oscillations. Note that the oscillations should be present only in $t_{pp} \geq t_0$ or in $t_{pp} \leq 0$ parts of the pump–probe data (see Equations (82) and (83)).
3. If the rwLSSA spectrum shows a presence of statistically meaningful oscillations in a reasonable range of frequencies, then the frequencies and the phases of the maximal amplitude signals from rwLSSA spectra can be used as initial guesses to fit the residuals of the PP(MC)³Fitting result with expression Equation (76) and basis functions B_o (Equation (82)) and B_{to} (Equation (83)). Since the coherent oscillations should correspond to the incoherent processes, the cross-correlations and decay times from the PP(MC)³Fitting results can be used.

The same procedure can also treat other unaccounted features in the experimental pump–probe data. One example is the presence of so-called coherent artifacts, which appear near the temporal overlap of the pump and probe pulses [69–72].

6.4. Cross-Correlation Time and Time Resolution

The last example we will show here is a demonstration of a concept rather than a PP(MC)³Fitting demonstration. Sometimes, statements are shared that the "time resolution of pump-probe experiments is limited by the cross-correlation time" [11]. This leads to frequent questions about whether the rate constants extracted from the fits are smaller than the cross-correlation times of the experiments. However, such a direct application of the idea that the cross-correlation time is the limit for the shortest obtainable decay lifetimes is questionable since there are a few common ways to express both the cross-correlation time (Equation (71)) and the decay time (Equation (30)). Therefore, in this section, we decided to use the MC sampling of the possible solutions with PP(MC)³Fitting to demonstrate the actual relation between the rate constants and the cross-correlation time.

To do that, we generated two sets of ten generic pump–probe data with a fixed decay time $\tau_r = 50$ fs, but with varied cross-correlation time $5 \leq \tau_{cc} \leq 500$ fs. The upper limit was determined by the capabilities of the standard methods to represent the transient function $B_t(t_{pp})$ (Equation (80)). The standard methods, using SciPy [73] packages, allow for stable and smooth data produced for the ratios τ_{cc}/τ_r up to $\tau_{cc}/\tau_r < 50$ (see Appendix D). Such implementation of the $B_t(t_{pp})$ function is used in PP(MC)³Fitting, but we limited ourselves to $\tau_{cc}/\tau_r \leq 10$.

All generated pump–probe data span the delays in the range $-1 \leq t_{pp} \leq +2$ ps with a step of 20 fs. Each set of data had a given signal-to-noise ratio (SN): one set had SN = 10 (high noise data), and the other SN = 100 (low noise data). Figure 14 shows an example of such data. Each of the datasets with a given τ_{cc} and SN parameters was fitted with PP(MC)³Fitting with the same function that it was generated with (Equation (80)), that is:

$$F(t_{pp}) = f_t \cdot B_t(t_{pp}). \tag{99}$$

Two fits were performed for each pump–probe dataset with a given τ_{cc} and SN: without regularization and with regularization for t_0 with $t_{0,reg} = 0 \pm 1$ fs (see Figure 14).

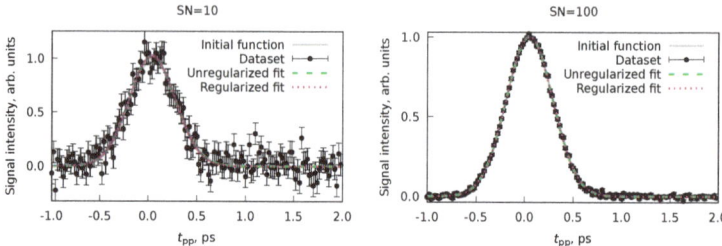

Figure 14. Examples of generated pump–probe datasets and their fits with PP(MC)³Fitting. The "initial function" is the original function, and the "dataset" is the generated dataset with the given SN level. The shown datasets were generated using Equation (80) with $\tau_{cc} = 335$ fs and $\tau_r = 50$ fs parameters.

The resulting trends for the three nonlinear parameters of the fit (t_0, τ_{cc}, and τ_r) are shown in Figure 15. As we can see, for the fully unregularized fit with SN = 10, the procedure gives reasonable results (see also Figure 15) up to $\tau_{cc} \sim 300$ fs ($\tau_{cc}/\tau_r = 6$). At a lower noise level (SN = 100), reasonable results extend to around $\tau_{cc} \sim 450$ fs ($\tau_{cc}/\tau_r = 9$). Upon applying the regularization, both the low noise level (SN = 100) and high noise level (SN = 10) fits were able to reproduce the preset parameters up to $\tau_{cc} = 500$ fs. Therefore, we can conclude that the decay times (Equation (30)) can be reliably fitted below the value of the cross-correlation time (Equations (5) and (75)) with a single dataset. The lower noise level and preliminary knowledge of the parameters, such as the position of the temporal overlap (t_0), allow shorter decay times to be reliably fitted. Combining the datasets in a global fit, as shown in Section 6.1, can also be used to increase the accuracy of the results.

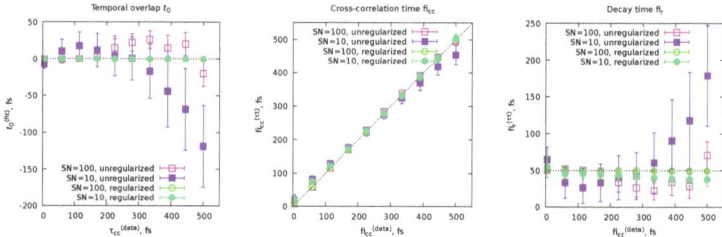

Figure 15. Results of fitting the various datasets with unregularized and regularized procedures. The dotted lines represent the actual defined parameters of the dataset. SN denotes the signal-to-noise ratio for the fitted dataset.

7. Conclusions

In this manuscript, we extensively discussed the inverse problem solution (fitting) for the pump–probe spectroscopic datasets. We examined the theoretical aspects of the inverse problem solution and the standard model used to fit the pump–probe data. Here, we provided rigorous proof that the classical set of model functions used to fit the pump–probe experimental data is sufficient to describe any pump–probe observables, given that only first-order reactions are possible (Appendix A.8). In addition, we have extended the standard set of the basis functions used to fit the pump–probe dynamics with two additional ones, describing coherent oscillations in the dataset (Equations (82) and (83)).

We proposed a general-purpose algorithm for treating the inverse problem of the pump–probe spectroscopy (outlined in Section 4.4). In short, it is based on separating the linear and nonlinear parameters. First, a global fit of the data is performed, and then the uncertainties of the fitted parameters are determined by the Markov chain Monte-Carlo routine. This approach was implemented in the Python software PP(MC)³Fitting, which can be used in a standardized fashion for various datasets. With the numerical examples, we highlighted the common ideas for application, such as global fits with shared parameters between different datasets, the presence of parasitic channels, and stepwise treatment of the coherent oscillations in the data. We also commented on a commonly misdiscussed issue of the time resolution in the pump–probe spectroscopy.

The presented PP(MC)³Fitting software is a complementary addition to the existing methods used to analyze experimental pump–probe data. Examples of such approaches include global and target analysis [32,38,74], KiMoPack [33], lifetime density maps analysis [75–77], and Maxwell–Bloch equation modeling [4,34]. Adding the PP(MC)³Fitting to the listed set of methods can be useful for the ultrafast community to robustly and effectively tackle complicated experimental pump–probe results with various dynamical observables (Supplementary Materials).

Supplementary Materials: The following supporting information can be downloaded at: https://www.mdpi.com/article/10.3390/photochem4010005/s1, The presented software is a complementary addition to the existing methods used to analyze experimental pump-probe data.

Author Contributions: Conceptualization, D.S.T. and M.S.; methodology, D.S.T. and D.G.; software, D.S.T. and D.G.; validation, D.S.T. and D.G.; formal analysis, D.S.T. and D.G.; investigation, D.S.T. and D.G.; resources, M.S.; data curation, D.S.T. and M.S.; writing—original draft preparation, D.S.T.; writing—review and editing, M.S. and D.G.; visualization, D.S.T.; supervision, M.S.; project administration, M.S.; funding acquisition, M.S. All authors have read and agreed to the published version of the manuscript.

Funding: This research received no external funding.

Data Availability Statement: The latest versions (at date of manuscript preparation) of the PPMC³Fitting code, manual, and numerical examples shown in the article are available in the Electronic Supporting Information. The latest version of the PPMC³Fitting code is available at https://gitlab.desy.de/denis.tikhonov/mcmcmcfitting/ (accessed on 18 January 2024).

Acknowledgments: This work has been supported by Deutsches Elektronen-Synchrotron DESY, a member of the Helmholtz Association (HGF).

Conflicts of Interest: The authors declare no conflicts of interest.

Abbreviations

The following abbreviations are used in this manuscript:

FEL	Free-electron laser
FT	Fourier transform
IR	Infrared
LSQ	Least-squares
(rw)LSSA	(Regularized weighted) least-squares spectral analysis
MC	Monte-Carlo
PAH	Polycyclic aromatic hydrocarbon
SN	Signal-to-noise ratio
XUV	Extreme ultraviolet

Appendix A. Detailed Derivations of Delta-Shaped Pump–Probe Dynamics

Appendix A.1. Solution of the First-Order Kinetics Equations

We are solving Equation (28) of the following form ($[A^*] = y$, $k_r = k$):

$$\dot{y} = \frac{dy}{dt} = -ky \tag{A1}$$

for function $y = y(t)$ with initial condition $y(0) = y_0$. Let us rewrite this equation as:

$$\frac{dy}{y} = -k\,dt$$

and integrate the left part from y_0 to $y(t)$ and the right part of the same equation from 0 to t as:

$$\int_{y_0}^{y(t)} \frac{dy}{y} = \ln\left(\frac{y(t)}{y_0}\right) = -k\int_0^t dt = -kt\,.$$

Rearranging this result, we obtain the final solution of the form:

$$y(t) = y_0 \cdot \exp(-kt)\,, \tag{A2}$$

as given in Equation (29).

Now, we integrate Equation (31) for the product of the decay ($[C] = z$) with substituted Equation (A2), which gives:

$$\dot{z} = ky = ky_0 \cdot \exp(-kt)\,. \tag{A3}$$

Such an equation can simply be integrated in time from 0 to t with the initial condition $z(0) = 0$ as:

$$\int_0^t \dot{z}\,dt = z(t) = ky_0 \int_0^t \exp(-kt)\,dt = y_0(1 - \exp(-kt))\,. \tag{A4}$$

Appendix A.2. Coherent Quantum Dynamics without Decay

We start from the von Neumann Equation (Equation (41)):

$$i\hbar\dot{\varrho} = [\mathcal{H}, \varrho] = \mathcal{H}\varrho - \varrho\mathcal{H}\,, \tag{A5}$$

where the density matrix ϱ (Equation (38)) and Hamiltonian matrix $\mathcal{H}$ (Equation (42)) are:

$$\varrho = \begin{pmatrix} \rho_{00} & \rho_{01} \\ \rho_{10} & \rho_{11} \end{pmatrix}, \quad \mathcal{H} = \hbar\omega \begin{pmatrix} 0 & 0 \\ 0 & 1 \end{pmatrix}.$$

where $\dot{\varrho}$ is a matrix, composed of elements $\dot{\rho}_{kl}$ ($k, l = 0, 1$). By computing two products of these matrices, we obtain:

$$\mathcal{H}\varrho = \hbar\omega \begin{pmatrix} 0 & 0 \\ \rho_{10} & \rho_{11} \end{pmatrix} \quad \text{and} \quad \varrho\mathcal{H} = \hbar\omega \begin{pmatrix} 0 & \rho_{01} \\ 0 & \rho_{11} \end{pmatrix}.$$

Substitution of these results in the initial equation results in a matrix equation:

$$i\hbar \begin{pmatrix} \dot{\rho}_{00} & \dot{\rho}_{01} \\ \dot{\rho}_{10} & \dot{\rho}_{11} \end{pmatrix} = \hbar\omega \begin{pmatrix} 0 & -\rho_{01} \\ +\rho_{10} & 0 \end{pmatrix}. \tag{A6}$$

We can rewrite this equation as a system of equations for corresponding individual matrix elements from the left and right sides:

$$\begin{cases} i\hbar\dot{\rho}_{00} = 0\,, \\ i\hbar\dot{\rho}_{01} = -\hbar\omega\rho_{01}\,, \\ i\hbar\dot{\rho}_{11} = 0\,, \\ i\hbar\dot{\rho}_{10} = +\hbar\omega\rho_{10}\,. \end{cases} \tag{A7}$$

By dividing the left and right sides by $i\hbar$, we arrive at Equations (43). First, let us integrate the equations, describing the dynamics of the diagonal elements (ρ_{00} and ρ_{11}) with initial conditions (Equation (40)) of $\rho_{00}(0) = 1 - p$ and $\rho_{11}(0) = p$:

$$\begin{cases} \int_{1-p}^{\rho_{00}(t)} \dot{\rho}_{00} dt = \rho_{00}(t) - (1-p) = 0 \int_0^t dt = 0 \Rightarrow \rho_{00}(t) = 1-p, \\ \int_p^{\rho_{11}(t)} \dot{\rho}_{11} dt = \rho_{11}(t) - p = 0 \int_0^t dt = 0 \Rightarrow \rho_{11}(t) = p. \end{cases}$$

For off-diagonal elements, the rate equations have the same form as Equation (A1) ($y = \rho_{01}, \rho_{10}$ and $k = \mp i\omega$). Thus, the solution is given with Equation (A2), that is:

$$\begin{cases} \rho_{01}(t) = \rho_{01}(0) \cdot \exp(+i\omega t), \\ \rho_{10}(t) = \rho_{10}(0) \cdot \exp(-i\omega t). \end{cases}$$

Taking into account the initial conditions (Equation (40)) of $\rho_{01}(0) = \sqrt{p \cdot (1-p)} \exp(+i\varphi)$ and $\rho_{10}(0) = \sqrt{p \cdot (1-p)} \exp(-i\varphi)$, we arrive at:

$$\begin{cases} \rho_{01}(t) = \sqrt{p \cdot (1-p)} \cdot \exp(+i\varphi + i\omega t), \\ \rho_{10}(t) = \sqrt{p \cdot (1-p)} \cdot \exp(-i\varphi - i\omega t). \end{cases}$$

Appendix A.3. Relation between Quantum and Classical Regimes

The classical incoherent regime in coherent regimes (Equation (51) or (57)) appears as a result of incoherent excitation. For instance, such an incoherent regime can appear when different molecules have different imprinted phases φ from the pump pulse. In this case, we need to average all these signals from all the incoherent molecules. Let us assume that the following expression gives the result of a molecule with an imprinted phase φ:

$$f(t_{pp}, \varphi) = F_0(t_{pp}) + F_1(t_{pp}) \cdot \cos(\omega t + \varphi),$$

where $F_0(t_{pp})$ is the phase-independent part of observable and $F_1(t_{pp})$ is the oscillation prefactor. Both Equations (51) and (57) can be reduced to such a form. The distribution of the molecules with various oscillation phases φ will be given by a probability distribution $P(\varphi)$, normalized as $\int_0^{2\pi} P(\varphi) d\varphi = 1$. In this case, the result of the ensemble observation will be:

$$f(t_{pp}) = \int_0^{2\pi} f(t_{pp}, \varphi) P(\varphi) d\varphi.$$

If we now assume that all phases are equally possible, i.e., $P(\varphi)$ is a uniform distribution for $\varphi \in [0; 2\pi)$ ($P(\varphi) = (2\pi)^{-1}$), then we get an averaging result:

$$f(t_{pp}) = \frac{1}{2\pi} \int_0^{2\pi} f(t_{pp}, \varphi) d\varphi =$$
$$= F_0(t_{pp}) \cdot \underbrace{\frac{1}{2\pi} \int_0^{2\pi} d\varphi}_{2\pi} + F_1(t_{pp}) \cdot \underbrace{\frac{1}{2\pi} \int_0^{2\pi} \cos(\omega t + \varphi) d\varphi}_{0} = F_0(t_{pp}).$$

In other words, the oscillating observables disappear, and only the nonoscillating incoherent part of the signal is left. This result can also be obtained by setting the nondiagonal elements to zero in the density matrix in Equation (47), which is known to lead to a classical regime of the quantum system [78,79].

Appendix A.4. Coherent Quantum Dynamics with Decay

The initial Linblad Equation (52) has the following form:

$$i\hbar\dot{\varrho} = [\mathcal{H}, \varrho] + i\hbar\gamma\left(\sigma^-\varrho\sigma^+ - \frac{1}{2}\{\sigma^+\sigma^-, \varrho\}\right) =$$
$$= \mathcal{H}\varrho - \varrho\mathcal{H} + i\hbar\gamma\left(\sigma^-\varrho\sigma^+ - \frac{1}{2}\sigma^+\sigma^-\varrho - \frac{1}{2}\varrho\sigma^+\sigma^-\right). \quad \text{(A8)}$$

The $\sigma^{\pm}$ matrices (Equation (53)) are the excitation/deexcitation operators. If we represent the system's wavefunction $|\psi\rangle = c_0|0\rangle + c_1|1\rangle$ with a vector of coefficients (c_0, c_1), and apply $\sigma^{\pm}$ for states $|0\rangle$ and $|1\rangle$, described as $(1,0)$ and $(0,1)$, respectively, then we get:

$$\sigma^+|0\rangle = \begin{pmatrix} 0 & 0 \\ 1 & 0 \end{pmatrix}\begin{pmatrix} 1 \\ 0 \end{pmatrix} = \begin{pmatrix} 0 \\ 1 \end{pmatrix} = |1\rangle$$

and

$$\sigma^-|1\rangle = \begin{pmatrix} 0 & 1 \\ 0 & 0 \end{pmatrix}\begin{pmatrix} 0 \\ 1 \end{pmatrix} = \begin{pmatrix} 1 \\ 0 \end{pmatrix} = |0\rangle .$$

To get the matrix equation for the system evolution, we need to calculate the additional term $i\hbar\gamma(\sigma^-\varrho\sigma^+ - (1/2)\cdot\{\sigma^+\sigma^-, \varrho\})$ of Equation (A5), which describes the decay in the quantum system. For this, we evaluate each of the following components:

$$\sigma^-\varrho\sigma^+ = \begin{pmatrix} 0 & 1 \\ 0 & 0 \end{pmatrix}\begin{pmatrix} \rho_{00} & \rho_{01} \\ \rho_{10} & \rho_{11} \end{pmatrix}\begin{pmatrix} 0 & 0 \\ 1 & 0 \end{pmatrix} = \begin{pmatrix} \rho_{11} & 0 \\ 0 & 0 \end{pmatrix},$$

$$\sigma^+\sigma^-\varrho = \begin{pmatrix} 0 & 0 \\ 1 & 0 \end{pmatrix}\begin{pmatrix} 0 & 1 \\ 0 & 0 \end{pmatrix}\begin{pmatrix} \rho_{00} & \rho_{01} \\ \rho_{10} & \rho_{11} \end{pmatrix} = \begin{pmatrix} 0 & 0 \\ \rho_{10} & \rho_{11} \end{pmatrix},$$

and

$$\varrho\sigma^+\sigma^- = \begin{pmatrix} \rho_{00} & \rho_{01} \\ \rho_{10} & \rho_{11} \end{pmatrix}\begin{pmatrix} 0 & 0 \\ 1 & 0 \end{pmatrix}\begin{pmatrix} 0 & 1 \\ 0 & 0 \end{pmatrix} = \begin{pmatrix} 0 & \rho_{01} \\ 0 & \rho_{11} \end{pmatrix}.$$

Substituting these matrices into Equation (A8) with the previously evaluated first part (Equation (A6)), we get:

$$i\hbar\begin{pmatrix} \dot{\rho}_{00} & \dot{\rho}_{01} \\ \dot{\rho}_{10} & \dot{\rho}_{11} \end{pmatrix} = \hbar\omega\begin{pmatrix} 0 & -\rho_{01} \\ +\rho_{10} & 0 \end{pmatrix} + i\hbar\gamma\begin{pmatrix} +\rho_{11} & -\frac{1}{2}\rho_{01} \\ -\frac{1}{2}\rho_{10} & -\rho_{11} \end{pmatrix}.$$

From this matrix equation, we can obtain a modified set of Equations (A7):

$$\begin{cases} i\hbar\dot{\rho}_{00} = +i\hbar\gamma\rho_{11}, \\ i\hbar\dot{\rho}_{01} = -\hbar\omega\rho_{01} - \frac{i\hbar\gamma}{2}\rho_{01}, \\ i\hbar\dot{\rho}_{10} = +\hbar\omega\rho_{10} - \frac{i\hbar\gamma}{2}\rho_{10}. \\ i\hbar\dot{\rho}_{11} = -i\hbar\gamma\rho_{11}. \end{cases}$$

Dividing these equations by $i\hbar$, for elements ρ_{01}, ρ_{10}, and ρ_{11}, we arrive to the decay differential equations (Equation (A1)) with $k = \gamma/2 - i\omega$, $k = \gamma/2 + i\omega$, and $k = \gamma$, respectively. This results in the solution given by Equation (A2) with corresponding decay rates. A similar procedure for ρ_{00} leads to an Equation (A3) with $k = \gamma$, which leads to the solution in Equation (A4). All these results are summarized in Equation (55) in the main text.

Appendix A.5. Reaction Scheme with Multiple Products

Let us consider the reaction scheme (a) from Figure 4, which can be represented with the following system of chemical reactions:

$$\begin{cases} M + \text{pump} \to A^* \\ A^* \xrightarrow{k_1} C_1 \\ A^* \xrightarrow{k_2} C_2 \\ \ldots \\ A^* \xrightarrow{k_N} C_N \\ A^* + \text{probe} \to B \end{cases}$$

It can be considered as an extension of scheme (27) with N various decay results $C_1, C_2, \ldots, C_N$. We will denote $[A^*] = y$ ($[A^*](0) = y(0) = y_0$) and $[C_i] = z_i$ ($i = 1, \ldots, N$). In this case, the rate equation for A^* will be:

$$\dot{y} = - \underbrace{\left(\sum_{i=1}^{N} k_i\right)}_{k_{\text{eff}}} y = -k_{\text{eff}} y .$$

This rate equation is the same Equation (A1) with solution (A2). For each of the products C_i, the rate equations are:

$$\dot{z}_i = +k_i \cdot y = k_i \cdot y_0 \cdot \exp(-k_{\text{eff}} t) ,$$

which is the same as Equation (A3). Thus, the solution for product C_i is given by Equation (A4). From this, the pump–probe dynamics of $[B]$ and $[C]$ are given by Equation (33). Since we can express rate constants through the decay lifetimes ($k_i = \tau_i^{-1}$, see Equation (30)), we can express the effective rate constant as:

$$\tau_r = \frac{1}{\underbrace{\left(\sum_{i=1}^{N} k_i\right)}_{k_{\text{eff}}}} = \left(\sum_i \frac{1}{\tau_i}\right)^{-1} .$$

Appendix A.6. Reaction Scheme with Sequential Metastable Intermediates

Let us consider the reaction scheme (b) from Figure 4, which can be represented with the following system of chemical reactions:

$$\begin{cases} M + \text{pump} \to A_1^* \\ A_1^* \xrightarrow{k_1} A_2^* \\ A_2^* \xrightarrow{k_2} \ldots \\ A_2^* + \text{probe} \to B \end{cases} ,$$

where we get the formation of the intermediate A_1^* by the pump, which is followed by the decay into A_2^*, which in turn decays further. The probing is done using B. First, we solve rate equation for $[A_1^*] = y_1$:

$$\dot{y}_1 = -k_1 y_1 ,$$

which is Equation (A1) with solution (Equation (A2)):

$$y_1(t) = \overbrace{y_1(0)}^{y_{10}} \cdot \exp(-k_1 t) = y_{10} \cdot \exp(-k_1 t) . \tag{A9}$$

The rate equation for $[A_2^*] = y_2$ looks as follows:

$$\dot{y}_2 = +k_1 y_1 - k_2 y_2 , \tag{A10}$$

which can be thought of as a combination of Equations (A1) and (A3). Thus, we can try to find the solution as a combination of (A2) and (A4) as:

$$y_2(t) = \alpha_1 \exp(-k_1 t) + \alpha_2 \exp(-k_2 t) , \tag{A11}$$

where α_1 and α_2 are coefficients. By applying a boundary condition $y_2(0) = 0$, we get $\alpha_1 = -\alpha_2$. Then, substituting (A11) and (A9) into (A10), we get:

$$\alpha_1 \cdot \exp(-k_1 t) \cdot (k_2 - k_1) = y_{10} \cdot k_1 \cdot \exp(-k_1 t) .$$

Dividing this equation by $\exp(-k_1 t) \cdot (k_2 - k_1)$, we get:

$$\alpha_1 = -\alpha_2 = \frac{y_{10} \cdot k_1}{k_2 - k_1} ,$$

which results in:

$$y_2(t) = y_{10} \cdot k_1 \frac{(\exp(-k_1 t) - \exp(-k_2 t))}{(k_2 - k_1)} .$$

The yield of B at $t_{pp} < 0$ is zero, and at $t_{pp} \geq 0$ it is $[B](t_{pp}) = p \cdot y_2(t_{pp})$, where p is the probing efficiency (similar to Equation (33)). This gives the following pump–probe yield of B:

$$[B](t_{pp}) = p \cdot \frac{y_{10} \cdot k_1}{(k_2 - k_1)} \cdot \theta(t_{pp}) \cdot (\exp(-k_1 t) - \exp(-k_2 t)) .$$

Appendix A.7. Reaction Scheme with Multiple Intermediates Forming a Single Product

Let us consider the reaction scheme (c) from Figure 4, which can be represented with the following system of chemical reactions:

$$\begin{cases} M + \text{pump} \to A_1^* \\ M + \text{pump} \to A_2^* \\ A_1^* \xrightarrow{k_1} C \\ A_2^* \xrightarrow{k_2} C \\ A_1^* + \text{probe} \to B_1 \\ A_2^* + \text{probe} \to B_2 \end{cases} .$$

The kinetic equations for $[A_i^*] = y_i$ ($i = 1, 2$) are the following ones:

$$\dot{y}_i = -k_i y_i ,$$

which are the decay Equation (A1), with the solution (Equation (A2)) given as:

$$y_i(t) = y_{i0} \cdot \exp(-k_i t) ,$$

where $y_{i0} = y_i(0)$ are the initial amounts of A_1^* and A_2^* created by the pump pulse. This solution (similar to Equation (33)) provides the amount of B_i as a function of the pump–probe delay t_{pp} to be:

$$[B_i](t_{pp}) = p_i \cdot y_{i0} \cdot \theta(t_{pp}) \cdot \exp(-k_i t) , \qquad (A12)$$

where p_i is the conversion efficiency of A_i^* into B_i by the probe pulse.

The rate equation for $[C] = z$ is as follows:

$$\dot{z} = +k_1 y_1 + k_2 y_2 = k_1 y_{10} \cdot \exp(-k_1 t) + k_2 y_{20} \cdot \exp(-k_2 t) ,$$

which can be solved by direct integration with boundary condition $z(0) = 0$ as:

$$\int_0^t \dot{z} dt = z(t) = k_1 y_{10} \cdot \int_0^t \exp(-k_1 t) dt + k_2 y_{20} \cdot \int_0^t \exp(-k_2 t) dt =$$
$$y_{10}(1 - \exp(-k_1 t)) + y_{20}(1 - \exp(-k_2 t)) .$$

At $t \to \infty$, $z(t) \to y_{10} + y_{20}$. At $t_{pp} < 0$, the yield of C is given as $[C](t_{pp}) = y_{10} + y_{20}$. At $t_{pp} \geq 0$, it is going to be the amount of full conversion ($y_{10} + y_{20}$) without the amount lost for B_1 and B_2 with a probe (Equation (A12)). Therefore, the total amount of C can be written (similar to Equation (33)) as:

$$[C](t_{pp}) = \sum_{i=1}^{2} \left(y_{i0} - p_i \cdot y_{i0} \cdot \theta(t_{pp}) \cdot \exp(-k_i t) \right).$$

Appendix A.8. General form of the Decay Dynamics Pump–Probe Equations

Now, we will consider the generalized version of the decay reaction dynamics, described in Section 3.2 of the main text and Appendixes A.5, A.6, and A.7 of the Appendix. Let us consider that we have N compounds $A_1, A_2, \ldots, A_N$, formed by the pump from the initial molecule M and/or those formed by the probe pulse from the species formed by the pump pulse. The amount of each of compound A_i will be denoted as y_i. An N-dimensional vector describing the current amounts of all compounds will be denoted as:

$$\mathbf{y} = \begin{pmatrix} y_1 \\ y_2 \\ \vdots \\ y_N \end{pmatrix},$$

where the pump pulse forms the initial distribution of states $\mathbf{y}_0$.

The system of reaction equations for free evolution of the system will be the following:

$$\begin{cases} A_1 \xrightarrow{k_{1 \to 2}} s_{1 \to 2} \cdot A_2 \\ A_1 \xrightarrow{k_{1 \to 3}} s_{1 \to 3} \cdot A_3 \\ \vdots \\ A_i \xrightarrow{k_{i \to j}} s_{i \to j} \cdot A_j \\ \vdots \\ A_N \xrightarrow{k_{N \to N-1}} s_{N \to N-1} \cdot A_{N-1} \end{cases}$$

where we consider all possible reactions of decay into each other ($1 \leq i, j \leq N$ and $i \neq j$) with rate constants $k_{i \to j} \geq 0$ and stoichiometric coefficients $s_{i \to j} \geq 0$, where $k_{i \to j} = 0$ and $s_{i \to j} = 0$ mean that this reaction is not happening. The free evolution of the system is, thus, given by the following rate equation:

$$\dot{\mathbf{y}} = -\mathcal{K}\mathbf{y}, \tag{A13}$$

where matrix $\mathcal{K}$ consists of the following elements. The diagonal elements K_{ii} are given as:

$$K_{ii} = +\sum_{j=1, j \neq i}^{N} k_{i \to j},$$

while the off-diagonal elements are defined as:

$$K_{ji} = -s_{i \to j} \cdot k_{i \to j}.$$

To obtain the solution of Equation (A13) [12], we will consider the solutions to the following eigenproblem:

$$\mathcal{K}\mathbf{u}_i = \gamma_i \mathbf{u}_i.$$

All eigenvalues should be nonnegative for physically-relevant reaction schemes ($\gamma_i \geq 0$). From a set of N orthonormal eigenvectors $\mathbf{u}_i$ ($i = 1, \ldots, N$ and $\mathbf{u}_i^T \mathbf{u}_j = \delta_{ij}$), we can form a matrix of orthogonal transformation:

$$\mathcal{U} = (\mathbf{u}_1, \mathbf{u}_2, \ldots, \mathbf{u}_N).$$

We can now diagonalize matrix $\mathcal{K}$ as:

$$\Gamma = \mathcal{U}\mathcal{K}\mathcal{U}^T = \mathrm{diag}(\gamma_1, \gamma_2, \ldots, \gamma_N).$$

Now, we can replace $\mathbf{y}$ and $\dot{\mathbf{y}}$ with $\mathbf{Y} = \mathcal{U}\mathbf{y}$ and $\dot{\mathbf{Y}} = \mathcal{U}\dot{\mathbf{y}}$ by multiplying Equation (A13) with $\mathcal{U}$ from the left side:

$$\dot{\mathbf{Y}} = \mathcal{U}\dot{\mathbf{y}} = -\mathcal{U}\mathcal{K}\overbrace{\mathcal{U}^T\mathcal{U}}^{\mathcal{E}}\mathbf{y} = -\Gamma \mathbf{Y},$$

where $\mathcal{E}$ is an $N \times N$ identity matrix. The solution for this equation is:

$$\mathbf{Y}(t) = \exp(-\Gamma \cdot t)\mathbf{Y}_0,$$

where $\mathbf{Y}_0 = \mathcal{U}\mathbf{y}_0$ is the initial condition and the exponential matrix $\exp(-\Gamma \cdot t)$ is:

$$\exp(-\Gamma \cdot t) = \mathrm{diag}(\exp(-\gamma_1 t), \exp(-\gamma_2 t), \ldots, \exp(-\gamma_N t)).$$

We can go back to $\mathbf{y} = \mathcal{U}^T\mathbf{Y}$ by multiplying this solution with $\mathcal{U}^T$ from the left, resulting in:

$$\mathbf{y}(t) = \overbrace{\mathcal{U}^T \exp(-\Gamma \cdot t)\mathcal{U}}^{\mathcal{T}(t)} \mathbf{y}_0 = \mathcal{T}(t)\mathbf{y}_0,$$

where $\mathcal{T}(t)$ is the free evolution operator. If we expand each of the components y_i, we get:

$$y_i(t) = \sum_{j=1}^{N} c_{ij} \exp(-\gamma_j t),$$

where the dynamics of each of the components is a sum of exponential decays with effective rate constants (for $\gamma_j > 0$) and static yields (for $\gamma_j = 0$) with constant coefficients c_{ij} that depend on the initial conditions $\mathbf{y}_0$ and on the reaction scheme.

To describe the probe process, we will also consider another set of rate equations:

$$\begin{cases} A_1 + \text{probe} \xrightarrow{p_{1\to 2}} s'_{1\to 2} A_2 \\ A_1 + \text{probe} \xrightarrow{p_{1\to 3}} s'_{1\to 3} A_3 \\ \vdots \\ A_i + \text{probe} \xrightarrow{p_{i\to j}} s'_{i\to j} A_j \\ \vdots \\ A_N + \text{probe} \xrightarrow{p_{N\to N-1}} s'_{N\to N-1} A_{N-1} \end{cases},$$

where $0 \leq p_{i\to j} \leq 1$ describe the probing efficiency of a given process and $s'_{i\to j} \geq 0$ are the stoichiometric coefficients ($p_{i\to j} = 0$ and $s'_{i\to j} = 0$ mean that this process does not happen). Therefore, we can describe probing at time $t > 0$ as an instant shuffling of the products as replacement current values $\mathbf{y}(t)$ with $\mathcal{P}\mathbf{y}(t)$, that is:

$$\mathbf{y}(t) \to \mathcal{P}\mathbf{y}(t) \tag{A14}$$

where the matrix $\mathcal{P}$ has the following nondiagonal elements:

$$P_{ji} = s'_{i \to j} \cdot p_{i \to j}$$

and diagonal elements:

$$P_{ii} = \sum_{j=1, j \neq i}^{N} (1 - p_{i \to j}).$$

At $t_{\text{pp}} < 0$, the system continues a free evolution until the detection. Therefore, we can describe the observable yield of all the components $\mathbf{f}(t_{\text{pp}})$ as:

$$\mathbf{f}(t_{\text{pp}}) = \lim_{t \to \infty} \mathbf{y}(t) = \overbrace{\left(\lim_{t \to \infty} \mathcal{T}(t)\right)}^{\mathcal{T}_\infty} \mathbf{y}_0 = \mathcal{T}_\infty \mathbf{y}_0, \quad (A15)$$

where $\mathcal{T}_\infty$ is given as:

$$\mathcal{T}_\infty = \mathcal{U}^{\text{T}} \overbrace{\text{diag}(\theta(-\gamma_1), \theta(-\gamma_2), \ldots, \theta(-\gamma_N))}^{\lim_{t \to \infty} \exp(-\Gamma \cdot t)} \mathcal{U},$$

where for $\gamma_i = 0$ we get $\theta(-\gamma_i) = 1$ and for $\gamma_i > 0$ we get $\theta(-\gamma_i) = 0$.

At $t_{\text{pp}} > 0$, we get a free evolution from the initial conditions $\mathbf{y}_0$, then the action of the probe according to Equation (A14), and then again free evolution until detection. We can write this (similar to Equation A15) as:

$$\mathbf{f}(t_{\text{pp}}) = \mathcal{T}_\infty \mathcal{P} \mathcal{T}(t_{\text{pp}}) \mathbf{y}_0. \quad (A16)$$

Combining Equations (A15) and (A16), we get the following equation describing all the pump–probe yields of all products:

$$\mathbf{f}(t_{\text{pp}}) = \mathcal{T}_\infty \cdot (\mathcal{E} + \theta(t_{\text{pp}}) \cdot (\mathcal{P}\mathcal{T}(t_{\text{pp}}) - \mathcal{E})) \mathbf{y}_0.$$

For each of the compounds A_i, their pump–probe yield can be described as:

$$f_i(t_{\text{pp}}) = q_0 + \sum_{j=1}^{N} q_i \cdot \theta(t_{\text{pp}}) \cdot \exp(-\gamma_i t_{\text{pp}}),$$

where q_j ($j = 0, 1, \ldots, N$) are the coefficients related to initial conditions $\mathbf{y}_0$, reaction scheme, and also probing efficiencies. Terms with exponents appear from the operator $\mathcal{T}(t_{\text{pp}})$ in the expression above.

Appendix B. Effects of the Duration of the Pulses and Experimental Setup Jitter

Appendix B.1. Sequential Convolution with Gaussian-Shaped Pulses

Let us consider a sequential convolution of the following form:

$$f \circledast p_1 \circledast p_2 =$$

$$= \frac{1}{\pi \tau_1 \tau_2} \int_{-\infty}^{+\infty} \left(\int_{-\infty}^{+\infty} f(t - t_1 - t_2) \cdot \exp\left(-\frac{t_1^2}{\tau_1^2}\right) dt_1 \right) \cdot \exp\left(-\frac{t_2^2}{\tau_2^2}\right) dt_2, \quad (A17)$$

where $f(t)$ is some arbitrary function, and $p_i(t)$ ($i = 1, 2$) are the normal distributions:

$$p_i(t) = \frac{1}{\sqrt{\pi} \cdot \tau_i} \cdot \exp\left(-\frac{t_i^2}{\tau_i^2}\right).$$

Trying to simplify Equation (A17), we can replace variables t_1 and t_2 with:

$$\begin{cases} T = t_1 + t_2, \\ q = t_2 - t_1, \end{cases} \quad \text{i.e.,} \quad \begin{cases} t_1 = \frac{T-q}{2}, \\ t_2 = \frac{T+q}{2}. \end{cases} \tag{A18}$$

The Jacobian J for such transformation is:

$$J = \det \begin{pmatrix} \frac{\partial t_1}{\partial T} & \frac{\partial t_2}{\partial T} \\ \frac{\partial t_1}{\partial q} & \frac{\partial t_2}{\partial q} \end{pmatrix} = \det \begin{pmatrix} +1/2 & +1/2 \\ -1/2 & +1/2 \end{pmatrix} = \frac{1}{2},$$

which means that $dt_1 dt_2 = J dT dq = \frac{1}{2} dT dq$. Substitution of coordinates from Equation (A18) into Equation (A17) results in:

$$f \circledast p_1 \circledast p_2 =$$
$$= \frac{1}{2\pi\tau_1\tau_2} \int_{-\infty}^{+\infty} f(t-T) \underbrace{\left(\int_{-\infty}^{+\infty} \exp\left(-\frac{(T-q)^2}{4\tau_1^2} - \frac{(T+q)^2}{4\tau_2^2} \right) dq \right)}_{I(T)} dT, \tag{A19}$$

and therefore, we only need to evaluate the integral $I(T)$. First, let us rearrange the expression inside the exponent as:

$$-\frac{(T-q)^2}{4\tau_1^2} - \frac{(T+q)^2}{4\tau_2^2} =$$

$$= -\frac{(\tau_1^2 + \tau_2^2)}{4\tau_1^2\tau_2^2} \cdot \left(T^2 + q^2 - 2 \overbrace{\frac{(\tau_2^2 - \tau_1^2)}{(\tau_1^2 + \tau_2^2)}}^{a} \cdot T \cdot q \right) =$$

$$= -\frac{(\tau_1^2 + \tau_2^2)}{4\tau_1^2\tau_2^2} \cdot \left(T^2 + q^2 - 2 \cdot a \cdot q + a^2 - a^2 \right) =$$

$$= -\frac{(\tau_1^2 + \tau_2^2)}{4\tau_1^2\tau_2^2} \cdot \left(1 - \frac{(\tau_2^2 - \tau_1^2)^2}{(\tau_1^2 + \tau_2^2)^2} \right) \cdot T^2 - \frac{(\tau_1^2 + \tau_2^2)}{4\tau_1^2\tau_2^2} \cdot (q - a)^2 =$$

$$= -\frac{T^2}{\tau_{12}^2} - \frac{(q-a)^2}{\tau_q^2}, \tag{A20}$$

where

$$\tau_{12}^2 = \tau_1^2 + \tau_2^2 \tag{A21}$$

and

$$\tau_q^2 = \frac{4\tau_1^2\tau_2^2}{(\tau_1^2 + \tau_2^2)} = \frac{4\tau_1^2\tau_2^2}{\tau_{12}^2}.$$

Substitution of the result from Equation (A20) into $I(T)$ from Equation (A19) results in:

$$I(T) = \int_{-\infty}^{+\infty} \exp\left(-\frac{(T-q)^2}{4\tau_1^2} - \frac{(T+q)^2}{4\tau_2^2} \right) dq =$$

$$= \exp\left(-\frac{T^2}{\tau_{12}^2} \right) \cdot \int_{-\infty}^{+\infty} \exp\left(-\frac{(q-a)^2}{\tau_q^2} \right) dq = \sqrt{\pi}\tau_q \cdot \exp\left(-\frac{T^2}{\tau_{12}^2} \right) =$$

$$= 2\sqrt{\pi} \frac{\tau_1 \tau_2}{\tau_{12}} \cdot \exp\left(-\frac{T^2}{\tau_{12}^2} \right),$$

which upon substitution into Equation (A19) and replacement of $T = t_{12}$ provides us a final answer of:

$$f \circledast p_1 \circledast p_2 = \frac{1}{\sqrt{\pi}\tau_{12}} \int_{-\infty}^{+\infty} f(t - t_{12}) \cdot \exp\left(-\frac{t_{12}^2}{\tau_{12}^2}\right) dt_{12} . \quad (A22)$$

In other words, convolution of function $f(t)$ with two Gaussian functions is equivalent to convolution with a single effective Gaussian function:

$$f_{12}(t) = \frac{1}{\sqrt{\pi}\tau_{12}} \exp\left(-\frac{t^2}{\tau_{12}^2}\right),$$

where the effective width τ_{12}^2 is given by Equation (A21). If we require the calculation of a triple convolution, we can apply this result sequentially as:

$$\overbrace{f \circledast p_1 \circledast p_2}^{f \circledast p_{12}} \circledast p_3 = f \circledast p_{123} ,$$

where the final effective Gaussian function $p_{123}(t)$ will have the effective width $\tau_{123}^2 = \tau_1^2 + \tau_2^2 + \tau_3^3$. Such a process can be continued for any arbitrary number of Gaussian functions, yielding the result listed in Equation (72).

Appendix B.2. Basis Functions for Fitting Observables with Finite Duration Pump/Probe Pulses and Experimental Jitter

Here, we explicitly calculate the convolution of the basis functions b_x (x = c, s, t, i, bo, to, given in Equations (61), (62), (63), (64), (65) and (66), respectively) for describing the instant pump–instant probe observables with the effective distribution $p(t) = (\sqrt{\pi}\tau_{cc})^{-1} \cdot \exp(-t^2/\tau_{cc}^2)$ (see Equation (73)) representing fluctuation of the pump–probe delay t_{pp}. The expression we will evaluate is the following (see Equations (72) and (76)):

$$B_x(t_{pp}) = b_x \circledast p = \frac{1}{\sqrt{\pi}\tau_{cc}} \int_{-\infty}^{+\infty} b_x(t_{pp} - t) \cdot \exp\left(-\frac{t^2}{\tau_{cc}^2}\right) dt . \quad (A23)$$

Appendix B.2.1. Constant Function

The first basis function is the constant ("c") function $b_c(t_{pp}) = 1$ (Equation (61)). Substituting it into Equation (A23) we get:

$$B_c(t_{pp}) = b_c \circledast p = \frac{1}{\sqrt{\pi}\tau_{cc}} \overbrace{\int_{-\infty}^{+\infty} \exp\left(-\frac{t^2}{\tau_{cc}^2}\right) dt}^{\sqrt{\pi}\tau_{cc}} = 1 .$$

Appendix B.2.2. Step Function

The second basis is the step ("s") function $b_s(t_{pp}) = \theta(t_{pp})$ (Equation (62)). Substituting it into Equation (A23) we get:

$$B_s(t_{pp}) = b_s \circledast p = \frac{1}{\sqrt{\pi}\tau_{cc}} \int_{-\infty}^{+\infty} \theta(t_{pp} - t) \exp\left(-\frac{t^2}{\tau_{cc}^2}\right) dt . \quad (A24)$$

The Heaviside step function $\theta(x)$ (Equation (24)) is nonzero only for values of argument $x \geq 0$ ($\theta(x) = 1$). Thus, the nonzero values of the integral are given by inequality $t_{pp} - t \geq 0 \Rightarrow t \leq t_{pp}$. With that, we can rewrite integral in Equation (A24) as:

$$\int_{-\infty}^{+\infty} \theta(t_{pp} - t) \exp\left(-\frac{t^2}{\tau_{cc}^2}\right) dt = \int_{-\infty}^{t_{pp}} \exp\left(-\frac{t^2}{\tau_{cc}^2}\right) dt =$$

$$\underbrace{\int_{-\infty}^{0} \exp\left(-\frac{t^2}{\tau_{cc}^2}\right) dt}_{\sqrt{\pi}\tau_{cc}/2} + \underbrace{\int_{0}^{t_{pp}} \exp\left(-\frac{t^2}{\tau_{cc}^2}\right) dt}_{\sqrt{\pi}\tau_{cc}\cdot\text{erf}(t_{pp}/\tau_{cc})/2} = \frac{\sqrt{\pi}\tau_{cc}}{2} \cdot \left(1 + \text{erf}\left(\frac{t_{pp}}{\tau_{cc}}\right)\right). \quad \text{(A25)}$$

Substitution of this result into Equation (A24) results in:

$$B_s(t_{pp}) = b_s \circledast p = \frac{1}{2} \cdot \left(1 + \text{erf}\left(\frac{t_{pp}}{\tau_{cc}}\right)\right).$$

Appendix B.2.3. Transient Function

The third basis is the transient ("t") function $b_t(t_{pp}) = \theta(t_{pp}) \exp(-kt_{pp})$ (Equation (63)), where $k = \tau_r^{-1}$ (Equation (30)). Substituting it into Equation (A23), we get:

$$B_t(t_{pp}) = b_t \circledast p = \frac{1}{\sqrt{\pi}\tau_{cc}} \int_{-\infty}^{+\infty} \theta(t_{pp} - t) \exp\left(-k \cdot (t_{pp} - t) - \frac{t^2}{\tau_{cc}^2}\right) dt. \quad \text{(A26)}$$

Removing the zero values of the Heaviside step function, similar to that in Equation (A25), we can rewrite the integral in Equation (A26) as:

$$\int_{-\infty}^{+\infty} \theta(t_{pp} - t) \exp\left(-k \cdot (t_{pp} - t) - \frac{t^2}{\tau_{cc}^2}\right) dt =$$

$$= \int_{-\infty}^{t_{pp}} \exp\left(-k \cdot (t_{pp} - t) - \frac{t^2}{\tau_{cc}^2}\right) dt = \exp(-kt_{pp}) \int_{-\infty}^{t_{pp}} \exp\left(-\frac{t^2}{\tau_{cc}^2} + kt\right) dt =$$

$$= \exp\left(\frac{k^2\tau_{cc}^2}{4}\right) \cdot \exp(-kt_{pp}) \int_{-\infty}^{t_{pp}} \exp\left(-\frac{1}{\tau_{cc}^2} \cdot \underbrace{\left(t - \frac{k\tau_{cc}^2}{2}\right)^2}_{q^2}\right) dt =$$

$$= \exp\left(\frac{k^2\tau_{cc}^2}{4}\right) \cdot \exp(-kt_{pp}) \int_{-\infty}^{t_{pp} - k\tau_{cc}^2/2} \exp\left(-\frac{q^2}{\tau_{cc}^2}\right) dq =$$

$$= \exp\left(\frac{k^2\tau_{cc}^2}{4}\right) \cdot \exp(-kt_{pp}) \cdot \left(\underbrace{\int_{-\infty}^{0} \exp\left(-\frac{q^2}{\tau_{cc}^2}\right) dq}_{\sqrt{\pi}\tau_{cc}/2} + \underbrace{\int_{0}^{t_{pp} - k\tau_{cc}^2/2} \exp\left(-\frac{t^2}{\tau_{cc}^2}\right) dt}_{\sqrt{\pi}\tau_{cc}\cdot\text{erf}(t_{pp}/\tau_{cc} - k\tau_{cc}/2)/2}\right) =$$

$$= \frac{\sqrt{\pi}\tau_{cc}}{2} \cdot \exp\left(\frac{k^2\tau_{cc}^2}{4}\right) \cdot \exp(-kt_{pp}) \cdot \left(1 + \text{erf}\left(\frac{t_{pp}}{\tau_{cc}} - \frac{k\tau_{cc}}{2}\right)\right)$$

Substitution of this result into Equation (A26) results in:

$$B_t(t_{pp}) = b_t \circledast p = \frac{1}{2} \exp\left(\frac{k^2\tau_{cc}^2}{4}\right) \cdot \exp(-kt_{pp}) \cdot \left(1 + \text{erf}\left(\frac{t_{pp}}{\tau_{cc}} - \frac{k\tau_{cc}}{2}\right)\right). \quad \text{(A27)}$$

The constant term $\exp\left(\frac{k^2\tau_{cc}^2}{4}\right)$ is preserved further, since it is crucial for keeping the values of this function from approaching zero in the whole pump–probe range.

Appendix B.2.4. Instant Increase Function

The fourth basis is the instant increase ("i") function $b_i(t_{pp}) = \delta(t_{pp})$ (Equation (64)). Substituting it into Equation (A23) we get (by definition of the Dirac delta function):

$$B_i(t_{pp}) = b_i \circledast p = \frac{1}{\sqrt{\pi}\tau_{cc}} \int_{-\infty}^{+\infty} \delta(t_{pp} - t) \exp\left(-\frac{t^2}{\tau_{cc}^2}\right) dt = \frac{1}{\sqrt{\pi}\tau_{cc}} \exp\left(-\frac{t_{pp}^2}{\tau_{cc}^2}\right).$$

The normalization factor $(\sqrt{\pi}\tau_{cc})^{-1}$ is later ignored to have the maximal value of this function fixed at one.

Appendix B.2.5. Nondecaying Coherent Oscillation Function

The fifth basis is the coherent oscillation ("o") function $b_o(t_{pp}) = \theta(t_{pp}) \cdot \cos(\omega t_{pp} + \varphi)$ (Equation (65)). Substituting it into Equation (A23), we get:

$$B_o(t_{pp}) = b_o \circledast p = \frac{1}{\sqrt{\pi}\tau_{cc}} \int_{-\infty}^{+\infty} \theta(t_{pp} - t) \cos(\omega \cdot (t_{pp} - t) + \varphi) \cdot \exp\left(-\frac{t^2}{\tau_{cc}^2}\right) dt. \quad (A28)$$

We can represent cosine as:

$$\cos(x) = \frac{\exp(ix) + \exp(-ix)}{2},$$

and thus, rewriting $b_o(t_{pp})$ as:

$$b_o(t_{pp}) =$$
$$= \frac{1}{2} \cdot \exp(+i\varphi) \cdot \overbrace{\theta(t_{pp}) \cdot \exp(+i\omega t_{pp})}^{b_+(t_{pp})} + \frac{1}{2} \cdot \exp(-i\varphi) \cdot \overbrace{\theta(t_{pp}) \cdot \exp(-i\omega t_{pp})}^{b_-(t_{pp})} =$$
$$= \frac{1}{2} \cdot \left(\exp(+i\varphi) \cdot b_+(t_{pp}) + \exp(-i\varphi) \cdot b_-(t_{pp})\right), \quad (A29)$$

where new functions $b_\pm(t_{pp})$ look the same as $b_t(t_{pp})$, but with the complex rate constants $k_\pm = \pm i\omega$. Thus, we can reduce Equation (A28) to a linear combination of Equations of type (A26). Applying this, we obtain the following result from Equation (A27):

$$B_o(t_{pp}) = \frac{1}{2} \cdot \left(\exp(+i\varphi) \cdot (b_+ \circledast p)(t_{pp}) + \exp(-i\varphi) \cdot (b_- \circledast p)(t_{pp})\right) =$$
$$= \frac{1}{4} \exp\left(-\frac{\omega^2 \tau_{cc}^2}{4}\right) \cdot \left[\exp(+i\omega t_{pp} + i\varphi) \cdot \left(1 + \text{erf}\left(\frac{t_{pp}}{\tau_{cc}} + \frac{i\omega \tau_{cc}}{2}\right)\right) + \right.$$
$$\left. + \exp(-i\omega t_{pp} - i\varphi) \cdot \left(1 + \text{erf}\left(\frac{t_{pp}}{\tau_{cc}} - \frac{i\omega \tau_{cc}}{2}\right)\right)\right] \quad (A30)$$

which cannot be properly simplified further. However, we can apply an approximation that $\omega \tau_{cc} \approx 0$, which is equivalent to the statement that the oscillation period $\tau_o = 2\pi/\omega$ is much larger than the cross-correlation time ($\tau_o \gg \tau_{cc}$). In this case, Equation (A30) can be rewritten as:

$$B_o(t_{pp}) = \frac{1}{2} \cos(\omega t_{pp} + \varphi) \cdot \left(1 + \text{erf}\left(\frac{t_{pp}}{\tau_{cc}}\right)\right).$$

Appendix B.2.6. Transient Coherent Oscillation Function

The sixth basis is the transient coherent oscillation ("to") function $b_{\text{to}}(t_{\text{pp}}) = \theta(t_{\text{pp}}) \cdot \cos(\omega t_{\text{pp}} + \varphi) \cdot \exp(-kt/2)$ (Equation (66)). By substituting it into Equation (A23), we obtain:

$$B_{\text{to}}(t_{\text{pp}}) = b_{\text{to}} \circledast p =$$
$$= \frac{1}{\sqrt{\pi}\tau_{\text{cc}}} \int_{-\infty}^{+\infty} \theta(t_{\text{pp}} - t) \cos(\omega \cdot (t_{\text{pp}} - t) + \varphi) \cdot \exp\left(-\frac{k}{2} \cdot (t_{\text{pp}} - t) - \frac{t^2}{\tau_{\text{cc}}^2}\right) dt \,. \quad \text{(A31)}$$

Similar to Equation (A29), we can rewrite the original function as:

$$b_{\text{to}}(t_{\text{pp}}) = \frac{1}{2} \cdot \left(\exp(+i\varphi) \cdot b_{+}(t_{\text{pp}}) + \exp(-i\varphi) \cdot b_{-}(t_{\text{pp}})\right),$$

where in this case:

$$b_{\pm}(t_{\text{pp}}) = \theta(t_{\text{pp}}) \cdot \exp(-k_{\pm}t_{\text{pp}})$$

with effective complex rate constants $k_{\pm} = (k/2) \mp i\omega$. By applying Equations (A27)–(A31) in the same fashion as in Equation (A30), and also applying approximation $\omega\tau_{\text{cc}} \approx 0$, we arrive at the final result:

$$B_{\text{to}}(t_{\text{pp}}) = \frac{1}{2} \exp\left(\frac{k^2 \tau_{\text{cc}}^2}{16}\right) \cdot \exp\left(-\frac{k t_{\text{pp}}}{2}\right) \cdot \cos(\omega t_{\text{pp}} + \varphi) \cdot \left(1 + \text{erf}\left(\frac{t_{\text{pp}}}{\tau_{\text{cc}}} - \frac{k \tau_{\text{cc}}}{4}\right)\right).$$

Appendix C. Regularized Weighted Least-Squares Spectral Analysis (rwLSSA)

The detailed derivation of the rwLSSA technique is provided in Ref. [68], focusing on the sine-FT. Here, we only outline the basic steps of the method for the general case of the FT. The time-dependent signal $y = y(t)$ is given as a discrete set of N points $\{t_n, y_i = y(t_n), \sigma_n\}_{n=1}^{N}$ with uncertainties σ_n for each point y_n. Our goal is to get an M-point spectral representation of these data with a set of points $\{\omega_m, f_m = f(\omega_m), \varsigma_m\}_{m=1}^{M}$, where ω_m refers to the points in the grid of angular frequencies, $f_m = A_m \cdot \exp(i\varphi_m)$ refers to the complex spectra with amplitude $A_m \in \mathbb{R}$ and phase $\varphi_m \in [-\pi; \pi)$, and ς_m refers to the m-th point's amplitude uncertainty.

The spectrum is computed through the following expression [68]:

$$\mathbf{f} = \Sigma_\alpha \left(\mathcal{S}^\dagger \mathcal{W}\right) \mathbf{y},$$

with the following components:

- $\mathbf{y} = (y_1, y_2, \ldots, y_N)$ is the N-dimensional vector of the data points;
- $\mathbf{f} = (f_1, f_2, \ldots, f_M)$ is the M-dimensional vector of spectral representation;
- $\mathcal{S}$ is the matrix of size $N \times M$ with elements $\mathcal{S}_{nm} = \exp(-i\omega_m t_n)$;
- $\mathcal{W} = \text{diag}(\sigma_1^{-2}, \sigma_2^{-2}, \ldots, \sigma_N^{-2})$ is the $N \times N$ diagonal matrix of weights;
- $\alpha \geq 0$ is the regularization parameter;
- Σ_α is the $M \times M$ covariance matrix defined as $\Sigma_\alpha^{-1} = \alpha \mathcal{E} + \mathcal{S}^\dagger \mathcal{W} \mathcal{S}$, where $\mathcal{E} = \text{diag}(1, 1, \ldots, 1)$ is the unit matrix of size $M \times M$.

The uncertainties of the spectral amplitudes ς_m are computed from the m-th diagonal elements $\Sigma_{\alpha,mm}$ and $(\mathcal{S}^\dagger \mathcal{W} \mathcal{S})_{mm}$ of the matrices Σ_α and $(\mathcal{S}^\dagger \mathcal{W} \mathcal{S})$, respectively, as [68]:

$$\varsigma_m = \sqrt{\Sigma_{\alpha,mm} \cdot (\alpha + (\mathcal{S}^\dagger \mathcal{W} \mathcal{S})_{mm}) / (\mathcal{S}^\dagger \mathcal{W} \mathcal{S})_{mm}} \,.$$

The regularization parameter was chosen automatically with an *a priori* regularization criterion from Ref. [68] as:

$$\alpha = \text{tr}\left(\mathcal{S}^\dagger \mathcal{W} \mathcal{S}\right) \cdot \left(M + \frac{N}{M} \cdot \text{tr}\left(\mathcal{S}^\dagger \mathcal{W} \mathcal{S}\right) \cdot \frac{(\mathbf{y}^T \mathcal{W} \mathbf{y})}{\text{tr}(\mathcal{W})}\right)^{-1}.$$

Appendix D. Issues with Numerical Implementation of the $B_t(t_{pp})$ Basis Function

The $B_t(t_{pp})$ basis function (Equation (80)), and also the $B_{to}(t_{pp})$ basis function (Equation (83)), have issues in the numerical usage, due to the unstable behavior in the $t_{pp} < 0$ region, where the near-zero $(1 + \text{erf}(x))$ function is being multiplied by an exponential function, which highlights all the small numerical noise effects.

To demonstrate it, we can calculate the $B_t(t_{pp})$ with different ratios τ_{cc}/τ_r. Here, we compare three alternative numerical representations of $B_t(t_{pp})$ with $\tau_r = 50$ fs and $\tau_r = 253, 457, 2390, 2593$ fs. The $B_t(t_{pp})$ was computed using the Gnuplot's implementation of $(1 + \text{erf}(x))$, and with two alternative Python implementations of $(1 + \text{erf}(x))$ from SciPy package [73]: using scipy.special.erf (definition #1) and using the cumulative distribution function of the normal distribution scipy.stats.norm.cdf (definition #2).

The results are shown in Figure A1. As one can see, at ratio $\tau_{cc}/\tau_r = 5$, all three numerical results are equivalent. At the ratio $\tau_{cc}/\tau_r = 9$, the Gnuplot and definition #1 start to fail, producing spurious oscillations near $t_{pp} = -0.5$ ps. At higher ratios, they eventually both fail, producing zeros. The definition #2 (through cumulative distribution function) is the most stable, allowing to reach $\tau_{cc}/\tau_r = 48$. However, at $\tau_{cc}/\tau_r = 52$ even definition #2 starts to fail, producing a cutoff at $t_{pp} = -1.8$, which is an abrupt break. Therefore, a stable ratio τ_{cc}/τ_r, where the definition #2 works stably is estimated to be $\tau_{cc}/\tau_r < 50$.

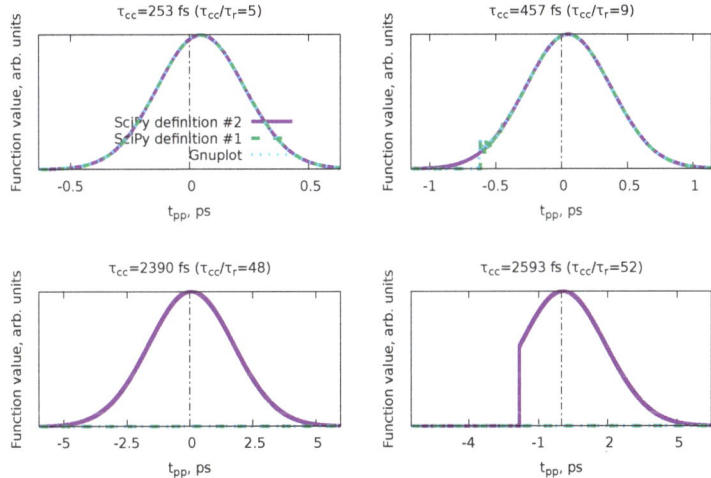

Figure A1. Illustration of numerical calculation of the $B_t(t_{pp})$ basis function (Equation (80)) with alternative numerical functions and various ratios of τ_{cc}/τ_r. Here, $\tau_r = 50$ fs and the region plotted is $-2.5 \cdot \tau_{cc} \leq t_{pp} \leq +2.5 \cdot \tau_{cc}$.

References

1. Zewail, A.H. Femtochemistry: Atomic-Scale Dynamics of the Chemical Bond Using Ultrafast Lasers (Nobel Lecture). *Angew. Chem. Int. Ed.* **2000**, *39*, 2586–2631. [CrossRef]
2. Young, L.; Ueda, K.; Gühr, M.; Bucksbaum, P.H.; Simon, M.; Mukamel, S.; Rohringer, N.; Prince, K.C.; Masciovecchio, C.; Meyer, M.; et al. Roadmap of ultrafast X-ray atomic and molecular physics. *J. Phys. At. Mol. Opt. Phys.* **2018**, *51*, 032003. [CrossRef]
3. Zettergren, H.; Domaracka, A.; Schlathölter, T.; Bolognesi, P.; Díaz-Tendero, S.; Łabuda, M.; Tosic, S.; Maclot, S.; Johnsson, P.; Steber, A.; et al. Roadmap on dynamics of molecules and clusters in the gas phase. *Eur. Phys. J.* **2021**, *75*, 152. [CrossRef]
4. Hertel, I.V.; Radloff, W. Ultrafast dynamics in isolated molecules and molecular clusters. *Rep. Prog. Phys.* **2006**, *69*, 1897. [CrossRef]
5. Scutelnic, V.; Tsuru, S.; Pápai, M.; Yang, Z.; Epshtein, M.; Xue, T.; Haugen, E.; Kobayashi, Y.; Krylov, A.I.; Møller, K.B.; et al. X-ray transient absorption reveals the 1Au (nπ*) state of pyrazine in electronic relaxation. *Nat. Commun.* **2021**, *12*, 5003. [CrossRef]
6. Lee, J.W.L.; Tikhonov, D.S.; Chopra, P.; Maclot, S.; Steber, A.L.; Gruet, S.; Allum, F.; Boll, R.; Cheng, X.; Düsterer, S.; et al. Time-resolved relaxation and fragmentation of polycyclic aromatic hydrocarbons investigated in the ultrafast XUV-IR regime. *Nat. Commun.* **2021**, *12*, 6107. [CrossRef] [PubMed]

7. Garg, D.; Lee, J.W.L.; Tikhonov, D.S.; Chopra, P.; Steber, A.L.; Lemmens, A.K.; Erk, B.; Allum, F.; Boll, R.; Cheng, X.; et al. Fragmentation Dynamics of Fluorene Explored Using Ultrafast XUV-Vis Pump-Probe Spectroscopy. *Front. Phys.* **2022**, *10*, 880793. [CrossRef]
8. Calegari, F.; Ayuso, D.; Trabattoni, A.; Belshaw, L.; Camillis, S.D.; Anumula, S.; Frassetto, F.; Poletto, L.; Palacios, A.; Decleva, P.; et al. Ultrafast electron dynamics in phenylalanine initiated by attosecond pulses. *Science* **2014**, *346*, 336–339. [CrossRef] [PubMed]
9. Wolf, T.J.A.; Paul, A.C.; Folkestad, S.D.; Myhre, R.H.; Cryan, J.P.; Berrah, N.; Bucksbaum, P.H.; Coriani, S.; Coslovich, G.; Feifel, R.; et al. Transient resonant Auger—Meitner spectra of photoexcited thymine. *Faraday Discuss.* **2021**, *228*, 555–570. [CrossRef] [PubMed]
10. Onvlee, J.; Trippel, S.; Küpper, J. Ultrafast light-induced dynamics in the microsolvated biomolecular indole chromophore with water. *Nat. Commun.* **2022**, *13*, 7462. [CrossRef]
11. Malý, P.; Brixner, T. Fluorescence-Detected Pump–Probe Spectroscopy. *Angew. Chem. Int. Ed.* **2021**, *60*, 18867–18875. [CrossRef]
12. Press, W.H.; Teukolsky, S.A.; Vetterling, W.T.; Flannery, B.P. *Numerical Recipes 3rd Edition: The Art of Scientific Computing*, 3rd ed.; Cambridge University Press: New York, NY, USA, 2007.
13. Mosegaard, K.; Tarantola, A. Monte Carlo sampling of solutions to inverse problems. *J. Geophys. Res. Solid Earth* **1995**, *100*, 12431–12447. [CrossRef]
14. Bingham, D.; Butler, T.; Estep, D. Inverse Problems for Physics-Based Process Models. *Annu. Rev. Stat. Its Appl.* **2024**, *11*. [CrossRef]
15. Tikhonov, A.N. Solution of incorrectly formulated problems and the regularization method. *Soviet Math. Dokl.* **1963**, *4*, 1035–1038.
16. Tikhonov, A.; Leonov, A.; Yagola, A. *Nonlinear Ill-Posed Problems*; Chapman & Hall: London, UK, 1998.
17. Hoerl, A.E.; Kennard, R.W. Ridge Regression: Biased Estimation for Nonorthogonal Problems. *Technometrics* **1970**, *12*, 55–67. [CrossRef]
18. Pedersen, S.; Zewail, A.H. Femtosecond real time probing of reactions XXII Kinetic description of probe absorption fluorescence depletion and mass spectrometry. *Mol. Phys.* **1996**, *89*, 1455–1502. [CrossRef]
19. Hadamard, J. Sur les problèmes aux dérivés partielles et leur signification physique. *Princet. Univ. Bull.* **1902**, *13*, 49–52.
20. Tikhonov, D.S.; Vishnevskiy, Y.V.; Rykov, A.N.; Grikina, O.E.; Khaikin, L.S. Semi-experimental equilibrium structure of pyrazinamide from gas-phase electron diffraction. How much experimental is it? *J. Mol. Struct.* **2017**, *1132*, 20–27. [CrossRef]
21. Hestenes, M.R.; Stiefel, E. Methods of conjugate gradients for solving linear systems. *J. Res. Natl. Bur. Stand.* **1952**, *49*, 409–436. [CrossRef]
22. Powell, M.J.D. An efficient method for finding the minimum of a function of several variables without calculating derivatives. *Comput. J.* **1964**, *7*, 155–162. [CrossRef]
23. Tibshirani, R. Regression Shrinkage and Selection via the Lasso. *J. R. Stat. Soc. Ser. (Methodol.)* **1996**, *58*, 267–288. [CrossRef]
24. Santosa, F.; Symes, W.W. Linear Inversion of Band-Limited Reflection Seismograms. *SIAM J. Sci. Stat. Comput.* **1986**, *7*, 1307–1330. [CrossRef]
25. Bauer, F.; Lukas, M.A. Comparing parameter choice methods for regularization of ill-posed problems. *Math. Comput. Simul.* **2011**, *81*, 1795–1841. [CrossRef]
26. Chiu, N.; Ewbank, J.; Askari, M.; Schäfer, L. Molecular orbital constrained gas electron diffraction studies: Part I. Internal rotation in 3-chlorobenzaldehyde. *J. Mol. Struct.* **1979**, *54*, 185–195. [CrossRef]
27. Baše, T.; Holub, J.; Fanfrlík, J.; Hnyk, D.; Lane, P.D.; Wann, D.A.; Vishnevskiy, Y.V.; Tikhonov, D.; Reuter, C.G.; Mitzel, N.W. Icosahedral Carbaboranes with Peripheral Hydrogen—Chalcogenide Groups: Structures from Gas Electron Diffraction and Chemical Shielding in Solution. *Chem.—Eur. J.* **2019**, *25*, 2313–2321. [CrossRef]
28. Vishnevskiy, Y.V.; Tikhonov, D.S.; Reuter, C.G.; Mitzel, N.W.; Hnyk, D.; Holub, J.; Wann, D.A.; Lane, P.D.; Berger, R.J.F.; Hayes, S.A. Influence of Antipodally Coupled Iodine and Carbon Atoms on the Cage Structure of 9,12-I2-closo-1,2-C2B10H10: An Electron Diffraction and Computational Study. *Inorg. Chem.* **2015**, *54*, 11868–11874. [CrossRef] [PubMed]
29. Metropolis, N.; Rosenbluth, A.W.; Rosenbluth, M.N.; Teller, A.H.; Teller, E. Equation of State Calculations by Fast Computing Machines. *J. Chem. Phys.* **2004**, *21*, 1087–1092. [CrossRef]
30. Hastings, W.K. Monte Carlo sampling methods using Markov chains and their applications. *Biometrika* **1970**, *57*, 97–109. [CrossRef]
31. Rosenbluth, M.N. Genesis of the Monte Carlo Algorithm for Statistical Mechanics. *AIP Conf. Proc.* **2003**, *690*, 22–30. [CrossRef]
32. Snellenburg, J.J.; Laptenok, S.; Seger, R.; Mullen, K.M.; van Stokkum, I.H.M. Glotaran: A Java-Based Graphical User Interface for the R Package TIMP. *J. Stat. Softw.* **2012**, *49*, 1–22. [CrossRef]
33. Müller, C.; Pascher, T.; Eriksson, A.; Chabera, P.; Uhlig, J. KiMoPack: A python Package for Kinetic Modeling of the Chemical Mechanism. *J. Phys. Chem.* **2022**, *126*, 4087–4099. [CrossRef] [PubMed]
34. Arecchi, F.; Bonifacio, R. Theory of optical maser amplifiers. *IEEE J. Quantum Electron.* **1965**, *1*, 169–178. [CrossRef]
35. Lindh, L.; Pascher, T.; Persson, S.; Goriya, Y.; Wärnmark, K.; Uhlig, J.; Chábera, P.; Persson, P.; Yartsev, A. Multifaceted Deactivation Dynamics of Fe(II) N-Heterocyclic Carbene Photosensitizers. *J. Phys. Chem.* **2023**, *127*, 10210–10222. [CrossRef] [PubMed]
36. Brückmann, J.; Müller, C.; Friedländer, I.; Mengele, A.K.; Peneva, K.; Dietzek-Ivanšić, B.; Rau, S. Photocatalytic Reduction of Nicotinamide Co-factor by Perylene Sensitized RhIII Complexes**. *Chem.—Eur. J.* **2022**, *28*, e202201931. [CrossRef] [PubMed]

37. Shchatsinin, I.; Laarmann, T.; Zhavoronkov, N.; Schulz, C.P.; Hertel, I.V. Ultrafast energy redistribution in C60 fullerenes: A real time study by two-color femtosecond spectroscopy. *J. Chem. Phys.* **2008**, *129*, 204308. [CrossRef] [PubMed]
38. van Stokkum, I.H.; Larsen, D.S.; van Grondelle, R. Global and target analysis of time-resolved spectra. *Biochim. Biophys. Acta (BBA)-Bioenerg.* **2004**, *1657*, 82–104. [CrossRef]
39. Connors, K. *Chemical Kinetics: The Study of Reaction Rates in Solution*; VCH Publishers, Inc.: New York, NY, USA, 1990.
40. Atkins, P.; Paula, J. *Atkins' Physical Chemistry*; Oxford University Press: Oxford, UK, 2008.
41. Rivas, D.E.; Serkez, S.; Baumann, T.M.; Boll, R.; Czwalinna, M.K.; Dold, S.; de Fanis, A.; Gerasimova, N.; Grychtol, P.; Lautenschlager, B.; et al. High-temporal-resolution X-ray spectroscopy with free-electron and optical lasers. *Optica* **2022**, *9*, 429–430. [CrossRef]
42. Debnath, T.; Mohd Yusof, M.S.B.; Low, P.J.; Loh, Z.H. Ultrafast structural rearrangement dynamics induced by the photodetachment of phenoxide in aqueous solution. *Nat. Commun.* **2019**, *10*, 2944. [CrossRef]
43. Atkins, P.; Friedman, R. *Molecular Quantum Mechanics*; OUP Oxford: Oxford, UK, 2011.
44. Manzano, D. A short introduction to the Lindblad master equation. *AIP Adv.* **2020**, *10*, 025106. [CrossRef]
45. Tikhonov, D.S.; Blech, A.; Leibscher, M.; Greenman, L.; Schnell, M.; Koch, C.P. Pump-probe spectroscopy of chiral vibrational dynamics. *Sci. Adv.* **2022**, *8*, eade0311. [CrossRef] [PubMed]
46. Sun, W.; Tikhonov, D.S.; Singh, H.; Steber, A.L.; Pérez, C.; Schnell, M. Inducing transient enantiomeric excess in a molecular quantum racemic mixture with microwave fields. *Nat. Commun.* **2023**, *14*, 934. [CrossRef] [PubMed]
47. Hatano, N. Exceptional points of the Lindblad operator of a two-level system. *Mol. Phys.* **2019**, *117*, 2121–2127. [CrossRef]
48. Schulz, S.; Grguraš, I.; Behrens, C.; Bromberger, H.; Costello, J.T.; Czwalinna, M.K.; Felber, M.; Hoffmann, M.C.; Ilchen, M.; Liu, H.Y.; et al. Femtosecond all-optical synchronization of an X-ray free-electron laser. *Nat. Commun.* **2015**, *6*, 5938. [CrossRef]
49. Savelyev, E.; Boll, R.; Bomme, C.; Schirmel, N.; Redlin, H.; Erk, B.; Düsterer, S.; Müller, E.; Höppner, H.; Toleikis, S.; et al. Jitter-correction for IR/UV-XUV pump-probe experiments at the FLASH free-electron laser. *New J. Phys.* **2017**, *19*, 043009. [CrossRef]
50. Schirmel, N.; Alisauskas, S.; Huülsenbusch, T.; Manschwetus, B.; Mohr, C.; Winkelmann, L.; Große-Wortmann, U.; Zheng, J.; Lang, T.; Hartl, I. Long-Term Stabilization of Temporal and Spectral Drifts of a Burst-Mode OPCPA System. In Proceedings of the Conference on Lasers and Electro-Optics, San Jose, CA, USA, 5–10 May 2019; p. STu4E.4. [CrossRef]
51. Lambropoulos, P. Topics on Multiphoton Processes in Atoms**Work supported by a grant from the National Science Foundation No. MPS74-17553. *Adv. At. Mol. Phys.* **1976**, *12*, 87–164. [CrossRef]
52. Kheifets, A.S. Wigner time delay in atomic photoionization. *J. Phys. At. Mol. Opt. Phys.* **2023**, *56*, 022001. [CrossRef]
53. Virtanen, P.; Gommers, R.; Oliphant, T.E.; Haberland, M.; Reddy, T.; Cournapeau, D.; Burovski, E.; Peterson, P.; Weckesser, W.; Bright, J.; et al. SciPy 1.0: Fundamental Algorithms for Scientific Computing in Python. *Nat. Methods* **2020**, *17*, 261–272. [CrossRef]
54. Chopra, P. Astrochemically Relevant Polycyclic Aromatic Hydrocarbons Investigated Using Ultrafast Pump-Probe Spectroscopy and Near-Edge X-ray Absorption Fine Structure Spectroscopy. Ph.D. Thesis, Christian-Albrechts-Universität zu Kiel, Kiel, Germany, 2022.
55. Garg, D. Electronic Structure and Ultrafast Fragmentation Dynamics of Polycyclic Aromatic Hydrocarbons. Ph.D. Thesis, Universität Hamburg, Hamburg, Germany, 2023.
56. Douguet, N.; Grum-Grzhimailo, A.N.; Bartschat, K. Above-threshold ionization in neon produced by combining optical and bichromatic XUV femtosecond laser pulses. *Phys. Rev. A* **2017**, *95*, 013407. [CrossRef]
57. Strüder, L.; Epp, S.; Rolles, D.; Hartmann, R.; Holl, P.; Lutz, G.; Soltau, H.; Eckart, R.; Reich, C.; Heinzinger, K.; et al. Large-format, high-speed, X-ray pnCCDs combined with electron and ion imaging spectrometers in a multipurpose chamber for experiments at 4th generation light sources. *Nucl. Instruments Methods Phys. Res. Sect. Accel. Spectrom. Detect. Assoc. Equip.* **2010**, *614*, 483–496. [CrossRef]
58. Erk, B.; Müller, J.P.; Bomme, C.; Boll, R.; Brenner, G.; Chapman, H.N.; Correa, J.; Düsterer, S.; Dziarzhytski, S.; Eisebitt, S.; et al. CAMP@FLASH: an end-station for imaging, electron- and ion-spectroscopy, and pump–probe experiments at the FLASH free-electron laser. *J. Synchrotron. Radiat.* **2018**, *25*, 1529–1540. [CrossRef]
59. Rossbach, J., FLASH: The First Superconducting X-ray Free-Electron Laser. In *Synchrotron Light Sources and Free-Electron Lasers: Accelerator Physics, Instrumentation and Science Applications*; Jaeschke, E., Khan, S., Schneider, J.R., Hastings, J.B., Eds.; Springer International Publishing: Cham, Switzerland, 2014; pp. 1–22. [CrossRef]
60. Beye, M.; Gühr, M.; Hartl, I.; Plönjes, E.; Schaper, L.; Schreiber, S.; Tiedtke, K.; Treusch, R. FLASH and the FLASH2020+ project—Current status and upgrades for the free-electron laser in Hamburg at DESY. *Eur. Phys. J. Plus* **2023**, *138*, 193. [CrossRef]
61. Garcia, J.D.; Mack, J.E. Energy Level and Line Tables for One-Electron Atomic Spectra∗. *J. Opt. Soc. Am.* **1965**, *55*, 654–685. [CrossRef]
62. Martin, W.C. Improved ^4i 1snl ionization energy, energy levels, and Lamb shifts for 1sns and 1snp terms. *Phys. Rev. A* **1987**, *36*, 3575–3589. [CrossRef] [PubMed]
63. Williams, T.; Kelley, C. Gnuplot 5.4: An Interactive Plotting Program. 2021. Available online: http://www.gnuplot.info (accessed on 20 January 2020).
64. LEVENBERG, K. A Method for the Solution of Certain Non-Linear Problems in Least Squares. *Q. Appl. Math.* **1944**, *2*, 164–168. [CrossRef]

65. Marquardt, D.W. An Algorithm for Least-Squares Estimation of Nonlinear Parameters. *J. Soc. Ind. Appl. Math.* **1963**, *11*, 431–441. [CrossRef]
66. Akaike, H. A new look at the statistical model identification. *IEEE Trans. Autom. Control* **1974**, *19*, 716–723. [CrossRef]
67. Vaníček, P. Approximate spectral analysis by least-squares fit. *Astrophys. Space Sci.* **1969**, *4*, 387–391. [CrossRef]
68. Tikhonov, D.S. Regularized weighted sine least-squares spectral analysis for gas electron diffraction data. *J. Chem. Phys.* **2023**, *159*, 174101. [CrossRef] [PubMed]
69. Ippen, E.P.; Shank, C.V., Techniques for Measurement. In *Ultrashort Light Pulses: Picosecond Techniques and Applications*; Shapiro, S.L., Ed.; Springer: Berlin/Heidelberg, Germany, 1977; pp. 83–122. [CrossRef]
70. Vardeny, Z.; Tauc, J. Picosecond coherence coupling in the pump and probe technique. *Opt. Commun.* **1981**, *39*, 396–400. [CrossRef]
71. Lebedev, M.V.; Misochko, O.V.; Dekorsy, T.; Georgiev, N. On the nature of "coherent artifact". *J. Exp. Theor. Phys.* **2005**, *100*, 272–282. [CrossRef]
72. Rhodes, M.; Steinmeyer, G.; Ratner, J.; Trebino, R. Pulse-shape instabilities and their measurement. *Laser Photonics Rev.* **2013**, *7*, 557–565. [CrossRef]
73. SciPy: Open Source Scientific Tools for Python. Available online: http://www.scipy.org/ (accessed on 18 January 2024).
74. van Stokkum, I.H.M.; Weißenborn, J.; Weigand, S.; Snellenburg, J.J. Pyglotaran: a lego-like Python framework for global and target analysis of time-resolved spectra. *Photochem. Photobiol. Sci.* **2023**, *22*, 2413–2431. [CrossRef]
75. Müller, M.G.; Niklas, J.; Lubitz, W.; Holzwarth, A.R. Ultrafast Transient Absorption Studies on Photosystem I Reaction Centers from Chlamydomonas reinhardtii. 1. A New Interpretation of the Energy Trapping and Early Electron Transfer Steps in Photosystem I. *Biophys. J.* **2003**, *85*, 3899–3922. [CrossRef]
76. Croce, R.; Müller, M.G.; Bassi, R.; Holzwarth, A.R. Carotenoid-to-Chlorophyll Energy Transfer in Recombinant Major Light-Harvesting Complex (LHCII) of Higher Plants. I. Femtosecond Transient Absorption Measurements. *Biophys. J.* **2001**, *80*, 901–915. [CrossRef]
77. Holzwarth, A.R.; Müller, M.G.; Niklas, J.; Lubitz, W. Ultrafast Transient Absorption Studies on Photosystem I Reaction Centers from Chlamydomonas reinhardtii. 2: Mutations near the P700 Reaction Center Chlorophylls Provide New Insight into the Nature of the Primary Electron Donor. *Biophys. J.* **2006**, *90*, 552–565. [CrossRef] [PubMed]
78. Zurek, W.H. Decoherence and the Transition from Quantum to Classical. *Phys. Today* **1991**, *44*, 36–44. [CrossRef]
79. Zurek, W.H. Decoherence and the Transition from Quantum to Classical–Revisited. *Los Alamos Sci.* **2002**, *27*, 86–109.

Disclaimer/Publisher's Note: The statements, opinions and data contained in all publications are solely those of the individual author(s) and contributor(s) and not of MDPI and/or the editor(s). MDPI and/or the editor(s) disclaim responsibility for any injury to people or property resulting from any ideas, methods, instructions or products referred to in the content.

Article

Spectral Optical Properties of Rabbit Brain Cortex between 200 and 1000 nm

Tânia M. Gonçalves [1], Inês S. Martins [1,2], Hugo F. Silva [1,2], Valery V. Tuchin [3,4,5] and Luís M. Oliveira [1,2,*]

1. Physics Department, School of Engineering, Polytechnic Institute of Porto, Rua Dr. António Bernardino de Almeida 431, 4249-015 Porto, Portugal; tania_02pinto@hotmail.com (T.M.G.); inessoraiamartins@gmail.com (I.S.M.); ugo96sylva@gmail.com (H.F.S.)
2. Center of Innovation in Engineering and Industrial Technology, ISEP, Rua Dr. António Bernardino de Almeida 431, 4249-015 Porto, Portugal
3. Science Medical Center, Saratov State University, 83 Astrakhanskaya Str., 410012 Saratov, Russia; tuchinvv@mail.ru
4. Interdisciplinary Laboratory of Biophotonics, National Research Tomsk State University, 36 Lenin's av., 634050 Tomsk, Russia
5. Laboratory of Laser Diagnostics of Technical and Living Systems, Institute of Precision Mechanics and Control Institute of the Russian Academy of Sciences, 24 Rabochaya Str., 410028 Saratov, Russia
* Correspondence: lmo@isep.ipp.pt

Abstract: The knowledge of the optical properties of biological tissues in a wide spectral range is highly important for the development of noninvasive diagnostic or treatment procedures. The absorption coefficient is one of those properties, from which various information about tissue components can be retrieved. Using transmittance and reflectance spectral measurements acquired from ex vivo rabbit brain cortex samples allowed to calculate its optical properties in the ultraviolet to the near infrared spectral range. Melanin and lipofuscin, the two pigments that are related to the aging of tissues and cells were identified in the cortex absorption. By subtracting the absorption of these pigments from the absorption of the brain cortex, it was possible to evaluate the true ratios for the DNA/RNA and hemoglobin bands in the cortex—12.33-fold (at 260 nm), 12.02-fold (at 411 nm) and 4.47-fold (at 555 nm). Since melanin and lipofuscin accumulation increases with the aging of the brain tissues and are related to the degeneration of neurons and their death, further studies should be performed to evaluate the evolution of pigment accumulation in the brain, so that new optical methods can be developed to aid in the diagnosis and monitoring of brain diseases.

Keywords: tissue spectroscopy; tissue optical properties; scattering coefficient; absorption coefficient; DNA content; blood content; pigment detection

Citation: Gonçalves, T.M.; Martins, I.S.; Silva, H.F.; Tuchin, V.V.; Oliveira, L.M. Spectral Optical Properties of Rabbit Brain Cortex between 200 and 1000 nm. *Photochem* **2021**, *1*, 190–208. https://doi.org/10.3390/photochem1020011

Academic Editor: Marcelo I. Guzman

Received: 25 June 2021
Accepted: 9 August 2021
Published: 11 August 2021

Publisher's Note: MDPI stays neutral with regard to jurisdictional claims in published maps and institutional affiliations.

Copyright: © 2021 by the authors. Licensee MDPI, Basel, Switzerland. This article is an open access article distributed under the terms and conditions of the Creative Commons Attribution (CC BY) license (https://creativecommons.org/licenses/by/4.0/).

1. Introduction

The optical properties of biological tissues are unique to those tissues and provide means for their identification. The estimation of their wavelength dependence allows the identification of biological components in tissues and the discrimination of pathologies [1]. There are various optical properties, but the most commonly used to characterize a biological tissue are the refractive index (RI), the absorption coefficient (μ_a), the scattering coefficient (μ_s) or the reduced scattering coefficient (μ'_s) and the scattering anisotropy (*g*) [2,3]. The estimation of tissue's optical properties is traditionally made using inverse simulations based on the Monte Carlo or the Adding–Doubling algorithms [3–5]. Due to the fact that such simulations estimate a set of optical properties for a single wavelength at a time, they are time consuming if we want to obtain such properties for a broad spectral range.

Although some diagnostic or treatment windows have been previously identified in the visible and infrared range of the electromagnetic spectrum [6,7], current biophotonics techniques can be applied within different spectral bands from the deep ultraviolet to the terahertz [8]. This means that a fast way to obtain the optical properties of a tissue for a wide spectral range is necessary. As recently published [1,9], if some spectral measurements from the tissue samples are available, certain equations can be used for a fast calculation of such properties. If the absorption properties of a tissue are considerably higher than its scattering properties, the Bouguer–Beer–Lambert law can be used to describe such an absorption-dominated medium [10]:

$$I(\lambda) = I_0(\lambda)e^{-A(\lambda)}, \quad (1)$$

where $A(\lambda)$ is the absorbance of the tissue as a function of wavelength, λ, which is defined as:

$$A(\lambda) = -ln\left(\frac{I(\lambda)}{I_0(\lambda)}\right) = \mu_a(\lambda) \times d, \quad (2)$$

with $I_0(\lambda)$ representing the excitation light intensity, $I(\lambda)$ representing the transmitted light intensity and d the sample thickness [10]. This equation can be used to calculate $\mu_a(\lambda)$ from the measured transmittance spectrum.

For any slab-form tissue sample that is irradiated by a light beam of intensity $I_0(\lambda)$, three optical phenomena will occur [2]: transmittance, reflectance and absorbance. If we can measure the total transmittance (T_t) and total reflectance (R_t) spectra from such a sample, its $A(\lambda)$ can be calculated using the following relation [1,9]:

$$T_t(\lambda) + R_t(\lambda) + A(\lambda) = 1. \quad (3)$$

Equation (3) can be used to calculate $A(\lambda)$ for any sample that contains major absorption, major scattering, or a combination of absorption and scattering properties. When combining Equations (2) and (3), it is possible to calculate $\mu_a(\lambda)$ from such T_t and R_t spectra that were measured from a tissue slab sample [1]:

$$\mu_a(\lambda) = \frac{1 - [T_t(\lambda) + R_t(\lambda)]}{d}, \quad (4)$$

where $T_t(\lambda)$ and $R_t(\lambda)$ are represented as ratios of total intensity of transmitted and reflected light measured with the corresponding integrating spheres to intensity of the incident light $I_0(\lambda)$, respectively; $\mu_a(\lambda)$ is calculated in cm^{-1} (or mm^{-1}), depending on the units used to measure d (cm or mm).

Since biological tissues are dominated by light scattering properties, once $\mu_a(\lambda)$ is calculated with Equation (4), and if collimated transmittance (T_c) spectra from the tissue are available, then the following form of the Bouguer–Beer–Lambert law can be used to calculate $\mu_s(\lambda)$ for the same wavelength range [1–3]:

$$\mu_s(\lambda) = -\frac{ln[T_c(\lambda)]}{d} - \mu_a(\lambda). \quad (5)$$

Spectral measurements do not usually allow to obtain μ'_s directly, but since its wavelength dependence is well described for the ultraviolet-near infrared (UV-NIR), discrete values can be estimated for that spectral range through inverse adding-doubling (IAD) simulations, and then fitted with a curve as described by Equation (6) [11].

$$\mu'_s(\lambda) = a \times \left(f_{Ray} \times \left(\frac{\lambda}{500\ nm}\right)^{-4} + (1 - f_{Ray}) \times \left(\frac{\lambda}{500\ nm}\right)^{-b_{Mie}} \right) \quad (6)$$

In Equation (6), a represents the value of μ'_s at 500 nm, f_{Ray} is the Rayleigh scattering fraction and b_{Mie} is the mean size of the Mie scatterers. These parameters can be obtained when the discrete μ'_s values that were generated through IAD are fitted with Equation (6). When performing the IAD estimations to obtain the discrete μ'_s values, the RI of the tissue (n_{tissue}) at the desired wavelengths is necessary. In general, the RI of tissues is measured at discrete wavelengths with multi-wavelength refractometers [12], or using the total internal reflection method with various lasers with emission within the desired wavelength range [13]. Once those RI values are measured, the tissue dispersion for that wavelength range is calculated by fitting the experimental RI data with equations such as the Cauchy (Equation (7)), the Conrady (Equation (8)) or the Cornu (Equation (9)) equations [13–15]:

$$n_{\text{tissue}}(\lambda) = A + \frac{B}{\lambda^2} + \frac{C}{\lambda^4}, \tag{7}$$

$$n_{\text{tissue}}(\lambda) = A + \frac{B}{\lambda} + \frac{C}{\lambda^{3.5}}, \tag{8}$$

$$n_{\text{tissue}}(\lambda) = A + \frac{B}{(\lambda - C)}, \tag{9}$$

where, A, B and C are the Cauchy, the Conrady or the Cornu parameters, which are obtained during the fitting of discrete experimental data. If the calculated tissue dispersion is not available for the entire wavelength range of interest, a broader dispersion can be calculated from $\mu_a(\lambda)$, using the Kramers–Kronig relations, which were developed for non-scattering materials [16]. In this calculation process, Equation (10) is the first one to be used to obtain the imaginary part of tissue dispersion ($\kappa(\lambda)$) [13,16]:

$$\kappa(\lambda) = \frac{\lambda}{4\pi}\mu_a(\lambda). \tag{10}$$

After obtaining $\kappa(\lambda)$, the dispersion that corresponds to the real part of RI can be calculated with Equation (11) [13,16,17]:

$$n_{\text{tissue}}(\lambda) = 1 + \frac{2}{\pi}\int_0^\infty \frac{\lambda_1}{\Lambda} \times \frac{\lambda_1}{\Lambda^2 - \lambda_1^2}\kappa(\Lambda)\mathrm{d}\Lambda, \tag{11}$$

where Λ represents the integrating variable over the wavelength domain and λ_1 is a fixed wavelength that can be adjusted for better vertical matching of the calculated dispersion to the one obtained from discrete experimental data.

Once the broad-band tissue dispersion is calculated through Equations (10) and (11), we can select discrete values from it to use in the IAD simulations. By running those simulations, the generated μ'_s values are then fitted with Equation (6) to obtain $\mu'_s(\lambda)$. Such spectrum can be combined with the μ_s spectrum that was calculated with Equation (5) to obtain the wavelength dependence of tissue scattering anisotropy ($g(\lambda)$) [1–3,18]:

$$g(\lambda) = 1 - \frac{\mu'_s(\lambda)}{\mu_s(\lambda)}. \tag{12}$$

In general, $g(\lambda)$ presents an increasing exponential behavior with increasing wavelength in the UV-NIR range [1], which can be described mathematically by Equation (13) [18] or Equation (14) [19].

$$g(\lambda) = a + b\left[1 - exp\left(\frac{\lambda - c}{d}\right)\right], \tag{13}$$

$$g(\lambda) = a \times exp(b \times \lambda) + c \times exp(d \times \lambda), \tag{14}$$

where a, b, c and d are parameters that are obtained during the fitting of g data.

Such calculation procedure, which relies only on IAD simulations to obtain discrete μ'_s values, is a fast way to obtain the wavelength dependency for the optical properties of a biological tissue, provided that spectral measurements from the tissue are available. A particular analysis of the calculated spectral optical properties provides information about tissue composition, contents of its biological chromophores and possibly indication of pathologies, as previously observed [1,20].

The brain is a complex organ, composed by different parts like the cortex and the cerebellum. The brain functions are complex and the incidence of brain diseases, such as Alzheimer, Parkinson or stroke have increased significantly in the past 30 years [21]. Such diseases occur as a natural consequence of the aging process, and are principally due to the degeneration and death of neurons [22–24]. A monitoring procedure to evaluate neuron aging, degeneration and death would be helpful to prevent these diseases. To develop such procedure with optical methods, the knowledge of the optical properties of the brain tissues is necessary. Only by knowing the spectral optical properties in a wide spectral range it will be possible to develop optimized optical imaging and spectroscopic methods to evaluate the health of superficial brain tissues, or to develop in-depth monitoring procedures with the combination of optical clearing treatments [2,25,26]. With the objective of obtaining the spectral optical properties of the brain cortex, we performed spectral measurements from ex vivo rabbit tissues. The methodology used in this study is described in Section 2 and the results are presented in Section 3.

2. Materials and Methods

The present study involves measurements from ex vivo animal tissues. The research follows the Declaration of Helsinki and was approved by the research review board in biomedical engineering of the Center of Innovation in Engineering and Industrial Technology (CIETI), in Porto, Portugal. Such approval has the number CIETI/Biomed_Research_2021_01.

In this study, brain cortex samples from recently sacrificed rabbits were prepared to conduct the experimental measurements. Such experimental studies consisted on RI measurements at discrete wavelengths and spectral measurements to calculate the optical properties between the UV and the NIR. Section 2.1 describes the tissue collection and preparation procedure, Section 2.2 describes the setup and measuring procedure used to obtain the discrete RI values and Section 2.3 describes the setups and measuring procedures to acquire the spectral measurements. Section 2.4 describes the calculations made to obtain tissue dispersion and the spectral optical properties of the brain cortex.

2.1. Tissue Collection and Sample Preparation

Five adult rabbits were acquired from a local breader that sells them for consumption. The animals were sacrificed on different days and the brain from each one was frozen for 12 h after dissecting it from within the skull. A cryostat (Leica^{TM}, Wetzlar, Germany, model CM1860 UV) was used to prepare tissue slices from the brain cortex. For the RI measurements, three samples were prepared in the cryostat with 3 mm thickness and an approximated square superficial area of about 1 cm × 1 cm. For the spectral measurements, ten samples were prepared with 0.5 mm thickness and an approximated circular form (~1 cm in diameter).

2.2. RI Measurements

To perform the RI measurements, the total internal reflection method was used [13,27–29]. The setup used for these measurements was constructed in our lab and is represented in Figure 1.

To obtain the RI values of the cortex at discrete wavelengths in the visible and NIR range, each of the three tissue samples that were prepared for this purpose was submitted to measurements with different lasers in the setup presented in Figure 1. The lasers used in these measurements had emission wavelengths at 401.4, 534.6, 626.6, 782.1, 820.8 and 850.7 nm. These lasers are laser diodes that were acquired from Edmund Optics, with the exception of the lasers that emit at 534.6 nm and at 626.6 nm, which were acquired from Kvant (Bratislava, Slovakia) and from Pasco (Roseville, CA, USA), respectively. The emmiting power of all lasers was 5 mW or less and their emmiting wavelengths were verified with a spectrometer from Avantes (Apeldoorn, Netherlands).

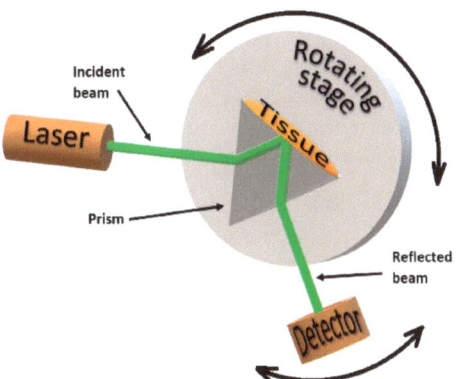

Figure 1. Total internal reflection setup for the RI measurements.

The dispersion prism presented in Figure 1 is a SCHOTT N-SF11 prism acquired from Edmund Optics, with an RI dependence on wavelength, as presented in Figure 2.

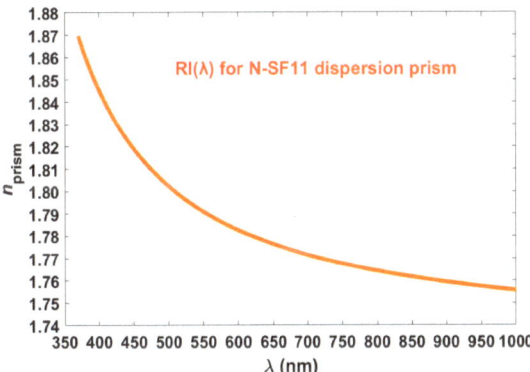

Figure 2. Wavelength dependence for the RI of the N-SF11 dispersion prism.

The curve represented in Figure 2 is described by the Sellmeier equation [30], at 20 °C:

$$n_{prism}^2 - 1 = \frac{K_1 \lambda^2}{\lambda^2 - L_1} + \frac{K_2 \lambda^2}{\lambda^2 - L_2} + \frac{K_3 \lambda^2}{\lambda^2 - L_3}, \quad (15)$$

where the Sellmeier coefficients have the following values for the SCHOTT N-SF11 glass: $K_1 = 1.7376$, $K_2 = 0.3137$, $K_3 = 1.8988$, $L_1 = 0.0132$, $L_2 = 0.023$, and $L_3 = 155.2363$. Acording to Ref. [30], since λ is represented in μm in Equation (15), K_1, K_2 and K_3 are dimentionless coefficients, while L_1, L_2 and L_3 are represented in μm². Three sets of measurements were made with each laser, one set per tissue sample. Temperature was kept at 20 ± 1 °C and the following procedure was followed for each set of measurements:

1. The sample was placed in perfect contact with the base of the prism (see Figure 1).
2. Illumination of the setup was made with the laser beam through one side of the prism.
3. The reflected beam was collected with a photocell (a laser power meter from Coherent with spectral resolution from 0.15 μm to 11 μm), connected to a voltmeter (from Wavetek Meterman) to read the electrical potential.

4. This measuring procedure was repeated for several incidence angles (α) between the incident laser beam and the normal to the air/prism interface. The angular resolution for these measurements was 1°.

Such procedure was repeated for the other lasers. To obtain the RI of the brain cortex at the laser wavelengths, the collected data from each set of measurements needed to be submitted to calculations, which are described in Section 2.4.

2.3. Spectral Measurements

To perform the spectral measurements, which are necessary to calculate the spectral optical properties of the brain cortex, ten ex vivo tissue samples were used. All these samples were sequentially submitted to measurements with the setups presented in Figure 3 to acquire $T_t(\lambda)$, $R_t(\lambda)$ and $T_c(\lambda)$.

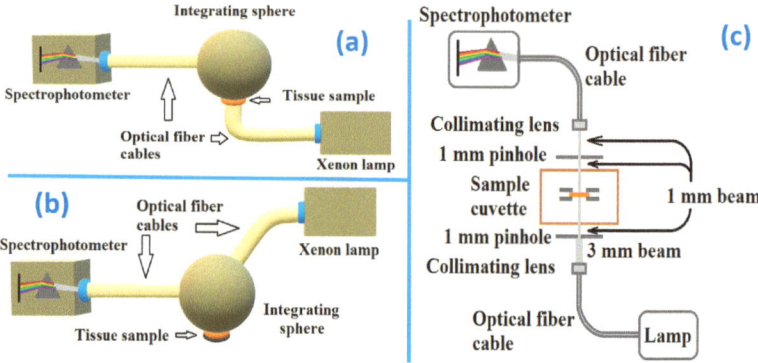

Figure 3. Experimental setups to measure: $T_t(\lambda)$ (a), $R_t(\lambda)$ (b) and $T_c(\lambda)$ (c).

In the measurements with the T_t setup, a collimated beam 6 mm in diameter from a Xenon lamp crosses the sample before entering the integrating sphere. The transmitted beam undergoes several reflections and integration inside the sphere, before being delivered to the spectrophotometer through an optical fiber cable. The R_t setup is similar to the one used to acquire T_t, but the beam from the lamp interacts with the tissue sample in a reflection mode. Such beam enters the integrating sphere at 8° with the vertical axis of the sphere. The reflected beam that leaves the tissue sample is reflected and integrated inside the sphere before being delivered to the spectrophotometer. In the case of T_c measurements, a beam from a deuterium-halogen lamp is delivered to the lower part of the sample cuvette through an optical fiber cable, a collimating lens and a pinhole, which collimate the beam and reduce its diameter to 1 mm. The unscattered transmitted beam is collected by another collimating lens into another optical fiber cable to be delivered to the spectrophotometer. Once ten spectra were collected from the brain cortex samples with all the setups presented in Figure 3, calculations, as described in Section 2.4, were performed to obtain the wavelength dependence of the optical properties of this tissue.

2.4. Calculations

After ending all experimental measurements, certain calculations were necessary to obtain the spectral optical properties of the brain cortex.

Regarding the RI measurements, and since the incident and reflected angles for the beam can only be measured outside the prism, at the air/prism interfaces, we needed to use the Snell–Descartes equation to convert the angle of the incident (or reflected) beam as measured outside the prism (α) to the incident (or reflected) angle at the prism/tissue interface (θ) [13,31]:

$$\theta = \beta - arcsin\left[\frac{1}{n_{prism}} \times sin(\alpha)\right], \tag{16}$$

where β represents the internal angle of the prism (60° for the prism we used) and n_{prism} represents the RI of the prism at the wavelength of the laser, which can be retrieved from Figure 2. To calculate the reflectance curve at the prism/tissue interface for each set of measurements, we performed the following calculation [13,15]:

$$R(\theta) = \frac{V(\theta) - V_{noise}}{V_{laser} - V_{noise}}, \tag{17}$$

with $V(\theta)$ representing the electrical potential measured for an incident (or reflected) angle θ at the prism/tissue interface, V_{noise} representing the potential measured with only background light and V_{laser} representing the potential measured directly from the laser.

The reflectance curves obtained from each set of measurements, as calculated with Equation (17), contain information about the critical angle of reflection, but to obtain such angle with precision, the first derivative of each curve needed to be calculated. Such calculation was made according to [13,15]:

$$deriv(\theta) = \frac{R(\theta_i) - R(\theta_{i-1})}{\theta_i - \theta_{i-1}}, \tag{18}$$

where θ_i and θ_{i-1} represent the consecutive angles of measurement and $R(\theta_i)$ and $R(\theta_{i-1})$ represent the corresponding reflectances measured at those angles. As previously observed in other studies [13,32,33], those derivatives present a strong peak, whose central angle corresponds to the critical angle of reflection (θ_c) at the prism/tissue interface.

For each set of measurements with a particular laser, by representing the curve for the first derivative of the reflectance as a function of the angle θ, we could identify the value of θ_c, which was then used in Equation (19) to calculate the RI of the tissue at the laser wavelength [13,14].

$$n_{tissue}(\lambda) = n_{prism}(\lambda) \times sin(\theta_c). \tag{19}$$

In Equation (19), $n_{tissue}(\lambda)$ represents the RI of the tissue at the laser wavelength and $n_{prism}(\lambda)$ represents the RI of the prism at the same wavelength as retrieved from Figure 2. Three sets of measurements were performed with each laser, meaning that three n_{tissue} values were obtained for each laser wavelength. The mean and standard deviation (SD) for the tissue RI were calculated for each laser wavelength to provide higher accuracy and data dispersion between the samples analyzed. The mean RI values of the rabbit cortex were fitted with Equations (7)–(9) to check which one provides a better fitting. It was verified that the Cauchy equation (Equation (7)) provides the best fit ($R^2 = 0.9793$) within the wavelength range of the lasers used (400–850 nm). When performing that fit, we obtained the A, B and C parameters indicated in Equation (7), as represented next:

$$n_{tissue}(\lambda) = 1.353 + \frac{5420}{\lambda^2} + \frac{0.3521}{\lambda^4}. \tag{20}$$

All graphs that result from these calculations are presented in Section 3.

Considering the spectral measurements, the first calculation was made with Equation (4) to obtain $\mu_a(\lambda)$. Such calculation was made ten times, using the ten pairs of T_t and R_t spectra that were measured from the samples, resulting in mean and SD for $\mu_a(\lambda)$.

To obtain the real part of the RI for the cortex as a function of wavelength, the ten μ_a spectra were first used in Equation (10) to calculate ten $\kappa(\lambda)$. The resulting imaginary dispersions were then used in Equation (11), which through adjustment of the λ_1 parameter allowed obtaining the real part of the RI for the cortex with optimized vertical matching to the curve described by Equation (20). Since ten n_{tissue} dispersions were obtained in this calculation, mean and SD data were calculated between 200 and 1000 nm.

After calculating $\mu_a(\lambda)$ and $n_{tissue}(\lambda)$ for the cortex, calculations were performed with Equation (5) to obtain the mean and SD for $\mu_s(\lambda)$. To obtain $\mu'_s(\lambda)$, ten IAD simulations with data from each individual sample were made at discrete wavelengths, with 50 nm increments between 200 and 1000 nm. For each of the ten sets of simulations (with data from a particular sample), the resulting μ'_s values were fitted by a curve described by Equation (6) to obtain its wavelength dependence. Ten calculated curves were used to obtain the mean and SD for $\mu'_s(\lambda)$.

Using the calculated values of $\mu_s(\lambda)$ and $\mu'_s(\lambda)$ in Equation (12), mean and SD of $g(\lambda)$ were obtained. Finally, using the individual μ_a and μ'_s spectra in Equation (21) [2,34], the mean and SD of light penetration depth in diffusion approximation ($\delta(\lambda)$) was obtained for the cortex:

$$\delta(\lambda) = \frac{1}{\sqrt{3\mu_a(\lambda)(\mu_a(\lambda) + \mu'_s(\lambda))}}. \tag{21}$$

All the results from these calculations are presented in Section 3.

3. Results

3.1. RI Measurements

As a result of the calculations described in Section 2.4, which are necessary to obtain the RI of the cortex, the first step consisted on obtaining the reflectance curves for each set of measurements with a particular laser. The second step consisted on calculating the 1st derivative curves of those reflectance curves. The reflectance curves were calculated for each set of measurements with Equation (17) and the 1st derivative curves were obtained with Equation (18). Figure 4 presents such results for the 782.1 nm laser.

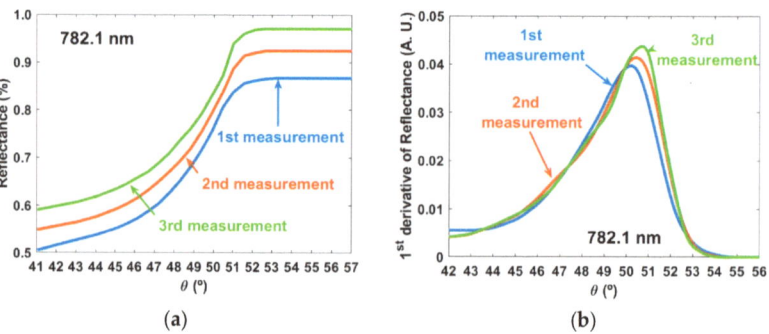

Figure 4. Curves calculated with the RI measurements for the 782.1 nm laser: reflectance curves (**a**) and 1st derivative curves (**b**).

Similar curves to the ones presented in panels of Figure 4 were calculated for the measurements with the other lasers. The θ_c values were retrieved from the curves in Figure 4b and others like those that were obtained for the other lasers. These θ_c values were used in Equation (19) to calculate the corresponding RI values of the brain cortex at the laser wavelengths. Table 1 presents the results of these calculations, their mean and SD.

Table 1. RI values of the brain cortex at the laser wavelengths.

Laser	n_{tissue}	Mean	SD
401.4 nm	1.3883 1.3877 1.3789	1.3850	0.0053
534.6 nm	1.3679 1.3735 1.3794	1.3736	0.0058
626.6 nm	1.3632 1.3686 1.3721	1.3680	0.0045
782.1 nm	1.3562 1.3609 1.3662	1.3611	0.0050
820.8 nm	1.3552 1.3566 1.3673	1.3597	0.0066
850 nm	1.3516 1.3583 1.3667	1.3589	0.0076

As indicated in Section 2.4, the mean RI values presented in Table 1 were submitted to fitting tests to check which of the curves described by Equations (7)–(9) provided a better dispersion curve. In these tests we verified that the best fitting was obtained with the Cauchy equation (Equation (7)). Equation (20), which is represented in Figure 5 along with the mean experimental RI values and the SD bars, describes such calculated dispersion for the rabbit brain cortex.

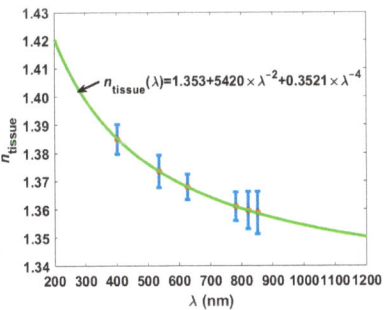

Figure 5. Experimental RI values for the rabbit cortex and calculated dispersion.

To compare our experimental data with other data for the brain cortex in literature and to use it later in the calculation of the cortex dispersion in the spectral range between 200 and 1000 nm from $\mu_a(\lambda)$, we extended the curve in Figure 5 to a lower wavelength of 200 nm and to a longer wavelength of 1200 nm. Ref. [35] indicates that the RI of human grey matter (cortex included) is 1.36 for the 456–1064 nm range. In our case and considering the same spectral range, we see that the rabbit brain cortex presents RI values between 1.35 and 1.38. On the other hand, it was reported by the authors of Ref. [36] that the rat brain cortex presents a RI of 1.3526 at 1100 ± 100 nm, and from our curve, we see that it is 1.3520 in that range. Authors of Ref. [37] reported that the cortex in rat brain presents an average RI value of 1.369, but no reference wavelength has been indicated for this value. We have obtained that value near 600 nm.

Considering that our experimental measurements to obtain the RI of rabbit brain cortex are accurate and also assuming that the calculated dispersion is valid outside the spectral range that corresponds to our measurements, we will now present the spectral measurements and calculations to obtain the spectral optical properties for this tissue.

3.2. Spectral Measurements and Calculated Spectral Optical Properties

We initiated the spectral measurements by acquiring the T_t spectra from all ten samples. As explained in Section 2.3, we used the setup presented in Figure 3a to perform these measurements. Using the setup presented in Figure 3b, we acquired the R_t spectra from the same ten samples, which were also submitted to measurements with the setup presented in Figure 3c to acquire the T_c spectra. The mean and SD for these measurements are presented in Figure 6.

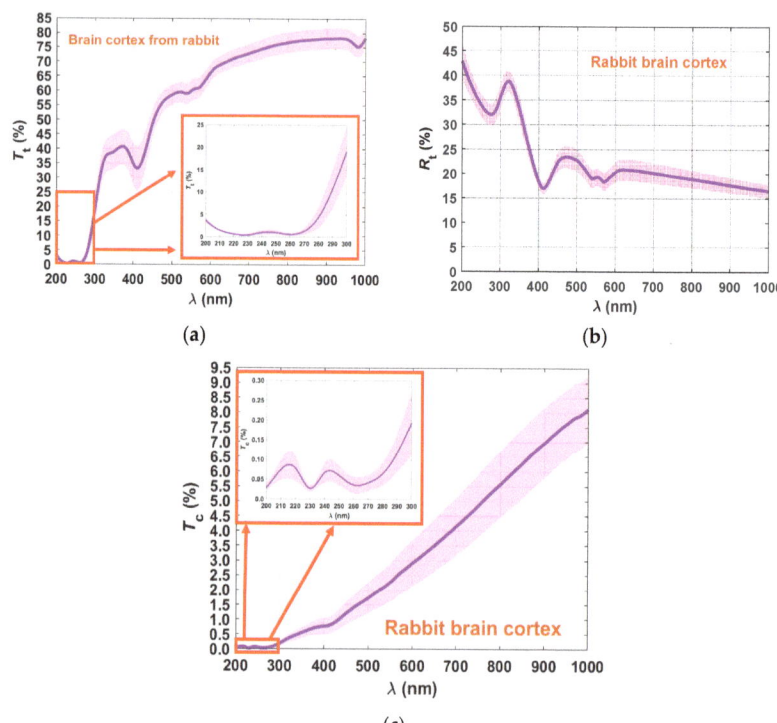

Figure 6. Mean spectra and SD of the rabbit brain cortex: $T_t(\lambda)$ (**a**), $R_t(\lambda)$ (**b**) and $T_c(\lambda)$ (**c**).

Although the R_t spectrum presented in Figure 6b shows small magnitude in the SD (less spreading between samples) for the entire spectral range, the tendency for T_t and T_c spectra is that the magnitude of the SD increases with increasing wavelength. Such a fact shows that different tissue samples present different transparency for the longer wavelengths, but for shorter wavelengths they are very similar.

Using the T_t and R_t spectra in Equation (4), we calculated $\mu_a(\lambda)$, and once it was calculated, Equations (10) and (11) were used to obtain $n_{\text{tissue}}(\lambda)$ for the cortex. Both these graphs are presented in Figure 7.

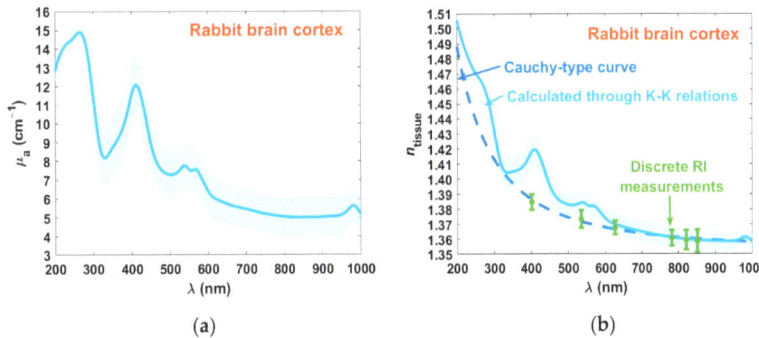

Figure 7. Spectral curves of $\mu_a(\lambda)$ (**a**) and $n_{tissue}(\lambda)$ (**b**) for the rabbit brain cortex.

As a result of the calculations made with Equation (4), $\mu_a(\lambda)$ presents increasing magnitude for the SD with increasing wavelength. In this graph (Figure 7a), we see several absorption peaks. The first one occurs at 230 nm, which correspond to amino acid connections of tyrosine and tryptophan in proteins [38,39]. The brain and especially the cortex contains several proteins, such as actin, albumin, α-tubulin, β-tubulin, neuron-specific enolase (NSE), and vimentin [40]. The second band occurs at 267 nm, showing a combination of the absorption band of DNA/RNA at 260 nm with the one of hemoglobin at 274 nm [41]. We see also the other absorption bands of oxygenated hemoglobin in the visible range—411 nm (Soret band) and 540/570 nm (Q bands) [41]. The graph in Figure 7a also shows the absorption band of water at 980 nm [41].

Considering the RI data presented in Figure 7b, we see that there is a good matching between the cortex dispersions that were calculated with the Kramers–Kronig relations and with the Cauchy equation in the entire spectral range. An exception to this good matching occurs in the locations that correspond to the absorption bands of oxygenated hemoglobin. Such poor matching at these spectral locations is due to the lack of hemoglobin sensitivity in the total reflection method that resulted in the Cauchy curve in Figure 7b.

The following step consisted on performing the IAD simulations to obtain discrete values of μ'_s. Such simulations were also performed ten times, considering the experimental measured spectra from the ten samples. On the other hand, using the calculated μ_a spectra in Equation (5), we calculated ten spectra for μ_s. The mean spectra and SD data for μ'_s and μ_s are presented in Figure 8.

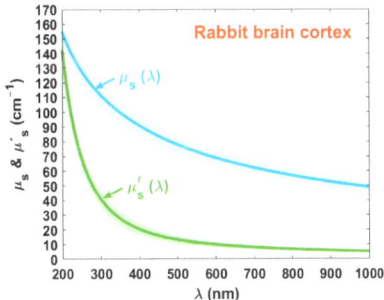

Figure 8. Spectral curves of $\mu'_s(\lambda)$ and $\mu_s(\lambda)$ for the rabbit brain cortex.

We can see from Figure 8 that as a result of fitting all μ'_s and μ_s estimations with Equation (6), both curves have a smooth decreasing exponential behavior with increasing wavelength. Both curves approach each other at lower wavelengths, and as the wavelength grows they tend to differ, as previously observed for other studies [1,9]. Considering the wavelengths 635, 671, and 808 nm, we see that the average μ'_s data in Figure 8 is 9.3, 8.6, and 6.8 cm^{-1}. The authors of Ref. [42] also presented μ'_s data for the rabbit brain after freezing at $-20\,°C$. Although the data in that study were not specifically referring to the brain cortex, their results are not too dissimilar from ours: 11.2, 9.2, and 5.7 cm^{-1}. Using the data in Figure 8, we calculated $g(\lambda)$ with Equation (12). Figure 9 presents the result of this calculation.

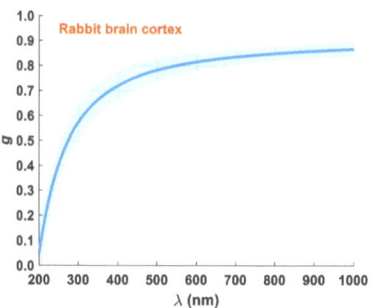

Figure 9. Spectral curve of $g(\lambda)$ for the rabbit brain cortex.

The wavelength dependence presented for the g-factor in Figure 9 shows the expected behavior for a biological tissue—g increases with increasing wavelength due to predominant Rayleigh scattering [1,9,11]. To finalize the calculation of the optical properties of the brain cortex, we used the μ_a and μ'_s spectra in Equation (21) to calculate $\delta(\lambda)$. Figure 10 presents the result of these calculations.

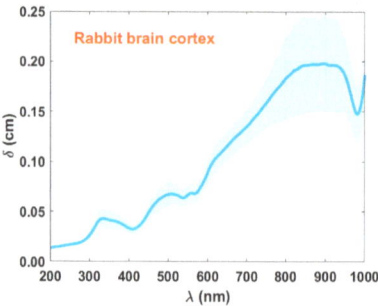

Figure 10. Spectral curve of $\delta(\lambda)$ for the rabbit brain cortex.

Once again, the wavelength dependence presented in Figure 10 is according to what is described in literature for biological tissues—δ increases with wavelength. We also see that such increase seems to saturate near 800 nm. The existence of absorption bands in the graph presented in Figure 10 is due to the dependence of δ on μ_a.

As a final discussion and interpretation of results, we performed an additional analysis on $\mu_a(\lambda)$, which is presented in Figure 7a. As previously described in the literature [1], by evaluating the ratios for the absorption bands in the μ_a spectrum, we can estimate a measure of the blood content and possibly of other absorbers in the tissue. Looking into Figure 7a, it seems that μ_a shows a baseline with some exponential decreasing dependency with increasing wavelength in the entire spectral range. Such behavior was previously observed for human colorectal tissues, where a discriminated content of a pigment (lipofuscin) was detected for normal and pathological mucosa [1,20].

The brain tissues are known to accumulate some particular pigments, such as hemosiderin [43], which is a microscopically visible granular iron pigment that emerges after rapid hemoglobin destruction or due to a fast and excessive iron deposition [44]. Another type of pigments that have been reported to accumulate in brain tissues are the carotenoid-type pigments, such as β-cryptoxanthin, lutein, and zeaxanthin, which contain oxygen atoms as hydroxyl, carbonyl, aldehyde, carboxylic, epoxide, and furanoxide groups in their molecules [45]. In addition to these pigments, the most common to accumulate in the brain, and in particular in the cortex are melanin and lipofuscin [46]. These two pigments originate from precursors, such as L-tyrosine, L-cysteine, and dopamine (in the case of melanin) [47] and from cell organelles, such as mitochondria, Golgi apparatus, and lysosomes (in the case of lipofuscin) [48–50]. Such precursors can easily associate with metals, particularly iron, to form melanin and lipofuscin [23].

Different mathematical equations that describe the wavelength dependence of μ_a for melanin and lipofuscin have been reported [51]:

$$\mu_{a-\text{melanin}}(\lambda) = A \times \lambda^{-B}, \tag{22}$$

$$\mu_{a-\text{lipofuscin}}(\lambda) = e^{(C-D \times \lambda)}, \tag{23}$$

where A, B, C and D are numerical parameters that can be determined when fitting experimental data. According to Ref. [51], these equations are valid for the range between the visible and the near infrared range (450–900 nm). In our case, and since we performed our studies between 200 and 1000 nm, Equations (22) and (23) might not be suitable to describe the wavelength dependencies for melanin and lipofuscin. Looking in the literature to find broader spectra for these pigments, we found graphical data [52,53], which we have reconstructed, as presented in Figure 11.

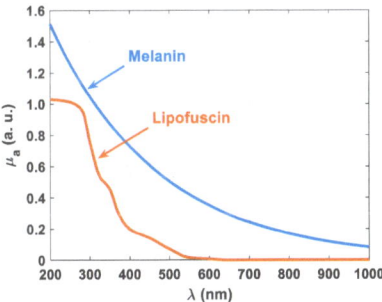

Figure 11. Spectral curves of $\mu_a(\lambda)$ for melanin and lipofuscin.

As represented in Figure 11, lipofuscin does not have a wavelength dependence as described by Equation (23), especially in the deep-UV range. Regarding melanin, we tried to fit the data in Figure 11 with a curve as described by Equation (22) and found that such an equation only provides a good fitting between 450 and 900 nm. Indeed, for an extended wavelength range, such as 200–1000 nm as we are using in our study, we need to avoid using Equations (22) and (23). This way, we used the numerical data from Figure 11 to reconstruct the accumulation of these pigments in the tissues used in our study. Such accumulation was found to be described as:

$$\mu_{a-\text{pigment}}(\lambda) = 3.5 \times M(\lambda) + 1.9 \times L(\lambda) \tag{24}$$

with $M(\lambda)$ and $L(\lambda)$ representing the wavelength dependencies of the absorption coefficient of melanin and lipofuscin, respectively, as presented in Figure 11. Equation (24) describes well the decreasing behavior of the baseline observed for the μ_a of the cortex that is presented in Figure 7a, but some contribution of other pigments or the precursors of melanin and lipofuscin may be included in that absorption spectrum. For instance, hemosiderin shows a wavelength dependence for μ_a, which is similar to the one of melanin that we have represented in Figure 11 [54]. Since there was no suspicion of rapid hemoglobin destruction or fast and excessive iron deposition in the cortex of the animals used in the study, we have neglected the absorption of hemosiderin in the calculation of $\mu_{a\text{-pigment}}(\lambda)$ in Equation (24). The carotenoid-type pigments, β-cryptoxanthin, lutein and zeaxanthin, all have a similar absorption spectrum, with low absorption in the UV and null absorption for wavelengths above ~500 nm. Between 400 and 500 nm, these carotenoid-type pigments present three absorption peaks, being β-cryptoxanthin the one with peaks of higher magnitude [55]. Since those absorption bands occur between 420 and 476 nm [55], they seem to have no influence in the resulting μ_a spectrum of the cortex. Looking into that spectrum in Figure 7a, between 400 and 500 nm we see only a well-defined Soret band, with center wavelength at 411 nm.

Regarding the melanin precursors, their absorption is mainly characterized by UV-bands, with zero absorption above 300 nm. In the case of L-cysteine, its absorption spectrum is the combination and overlapping of two bands, the one with higher-magnitude centered near 230 nm and the one with lower-magnitude centered near 280 nm [56]. This means that the absorption band that we see in Figure 7a at 230 nm may not be only due to the amino acid connections in proteins. It may also indicate the presence of L-cysteine in the cortex samples we used. No evidence of the 280 nm band was observed in the spectrum in Figure 7a, but due to the strong band at 267 nm that results from the combination of the DNA/RNA band at 260 nm and the hemoglobin band at 274 nm, the 280 nm band of L-cysteine can be masked. The absorption spectrum of L-tyrosine is the combination of two bands with a small overlapping, the higher-magnitude being centered at a wavelength smaller than 240 nm and the one with lower-magnitude being centered near 285 nm [57]. In the case of dopamine, a single absorption band occurs, with a central wavelength near 280 nm [58]. For the same reasons presented above for L-cysteine, the absorption bands of L-tyrosine and dopamine can be masked by the absorption bands of the amino acid connections in proteins and the one that results from the combination of DNA/RNA and hemoglobin. This means that melanin precursors may also be present in the samples used in this study.

For the case of the lipofuscin precursors, we found only the absorption spectrum of mitochondria. Similar to the case of the melanin precursors, such absorption is limited in the UV, with zero absorption for wavelengths longer than 400 nm. Mitochondria present a single absorption band between 300 and 400 nm, with a central wavelength near 360 nm [59]. Looking into the absorption spectrum presented in Figure 7a, the absorption band of mitochondria can be masked by the Soret band, which extends to a lower wavelength close to 320 nm.

Considering this discussion and assuming that only melanin and lipofuscin contribute the decaying baseline observed in the absorption spectrum of the brain cortex, we subtracted the curve described by Equation (24) to that spectrum. Figure 12 presents the $\mu_a(\lambda)$ of the rabbit brain cortex (the same as in Figure 7a), the $\mu_{a\text{-pigment}}(\lambda)$ as described by Equation (24) and a third curve, which represents the difference of the previous two. We have also presented in Figure 12 the ratios for the most significant absorption bands—DNA/RNA (260 nm) and oxygenated hemoglobin (Soret and Q-bands at 411 and 555 nm). The ratios that are presented in blue were calculated for the peaks in the original $\mu_a(\lambda)$, considering the $\mu_{a\text{-pigment}}(\lambda)$ curve as baseline. The ratios that are presented in green were calculated in the same manner, but now considering the peaks in $\mu_a(\lambda)$ after subtracting the curve for $\mu_{a\text{-pigment}}(\lambda)$. In this case, the baseline is the horizontal black line that was fixed at the minimum value of the green curve, which occurs at 804 nm.

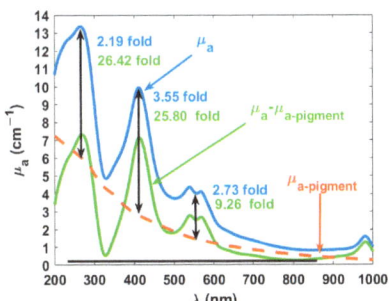

Figure 12. Wavelength dependencies of μ_a for the melanin and lipofuscin pigments and for the rabbit brain cortex, before (blue line) and after (green line) subtracting the absorption of the pigments.

In the first analysis of the data in Figure 12, we see that by subtracting the absorption of the pigments from the calculated μ_a, a decrease occurred in the entire spectral range. Performing a similar comparison with literature data as the one made for the μ'_s data in Figure 8, we see some similarity between the corrected μ_a data in Figure 12 and that reported in Ref. [42] for rabbit brain after freezing at $-20\ °C$. Considering once again the wavelengths 635, 671, and 808 nm, the corrected μ_a spectrum in Figure 12 presents the values: 0.8, 0.6, and 0.3 cm^{-1}, while in Ref. [42] the values of 0.8, 0.5, and 0.4 cm^{-1} were reported at those wavelengths.

Considering now the ratios presented in Figure 12 and comparing between the values obtained before and after subtracting the absorption of the pigments, we see that both melanin and lipofuscin were camouflaging the real content of both DNA/RNA and hemoglobin in the rabbit brain. Such contents were not only low but deceiving, since from the blue curve in Figure 12 we see a higher ratio at the Soret band than at the DNA/RNA band. After subtracting the absorption of the pigments from the μ_a of the cortex, we see that in reality the DNA/RNA band presents a higher ratio than the one observed at the Soret band.

According to Double et al. [23], both melanin and lipofuscin are two pigments that are produced within the brain and accumulate there at a slow pace from the early years of life. Lipofuscin accumulation in the brain has been associated with the brain cell aging process [23,60–62], while melanin has been suggested to be produced by brain cells as an iron-regulatory molecule with protective functions. A more recent review was published by Moreno-García et al. [24], where new findings on the melanin and lipofuscin accumulation in the brain and their interrelation is discussed. According to these authors, and considering the human brain, catecholaminergic neurons are characterized by an age-related accumulation of melanin. When a neuron degenerates, it releases melanin and other products such undegraded cell components or lipids, a process that contributes to initiating and worsening an eventual immune response, which ultimately leads to the neurodegenerative process. These authors also reported that the accumulation of lipofuscin aggregates in the brain during the normal aging process leads to striking morphological changes in neurons. Lipofuscin is described as an undigested bioproduct of central processes of cellular detoxification of autophagy, which is associated with both the aging and the neurodegeneration processes in the brain. It is also noted that the colocalization of both melanin and lipofuscin may produce a redox crosstalk between them. Such crosstalk, under certain conditions, may induce the production of melanin from melanized lipofuscin and under the presence of ferrous sulfide, lipofuscin can be transformed into melanin after a pseudoperoxidation process [24]. We remember that the tissue samples used for our study were retrieved from adult rabbits, meaning that the accumulation of such pigments should be expected in our study. Other authors have also reported that such pigment accumulation is related to tissue and cell degeneration as a consequence of the aging process [60,63,64]. It has also been reported that lipofuscin is able to aggregate transition metals, such as iron and copper, which turn this pigment cytotoxic, changing the oxidation of cellular components such as proteins, lipids and RNA/DNA [62]. In the case of brain tissues, such cellular changes may be at the basis of the degeneration and death of neurons has reported in Ref. [22], and may be the primary responsible for the occurrence of Alzheimer, Parkinson, and stroke diseases.

To prevent the development of such diseases, it is necessary to develop a noninvasive method to evaluate the accumulation of such age-related pigments in brain tissues. Since in our study we were able to detect the presence of such pigments from the $\mu_a(\lambda)$ of the brain cortex, possibly the application of noninvasive diffuse reflectance (R_d) measurements can be used to develop such method. As recently reported [20], such measurements can be combined with a machine learning approach to reconstruct $\mu_a(\lambda)$ and allow to monitor pigment accumulation.

4. Conclusions

In this study, we used spectral measurements from ex vivo brain cortex samples from adult rabbits to calculate the wavelength dependencies of tissues optical properties. The results obtained in this study are accordingly to others that have been reported for different biological tissues. An analysis on the absorption of the cortex showed that it contains some pigments, namely melanin and lipofuscin, which hides the true content of other biological absorbers that the tissue contains. By reconstructing the exponential decreasing absorption with increasing wavelength of these pigments we were able to subtract it from the calculated absorption coefficient of the cortex and such procedure permitted the correct evaluation of the blood and DNA/RNA contents in the rabbit brain.

The accumulation of melanin and lipofuscin has been described to be related to the degeneration processes that occur in tissues and cells as a consequence of the ageing process. In the case of the brain, such degeneration and consequent neuron death has been connected to the development of certain pathologies, such as Alzheimer, Parkinson, and stroke. Due to this relation, it would be interesting to evaluate the content of melanin and lipofuscin in brain tissues from animal specimens at different ages to monitor its progress. Such evaluation could also be performed in humans to identify the pigment accumulation as a function of age and try to relate it with the development of brain diseases. To perform such evaluation in humans in vivo using a noninvasive procedure, diffuse reflectance spectra could be acquired and processed with machine learning algorithms to reconstruct the absorption spectra of the brain tissues. There are several ethical and experimental issues associated with such research in humans, but if the absorption spectra that quantifies pigment accumulation can be calculated from animal models, a method could be developed to diagnose and monitor the development of age-related brain diseases.

Author Contributions: Conceptualization, L.M.O. and V.V.T.; methodology, L.M.O.; software, T.M.G. and I.S.M.; validation, L.M.O.; investigation, T.M.G., H.F.S. and I.S.M.; writing—original draft preparation, L.M.O.; writing—review and editing, T.M.G., H.F.S., I.S.M. and V.V.T. All authors have read and agreed to the published version of the manuscript.

Funding: The work of L.M.O. was supported by the Portuguese Science Foundation, Grant No. FCT-UIDB/04730/2020. The work of V.V.T. was supported by the Government of the Russian Federation (project No. 075-15-2019-1885 to support scientific research projects implemented under the supervision of leading scientists at Russian institutions, Russian institutions of higher education).

Institutional Review Board Statement: The present study involves measurements from ex vivo animal tissues. The research follows the Declaration of Helsinki and was approved by the research review board in biomedical engineering of the Center of Innovation in Engineering and Industrial Technology (CIETI), in Porto, Portugal. Such approval has the number CIETI/Biomed_Research_2021_01.

Informed Consent Statement: Not applicable.

Data Availability Statement: The data that supports the findings of this study are available from the corresponding author upon reasonable request.

Conflicts of Interest: The authors declare no conflict of interest.

References

1. Carvalho, S.; Carneiro, I.; Henrique, R.; Tuchin, V.V.; Oliveira, L. Lipofuscin-type pigment as a marker of colorectal cancer. *Electronics* **2020**, *9*, 1805. [CrossRef]
2. Oliveira, L.M.; Tuchin, V.V. *The Optical Clearing Method—A New Tool for Clinical Practice and Biomedical Engineering*; Springer: Cham, Switzerland, 2019; pp. 1–106.
3. Tuchin, V.V. *Tissue Optics: Light Scattering Methods and Instruments for Medical Diagnostics*, 3rd ed.; SPIE Press: Bellingham, WA, USA, 2015; pp. 245–358.
4. Wang, L.; Jacques, S.L.; Zheng, L. MCML-Monte Carlo modeling of light transport in multi-layered tissues. *Comp. Method Prog. Biomed.* **1995**, *47*, 131–146. [CrossRef]
5. Prahl, S.A.; van Gemert, M.J.C.; Welch, A.J. Determining the optical properties of turbid media by using the adding-doubling method. *Appl. Opt.* **1993**, *32*, 559–568. [CrossRef]
6. Sordillo, D.C.; Sordillo, L.A.; Sordillo, P.P.; Shi, L.; Alfano, R.R. Short wavelength infrared optical windows for evaluation of benign and malignant tissues. *J. Biomed. Opt.* **2017**, *22*, 45002. [CrossRef]
7. Shi, L.; Sordillo, L.A.; Rodríguez-Contreras, A.; Alfano, R. Transmission in near-infrared optical Windows for deep-brain imaging. *J. Biophotonics* **2016**, *9*, 38–43. [CrossRef]
8. Oliveira, L.M.; Zaytsev, K.I.; Tuchin, V.V. Improved biomedical imaging over a wide spectral range from UV to THz towards multimodality. In Proceedings of the Third International Conference on Biophotonics Riga 2020, Riga, Latvia, 24–25 August 2020; Spigulis, J., Ed.; SPIE: Bellingham, WA, USA, 2020; Volume 11585, p. 11585.
9. Carneiro, I.; Carvalho, S.; Henrique, R.; Oliveira, L.; Tuchin, V. Measurement of optical properties of normal and pathological human liver tissue from deep-UV to NIR. In Proceedings of the Photonics Europe: Tissue Optics and Photonics, Online Conference, 2 April 2020; Tuchin, V.V., Blondel, W., Zalevsky, Z., Eds.; SPIE: Bellingham, WA, USA, 2020; Volume 11363, p. 113630G.
10. Backman, B.; Wax, A.; Zhang, H. *A Laboratory Manual in Biophotonics*; CRC Press: Boca Raton, FL, USA, 2018; p. 100.
11. Jacques, S.L. Optical properties of biological tissues: A review. *Phys. Med. Biol.* **2013**, *58*, R37–R61. [CrossRef]
12. Yanina, I.Y.; Lazareva, E.N.; Tuchin, V.V. Refractive index of adipose tissue and lipid droplet measured in wide spectral and temperature ranges. *Appl. Opt.* **2018**, *57*, 4839. [CrossRef]
13. Martins, I.; Silva, H.; Tuchin, V.V.; Oliveira, L. Estimation of rabbit pancreas dispersion between 400 and 1000 nm. *J. Biomed. Phot. Eng.* **2021**, *7*, 020303.
14. Lazareva, E.N.; Oliveira, L.; Yanina, I.Y.; Chernomyrdin, N.V.; Musina, G.R.; Tuchina, D.K.; Bashkatov, A.N.; Zaytsev, K.I.; Tuchin, V.V. Refractive index measurements of tissue and blood and OCAs in a wide spectral range. In *Tissue Optical Clearing: New Prospects in Optical Imaging*; Zhu, D., Genina, E., Tuchin, V., Eds.; CRC Press: Boca Raton, FL, USA, to be published.
15. Carneiro, I.; Carvalho, S.; Henrique, R.; Oliveira, L.; Tuchin, V.V. Simple multimodal optical technique for evaluation of free/bound water and dispersion of human liver tissue. *J. Biomed. Opt.* **2017**, *22*, 125002. [CrossRef]
16. Sydoruk, O.; Zhernovaya, O.; Tuchin, V.V.; Douplik, A. Refractive index of solutions of human hemoglobin from the near-infrared to the ultraviolet range: Kramers-Kronig analysis. *J. Biomed. Opt.* **2012**, *17*, 115002. [CrossRef]
17. Gienger, J.; Groβ, H.; Neukammer, J.; Bär, M. Determining the refractive index of human hemoglobin solutions by Kramers-Kronig relations with an improved absorption model. *Appl. Opt.* **2016**, *55*, 8951–8961. [CrossRef]
18. Bashkatov, A.N.; Genina, E.A.; Kozintseva, M.D.; Kochubei, V.I.; Gorodkov, S.Y.; Tuchin, V.V. Optical properties of peritoneal biological tissues in the spectral range of 350–2500 nm. *Opt. Spectrosc.* **2016**, *120*, 6–14. [CrossRef]
19. Carvalho, S.; Gueiral, N.; Nogueira, E.; Henrique, R.; Oliveira, L.; Tuchin, V. Comparative study of the optical properties of colon mucosa and colon precancerous polyps between 400 and 1000 nm. In Proceedings of the BIOS-Photonics West 2017: Dynamics and Fluctuations in Biomedical Photonics, San Francisco, CA, USA, 28 January–2 February 2017; Tuchin, V.V., Larin, K.V., Leahy, M.J., Wang, R.K., Eds.; SPIE: Bellingham, WA, USA, 2017; Volume 10063, p. 100631L.
20. Fernandes, L.; Carvalho, S.; Carneiro, I.; Henrique, R.; Tuchin, V.V.; Oliveira, H.; Oliveira, L. Diffuse reflectance and machine learning techniques to differentiate colorectal cancer ex vivo. *Chaos* **2021**, *31*, 053118. [CrossRef] [PubMed]
21. GBD 2016 Neurology Collaborators. Global, regional, and national burden of neurological disorders, 1990–2016: A systematic analysis for the global burden of disease study 2016. *Lancet Neurol.* **2019**, *18*, 459–480. [CrossRef]
22. Mattson, M.P.; Duan, W.; Pedersen, W.A.; Culmsee, C. Neurodegenerative disorders and ischemic brain diseases. *Apoptosis* **2001**, *6*, 69–81. [CrossRef] [PubMed]
23. Double, K.L.; Dedov, V.N.; Fedorow, H.; Kettle, E.; Halliday, G.M.; Garner, B.; Brunk, U.T. The comparative biology of neuromelanin and lipofuscin in the brain. *Cell Mol. Life Sci.* **2008**, *65*, 1669–1682. [CrossRef] [PubMed]
24. Moreno-García, A.; Kun, A.; Calero, M.; Calero, O. The neuromelanin paradox and its role in oxidative stress and neurodegeneration. *Antioxidants* **2021**, *10*, 124. [CrossRef]
25. Tuchin, V.V. *Optical Clearing of Tissues and Blood*; SPIE Press: Bellingham, WA, USA, 2005; p. 79.
26. Sdobnov, A.Y.; Darvin, M.; Genina, E.A.; Bashkatov, A.N.; Lademann, J.; Tuchin, V.V. Recent progress in tissue optical clearing for spectroscopic application. *Spectrochim. Acta Part A Biomol. Spectrosc.* **2018**, *197*, 216–229. [CrossRef]
27. Li, H.; Xie, S. Measurement method of the refractive index of biotissue by total internal reflection. *Appl. Opt.* **1996**, *35*, 1793–1975. [CrossRef]
28. Ding, H.; Lu, J.Q.; Jacobs, K.M.; Hu, X.H. Determination of refractive indices of porcine skin tissues and intralipid at eight wavelengths between 325 and 1557 nm. *J. Opt. Soc. Am. A* **2005**, *22*, 1151–1157. [CrossRef]

29. Deng., Z.; Wang, J.; Ye, Q.; Sun, T.; Zhou, W.Y.; Mei, J.; Zhang, C.; Tian, J. Determination of continuous complex refractive index dispersion of biotissue based on internal reflection. *J. Biomed. Opt.* **2016**, *21*, 015004. [CrossRef]
30. Refractive Index Database. Available online: https://refractiveindex.info/ (accessed on 1 June 2021).
31. Song, Q.W.; Ku, C.Y.; Zhang, C.; Gross, R.B.; Birge, R.R.; Michalak, R. Modified critical angle method for measuring the refractive index of bio-optical materials and its application to bacteriorhodopsin. *J. Opt. Soc. Am. B* **1995**, *12*, 797–803. [CrossRef]
32. Carneiro, I.; Carvalho, S.; Henrique, R.; Oliveira, L.; Tuchin, V.V. Water content and scatterers dispersion evaluation in colorectal tissues. *J. Biomed. Phot. Eng.* **2017**, *3*, 040301. [CrossRef]
33. Carvalho, S.; Gueiral, N.; Nogueira, E.; Henrique, R.; Oliveira, L.; Tuchin, V.V. Wavelength dependence of the refractive index of human colorectal tissues: Comparison between healthy mucosa and cancer. *J. Biomed. Phot. Eng.* **2016**, *2*, 040307. [CrossRef]
34. Bashkatov, A.N.; Genina, E.A.; Kochubey, V.I.; Tuchin, V.V. Optical properties of the subcutaneous adipose tissue in the spectral range 400–2500 nm. *Opt. Spectrosc.* **2005**, *99*, 836–842. [CrossRef]
35. Roggan, A.; Dörschel, K.; Minet, O.; Wolff, D.; Müller, G. The optical properties of biological tissue in the near infrared wavelength range—Review and measurements. In *Laser-Induced Interstitial Thermotherapy*; Müller, G., Roggan, A., Eds.; SPIE Press: Bellingham, WA, USA, 1995; pp. 10–44.
36. Binding, J.; Arous, J.B.; Léger, J.F.; Gigan, S.; Boccara, C.; Bourdieu, L. Brain refractive index measured in vivo with high-NA defocus-corrected full-field OCT and consequences for two-photon microscopy. *Opt. Express* **2011**, *19*, 4833–4847. [CrossRef]
37. Sun, J.; Lee, S.J.; Wu, L.; Sarntinoranont, M.; Xie, H. Refractive index measurement of acute rat brain tissue slices using optical coherence tomography. *Opt. Express* **2012**, *20*, 1084–1095. [CrossRef]
38. Fasman, G.D. Ultraviolet spectra of derivatives of cysteine, cysteine, histidine, phenylalanine, tyrosine, and tryptophan. In *Handbook of Biochemistry and Molecular Biology*, 3rd ed.; Fasman, G.D., Ed.; CRC Press: Boca Raton, FL, USA, 2018; Volume 1, Chapter 17.
39. Wetlaufer, D.B. Ultraviolet spectra of proteins and amino acids. In *Advances in Protein Chemistry*; Afinsen, C.B., Jr., Ed.; Academic Press: London, UK, 1962; Chapter 6; Volume 17.
40. Narayan, R.K.; Heydon, W.E.; Creed, G.J.; Jacobowitz, D.M. Identification of major proteins in human cerebral cortex and brain tumors. *J. Protein Chem.* **1985**, *4*, 375–389. [CrossRef]
41. Zhou, Y.; Yao, J.; Wang, L.V. Tutorial on photoacoustic tomography. *J. Biomed. Opt.* **2016**, *21*, 061007. [CrossRef]
42. Pitzschke, A.; Lovisa, B.; Seydoux, O.; Haenggi, M.; Oertel, M.F.; Zellweger, M.; Tardy, Y.; Wagnières, G. Optical properties of rabbit brain in the red and near-infrared: Changes observed under in vivo, postmortem, frozen, and formalin-fixated conditions. *J. Biomed. Opt.* **2015**, *20*, 025006. [CrossRef] [PubMed]
43. Koeppen, A.H.; Dentinger, M.P. Brain hemosiderin and superficial siderosis of the central nervous system. *J. Neuropathol. Exp. Neurol.* **1988**, *47*, 249–270. [CrossRef]
44. Granick, S. Ferritin: Its properties and significance for iron metabolism. *Chem. Rev.* **1946**, *38*, 379–403. [CrossRef] [PubMed]
45. Maoka, T. Carotenoids as natural functional pigments. *J. Nat. Med.* **2020**, *74*, 1–16. [CrossRef]
46. Johansson, J. Spectroscopic method for determination of the absorption coefficient in brain tissue. *J. Biomed. Opt.* **2010**, *15*, 057005. [CrossRef] [PubMed]
47. Wakamatsu, K.; Murase, T.; Zucca, F.A.; Zecca, L.; Ito, S. Biosynthetic pathway to measure neuromelanin and its aging process. *Pigment. Cell Melanoma Res.* **2012**, *25*, 792–803. [CrossRef] [PubMed]
48. Gilissen, E.P.; Staneva-Dobrovski, L. Distinct types of lipofuscin pigment in the hippocampus and cerebellum of aged cheirogaleid primates. *Anat. Rec.* **2013**, *296*, 1895–1906. [CrossRef]
49. Heinsen, H. Lipofuscin in the cerebellar cortex of albino rats: An electron microscopic study. *Anat. Embryol.* **1979**, *115*, 333–345. [CrossRef]
50. Ivy, G.O.; Kanai, S.; Ohta, M.; Smith, G.; Sato, Y.; Kobayashi, M.; Kitani, K. Lipofucin-like substances accumulate rapidly in brain, retina and internal organs with cysteine protease inhibition. *Adv. Exp. Med. Biol.* **1989**, *266*, 31–45.
51. Johansson, J.D.; Wårdell, K. Intracerebral quantitative chromophore estimation from reflectance spectra captured during deep brain stimulation implantation. *J. Biophot.* **2013**, *6*, 435–445. [CrossRef] [PubMed]
52. Różanowska, M.B.; Pawlak, A.; Różanowski, B. Products of docosahexaenoate oxidation as contributors to photosensitizing properties of retinal lipofuscin. *Int. J. Mol. Sci.* **2021**, *22*, 3525. [CrossRef]
53. Zonios, G.; Dimou, A.; Bassukas, I.; Galaris, D.; Tsolakidis, A.; Kaxiras, E. Melanin absorption spectroscopy: New method for noninvasive skin investigation and melanoma detection. *J. Biomed. Opt.* **2008**, *13*, 014017. [CrossRef] [PubMed]
54. Zhang, L.; Zou, X.; Zhang, B.; Cui, L.; Zhang, J.; Mao, Y.; Chen, L.; Ji, M. Label-free imaging of hemoglobin degradation and hemosiderin formation in brain tissues with femtosecond pump-probe microscopy. *Theranostics* **2018**, *8*, 4129–4140. [CrossRef] [PubMed]
55. Zaghdoudi, K.; Ngomo, O.; Vanderesse, R.; Anoux, P.; Myrzakhmetov, B.; Frochot, C.; Guiavarch, Y. Extraction, identification and photo-physical characterization of persimmon (*Diospyros kaki* L.) carotenoids. *Foods* **2017**, *6*, 4. [CrossRef] [PubMed]
56. Raja, A.S.; Sathiyabama, J.; Venkatesan, R.; Prathipa, V. Corrosion control of carbon steel by eco-friendly inhibitor L-Cysteine-Zn2+ system in acqueous medium. *J. Chem. Biol. Phys. Sci.* **2014**, *4*, 3182–3189.
57. Hazra, C.; Samanta, T.; Mahalingam, V. A resonance energy transfer approach for the selective detection of aromatic amino acids. *J. Mater. Chem. C* **2014**, *2*, 10157–10163. [CrossRef]

58. Lin, J.-H.; Yu, C.-J.; Yang, Y.-C.; Tseng, W.-L. Formation of fluorescence polydopamine dots from hydroxyl radical-induced degradation of polydopamine nanoparticles. *Phys. Chem. Chem. Phys.* **2015**, *17*, 15124. [CrossRef]
59. Feng, S.; Harayama, T.; Montessuit, S.; David, F.P.; Winssinger, N.; Martinou, J.-C.; Riezman, H. Mitochondria-specific photoactivation to monitor local sphingosine metabolism and function. *Elife* **2018**, *7*, e34555. [CrossRef]
60. Jung, T.; Bader, N.; Grune, T. Lipofuscin: Formation, distribution, and metabolic consequences. *Ann. N. Y. Acad. Sci.* **2007**, *1119*, 97–111. [CrossRef]
61. Katz, M.L.; Robinson, W.G., Jr. What is lipofuscin? Defining characteristics and differentiation from other autofluorescent lysosomal storage bodies. *Arch. Gerontol. Geriatr.* **2002**, *34*, 169–184. [CrossRef]
62. Höhn, A.; Jung, T.; Grimm, S.; Grune, T. Lipofuscin-bound iron is a major intracellular source of oxidants: Role in senescent cells. *Free Rad. Biol. Med.* **2010**, *48*, 1100–1108. [CrossRef]
63. Hunter, J.J.; Morgan, J.I.W.; Merigan, W.H.; Sliney, D.H.; Sparrow, J.R.; Williams, D.R. The susceptibility of the retina to photochemical damage from visible light. *Prog. Retin Eye Res.* **2012**, *31*, 28–42. [CrossRef]
64. Gosnell, M.E.; Anwer, A.G.; Cassano, J.C.; Sue, C.M.; Goldys, E.M. Functional hyperspectral imaging captures subtle details of cell metabolism in olfactory neurosphere cells, disease-specific models of neurodegenerative disorders. *Biochim. Biophys. Acta* **2016**, *1863*, 56–63. [CrossRef]

Article

Photoprotective Properties of Eumelanin: Computational Insights into the Photophysics of a Catechol:Quinone Heterodimer Model System

Victoria C. Frederick [1], Thomas A. Ashy [1], Barbara Marchetti [1], Michael N. R. Ashfold [2] and Tolga N. V. Karsili [1,*]

[1] Department of Chemistry, University of Louisiana at Lafayette, Lafayette, LA 70503, USA; victoria.frederick1@louisiana.edu (V.C.F.); thomas.ashy1@louisiana.edu (T.A.A.); barbara.marchetti1@louisiana.edu (B.M.)
[2] School of Chemistry, University of Bristol, Bristol BS8 1TS, UK; mike.ashfold@bris.ac.uk
* Correspondence: tolga.karsili@louisiana.edu

Citation: Frederick, V.C.; Ashy, T.A.; Marchetti, B.; Ashfold, M.N.R.; Karsili, T.N.V. Photoprotective Properties of Eumelanin: Computational Insights into the Photophysics of a Catechol:Quinone Heterodimer Model System. *Photochem* **2021**, *1*, 26–37. https://doi.org/10.3390/photochem1010003

Academic Editor: Marcelo I. Guzman

Received: 19 February 2021
Accepted: 7 March 2021
Published: 10 March 2021

Publisher's Note: MDPI stays neutral with regard to jurisdictional claims in published maps and institutional affiliations.

Copyright: © 2021 by the authors. Licensee MDPI, Basel, Switzerland. This article is an open access article distributed under the terms and conditions of the Creative Commons Attribution (CC BY) license (https://creativecommons.org/licenses/by/4.0/).

Abstract: Melanins are skin-centered molecular structures that block harmful UV radiation from the sun and help protect chromosomal DNA from UV damage. Understanding the photodynamics of the chromophores that make up eumelanin is therefore paramount. This manuscript presents a multi-reference computational study of the mechanisms responsible for the experimentally observed photostability of a melanin-relevant model heterodimer comprising a catechol (C)–benzoquinone (Q) pair. The present results validate a recently proposed photoinduced intermolecular transfer of an H atom from an OH moiety of C to a carbonyl-oxygen atom of the Q. Photoexcitation of the ground state C:Q heterodimer (which has a π-stacked "sandwich" structure) results in population of a locally excited ππ* state (on Q), which develops increasing charge-transfer (biradical) character as it evolves to a "hinged" minimum energy geometry and drives proton transfer (i.e., net H atom transfer) from C to Q. The study provides further insights into excited state decay mechanisms that could contribute to the photostability afforded by the bulk polymeric structure of eumelanin.

Keywords: photophysics; photoprotection; photostability; eumelanin; catechol; benzoquinone; ultraviolet; conical intersection

1. Introduction

The fundamental photochemistry of prototypical organic and biological chromophores is attracting ever more attention [1–3]—driven, in part, by ambitions to advance understanding (and prevention) of photoinduced damage in biomolecules [4–11] and to improve the photoprotection offered by sunscreen molecules [12–15]. Ultraviolet (UV) excitation of any given molecule increases its total energy, typically to values in excess of many of the energy barriers associated with reaction on the ground state potential energy (PE) surface. The excited state molecules formed upon UV absorption may decay in a number of ways that have traditionally been illustrated using a Jablonski diagram. Extremes of these behaviors include:

(i) Photoreaction, by photodissociation, which constitutes the dominant decay mechanism for many small heterocyclic molecules in the gas phase. Much studied examples include phenol [2,4,16–25], pyrrole [26–34] and indole [35–38].

(ii) Photostability, wherein the photoexcited molecule decays back to the ground state, rapidly and with high efficiency, without any permanent chemical transformation. Such non-radiative decay (generically termed internal conversion) is the desired photophysical response for the DNA/RNA nucleobases [4,39–65] and, for example, for derivatives of *p*-aminobenzoates, cinnamates, salicylates, anthranilates, camphor, dibenzoyl methanes and/or benzophenones used in commercial sunscreens [13,14]. Internal conversion processes are mediated by conical intersections (CIs) — regions of the PE surfaces where distinct electronic states become energetically degenerate [66,67]. These points of degeneracy develop into CIs when orthogonal motions are considered, which facilitate non-adiabatic coupling (i.e., the funneling of population) from the photoexcited state to a lower (e.g., the ground) state.

Sunscreen molecules are chosen on the basis of good photostability. Those that find use in commercial sunscreen products have been chosen/engineered to absorb in the UV-A/B regions of the electromagnetic spectrum and, following photoexcitation, to undergo efficient non-radiative decay back to the ground state—releasing the excess energy as local heating in the formulation of which they are a part. In oxybenzone, for example, $\pi^* \leftarrow \pi$ excitation of the dominant (in the ground (S_0) state) enol-conformer populates the strongly absorbing $^1\pi\pi^*$ state, which relaxes—either directly or via an (optically dark) $^1n\pi^*$ state—toward its minimum energy keto-configuration (i.e., an intramolecular H atom transfer (HAT) process) and onwards towards the CI with the S_0 state at non-planar geometries (one ring twists relative to the other about the central aliphatic C–C bond). A reverse H atom transfer process on the S_0 PES and vibrational energy transfer to the surrounding solvent results in (efficient, but not complete) reformation of the original enol-conformer with a (solvent-dependent) time constant [32,68–72].

Many sunscreen molecules occur naturally on UV-exposed regions of biological systems. In mammals, for example, natural molecular sunscreens are localized in the skin (eumelanin and pheomelanin), sweat glands (urocanic acid) and the cornea of the eye (kynurenines). As with the DNA/RNA nucleobases, natural sunscreen molecules have evolved to cope with exposure to UV radiation, which, in the case of the skin, means protecting the lower epidermis from UV-induced DNA damage. Eumelanin, the most abundant melanin found in humans, provides many beneficial functions including serving as a naturally occurring sunscreen [73–75]. Eumelanin is produced via melanogenesis, wherein tyrosine is oxidized and polymerized, resulting in a heterogeneous pigment composed of cross-linked 5,6-dihydroxyindole (DHI) and 5,6-dihydroxyindole-2-carboxylic acid (DHICA) based polymers. The mechanism(s) underpinning its photostability remain active research topics, however [76–78].

Several studies have sought to address these issues by exploring aspects of the photophysics prevailing within individual molecular components of eumelanin—including DHI [79–82] and DHICA [75,83–85], various monohydroxyindoles [86,87] and catechol [88–93] in the gas and/or solution phase. Building on earlier work by Van Anh and Williams [94], Kohler and co-workers [95] recently reported ultrafast transient absorption studies following UV photoexcitation of non-covalently bonded heterodimers based on *ortho*-positioned dihydroxyphenol (catechol) and 1,2-benzoquinone groups. For experimental reasons, it was necessary to work with the chemically stable 3,5-di-*tert*-butyl substituted C and Q molecules. The experimental data provided rather convincing evidence that intermolecular HAT from the (acidic) O–H proton donor on catechol to a carbonyl oxygen (proton acceptor) group on the quinone could constitute another decay pathway and thus another source of photostability following UV photoexcitation. Here we report multi-reference computational studies designed to explore the photophysics of this model catechol (C), 1,2-benzoquinone (Q) (henceforth C:Q) heterodimer (without the *tert*-butyl substituents, for computational simplicity), which provide mechanistic insight in support of these recent experiments [95].

2. Computational Methods

The ground state minimum energy geometry of the π-stacked configuration of the C:Q heterodimer was optimized using the ωB97XD functional of Density Functional Theory [96], coupled to the 6-311+G(*d*, *p*) Pople basis set [97]. This functional was chosen since it is capable of describing the long-range correlation effects inherent to charge-separated configurations, as well as the dispersion interactions between the individual chromophores. Optimizations of other plausible side-on, hydrogen-bonded and π-stacked configurations were undertaken but, in all cases, these alternative starting structures converged to the π-stacked configuration shown below.

The present study focused on the net transfer of an H atom, so the tautomer formed by HAT was next optimized. In the closed-shell singlet configuration, this tautomer should represent an unstable high energy point on the diabatic ground state PE surface due to the long-range attractive interaction that encourages reformation of the parent heterodimer. Thus, the corresponding triplet-spin configuration of the C:Q biradical tautomer (which can also be viewed as a semiquinone, or catechoxyl, radical pair) was optimized in order to maintain a long-range minimum energy structure. A similar methodology was used recently to follow the photoinduced evolution of a cyclopropenone-containing enediyne system through to a biradical configuration [98].

Vertical excitation energies and transition dipole moments of the C:Q heterodimer were calculated using complete active space self-consistent field (CASSCF) and complete active space second-order perturbation theory (CASPT2) methods. The CASSCF calculation was state-averaged with the lowest four singlet and the lowest four triplet states and employed an active space of ten electrons in ten orbitals (10/10)—comprising the five highest valence orbitals and five lowest unoccupied molecular orbitals (shown later).

PE profiles associated with the HAT coordinate were then constructed by interpolating the geometries between the ground state minimum of the C:Q heterodimer and the biradical tautomer, using a linear interpolation in internal coordinates (LIIC). Using the CASPT2 method [99] and a cc-pVDZ basis set, PE values at each point along the LIIC were calculated for the lowest four singlet states using a state-averaged CASSCF reference wavefunction. The same (10/10) active space was used, along with an imaginary level shift of 0.5 E_H to aid convergence and circumvent the involvement of intruder states.

All DFT calculations were calculated using the Gaussian 16 computational package [100], whilst all CASPT2 calculations used the MOLPRO computational package [101].

3. Results and Discussion

An isolated gas-phase multi-reference computational study of the UV photoinduced chemistry of C:Q heterodimers is presented, the results of which support and extend conclusions reached in recent transient absorption studies of this system in a weakly interacting solvent (cyclohexane) [95]. This section is sub-divided into sections addressing the minimum energy structures of the heterodimer and its biradical tautomer, the electronic spectroscopy of the former and then the topography of the PE surfaces sampled following photoexcitation of the heterodimer.

3.1. Minimum Energy Geometries of the Ground State Heterodimer and Its Biradical Tautomer

As a reminder, the ωB97XD functional was used in order to achieve an appropriately balanced description of the dominant π-π interactions between the C and Q chromophores and the long-range correlation effects. As Figure 1a,b show, the ground state minimum energy geometry of the C:Q heterodimer exhibits a π-stacked configuration. (The Cartesian coordinates of all atoms in this minimum energy structure are provided as Supplementary Materials, as are (harmonic) normal mode wavenumbers for the ground state C:Q heterodimer and the bare C monomer.) Alternative side-on hydrogen bonding could also be expected to provide a strong intermolecular interaction, but the π-stacked ground state configuration shown in Figure 1 offers both π-π and hydrogen-bonding, and the stability of this π-stacked structure can be understood by recognizing two stabilizing interactions. One is a π–π interaction between bonding π electrons on the catechol moiety and the antibonding π^* orbital localized on the benzoquinone. These orbitals are reasonably well-matched in energy. The second is the inter-chromophore hydrogen-bonding between an O–H donor, local to catechol, and a carbonyl oxygen acceptor, localized on the benzoquinone chromophore. This deduced C:Q heterodimer structure, involving one intra- and one intermolecular H-bond, is fully consistent with that derived by analysis of the Fourier transform infrared spectrum of mixed solutions of (the di-*tert*-butyl substituted forms of) C and Q in cyclohexane [95].

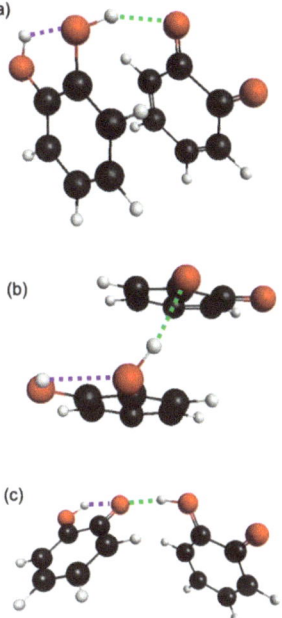

Figure 1. (a) Side and (b) top view of the optimized structure of the catechol (C), 1,2-benzoquinone (Q) (C:Q) heterodimer in its ground electronic state. (c) Optimized geometry of the biradical tautomer formed by H atom transfer. Intramolecular and intermolecular H-bonds are indicated by, respectively, dashed purple and green lines.

Figure 1c shows the minimum energy geometry of the biradical formed upon HAT. (The Cartesian coordinates of all atoms in this minimum energy structure are also provided in the Supplementary Materials). The "hinged" structural arrangement of the C and Q chromophores is very different from the π-stacked configuration of the ground state heterodimer, though it again displays one intermolecular and one intramolecular hydrogen bond. The breakdown of the π-stacking upon biradical formation can be understood by recognizing that the lowest energy biradical configuration has ππ* character, wherein the π- and π*-orbitals are localized on, respectively, the C and Q moieties (vide infra). The ensuing electron–electron repulsion destroys the π–π interaction inherent to the ground state parent structure, leaving inter-chromophore H-bonding as the dominant non-covalent interaction in the biradical tautomer.

We note that the experimentally studied C:Q heterodimer contains bulky *tert*-butyl substituents which may affect the π-stacking. That said, we do not expect this to have a serious impact on the excited state photophysics deduced here, as the *tert*-butyl group is a σ-perturbing substituent, while the dominant effects observed in the photophysics of C:Q are π-centered.

3.2. The Electronic Spectrum of the C:Q Heterodimer

Table 1 lists the vertical excitation energies (VEEs) to the first few singlet and triplet excited states of the C:Q heterodimer from the π-stacked minimum energy configuration. The active space orbitals used in the CASPT2 computations, shown in Figure 2, may be used along with Table 1 to identify the dominant orbital promotions involved in preparing these various excited states.

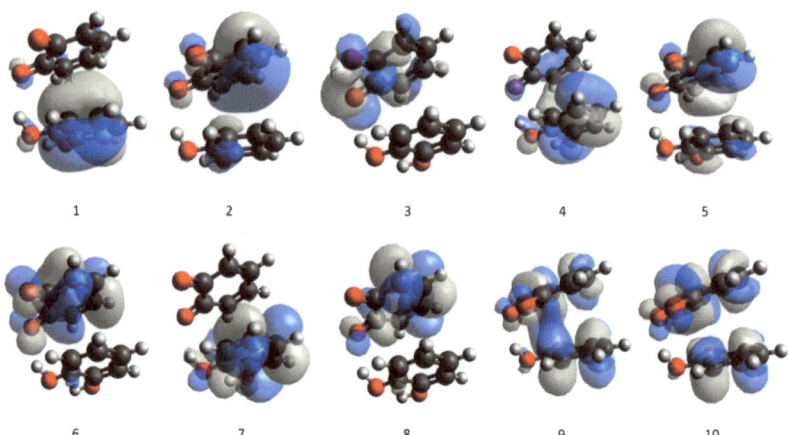

Figure 2. Active space orbitals used in the CASSCF/CASPT2 calculations of the C:Q heterodimer. Q sits above C in all depictions, but the orientation of the heterodimer structure is varied to allow better visualization of the various occupied (1–5) and virtual (6–10) orbitals. The orbital numbering aligns with that used to describe the dominant promotions associated with forming the various excited states in Table 1.

Table 1. Calculated vertical excitation energies (VEE) and oscillator strengths (f) to the lowest singlet and lowest triplet excited states of the C:Q heterodimer. The entries in the "Character" column show the dominant electron promotions between the active space orbitals shown in Figure 2, with the respective contributions (i.e., the squares of the associated coefficients) shown in parentheses.

Electronic State	Character	VEE/eV	f
S_1-S_0	6←3 (0.71) 6←3 + 6←5 (0.04)	1.93	0.001
S_2-S_0	6←5 (0.51) 6←4 (0.34)	3.23	0.0541
T_1-S_0	6←5 (0.72) 8←3 (0.04)	1.64	-
T_2-S_0	6←5 (0.73) 8←5 (0.04)	2.25	-

Vertical excitation to the S_1 state from the π-stacked ground state minimum energy geometry is dominated by electron promotion from a largely non-bonding (n) orbital, localized on the carbonyl oxygen atom, to an antibonding π^* orbital, both of which are localized on the benzoquinone moiety. The S_1 state is best viewed as a locally excited state (i.e., the excitation is concentrated on a common chromophore) with $n\pi^*$ character, and optically "dark" (i.e., the S_1–S_0 transition has a low oscillator strength—reflecting the poor spatial overlap of the n and π^* orbitals).

The S_2 state is best described by a mixture of two configurations. The 6←5 orbital promotion involves excitation of an electron from the π highest occupied molecular orbital (HOMO) (which is mainly localized on Q but extends over the C moiety also) to the π^* lowest unoccupied molecular orbital (LUMO) localized on Q. 6←4 promotion, in contrast, involves excitation from a bonding π orbital, largely localized on the C chromophore, to the Q-localized π^* antibonding orbital. Both promotions can be pictured as $\pi^* \leftarrow \pi$ transitions: The S_2 state is thus best viewed as having $\pi\pi^*$ character but, even in the vertical region, formation of the S_2 state of the heterodimer involves some electron transfer from C to Q—which likely contributes to the high oscillator strength reported in Table 1. For completeness, excitation energies to the first two triplet excited states, both of which are also best described as locally excited $\pi\pi^*$ states, are also included in Table 1.

The predicted oscillator strengths and VEEs for the (weak) S_1–S_0 and (strong) S_2–S_0 transitions reported in Table 1 match well with the maxima evident (at λ ~595 nm and ~400 nm) in the UV absorption spectrum of the (*t*-butyl substituted) C:Q heterodimer in cyclohexane [95], lending further support to our expectation that the *tert*-butyl substituents have little effect on the electronic properties (and excited state photophysics) of the heterodimer.

3.3. Photophysics of the C:Q Heterodimer

We now consider the possible fate(s) of the C:Q heterodimer following UV photoexcitation and the potential role of such photophysics in explaining the photostability of eumelanin. Motivated by the work of Kohler and co-workers [95], the ensuing discussion focuses on the mechanism of photoinduced HAT. Figure 3a displays PE profiles of the ground and first two excited singlet electronic states along the LIIC between the ground state minimum energy geometry and that of the optimized biradical tautomer formed by HAT (henceforth Q_{LIIC})—the structures of both of which are reproduced again as insets in Figure 3a. We caution that the use of a LIIC almost inevitably means that the present calculations do not capture the true minimum energy path from reactant to product, but they are expected to identify key topographical features of the PE surfaces under study. The ground state (black) PE profile increases, reaching a maximum at Q_{LIIC} ~0.5, beyond which it decreases en route to the biradical tautomer. The electronic wavefunction of the adiabatic ground state switches at Q_{LIIC} ~0.5, as illustrated in Figure 3b,c, which illustrates the increasing $\pi\pi^*$ character of the ground state configuration of the biradical tautomer (which correlates diabatically with the S_2 state of the parent C:Q heterodimer). The stability of the biradical structure can be understood by considering the change in electronic character. In the $\pi\pi^*$ configuration, the O-atom donor of the pre-existing OH moiety contains a doubly occupied *p* orbital which, when viewed from the biradical minimum, provides a long-range repulsive interaction in the reverse HAT direction. As Figure 3b,c show, the singly occupied molecular orbitals in the S_0 state of the biradical are localized on different chromophores: the S_0 state at $Q_{LIIC} > 0.5$ is best described as a charge-separated (or charge transfer) state.

The S_1 state (red in Figure 3a) has $n\pi^*$ character in the vertical region and is bound with respect to initial motion along Q_{LIIC}—reflecting the fact that the $\pi^* \leftarrow n$ transition is localized on the benzoquinone moiety and shows no net driving force for HAT. The S_2 state, in contrast, has $\pi\pi^*$ character at the Franck–Condon geometry and shows net reactivity with respect to the "hinge-like" geometry change along Q_{LIIC}. This can be understood by recognizing that the transition involves π and π^* orbitals that are initially largely localized on a single chromophore but then develop increasing charge-separated character, as shown in Figure 3b,c. The diabatic $^1\pi\pi^*$ state progressively develops charge transfer (CT) character as $Q_{LIIC} \to 1$, which is neutralized by proton transfer from the C to the Q moiety. Such photoinduced HATs are also frequently termed proton-coupled electron transfer (PCET) or electron-driven proton transfer (EDPT) processes. Upon increasing Q_{LIIC} from the Franck–Condon region, the diabatic CT state crosses both the $^1n\pi^*$ state and the ground state. This is the origin of the evolution of the ground state electronic wavefunction described above. As in many related systems [4,102,103], these diabatic crossing points will surely be CIs when motion along orthogonal modes are considered, and represent regions of configuration space where internal conversion between electronic states is favorable (i.e., where population is funneled efficiently to the lower PE surface).

Given the foregoing descriptions of the various electronic states of the C:Q heterodimer, the reported photophysics can be rationalized as follows: Photoexcitation populates the "bright" $^1\pi\pi^*$ state, which is initially largely localized on Q but evolves spontaneously along the coordinate associated with HAT. Internal conversion is likely to occur at both the S_2/S_1 and S_1/S_0 CIs (see Figure 3a). The former may well lead to some population becoming temporarily trapped in the $^1n\pi^*$ state—as has been proposed in the case of oxybenzone [72]—while non-adiabatic interaction at the latter CI will promote efficient internal conversion back to the S_0 state. Having accessed the S_0 PE surface, population may bifurcate to reform the ground state heterodimer (thereby demonstrating photostability) or evolve towards the biradical tautomer and thence to two (potentially harmful) semiquinone free radical species. Thus, the extent to which the C:Q heterodimer offers photoprotection and photostability will be sensitively dependent on the non-adiabatic dynamics prevailing at the S_1/S_0 CI—which will be sensitive to the detailed topography of the CI and the nuclear momenta within the evolving population. Such details, in turn, are likely to be sensitively dependent upon the natures of any (less benign than *tert*-butyl) substituents within the C and Q moieties and, in any condensed phase application, to the prevailing solvent [104]. Extrapolating to eumelanin itself, any such competition between reformation of the minimum energy ground state structure and biradical (and thence radical) formation might well be influenced by the extent (or otherwise) to which the system is able to distort away from any structural layering imposed by more extensive π-stacking between polymer strands.

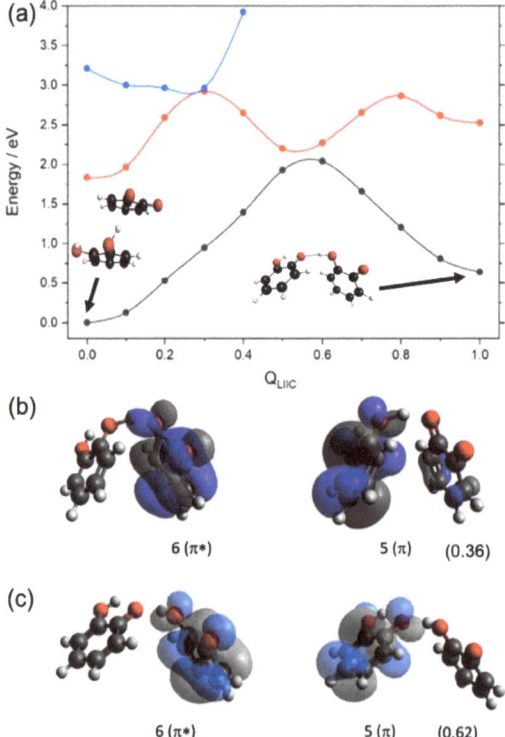

Figure 3. (**a**) Adiabatic potential energy (PE) profiles of the S_0 (black), S_1 (red) and S_2 (blue) states of the C:Q heterodimer plotted as a function of the linear interpolation in internal coordinates (LIIC) linking the ground state minimum energy geometry (at $Q_{LIIC} = 0$) with that of the optimized biradical tautomer (at $Q_{LIIC} = 1.0$). Representations of the evolving structure of the ground state heterodimer and of orbitals 5 and 6 are shown below for Q_{LIIC} = (**b**) 0.6 and (**c**) 1.0 (with, in each case, the square of the coefficient associated with this ππ* contribution to the S_0 state configuration shown in parenthesis).

4. General Discussion and Conclusions

This study, which is limited to the isolated heterodimer only, adds to the growing body of computational research aimed at exploring possible excited state decay paths in organic acid-base heterodimers. The present results support earlier suggestions, from analysis of transient absorption measurements in a weakly interacting solvent [95], that HAT from an OH moiety of a catechol sub-unit to the carbonyl-oxygen atom of a quinone unit arranged in a π-stacked C:Q heterodimer could contribute to the pool of photoprotection mechanisms available to eumelanin upon exposure to UV radiation.

From the photophysical perspective, the π-stacked chromophores in the C:Q heterodimer exhibit similarities and differences with the excited state decay mechanisms identified for the chromophores in double-stranded DNA. UV excitation of a, base pair starts with a $\pi^* \leftarrow \pi$ promotion localized on the purine (adenine (A) or guanine (G)), which is dissipated by PCET to the pyrimidine (thymine (T) or cytosine (C)) partner and subsequent coupling via a CI to the S_0 state [105–109]. The H atom in these cases is transferred within an H-bonded base pair wherein the individual bases are parts of complementary strands (i.e., an *inter*-strand HAT process). UV photoinduced *intra*-strand electron transfer between stacked nucleobases—more reminiscent of the present situation—has been identified also, but the subsequent charge-separation (and ultimate photostability) is again achieved by an *inter*-strand proton transfer in the resulting radical anion base-pair [110,111].

As noted in the Introduction, eumelanin is a heterogeneous macromolecule, and much remains to be learned both about its exact structure and the mechanisms of the photoprotection it affords. Several studies of intramolecular processes contributing to the decay of excited states of monomers (and oligomers) of various of the proposed key sub-units of eumelanin, like DHI and DHICA, have been reported [75,82,84], along with some studies of their intermolecular interactions with solvent molecules [104]. The present work supports another inter-chromophore excited state decay pathway wherein HAT facilitates non-radiative coupling to, and reformation of, the ground state C:Q heterodimer. But, as experimental studies of (the di-*tert*-butyl substituted form of) this heterodimer also show, the biradical structure at the asymptote of the HAT coordinate can decompose to two semiquinone radicals [95]. While it is notable from an energetic perspective that absorption of one photon with an energy less than that required to break an O–H bond in bare catechol [88] could result in the formation of two semiquinone radicals, it is unlikely that nature would have adopted eumelanin as a skin pigment if such heterodimers could act as significant light-driven radical generation centers. Clearly, much further work will be needed in order to establish the importance (or otherwise) of the excited state decay pathways identified thus far for small constituent parts to the overall photoprotection afforded by bulk eumelanin.

Supplementary Materials: The following are available online at https://www.mdpi.com/2673-7256/1/1/2637/s1, Cartesian coordinates associated with the various optimized structures of C:Q and (harmonic) normal mode wavenumbers for the ground states of bare C and the C:Q heterodimer.

Author Contributions: Conceptualization, T.N.V.K., B.M. and M.N.R.A.; methodology, T.N.V.K.; validation, V.C.F., B.M. and T.N.V.K.; formal analysis, V.C.F., T.A.A., B.M. and T.N.V.K.; investigation, V.C.F., T.A.A., B.M.; data curation, V.C.F., T.A.A., B.M.; writing—original draft preparation, T.N.V.K. and M.N.R.A.; writing—review and editing, T.N.V.K. and M.N.R.A.; T.N.V.K. and M.N.R.A.; project administration, T.N.V.K. and M.N.R.A. All authors have read and agreed to the published version of the manuscript.

Funding: T.N.V.K. acknowledges the University of Louisiana at Lafayette for start-up funds.

Data Availability Statement: The data supporting this study are available from the corresponding author on reasonable request.

Conflicts of Interest: The authors declare no conflict of interest.

References

1. Sobolewski, A.L.; Domcke, W.; Dedonder-Lardeux, C.; Jouvet, C. Excited-state hydrogen detachment and hydrogen transfer driven by repulsive $^1\pi\sigma^*$ states: A new paradigm for nonradiative decay in aromatic biomolecules. *Phys. Chem. Chem. Phys.* **2002**, *4*, 1093–1100. [CrossRef]
2. Ashfold, M.N.R.; King, G.A.; Murdock, D.; Nix, M.G.D.; Oliver, T.A.A.; Sage, A.G. πσ* excited states in molecular photochemistry. *Phys. Chem. Chem. Phys.* **2010**, *12*, 1218–1238. [CrossRef] [PubMed]

3. Roberts, G.M.; Stavros, V.G. The role of πσ* states in the photochemistry of heteroaromatic biomolecules and their subunits: Insights from gas-phase femtosecond spectroscopy. *Chem. Sci.* **2014**, *5*, 1698. [CrossRef]
4. Marchetti, B.; Karsili, T.N.V.; Ashfold, M.N.R.; Domcke, W. A 'bottom up', ab initio computational approach to understanding fundamental photophysical processes in nitrogen containing heterocycles, DNA bases and base pairs. *Phys. Chem. Chem. Phys.* **2016**, *18*, 20007–20027. [CrossRef] [PubMed]
5. Barbatti, M.; Aquino, A.J.A.; Szymczak, J.J.; Nachtigallová, D.; Hobza, P.; Lischka, H. Relaxation mechanisms of UV-photoexcited DNA and RNA nucleobases. *Proc. Natl. Acad. Sci. USA* **2010**, *107*, 21453–21458. [CrossRef] [PubMed]
6. Canuel, C.; Mons, M.; Piuzzi, F.; Tardivel, B.; Dimicoli, I.; Elhanine, M. Excited states dynamics of DNA and RNA bases: Characterization of a stepwise deactivation pathway in the gas phase. *J. Chem. Phys.* **2005**, *122*, 74316. [CrossRef]
7. Crespo-Hernández, C.E.; Cohen, B.; Hare, P.M.; Kohler, B. Ultrafast Excited-State Dynamics in Nucleic Acids. *Chem. Rev.* **2004**, *104*, 1977–2020. [CrossRef] [PubMed]
8. Gustavsson, T.; Improta, R.; Markovitsi, D. DNA/RNA: Building Blocks of Life under UV Irradiation. *J. Phys. Chem. Lett.* **2010**, *1*, 2025–2030. [CrossRef]
9. Middleton, C.T.; de la Harpe, K.; Su, C.; Law, Y.K.; Crespo-Hernández, C.E.; Kohler, B. DNA Excited-State Dynamics: From Single Bases to the Double Helix. *Annu. Rev. Phys. Chem.* **2009**, *60*, 217–239. [CrossRef]
10. Ullrich, S.; Schultz, T.; Zgierski, M.Z.; Stolow, A. Electronic relaxation dynamics in DNA and RNA bases studied by time-resolved photoelectron spectroscopy. *Phys. Chem. Chem. Phys.* **2004**, *6*, 2796–2801. [CrossRef]
11. Improta, R.; Santoro, F.; Blancafort, L. Quantum Mechanical Studies on the Photophysics and the Photochemistry of Nucleic Acids and Nucleobases. *Chem. Rev.* **2016**, *116*, 3540–3593. [CrossRef] [PubMed]
12. Karsili, T.N.V.; Marchetti, B.; Ashfold, M.N.R.; Domcke, W. Ab initio study of potential ultrafast internal conversion routes in oxybenzone, caffeic acid, and ferulic acid: Implications for sunscreens. *J. Phys. Chem. A* **2014**, *118*, 11999–12010. [CrossRef]
13. Baker, L.A.; Marchetti, B.; Karsili, T.N.V.; Stavros, V.G.; Ashfold, M.N.R. Photoprotection: Extending lessons learned from studying natural sunscreens to the design of artificial sunscreen constituents. *Chem. Soc. Rev.* **2017**, *46*, 3770–3791. [CrossRef]
14. Holt, E.L.; Stavros, V.G. Applications of ultrafast spectroscopy to sunscreen development, from first principles to complex mixtures. *Int. Rev. Phys. Chem.* **2019**, *38*, 243–285. [CrossRef]
15. Woolley, J.M.; Losantos, R.; Sampedro, D.; Stavros, V.G. Computational and experimental characterization of novel ultraviolet filters. *Phys. Chem. Chem. Phys.* **2020**, *22*, 25390–25395. [CrossRef] [PubMed]
16. Harris, S.J.; Karsili, T.N.V.; Murdock, D.; Oliver, T.A.A.; Wenge, A.M.; Zaouris, D.K.; Ashfold, M.N.R.; Harvey, J.N.; Few, J.D.; Gowrie, S.; et al. A Multipronged Comparative Study of the Ultraviolet Photochemistry of 2-, 3-, and 4-Chlorophenol in the Gas Phase. *J. Phys. Chem. A* **2015**, *119*, 6045–6056. [CrossRef] [PubMed]
17. Lin, Y.-C.; Lee, C.; Lee, S.-H.; Lee, Y.-Y.; Lee, Y.T.; Tseng, C.-M.; Ni, C.-K. Excited-state dissociation dynamics of phenol studied by a new time-resolved technique. *J. Chem. Phys.* **2018**, *148*, 74306. [CrossRef]
18. Nix, M.G.D.; Devine, A.L.; Cronin, B.; Dixon, R.N.; Ashfold, M.N.R. High resolution photofragment translational spectroscopy studies of the near ultraviolet photolysis of phenol. *J. Chem. Phys.* **2006**, *125*, 133318. [CrossRef]
19. Roberts, G.M.; Chatterley, A.S.; Young, J.D.; Stavros, V.G. Direct observation of hydrogen tunneling dynamics in photoexcited phenol. *J. Phys. Chem. Lett.* **2012**, *3*, 348–352. [CrossRef]
20. Iqbal, A.; Cheung, M.S.Y.; Nix, M.G.D.; Stavros, V.G. Exploring the time-scales of H-atom detachment from photoexcited phenol-h6 and phenol-d5: Statistical vs. Nonstatistical decay. *J. Phys. Chem. A* **2009**, *113*, 8157–8163. [CrossRef]
21. Ashfold, M.N.R.; Devine, A.L.; Dixon, R.N.; King, G.A.; Nix, M.G.D.; Oliver, T.A.A. Exploring nuclear motion through conical intersections in the UV photodissociation of phenols and thiophenol. *Proc. Natl. Acad. Sci. USA* **2008**, *105*, 12701–12706. [CrossRef]
22. Vieuxmaire, O.P.J.; Lan, Z.; Sobolewski, A.L.; Domcke, W. Ab initio characterization of the conical intersections involved in the photochemistry of phenol. *J. Chem. Phys.* **2008**, *129*, 224307. [CrossRef] [PubMed]
23. Karsili, T.N.V.; Wenge, A.M.; Harris, S.J.; Murdock, D.; Harvey, J.N.; Dixon, R.N.; Ashfold, M.N.R. O-H bond fission in 4-substituted phenols: S_1 state predissociation viewed in a Hammett-like framework. *Chem. Sci.* **2013**, *4*, 2434–2446. [CrossRef]
24. Iqbal, A.; Pegg, L.J.; Stavros, V.G. Direct versus indirect H atom elimination from photoexcited phenol molecules. *J. Phys. Chem. A* **2008**, *112*, 9531–9534. [CrossRef] [PubMed]
25. Karsili, T.N.V.; Wenge, A.M.; Marchetti, B.; Ashfold, M.N.R. Symmetry matters: Photodissociation dynamics of symmetrically versus asymmetrically substituted phenols. *Phys. Chem. Chem. Phys.* **2014**, *16*, 588–598. [CrossRef] [PubMed]
26. Karsili, T.N.V.; Marchetti, B.; Moca, R.; Ashfold, M.N.R. UV photodissociation of pyrroles: Symmetry and substituent effects. *J. Phys. Chem. A* **2013**, *117*, 12067–12074. [CrossRef]
27. Vallet, V.; Lan, Z.; Mahapatra, S.; Sobolewski, A.L.; Domcke, W. Photochemistry of pyrrole: Time-dependent quantum wave-packet description of the dynamics at the $^1\pi\sigma^*$-S_0 conical intersections. *J. Chem. Phys.* **2005**, *123*, 144307. [CrossRef] [PubMed]
28. Vallet, V.; Lan, Z.; Mahapatra, S.; Sobolewski, A.L.; Domcke, W. Time-dependent quantum wave-packet description of the $^1\pi\sigma^*$ photochemistry of pyrrole. *Faraday Discuss.* **2004**, *127*, 283–293. [CrossRef]
29. Roberts, G.M.; Williams, C.A.; Yu, H.; Chatterley, A.S.; Young, J.D.; Ullrich, S.; Stavros, V.G. Probing ultrafast dynamics in photoexcited pyrrole: Timescales for $^1\pi\sigma^*$ mediated H-atom elimination. *Faraday Discuss.* **2013**, *163*, 95–116. [CrossRef]
30. Sobolewski, A.L.; Domcke, W. Conical intersections induced by repulsive $^1\pi\sigma^*$ states in planar organic molecules: Malonaldehyde, pyrrole and chlorobenzene as photochemical model systems. *Chem. Phys.* **2000**, *259*, 181–191. [CrossRef]

31. Cronin, B.; Nix, M.G.D.; Qadiri, R.H.; Ashfold, M.N.R. High resolution photofragment translational spectroscopy studies of the near ultraviolet photolysis of pyrrole. *Phys. Chem. Chem. Phys.* **2004**, *6*, 5031–5041. [CrossRef]
32. Staniforth, M.; Young, J.D.; Cole, D.R.; Karsili, T.N.V.; Ashfold, M.N.R.; Stavros, V.G. Ultrafast excited-state dynamics of 2,4-dimethylpyrrole. *J. Phys. Chem. A* **2014**, *118*, 10909–10918. [CrossRef]
33. Cole-Filipiak, N.C.; Staniforth, M.; Rodrigues, N.D.N.; Peperstraete, Y.; Stavros, V.G. Ultrafast Dissociation Dynamics of 2-Ethylpyrrole. *J. Phys. Chem. A* **2017**, *121*, 969–976. [CrossRef]
34. Blank, D.A.; North, S.W.; Lee, Y.T. The ultraviolet photodissociation dynamics of pyrrole. *Chem. Phys.* **1994**, *187*, 35–47. [CrossRef]
35. Iqbal, A.; Stavros, V.G. Exploring the Time Scales of H-Atom Elimination from Photoexcited Indole. *J. Phys. Chem. A* **2010**, *114*, 68–72. [CrossRef]
36. Sobolewski, A.L.; Domcke, W. Ab initio investigations on the photophysics of indole. *Chem. Phys. Lett.* **1999**, *315*, 293–298. [CrossRef]
37. Nix, M.G.D.; Devine, A.L.; Cronin, B.; Ashfold, M.N.R. High resolution photofragment translational spectroscopy of the near UV photolysis of indole: Dissociation via the $^1\pi\sigma^*$ state. *Phys. Chem. Chem. Phys.* **2006**, *8*, 2610–2618. [CrossRef]
38. Lin, M.-F.; Tseng, C.-M.; Lee, Y.T.; Ni, C.-K. Photodissociation dynamics of indole in a molecular beam. *J. Chem. Phys.* **2005**, *123*, 124303. [CrossRef] [PubMed]
39. Matsika, S. Radiationless Decay of Excited States of Uracil through Conical Intersections. *J. Phys. Chem. A* **2004**, *108*, 7584–7590. [CrossRef]
40. Lan, Z.; Fabiano, E.; Thiel, W. Photoinduced Nonadiabatic Dynamics of 9H-Guanine. *ChemPhysChem* **2009**, *10*, 1225–1229. [CrossRef] [PubMed]
41. Kistler, K.A.; Matsika, S. Photophysical pathways of cytosine in aqueous solution. *Phys. Chem. Chem. Phys.* **2010**, *12*, 5024–5031. [CrossRef]
42. Delchev, V.B.; Sobolewski, A.L.; Domcke, W. Comparison of the non-radiative decay mechanisms of 4-pyrimidinone and uracil: An ab initio study. *Phys. Chem. Chem. Phys.* **2010**, *12*, 5007–5015. [CrossRef] [PubMed]
43. Barbatti, M.; Szymczak, J.J.; Aquino, A.J.A.; Nachtigallov, D.; Lischka, H. The decay mechanism of photoexcited guanine—A nonadiabatic dynamics study. *J. Chem. Phys.* **2011**, *134*, 14304. [CrossRef]
44. Nachtigallovaí, D.; Aquino, A.J.A.; Szymczak, J.J.; Barbatti, M.; Hobza, P.; Lischka, H. Nonadiabatic dynamics of uracil: Population split among different decay mechanisms. *J. Phys. Chem. A* **2011**, *115*, 5247–5255. [CrossRef]
45. Perun, S.; Sobolewski, A.L.; Domcke, W. Ab initio studies on the radiationless decay mechanisms of the lowest excited singlet states of 9H-adenine. *J. Am. Chem. Soc.* **2005**, *127*, 6257–6265. [CrossRef] [PubMed]
46. Serrano-Andrés, L.; Merchán, M.; Borin, A.C. A Three-State Model for the Photophysics of Adenine. *Chem. A Eur. J.* **2006**, *12*, 6559–6571. [CrossRef] [PubMed]
47. Barbatti, M.; Lischka, H. Nonadiabatic deactivation of 9H-adenine: A comprehensive picture based on mixed quantum-classical dynamics. *J. Am. Chem. Soc.* **2008**, *130*, 6831–6839. [CrossRef] [PubMed]
48. Conti, I.; Garavelli, M.; Orlandi, G. Deciphering low energy deactivation channels in adenine. *J. Am. Chem. Soc.* **2009**, *131*, 16108–16118. [CrossRef] [PubMed]
49. Barbatti, M.; Lan, Z.; Crespo-Otero, R.; Szymczak, J.J.; Lischka, H.; Thiel, W. Critical appraisal of excited state nonadiabatic dynamics simulations of 9H-adenine. *J. Chem. Phys.* **2012**, *137*, 22A503. [CrossRef] [PubMed]
50. Zgierski, M.Z.; Patchkovskii, S.; Fujiwara, T.; Lim, E.C. On the Origin of the Ultrafast Internal Conversion of Electronically Excited Pyrimidine Bases. *J. Phys. Chem. A* **2005**, *109*, 9384–9387. [CrossRef]
51. Triandafillou, C.G.; Matsika, S. Excited-state tautomerization of gas-phase cytosine. *J. Phys. Chem. A* **2013**, *117*, 12165–12174. [CrossRef]
52. Plasser, F.; Crespo-Otero, R.; Pederzoli, M.; Pittner, J.; Lischka, H.; Barbatti, M. Surface hopping dynamics with correlated single-reference methods: 9H-adenine as a case study. *J. Chem. Theory Comput.* **2014**, *10*, 1395–1405. [CrossRef]
53. Marian, C.M. The guanine tautomer puzzle: Quantum chemical investigation of ground and excited states. *J. Phys. Chem. A* **2007**, *111*, 1545–1553. [CrossRef] [PubMed]
54. Matsika, S.; Krause, P. Nonadiabatic events and conical intersections. *Ann. Rev. Phys. Chem.* **2011**, *62*, 621–643. [CrossRef] [PubMed]
55. Prokhorenko, V.I.; Picchiotti, A.; Pola, M.; Dijkstra, A.G.; Miller, R.J.D. New Insights into the Photophysics of DNA Nucleobases. *J. Phys. Chem. Lett.* **2016**, *7*, 4445–4450. [CrossRef] [PubMed]
56. Pepino, A.J.; Segarra-Martí, J.; Nenov, A.; Improta, R.; Garavelli, M. Resolving Ultrafast Photoinduced Deactivations in Water-Solvated Pyrimidine Nucleosides. *J. Phys. Chem. Lett.* **2017**, *8*, 1777–1783. [CrossRef]
57. Francés-Monerris, A.; Gattuso, H.; Roca-Sanjuán, D.; Tuñón, I.; Marazzi, M.; Dumont, E.; Monari, A. Dynamics of the excited-state hydrogen transfer in a (dG)·(dC) homopolymer: Intrinsic photostability of DNA. *Chem. Sci.* **2018**, *9*, 7902–7911. [CrossRef] [PubMed]
58. Pilles, B.M.; Maerz, B.; Chen, J.; Bucher, D.B.; Gilch, P.; Kohler, B.; Zinth, W.; Fingerhut, B.P.; Schreier, W.J. Decay Pathways of Thymine Revisited. *J. Phys. Chem. A* **2018**, *122*, 4819–4828. [CrossRef] [PubMed]
59. Perun, S.; Sobolewski, A.L.; Domcke, W. Conical Intersections in Thymine. *J. Phys. Chem. A* **2006**, *110*, 13238–13244. [CrossRef] [PubMed]
60. Ismail, N.; Blancafort, L.; Olivucci, M.; Kohler, B.; Robb, M.A. Ultrafast decay of electronically excited singlet cytosine via a π,π^* to n_0,π^* state switch. *J. Am. Chem. Soc.* **2002**, *124*, 6818–6819. [CrossRef]

61. Blancafort, L.; Cohen, B.; Hare, P.M.; Kohler, B.; Robb, M.A. Singlet excited-state dynamics of 5-fluorocytosine and cytosine: An experimental and computational study. *J. Phys. Chem. A* **2005**, *109*, 4431–4436. [CrossRef]
62. Kistler, K.A.; Matsika, S. Radiationless decay mechanism of cytosine: An ab initio study with comparisons to the fluorescent analogue 5-methyl-2-pyrimidinone. *J. Phys. Chem. A* **2007**, *111*, 2650–2661. [CrossRef]
63. Zechmann, G.; Barbatti, M. Photophysics and deactivation pathways of thymine. *J. Phys. Chem. A* **2008**, *112*, 8273–8279. [CrossRef] [PubMed]
64. Asturiol, D.; Lasorne, B.; Robb, M.A.; Blancafort, L. Photophysics of the π,π and n,π states of thymine: MS-CASPT2 minimum-energy paths and CASSCF on-the-fly dynamics. *J. Phys. Chem. A* **2009**, *113*, 10211–10218. [CrossRef]
65. Yamazaki, S.; Domcke, W.; Sobolewski, A.L. Nonradiative Decay Mechanisms of the Biologically Relevant Tautomer of Guanine. *J. Phys. Chem. A* **2008**, *112*, 11965–11968. [CrossRef] [PubMed]
66. Domcke, W.; Yarkony, D.R.; Koeppel, H. *Conical Intersections: Theory, Computation and Experiment*; World Scientific: Singapore, 2011.
67. Domcke, W.; Yarkony, D.R.; Koeppel, H. *Conical Intersections: Electronic Structure, Dynamics & Spectroscopy*; World Scientific: Singapore, 2004.
68. Baker, L.A.; Horbury, M.D.; Greenough, S.E.; Ashfold, M.N.R.; Stavros, V.G. Broadband Ultrafast Photoprotection by Oxybenzone Across the UVB and UVC Spectral Regions. *Photochem. Photobiol. Sci.* **2015**, *14*, 1814–1820. [CrossRef] [PubMed]
69. Baker, L.A.; Horbury, M.D.; Greenough, S.E.; Coulter, P.M.; Karsili, T.N.V.; Roberts, G.M.; Orr-Ewing, A.J.; Ashfold, M.N.R.; Stavros, V.G. Probing the Ultrafast Energy Dissipation Mechanism of the Sunscreen Oxybenzone after UVA Irradiation. *J. Phys. Chem. Lett.* **2015**, *6*, 1363–1368. [CrossRef] [PubMed]
70. Baker, L.A.; Grosvenor, L.C.; Ashfold, M.N.R.; Stavros, V.G. Ultrafast Photophysical Studies of a Multicomponent Sunscreen: Oxybenzone—Titanium Dioxide Mixtures. *Chem. Phys. Lett.* **2016**, *664*, 39–43. [CrossRef]
71. Deng, Z.; Sun, S.; Zhou, M.; Huang, G.; Pang, J.; Dang, L.; Li, M.-D. Revealing Ultrafast Energy Dissipation Pathway of Nanocrystalline Sunscreens Oxybenzone and Dioxybenzone. *J. Phys. Chem. Lett.* **2019**, *10*, 6499–6503. [CrossRef]
72. Li, C.-X.; Guo, W.-W.; Xie, B.-B.; Cui, G. Photodynamics of oxybenzone sunscreen: Nonadiabatic dynamics simulations. *J. Chem. Phys.* **2016**, *145*, 74308. [CrossRef] [PubMed]
73. McGrath, J.A.; Eady, R.A.J.; Pope, F.M. *Rook's Textb. Dermatology*; John Wiley & Sons: Hoboken, NJ, USA, 2004.
74. Solano, F. Photoprotection and skin pigmentation: Melanin-related molecules and some other new agents obtained from natural sources. *Molecules* **2020**, *25*, 1537. [CrossRef] [PubMed]
75. Sobolewski, A.L.; Domcke, W. Photophysics of eumelanin: Ab initio studies on the electronic spectroscopy and photochemistry of 5,6-dihydroxyindole. *Chemphyschem* **2007**, *8*, 756–762. [CrossRef] [PubMed]
76. Corani, A.; Huijser, A.; Iadonisi, A.; Pezzella, A.; Sundström, V.; d'Ischia, M. Bottom-Up Approach to Eumelanin Photoprotection: Emission Dynamics in Parallel Sets of Water-Soluble 5,6-Dihydroxyindole-Based Model Systems. *J. Phys. Chem. B* **2012**, *116*, 13151–13158. [CrossRef] [PubMed]
77. Corani, A.; Huijser, A.; Gustavsson, T.; Markovitsi, D.; Malmqvist, P.-Å.; Pezzella, A.; d'Ischia, M.; Sundström, V. Superior Photoprotective Motifs and Mechanisms in Eumelanins Uncovered. *J. Am. Chem. Soc.* **2014**, *136*, 11626–11635. [CrossRef] [PubMed]
78. Marchetti, B.; Karsili, T.N.V. Theoretical insights into the photo-protective mechanisms of natural biological sunscreens: Building blocks of eumelanin and pheomelanin. *Phys. Chem. Chem. Phys.* **2016**, *18*, 3644–3658. [CrossRef] [PubMed]
79. Datar, A.; Hazra, A. Pathways for Excited-State Nonradiative Decay of 5,6-Dihydroxyindole, a Building Block of Eumelanin. *J. Phys. Chem. A* **2017**, *121*, 2790–2797. [CrossRef] [PubMed]
80. Ghosh, P.; Ghosh, D. Non-radiative decay of an eumelanin monomer: To be or not to be planar. *Phys. Chem. Chem. Phys.* **2019**, *21*, 6635–6642. [CrossRef] [PubMed]
81. Ghosh, P.; Ghosh, D. Effect of microsolvation on the non-radiative decay of the eumelanin monomer. *Phys. Chem. Chem. Phys.* **2019**, *21*, 26123–26132. [CrossRef] [PubMed]
82. Crane, S.W.; Ghafur, O.; Cowie, T.Y.; Lindsay, A.G.; Thompson, J.O.F.; Greenwood, J.B.; Bebbington, M.W.P.; Townsend, D. Dynamics of electronically excited states in the eumelanin building block 5,6-dihydroxyindole. *Phys. Chem. Chem. Phys.* **2019**, *21*, 8152–8160. [CrossRef]
83. Gauden, M.; Pezzella, A.; Panzella, L.; Napolitano, A.; d'Ischia, M.; Sundström, V. Ultrafast Excited State Dynamics of 5,6-Dihydroxyindole, A Key Eumelanin Building Block: Nonradiative Decay Mechanism. *J. Phys. Chem. B* **2009**, *113*, 12575–12580. [CrossRef] [PubMed]
84. Huijser, A.; Pezzella, A.; Hannestad, J.K.; Panzella, L.; Napolitano, A.; d'Ischia, M.; Sundström, V. UV-dissipation mechanisms in the eumelanin building block DHICA. *Chemphyschem* **2010**, *11*, 2424–2431. [CrossRef]
85. Corani, A.; Pezzella, A.; Pascher, T.; Gustavsson, T.; Markovitsi, D.; Huijser, A.; d'Ischia, M.; Sundström, V. Excited-State Proton-Transfer Processes of DHICA Resolved: From Sub-Picoseconds to Nanoseconds. *J. Phys. Chem. Lett.* **2013**, *4*, 1383–1388. [CrossRef] [PubMed]
86. Oliver, T.A.A.; King, G.A.; Ashfold, M.N.R. Position matters: Competing O–H and N–H photodissociation pathways in hydroxy- and methoxy-substituted indoles. *Phys. Chem. Chem. Phys.* **2011**, *13*, 14646–14662. [CrossRef] [PubMed]
87. Crane, S.W.; Ghafur, O.; Saalbach, L.; Paterson, M.J.; Townsend, D. The influence of substituent position on the excited state dynamics operating in 4-, 5- and 6-hydroxyindole. *Chem. Phys. Lett.* **2020**, *738*, 136870. [CrossRef]

88. King, G.A.; Oliver, T.A.A.; Dixon, R.N.; Ashfold, M.N.R. Vibrational energy redistribution in catechol during ultraviolet photolysis. *Phys. Chem. Chem. Phys.* **2012**, *14*, 3338–3345. [CrossRef]
89. Chatterley, A.S.; Young, J.D.; Townsend, D.; Żurek, J.M.; Paterson, M.J.; Roberts, G.M.; Stavros, V.G. Manipulating dynamics with chemical structure: Probing vibrationally-enhanced tunnelling in photoexcited catechol. *Phys. Chem. Chem. Phys.* **2013**, *15*, 6879–6892. [CrossRef]
90. Young, J.D.; Staniforth, M.; Paterson, M.J.; Stavros, V.G. Torsional Motion of the Chromophore Catechol following the Absorption of Ultraviolet Light. *Phys. Rev. Lett.* **2015**, *114*, 233001. [CrossRef] [PubMed]
91. Grieco, C.; Kohl, F.R.; Zhang, Y.; Natarajan, S.; Blancafort, L.; Kohler, B. Intermolecular Hydrogen Bonding Modulates O-H Photodissociation in Molecular Aggregates of a Catechol Derivative. *Photochem. Photobiol.* **2019**, *95*, 163–175. [CrossRef] [PubMed]
92. Grieco, C.; Hanes, A.T.; Blancafort, L.; Kohler, B. Effects of Intra- and Intermolecular Hydrogen Bonding on O-H Bond Photodissociation Pathways of a Catechol Derivative. *J. Phys. Chem. A* **2019**, *123*, 5356–5366. [CrossRef]
93. Turner, M.A.P.; Turner, R.J.; Horbury, M.D.; Hine, N.D.M.; Stavros, V.G. Examining solvent effects on the ultrafast dynamics of catechol. *J. Chem. Phys.* **2019**, *151*, 84305. [CrossRef]
94. Anh, N.V.; Williams, R.M. Bis-semiquinone (bi-radical) formation by photoinduced proton coupled electron transfer in covalently linked catechol–quinone systems: Aviram's hemiquinones revisited. *Photochem. Photobiol. Sci.* **2012**, *11*, 957–961. [CrossRef] [PubMed]
95. Grieco, C.; Empey, J.M.; Kohl, F.R.; Kohler, B. Probing eumelanin photoprotection using a catechol:quinone heterodimer model system. *Faraday Discuss.* **2019**, *216*, 520–537. [CrossRef]
96. Mardirossian, N.; Head-Gordon, M. ωb97X-D: A 10-parameter, range-separated hybrid, generalized gradient approximation density functional with nonlocal correlation, designed by a survival-of-the-fittest strategy. *Phys. Chem. Chem. Phys.* **2014**, *16*, 9904–9924. [CrossRef]
97. Hehre, W.J.; Stewart, R.F.; Pople, J.A. Self-Consistent Molecular-Orbital Methods. I. Use of Gaussian Expansions of Slater-Type Atomic Orbitals. *J. Chem. Phys.* **1969**, *51*, 2657–2664. [CrossRef]
98. Léger, S.J.; Marchetti, B.; Ashfold, M.N.R.; Karsili, T.N.V. The Role of Norrish Type-I Chemistry in Photoactive Drugs: An ab initio Study of a Cyclopropenone-Enediyne Drug Precursor. *Front. Chem.* **2020**, *8*, 1200. [CrossRef]
99. Andersson, K.; Malmqvist, P.; Roos, B.O. Second-order perturbation theory with a complete active space self-consistent field reference function. *J. Chem. Phys.* **1992**, *96*, 1218–1226. [CrossRef]
100. Frisch, M.J.; Trucks, G.W.; Schlegel, H.B.; Scuseria, G.E.; Robb, M.A.; Cheeseman, J.R.; Scalmani, G.; Barone, V.; Petersson, G.A.; Nakatsuji, H.; et al. *Gaussian 09, Revision C.01*; Gaussian Inc.: Wallingford, CT, USA, 2016.
101. Werner, H.-J.; Knowles, P.J.; Knizia, G.; Manby, F.R.; Schütz, M. Molpro: A general-purpose quantum chemistry program package. *WIREs Comput. Mol. Sci.* **2012**, *2*, 242–253. [CrossRef]
102. Banyasz, A.; Ketola, T.; Martinez-Fernandez, L.; Improta, R.; Markovitsi, D. Adenine radicals generated in alternating AT duplexes by direct absorption of low-energy UV radiation. *Faraday Discuss.* **2018**, *207*, 181–197. [CrossRef] [PubMed]
103. Perun, S.; Sobolewski, A.L.; Domcke, W. Role of electron-driven proton-transfer processes in the excited-state deactivation of the adenine-thymine base pair. *J. Phys. Chem. A* **2006**, *110*, 9031–9038. [CrossRef]
104. Nogueira, J.J.; Corani, A.; el Nahhas, A.; Pezzella, A.; d'Ischia, M.; González, L.; Sundström, V. Sequential Proton-Coupled Electron Transfer Mediates Excited-State Deactivation of a Eumelanin Building Block. *J. Phys. Chem. Lett.* **2017**, *8*, 1004–1008. [CrossRef]
105. Sobolewski, A.L.; Domcke, W.; Hättig, C. Tautomeric selectivity of the excited-state lifetime of guanine/cytosine base pairs: The role of electron-driven proton-transfer processes. *Proc. Natl. Acad. Sci. USA* **2005**, *102*, 17903–17906. [CrossRef] [PubMed]
106. Gobbo, J.P.; Saurí, V.; Roca-Sanjuán, D.; Serrano-Andrés, L.; Merchán, M.; Borin, A.C. On the Deactivation Mechanisms of Adenine–Thymine Base Pair. *J. Phys. Chem. B* **2012**, *116*, 4089–4097. [CrossRef] [PubMed]
107. Groenhof, G.; Schäfer, L.V.; Boggio-Pasqua, M.; Goette, M.; Grubmüller, H.; Robb, M.A. Ultrafast Deactivation of an Excited Cytosine−Guanine Base Pair in DNA. *J. Am. Chem. Soc.* **2007**, *129*, 6812–6819. [CrossRef] [PubMed]
108. Sauri, V.; Gobbo, J.P.; Serrano-Pérez, J.J.; Lundberg, M.; Coto, P.B.; Serrano-Andrés, L.; Borin, A.C.; Lindh, R.; Merchán, M.; Roca-Sanjuán, D. Proton/Hydrogen Transfer Mechanisms in the Guanine–Cytosine Base Pair: Photostability and Tautomerism. *J. Chem. Theory Comput.* **2013**, *9*, 481–496. [CrossRef] [PubMed]
109. Röttger, K.; Marroux, H.J.B.; Grubb, M.P.; Coulter, P.M.; Böhnke, H.; Henderson, A.S.; Galan, M.C.; Temps, F.; Orr-Ewing, A.J.; Roberts, G.M. Ultraviolet Absorption Induces Hydrogen-Atom Transfer in G·C Watson–Crick DNA Base Pairs in Solution. *Angew. Chemie Int. Ed.* **2015**, *54*, 14719–14722. [CrossRef]
110. Crespo-Hernández, C.E.; Cohen, B.; Kohler, B. Base stacking controls excited-state dynamics in A·T DNA. *Nature* **2005**, *436*, 1141–1144. [CrossRef]
111. Zhang, Y.; de la Harpe, K.; Beckstead, A.A.; Improta, R.; Kohler, B. UV-Induced Proton Transfer between DNA Strands. *J. Am. Chem. Soc.* **2015**, *137*, 7059–7062. [CrossRef]

Article

Modelling Photoionisations in Tautomeric DNA Nucleobase Derivatives 7H-Adenine and 7H-Guanine: Ultrafast Decay and Photostability

Javier Segarra-Martí *,†, Sara M. Nouri and Michael J. Bearpark *

Department of Chemistry, Molecular Sciences Research Hub, White City Campus, Imperial College London, 82 Wood Lane, London W12 0BZ, UK; sara.nouri16@imperial.ac.uk
* Correspondence: javier.segarra@uv.es (J.S.-M.); m.bearpark@imperial.ac.uk (M.J.B.)
† Present address: Instituto de Ciencia Molecular, Universitat de Valencia, P.O. Box 22085, 46071 Valencia, Spain.

Abstract: The study of radiation effects in DNA is a multidisciplinary endeavour, connecting the physical, chemical and biological sciences. Despite being mostly filtered by the ozone layer, sunlight radiation is still expected to (photo)ionise DNA in sizeable yields, triggering an electron removal process and the formation of potentially reactive cationic species. In this manuscript, photoionisation decay channels of important DNA tautomeric derivatives, 7H-adenine and 7H-guanine, are characterised with accurate CASSCF/XMS-CASPT2 theoretical methods. These simulation techniques place the onset of ionisation for 7H-adenine and 7H-guanine on average at 8.98 and 8.43 eV, in line with recorded experimental evidence when available. Cationic excited state decays are analysed next, uncovering effective barrierless deactivation routes for both species that are expected to decay to their (cationic) ground state on ultrafast timescales. Conical intersection topographies reveal that these photoionisation processes are facilitated by sloped single-path crossings, known to foster photostability, and which are predicted to enable the (VUV) photo-protection mechanisms present in these DNA tautomeric species.

Keywords: photoionisation; CASSCF/CASPT2; photostability; DNA/RNA; UV/Vis spectroscopy; excited states; ionisation potentials; conical intersections

Citation: Segarra-Martí, J.; Nouri, S.M.; Bearpark, M.J. Modelling Photoionisations in Tautomeric DNA Nucleobase Derivatives 7H-Adenine and 7H-Guanine: Ultrafast Decay and Photostability. *Photochem* **2021**, *1*, 287–301. https://doi.org/10.3390/photochem1020018

Academic Editor: Marcelo I. Guzman

Received: 9 August 2021
Accepted: 29 August 2021
Published: 10 September 2021

Publisher's Note: MDPI stays neutral with regard to jurisdictional claims in published maps and institutional affiliations.

Copyright: © 2021 by the authors. Licensee MDPI, Basel, Switzerland. This article is an open access article distributed under the terms and conditions of the Creative Commons Attribution (CC BY) license (https://creativecommons.org/licenses/by/4.0/).

1. Introduction

The study of radiation induced DNA damage is a topic of utmost interest to a broad range of disciplines as it is known to mediate in the formation of lesions [1] that lead to mutations and to healthcare concerns such as skin cancer melanoma [2]. An enhanced understanding of these processes has been achieved over the years by studying DNA excited state dynamics triggered upon radiation exposure over a wide range of derivatives: from its monomers the nucleobases (and nucleosides/tides), [3] to dimers [4–6], single- and double-strands [7,8] and more recently other relevant motifs like guanine quadruplexes [9,10]. This has helped uncover an intricate excited state reactivity landscape triggered in our genetic material upon light absorption that is mediated by a mixture of localised and excitonic (delocalised) states [11], and that spreads across spin multiplicities [3,12].

Work reported in the literature thus far has focused on the study of DNA reactivity upon UV-light exposure [13,14], as it is often favoured due to its relation with sunlight-mediated lesions [15,16].

Photoionisation, on the other hand, is less studied but has recently gained attention due to the development of light sources capable of generating and monitoring photoionised (cationic) species [17,18]. Cationic states are known to be formed not only directly upon photoionisation but also in many charge transfer and separation processes mediating DNA damage [19–21] and repair [22]. Moreover, recent experiments with guanine-rich sequences have revealed the formation of significant cationic yields even when irradiating with UV-B light sources [10,23], which possess longer wavelength than those required to ionise DNA nucleobases. This means the onset of ionisation might be energetically well below the ionisation potential of the nucleobases [24], making them a more widespread event than initially thought and with sizeable involvement in UV-induced DNA photophysics [19–21]. Uncovering how cationic species evolve and react is thus of interest to understand DNA photo-processes occurring upon radiation exposure and is the main goal of this work.

The reactivity of the ensuing DNA nucleobase cationic species still remains vastly unexplored, despite recent efforts with state-of-the-art instrumentation on this front [25,26]. Whereas pyrimidine nucleobases have been studied in relative detail [27–30], purines are less explored perhaps due to their more complex electronic structure [31], studies on their tautomeric forms being even scarcer.

DNA non-canonical nucleobases, like the 7H tautomeric forms of DNA nucleobases adenine and guanine studied in this work (see Figure 1), feature in non-negligible yields in our genetic material and possess markedly different electronic properties, which are known to affect their reactivity [32]. These species are rarely studied in the literature in their singlet form [32], and even less so in the cationic (doublet) manifold, despite being important to describe some of the photo-processes underpinning DNA damage and repair. The study of these species is however fundamental to understand how prebiotic extreme UV-light exposure in DNA nucleobases, and their ability to withstand such radiation, may have played a role in selecting our genetic lexicon [13,14,33].

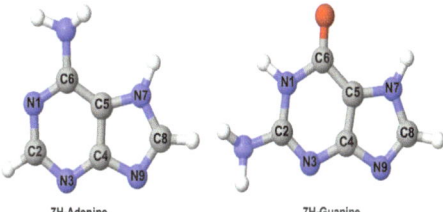

Figure 1. Molecular structure of 7H-adenine (**left**) and 7H-guanine (**right**) and their atom labelling.

This is particularly relevant in 7H-guanine, as it has been reported to be the most stable guanine tautomer in the gas phase [34,35]. This higher stability of the 7H tautomer of guanine in the gas phase is something that has, however, been contested over the years: in [36], a comparison of theoretical estimates with photoelectron spectra appears to support a 7H-guanine assignment as the most stable tautomer [35,37,38], but experiments in molecular beams appear to favour the 9H tautomer based on recorded IR spectroscopic signals instead [36,39]. This suggests the specific conditions in which guanine is measured are critical in this tautomeric equilibrium.

In this work photoionisation processes on DNA purine tautomeric 7H species (7H-adenine and 7H-guanine, Figure 1) are studied (to our knowledge) for the first time using accurate multireference perturbation theory (CASPT2) modelling techniques. We focus on studying the 7H tautomers of both adenine and guanine, as these feature several (accessible) cationic electronic excited states upon ionisation, making the comparison between analogous tautomers featuring different purine-substituted frames more interesting in terms of understanding how chemical substitution impacts their photophysics. These systems are nevertheless of interest due to their strong resemblance with our current genetic lexicon, and studying their photo-protection mechanisms may eventually help us understand how our current nucleobases were chosen during prebiotic times and whether photostability played a critical role in it [13,14]. We analyse their ionisation potentials, where we observe systematic blue-shifts in the values for $^2\pi^+$ and red-shifts for $^2n^+$ states, the latter being closer to the experimental evidence, and estimate the onset of ionisation to be at 8.98 and 8.43 eV for 7H-adenine and 7H-guanine, respectively. Cationic excited state decays are predicted to be effectively barrierless and thus very short-lived (ultrafast) for both systems, though the details of the decay mechanisms are different. 7H-guanine$^+$ displays localised distortions in the 5-membered right along the decay, whereas 7H-adenine$^+$ features delocalised motions across the whole molecular scaffold and presents an accessible conical intersection less than a tenth of an eV from the cationic ground state, which is expected to facilitate a potential chemiexcitation channel. The characterised sloped and single-path conical intersection topographies support ultrafast deactivation channels present for both systems, which are expected to be central in fostering photostability, and that suggest (VUV) photo-protection mechanisms in DNA nucleobases might also be available to (at least some of) their tautomeric derivatives.

2. Results

The results are laid out as follows: we analyse the ionisation potentials of 7H-adenine and 7H-guanine first and move next to the study of their cationic excited state decay upon photoionisation.

2.1. Ionisation Potentials

Figure 2 displays the ionisation potentials obtained for the 7H tautomeric species considered in this study, with a comparison to recorded experimental evidence where appropriate and available.

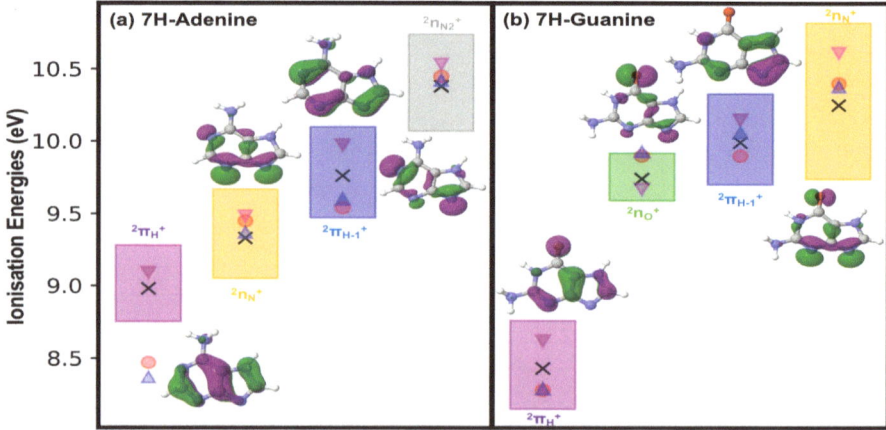

Figure 2. Schematic representation of the gas phase vertical ionisation potentials of (**a**) 7H-adenine and (**b**) 7H-guanine, computed with a range of zeroth-order CASPT2 Hamiltonians. The uncertainty range spanned by the different zeroth-order Hamiltonians is provided. CASPT2 average values are given as black crosses, magenta inverted triangles represent XMS-CASPT2(IPEA = 0.0) estimates used in Section 2.2, red dots denote experimental evidence found in the literature for guanine [37,40] and 9H-adenine [41], and blue triangles refer to EOM-IP/CCSD/cc-pVTZ-dff//RI-MP2/cc-pVTZ-dff theoretical estimates of Bravaya et al. [35]. Purple squares denote ionisation energy ranges for $^2\pi_H^+$, green for $^2n_O^+$, blue for $^2\pi_{H-1}^+$, orange for $^2n_N^+$ and grey for $^2n_{N_2}^+$ states. Associated singly occupied molecular orbitals (SOMOs) depicting the different cationic states are also provided. Specific values for each CASPT2 formulation are provided in Tables S1 and S2 in the Supplementary Materials.

The first cationic state in 7H-purine derivatives is characterised by the $^2\pi_H^+$ state, which corresponds to an unpaired electron in the singly occupied molecular orbital (SOMO). This is in line with what has been observed for other DNA/RNA nucleobase systems, where the SOMO is always the one embodying the first (i.e., energetically lowest) electron removal upon ionising radiation exposure [24,27–30]. This state is placed at 8.98 and 8.43 eV on average for 7H-adenine and 7H-guanine respectively, and feature similar standard deviations, with respect to the choice of CASPT2 zeroth order Hamiltonian, of 0.19 and 0.22 eV. These average ionisation values compare well with the available experimental evidence reported for guanine [37,40], which several works attribute to signals coming from the 7H tautomer [35,38]. On the other hand, it differs more significantly from estimates available for adenine that are however referred to the 9H canonical species in both theory [35] and experiment [41]. Particularly the average CASPT2 value obtained for 7H-guanine, the more comparable case, is in agreement with those reported experimentally [40] and theoretically [35] in the literature, being blue-shifted by around a tenth of an eV off those estimates.

The second ionisation potential embodies in both cases lone pair (also referred to as σ) [35] states, of $^2n_N^+$ character for 7H-adenine and (mostly) $^2n_O^+$ for 7H-guanine, which are placed on average at 9.33 and 9.74 eV, respectively. As can be seen in the SOMOs provided in Figure 2, these states feature delocalisations over several lone-pair sites, which affect particularly 7H-guanine as it leads to a state of $^2n_N^+$ and $^2n_O^+$ mixed character, even if $^2n_O^+$ is more prominent and thus used to label the state. The associated variances arising from different CASPT2 formulations seem to differ more prominently, leading to 0.22 eV for 7H-adenine and a smaller 0.13 eV for 7H-guanine. The (averaged) ionisation potential obtained for 7H-guanine is red-shifted by ~0.2 eV compared to experimental [40] and theoretical reference estimates [35], and red-shifted as well for 7H-adenine by ~0.1 eV. This suggests changes in the 7H tautomer appear to affect less prominently $^2n_N^+$ states, as theoretical [35] and experimental [41] estimates for the 9H-tautomer seem to fit those obtained here for the 7H species.

The third cationic state corresponds to electron removal from $^2\pi_{H-1}^+$ and is placed on average at 9.76 and 9.99 eV for 7H-adenine and 7H-guanine, respectively, with an associated standard deviation of 0.23 eV in both cases. Interestingly, averaged values are in this case are blue-shifted from experiment, by 0.18 eV in 7H-adenine and by 0.09 eV in 7H-guanine, being close to experimental [40,41] and reference theoretical [35] estimates.

The fourth and last cationic state analysed corresponds to ionisation in lone pair states, of n_N character for both systems, leading to $^2n_{N2}^+$ for 7H-adenine and $^2n_N^+$ for 7H-guanine. These are predicted to be placed (on average) at 10.38 and 10.25 eV for 7H-adenine and 7H-guanine, respectively, with associated standard deviations of 0.24 and 0.40 eV. These ionisation energies are red-shifted with respect to experimental [40,41] and theoretical [35] evidence by less than 0.1 eV in 7H-adenine, keeping in mind the experiment actually refers to 9H-adenine [41], and by ~0.15 eV in 7H-guanine.

Upon inspection of the different estimates obtained with diverse zeroth-order Hamiltonians used (and reported in Tables S1 and S2), we select the XMS-CASPT2(IPEA = 0.0) formulation for subsequent calculations in Section 2.2: it shows qualitative agreement (slightly blue-shifted estimates) with available experiments as seen in the magenta inverted triangles in Figure 2, and it has been reported to be essential when accurately describing interstate crossings [42].

2.2. Excited State Decays

It has been recently observed that, upon ionisation, an electron might be removed from different molecular orbitals creating diverse cationic species with similar probabilities [43,44]. This makes the choice of the initial state created in our model somewhat ambiguous, as any of the states (or linear combinations of them) are in principle possible. We here assume an electron removal from the highest-lying cationic state considered as done in previous studies [29,30,45]. This enables exploring all critical structures along the potential energy surface until reaching the cationic ground state thus providing an overview of the different pathways potentially triggered in this complex photo-process.

All critical structures found along the cationic excited state decay of 7H-adenine and guanine remain in-plane. This is shared with what we and others have found for pyrimidine nucleobases and derivatives [28–30,46], and justifies focusing on evaluating distortions along the different bond lengths affected during the photo-process.

Figure 3 displays the potential energy profiles obtained for 7H-adenine$^+$. Upon ionisation of the highest-lying $^2n_{N2}^+$ state, a rapid decay is predicted with an associated

~0.7 eV stabilisation, with significant 0.03 Å elongations of N1-C6, C4-C5 and C8-N9 and shortening of the C8-C9 bonds until reaching $(^2n_{N2}{}^+/^2\pi_{H-1}^+)_{CI}$. At this crossing, population is expected to transfer to the $^2\pi_{H-1}^+$ state and further decay to its shallow minimum $((^2\pi_{H-1}^+)_{min})$, which entails a slight 0.02 Å contraction of N1-C2 and an elongation of C7-N9 and that is placed at 0.64 eV adiabatically from the cationic ground state minimum $((^2\pi_H^+)_{min})$.

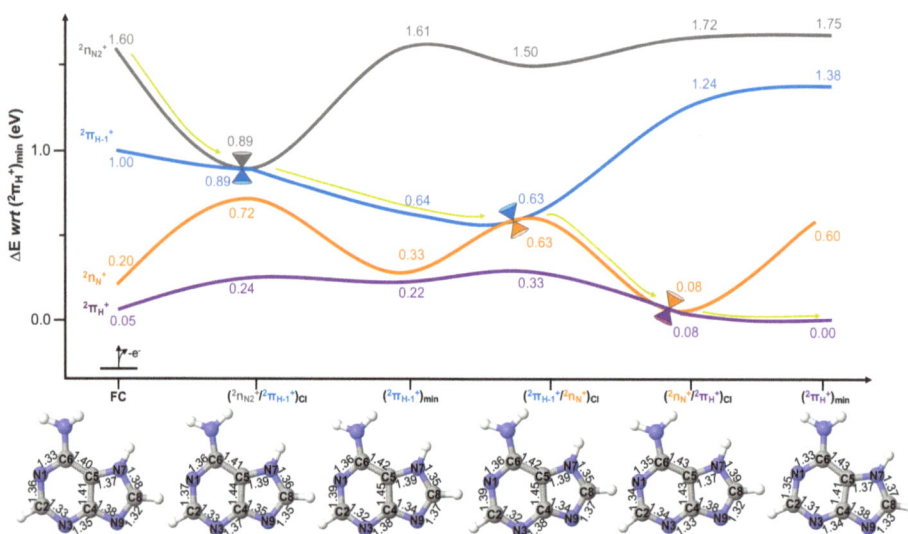

Figure 3. Potential energy surfaces of 7H-adenine$^+$ computed at the XMS-CASPT2 level of theory. All energies are given in eV with respect to $(^2\pi_H^+)_{min}$. Yellow arrows represent the evolution of the excited state population assuming initial activation of the highest-lying $^2n_{N2}{}^+$ (grey) state. XMS-CASPT2 optimised structures are provided to display the nuclear displacements along the photo-reaction, with bond distances given in Å.

From $(^2\pi_{H-1}^+)_{min}$ the $(^2\pi_{H-1}^+/^2n_N^+)_{CI}$ can be reached barrierlessly by relaxing just a further 0.01 eV, being almost identical in structure to $(^2\pi_{H-1}^+)_{min}$, and displaying changes in their structure only beyond the pm. $(^2\pi_{H-1}^+/^2n_N^+)_{CI}$ funnels the excited state population to the $^2n_N^+$ state, that does not feature a minimum and is thus predicted to decay swiftly to $(^2n_N^+/^2\pi_H^+)_{CI}$ mediating the decay to the ground state. This crossing, which mediates the final part of the decay, is placed at just 0.08 eV from the cationic ground state minimum, featuring pronounced 0.05 Å N1-C2, N3-C4 and C8-N9 bond contractions and a 0.04 Å N7-C8 elongation from the previous crossing $(^2\pi_{H-1}^+/^2n_N^+)_{CI}$.

The population is then finally transferred to the cationic ground state $(^2\pi_H^+)_{min}$, which entails a further 0.03 Å shortening of C2-N3 and 0.02 of C4-C5 and N7-C8 bonds. These small distortions encompass the strong structural resemblance with $(^2n_N^+/^2\pi_H^+)_{CI}$, as well as the energetic proximity: this implies re-crossings are possible just from available thermal energy potentially enabling chemiexcitation [47] events to populate back the cationic excited state.

The potential energy curves depicting photoionised excited state decay in 7H-guanine$^+$ are given in Figure 4. As can be seen, upon ionisation to the highest-lying computed $(^2n_N^+)$ state, situated at 2.07 eV adiabatically from the cationic ground state, a swift decay is predicted to reach $(^2n_N^+/^2\pi_{H-1}^+)_{CI}$ placed at 1.80 eV and transfer population to the $^2\pi_{H-1}^+$ state. This decay is mediated by very small (~0.02 Å) in-plane bond length rearrangements across the whole structure, which indicates the proximity of this structure to the equilibrium Franck–Condon (FC) region. It is also interesting to note that $^2n_N^+$ and $^2\pi_{H-1}^+$ states are practically degenerate energetically at the FC equilibrium region, and either state could thus be expected to initially be populated. However, they are both expected to lead to the population of the $^2\pi^+_{H-1}$ state, either direct or indirectly (via $(^2n^+_N/^2\pi^+_{H-1})_{CI}$) upon ionising radiation exposure.

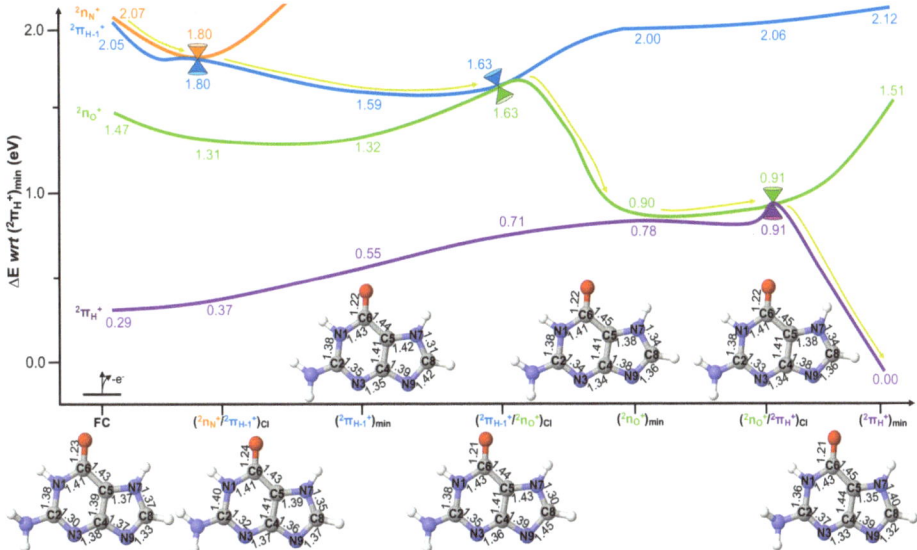

Figure 4. Potential energy surfaces of 7H-guanine$^+$ computed at the XMS-CASPT2 level of theory. All energies are given in eV with respect to $(^2\pi^+_H)_{min}$. Yellow arrows represent the evolution of the excited state population assuming initial activation of the highest-lying $^2n^+_N$ (orange) state. XMS-CASPT2 optimised structures are provided to display the nuclear displacements along the photo-reaction, with bond distances given in Å.

Upon populating $^2\pi^+_{H-1}$, further relaxation is predicted until reaching $(^2\pi^+_{H-1})_{min}$, placed at 1.59 eV and encompassing significant changes that are in this case largely localised in the 5-member ring: elongations of 0.05 Å for C8-N9 and 0.03 for C4-N9 and C5-N7 bonds and shortenings of 0.04 Å for N7-C8 and 0.03 for C4-N9 bonds. This is in line with the more localised character of the SOMO describing the $^2\pi^+_{H-1}$ state over the five-member ring, which is depicted in Figure 2.

From $(^2\pi^+_{H-1})_{min}$ a very small (0.04 eV) barrier is predicted to be surmounted to reach $(^2\pi^+_{H-1}/^2n^+_O)_{CI}$ mediating population transfer to the $^2n^+_O$ state and placed at 1.63 eV. This crossing is characterised by a further (0.03 Å) elongation of the C8-N9 bond.

Upon populating $^2n^+_O$, very significant relaxation (~0.7 eV) is predicted until reaching $(^2n^+_O)_{min}$, placed at 0.90 eV adiabatically from the cationic ground state. This minimum encompasses further (and very pronounced) localised distortions in the 5-member ring: C5-N7 and C8-N9 shorten by 0.05 and 0.09 Å, respectively, an an elongation of the N7-C8 bond of 0.04 Å. It is important to note that this minimum structure features a very small energy difference with the $^2\pi^+_H$ ground cationic state, which could potentially be part of the $(^2n^+_O/^2\pi^+_H)_{CI}$ intersection seam. Indeed, this conical intersection is separated from $(^2n^+_O)_{min}$ by a mere 0.01 eV featuring an almost identical structure.

The $(^2n^+_O/^2\pi^+_H)_{CI}$ crossing is expected to be reached almost barrierlessly (i.e., there is a negligible 0.01 eV barrier to be surmounted) and to mediate an efficient excited state decay to the cationic ground state, which is predicted to rapidly decay to $(^2\pi^+_H)_{min}$ where population is predicted to remain. The cationic ground state minimum encompasses significant structural rearrangements which are again partly localised in the five-member ring but that in this case expand across the molecular scaffold. This is encompassed by significant 0.06 Å elongations in N7-C8, 0.04 C2-N3 and 0.03 for the C4-C5 bonds, and shortenings of 0.04 Å for C8-N9 and 0.03 for C5-N7 bonds.

Conical Intersection Characterisation

The topography of the characterised conical intersections is analysed next, to provide complementary insights into the excited state decay mechanisms of 7H-purine DNA photoionisations.

To assess topographies of conical intersections and compare them we employ a series of parameters used by Fdez Galván et al. in a recent study [48]. Table 1 features estimates for pitch (δ), asymmetry (Δ), relative tilt (σ), and tilt heading (θ_s), computed from the gradient difference and non-adiabatic coupling vectors obtained at the converged conical intersection geometries. For intersection analyses we use branching plane vectors so that $\Delta \geq 0$ and $\theta_s \in [0, \frac{\pi}{2}]$, as this makes them comparable to one another [48]. The values obtained are employed to classify the topography as peaked/sloped, and bifurcating/single-path, based on $\mathcal{P}$ and $\mathcal{B}$ values [48] which are related to those originally proposed by Ruedenberg and co-workers [49]. Values of $\mathcal{P} < 1$ lead to peaked topographies and >1 to sloped ones, whereas $\mathcal{B} < 1$ refers to bifurcating character and >1 to single-path [48].

As can be seen in Table 1, the conical intersection mediating decay to the cationic ground state in 7H-adenine$^+$ and 7H-guanine$^+$, which refer to $(^2n_N^+/^2\pi_H^+)_{CI}$ and $(^2n_O^+/^2\pi_H^+)_{CI}$ respectively, feature both sloped and single-path character that have been suggested to foster photostability [50,51]. These conical intersection structures have similar associated pitch values and large asymmetries while featuring negligible tilt heading angles, which supports distortions mostly arising from in-plane motions as those described above.

Table 1. Characterised conical intersection parameters for the different crossings of 7H-adenine$^+$ and 7H-guanine$^+$ obtained at the XMS-CASPT2 level of theory. Pitch (δ), asymmetry (Δ) and relative tilt (σ) are given in atomic units, whereas tilt headings (θ_s) are provided in degrees.

	δ	Δ	σ	θ_s	$\mathcal{P}$	$\mathcal{B}$	Intersection Type [48]
7H-adenine$^+$							
$(^2n_N^+/^2\pi_H^+)_{CI}$	0.060	0.784	1.486	0.005	1.237	1.171	Sloped, single-path
$(^2\pi_{H-1}^+/^2n_N^+)_{CI}$	0.129	0.904	1.374	3.693	1.069	1.092	Sloped, single-path
$(^2n_{N2}^+/^2\pi_{H-1}^+)_{CI}$	0.054	0.594	0.459	1.712	0.132	0.657	Peaked, bifurcating
7H-guanine$^+$							
$(^2n_O^+/^2\pi_H^+)_{CI}$	0.059	0.866	5.780	0.018	17.899	2.754	Sloped, single-path
$(^2\pi_{H-1}^+/^2n_O^+)_{CI}$	0.056	0.572	3.906	0.020	9.709	2.646	Sloped, single-path
$(^2n_N^+/^2\pi_{H-1}^+)_{CI}$	0.063	0.934	0.239	0.297	0.029	0.319	Peaked, bifurcating

The second conical intersection, $(^2\pi_{H-1}^+/^2n_N^+)_{CI}$, which mediates $^2\pi_{H-1}^+ \rightarrow ^2n_N^+$ population transfer in 7H-adenine$^+$, also displays a sloped and single-path character. This crossing features a large asymmetry close to 1 (that would refer to completely asymmetric) and a slight tilt angle, with borderline $\mathcal{P}$ and $\mathcal{B}$ values that make it sloped and single-path. $(^2\pi_{H-1}^+/^2n_O^+)_{CI}$ in 7H-guanine mediates $^2\pi_{H-1}^+ \rightarrow ^2n_O^+$ decay and strongly resembles the sloped and single-path topography displayed by $(^2n_O^+/^2\pi_H^+)_{CI}$. These further support ultrafast deactivation mechanisms in both 7H-adenine$^+$ and 7H-guanine$^+$, which in turn support the photo-protection mechanisms associated to these species [13,14,33].

The last conical intersection considered in this study connects the initially accessed $^2n_{N2}^+$ ($^2n_N^+$) with the $^2\pi_{H-1}^+$ state in 7H-adenine$^+$ (7H-guanine$^+$), and mediates the initial part of the photoionisation process. These conical intersections present more significant changes, featuring the smallest relative tilts and leading to a peaked and bifurcating character. This suggests the emergence of potential re-crossings (and population transfers back and forth) leading to a wave packet splitting on two different directions, which is expected to embody the slower component of the decay.

3. Discussion

A few trends emerge when considering the simulation of ionisation potentials for 7H-adenine and 7H-guanine using different multireference perturbation theory approaches as in the present study: the averaged (CASPT2) values obtained are blue-shifted with respect to experimental evidence and other more correlated (EOM-IP-CCSD) theoretical approaches when considering $^2\pi$ states, while they are red-shifted for ^2n lone-pair states. There also seems to be a better agreement between the simulated ^2n states and recorded experimental evidence, which might be due to the lesser reliance of (covalent) lone-pair states on the dynamic electron correlation included in the model as compared to the strong dependence of π delocalised (ionic) states [52–54]. It is also worth noting that the computed standard deviation of the $^2n_N^+$ state in 7H-guanine is twice as big as that modelled for any other state considered in this study. This is partly due to the mixed character of the $^2n_O^+$ state, which can be easily seen by the delocalisation of the SOMO orbital over both n_O and n_N sites (see Figure 2), and which leads to large off-diagonal elements in multistate and extended multistate formulations of CASPT2 [42], as well as to lower weight references in its single-state variant that also affect the overall accuracy [55]. The values of the off-diagonal elements, together with the leading configuration state functions characterising the different wave functions, are provided in Tables S3–S8 in the supporting information. Significant couplings are observed between $^2\pi_{H-1}^+$ and $^2n_N^+$ states, which can also be rationalised by the different weights contributing to the diverse cationic states studied featuring a 50%/20% n_N/n_O character at the multistate (as well as single-state) level but that is uncoupled upon employing the extended multistate variant. This, together with the closer ionisation potentials obtained for XMS-CASPT2(IPEA = 0.0) with respect to both CASPT2 average and the recorded evidence (see Table S2 in the Supplementary Materials) justify its choice as the method used throughout for optimisations.

It should be stated that more accurate approaches than those of Bravaya et al. [35] used here as reference, have been reported for modelling the ionisation potentials of these species in the literature: the work of Ortiz and co-workers employing the P3 method [56] in 9H-adenine [57] and guanine [38] has been shown to match experiments within under a tenth of an eV. The EOM-IP-CCSD approach of Bravaya et al. [35] was however deemed a better comparison for this study as it is grounded on many body perturbation theory [58,59], as is the CASPT2 method [60], while including more correlation as well as featuring a larger basis set. This reveals a lack of correlation in the model when describing $^2\pi^+$ (ionic) states, which has been shown to be partially corrected upon increasing the zeroth-order Hamiltonian (i.e., the active space size) [61,62], and that explains the larger deviations registered when comparing with experiment particularly for $^2\pi^+$ states. Additionally, it should be noted that more recent experimental results were found compared to those of Lin et al. [40] used as reference in Figure 2: the recorded evidence by Zaytseva et al. [37] measured almost four decades later, qualitatively and almost quantitatively (all photoelectron measurements being within 0.05 eV on average with the largest deviation being below 0.1 eV) agrees with the original measures of Lin et al. [40] thus validating the experimental reference employed.

Whereas all values used (as well as those computed) in this study refer to the 7H tautomer of guanine, we could not find data on the ionisation potentials of 7H-adenine so we used those available for 9H-adenine instead [41], which makes comparisons with the computed 7H-adenine data much less straight forward. As can be seen in Figure 2, differences in the ionisation potentials of 7H- (average CASPT2 values in black crosses) and 9H-adenine (blue triangles (theory) [35] and red circles (experiment) [41]) seem to be small for all but the lowest-lying $^2\pi_H^+$ state, where a significant ~0.5 eV difference is observed. This is however in line with previous theoretical estimates found in the literature using density functional theory and the B3LYP functional, which predict a ~0.4 eV blue-shift on the $^2\pi_H^+$ state when going from 9H- to 7H-adenine [63]. This large deviation between the computed values and the experimental evidence [41] thus arises due to different tautomeric species being measured, rather than the lack of accuracy of the model.

An aspect not considered in this study is how the solvent affects ionisation in these species. Experimental and theoretical estimates for DNA nucleobases and nucleosides recorded in the literature point towards substantial red-shifts of the ionisation energies when in water solution [64,65]. To assess this, not only on ionisation potentials but also along the cationic excited state decays. Moreover, different mechanisms such as proton-coupled electron transfer (PCET) with the environment are expected to dominate ionisation in solution, having been summarised elsewhere [66]. More sophisticated approaches are thus required [67–69] to study these processes, which are out of the scope of this manuscript and that will be considered in future work.

Upon photoionisation, it is predicted that 7H-adenine$^+$ will undergo an ultrafast step-wise $^2n_{N2}^+ \rightarrow {}^2\pi_{H-1}^+ \rightarrow {}^2n_N^+$ decay, which is analogous to what has been observed for other DNA species [28–30]. A major difference, however, is brought by the very small energies (< 0.1 eV) separating $(^2n_N^+/^2\pi_H^+)_{CI}$ and $(^2\pi_H^+)_{min}$ critical structures (see Figure 3), which effectively enable a $^2n_N^+ \rightleftharpoons {}^2\pi_H^+$ equilibrium where both states could in principle present significant populations after decay to the ground state. Given the small energies separating them, this is likely to occur even when very little thermal energy is available and despite the $^2n_N^+$ state not presenting a well-defined minimum, thus paving the way for a potential population of the excited cationic state via chemiexcitation [47]. It is worth noting that, due to the elevated computational cost of XMS-CASPT2 gradients and non-adiabatic couplings, a reduced active space had to be used. Upon inspection of the obtained off-diagonal elements (see Tables S3 and S7 in the Supplementary Materials), we observe a relative artificial increase of this magnitude due to the smaller zeroth-order reference function, which, however, remains small enough to not bias the ensuing optimisations.

Interestingly, the decay mechanism outlined above (aside from the chemiexcitation channel accessible at the cationic ground state minimum) is similar to what has been found recently by Zhao et al. on the deactivation of 9H-adenine$^+$: [70] all low-lying cationic states appear to be vibronically coupled, which might explain the ultrafast timescales observed in the 9H species, and that are here suggested to apply also to the 7H tautomer.

7H-guanine$^+$, on the other hand, presents a predicted step-wise and ultrafast $^2n_N^+ \rightarrow {}^2\pi_{H-1}^+ \rightarrow {}^2n_O^+ \rightarrow {}^2\pi_H^+$ decay to the (cationic) ground state. This swift decay route is expected despite the associated potential energy barriers predicted along $^2\pi_{H-1}^+$ and $^2n_O^+$ state deactivations, which are however too small (0.04 eV for the former and 0.01 eV for the latter) to hamper decay.

Interestingly, despite involving ionisations mostly localised in the carbonyl moiety (i.e., the $^2n_O^+$ state), 7H-guanine$^+$ does not display significant C=O bond elongations, which is the main reaction coordinate in cationic pyrimidine nucleobases [27–30]. Another distinctive feature of 7H-guanine$^+$ and its decay is that geometrical distortions along the deactivation, at a difference from 7H-adenine$^+$, localise mostly over the five-member ring: this is a feature that has also been observed in other guanine derivatives such as thienoguanosine while studying their (singlet) excited state decays [71].

Reactivity in the cationic manifold is markedly different from what has been observed in the neutral singlet species: whereas 7H-adenine [72,73] and 7H-guanine [74] are reported to feature a potential energy barrier along the reaction coordinate hampering decay, which reflects experimentally in longer decay times [75], cations are barrierless and thus expected to decay rapidly to the (cationic) ground state. It is also worth noting that, whereas 7H DNA purines (as well as their biologically relevant 9H counterparts) feature strong out-of-plane motions along the decay in their singlet manifold [76], cations remain planar in line with what has been previously observed for other canonical DNA species [27–30].

The optimised conical intersection topographies highlight that despite featuring different lone pair states (i.e., localised on different atoms), the lowest-lying crossings of both 7H-adenine$^+$ and 7H-guanine$^+$ nevertheless display very strong similarities. This is also observed for the highest-lying crossing studied, $(^2n_{N2}^+/^2\pi_{H-1}^+)_{CI}$ and $(^2n_N^+/^2\pi_{H-1}^+)_{CI}$

for 7H-adenine$^+$ and 7H-guanine$^+$ respectively, which again entail different lone pair cationic states but that lead to analogous intersection topographies that strongly resemble one another, thus pointing at symmetry (i.e., π vs σ) as one of the main factor governing these crossings and thus their efficient decay [77,78].

The results also seem to point at the ability of the ground cationic state to lead to sloped single-path intersections, which are known to foster photostability [50,51], while other $^2\pi^+$ states like $^2\pi^+_{H-1}$ lead upon crossing with different symmetry lone-pair states to peaked and bifurcating topographies instead.

Photostability refers in this case to the unique ability shown by the nucleobases [3] to divert the (cationic) excited state wave packet created upon absorption to the ground state in ultrafast timescales, reducing the time spent in the more reactive electronic excited states that may lead to photo-damaging reactions [1]. This is supported by the topographies of the conical intersections mediating decay to the ground state, as they feature the (only and) preferential decay direction towards the regeneration of the (cationic) ground state structure thus helping avoid photodegradation [13,14]. This is a well-known feature in the singlet manifold [3,13,14,79–81], which is relatively unique to DNA nucleobases and that justifies their resilience to radiation, and has also been observed in other DNA cationic systems [29,30]. It is important to note, however, that ultrafast deactivations in the cationic manifold might be more frequent in organic molecular systems than in the singlet due to the smaller energy gaps between states [50,51], its role towards DNA photostability thus being less evident and a possibility that we are currently investigating.

In this study, we have favoured a thorough study of the potential energy landscape, characterising the different accessible conical intersections mediating decay, as that allowed us to employ accurate XMS-CASPT2 methodologies to uncover the qualitative traits of these photo-processes. Another option would have been to use non-adiabatic molecular dynamics, such as those based on cost-effective surface hopping approaches. Non-adiabatic molecular dynamics would thus provide temporal estimates to the photo-processes studied, but at present we would have to rely on simpler (i.e., less accurate) electronic structure potential energy surfaces due to their elevated cost.

The present study provides an adequate reference for the decay channels available to both 7H adenine and guanine species, with the caveat that only conical intersection topographies are considered and not how the intersection themselves are reached [82]. This is known to significantly influence decay [83], and is an important aspect that can be extracted from dynamics simulations and that will be explored in future work.

In summary, we observe markedly different decay pathways in 7H-adenine$^+$ and 7H-guanine$^+$, which however are expected to lead in both cases to ultrafast deactivations that are ultimately (i.e., on the last step of the decay) mediated by sloped and single-path conical intersections with the cationic ground state known to foster photostability [50,51]. These decay mechanisms differ in the specifics but lead to the same conclusions as those previously characterised for pyrimidine nucleobase cations [29,30], which highlight the resilience of our genetic material against VUV light radiation that we here tentatively also extend to purine tautomeric DNA bases.

4. Computational Details

OpenMolcas [84,85] was used to model the ionisation potentials reported, making use of the complete active space self-consistent field (CASSCF) [86,87] and its second-order perturbation theory extension (CASPT2) [60]. For these simulations, an Atomic Natural Orbital (ANO) large basis set with a double-ζ polarised contraction was used [88,89].

The active spaces for modelling the ionisation potentials contained all valence π and n lone pair orbitals, as well as all valence unoccupied π^* orbitals, totalling 18 electrons in 13 orbitals for 7H-adenine (17 in 13 for cations) and 20 electrons in 14 orbitals for 7H-guanine (19 in 14 for cations). CASSCF wave functions were averaged over five doublet states and were subsequently used for single-point CASPT2 energy corrections employing the single-state (SS) [90,91], multistate (MS) [92], and extended multistate (XMS) [93] variants. An imaginary level shift of 0.2 a.u. was employed in the perturbative step to avoid the presence of intruder states [94], and IPEA shifts [95] of 0.0 and 0.25 a.u. were both tested, resulting in 6 different CASPT2 zeroth-order Hamiltonians: SS-CASPT2(IPEA = 0.0), SS-CASPT2(IPEA = 0.25), MS-CASPT2(IPEA = 0.0), MS-CASPT2(IPEA = 0.25), XMS-CASPT2(IPEA = 0.0) and XMS-CASPT2(IPEA = 0.25).

To aid the discussion of ionisation energies, we have chosen to average the six different CASPT2 formulations as this provides the mean value as well as the standard deviation expected by modifying the zeroth-order Hamiltonian.

Relaxation pathways (cationic ground and excited state minima, as well as conical intersections) were characterised using BAGEL [96,97] and a cc-pVDZ basis set with its density fitting auxiliary basis using the XMS-CASPT2(IPEA = 0.0) method. This choice was motivated by XMS-CASPT2(IPEA = 0.0) featuring the closest ionisation potential estimates to the recorded evidence, and by providing the best balance for the simultaneous description of covalent and ionic excited states [93], being more reliable at or nearby crossing regions [42]. A reduced active space had to be employed due to the elevated cost of these simulations. 7H-adenine comprised the full π valence occupied space and one unoccupied virtual π^* orbital plus two n_N lone pair orbitals to account for $^2n_N^+$ states, leading to 16 electrons in nine orbitals (15 in 9 for cations). 7H-Guanine, on the other hand, featured all π valence occupied orbitals minus the fully in-phase (least contributing, that is, 1.99 occupation number) orbital and one unoccupied virtual π^* orbital plus two n lone pairs (n_O and n_N) to describe $^2n_O^+$ and $^2n_N^+$ states, totalling also 16 electrons in 9 orbitals (15 in 9 for cations).

The characterised cationic ground and excited state minima, as well as the different low-lying conical intersections, were optimised with XMS-CASPT2 using the single-state single reference (i.e., fully internally contracted) algorithm and no IPEA shift corrections as implemented in BAGEL [96,97]. Minima and CI optimisations (using the projection method of Bearpark et al.[98]) made use of analytical gradients [99–101] and non-adiabatic couplings [102]. Orbital visualisation was performed with Molden [103].

Conical intersection parameters were obtained from the XMS-CASPT2 gradient difference and non-adiabatic coupling vectors obtained at the characterised minimum energy conical intersections. These vectors were then used as described by Galván et al. to obtain pitch (δ), asymmetry (Δ) and relative tilt (σ), as well as $\mathcal{P}$ and $\mathcal{B}$ parameters to define the intersection types [48].

Supplementary Materials: The following are available online at https://www.mdpi.com/article/10.3390/photochem1020018/s1: Tables S1–S8 with estimates for the ionisation potentials with the different CASPT2 approaches and reference weights and off-diagonal terms of the MS and XMS-CASPT2 effective Hamiltonians.

Author Contributions: Conceptualization, J.S.-M. and M.J.B.; methodology, J.S.-M. and M.J.B.; software, J.S.-M.; validation, J.S.-M., S.M.N. and M.J.B.; formal analysis, J.S.-M. and S.M.N.; investigation, J.S.-M. and S.M.N.; resources, M.J.B.; data curation, J.S.-M. and S.M.N.; writing–original draft preparation, J.S.-M.; writing–review and editing, J.S.-M., S.M.N. and M.J.B.; visualization, J.S.-M. and S.M.N.; supervision, M.J.B.; project administration, J.S.-M. and M.J.B.; funding acquisition, J.S.-M. All authors have read and agreed to the published version of the manuscript.

Funding: The project that gave rise to these results received the support of a fellowship from "La Caixa" Foundation (ID 100010434) and from the European Union's Horizon 2020 research and innovation programme under the Marie Skłodowska-Curie grant agreement No 847648, fellowship code "LCF/BQ/PI20/11760022" (J.S.-M.).

Acknowledgments: We thankfully acknowledge computing resources at *Tirant* and technical support provided by Servei d'Informatica de la Universitat de Valencia. We also thank the use and support provided by the Imperial College Research Computing Service (DOI: 10.14469/hpc/2232).

Conflicts of Interest: The authors declare no conflict of interest. The funders had no role in the design of the study; in the collection, analyses, or interpretation of data; in the writing of the manuscript, or in the decision to publish the results.

References

1. Schreier, W.J.; Gilch, P.; Zinth, W. Early Events of DNA Photodamage. *Annu. Rev. Phys. Chem.* **2015**, *66*, 497–519. [CrossRef]
2. Noonan, F.P.; Zaidi, M.R.; Wolnicka-Glubisz, A.; Anver, M.R.; Bahn, J.; Wielgus, A.; Cadet, J.; Douki, T.; Mouret, S.; Tucker, M.A.; et al. Melanoma induction by ultraviolet A but not ultraviolet B radiation requires melanin pigment. *Nat. Commun.* **2012**, *3*, 884. [CrossRef] [PubMed]
3. Improta, R.; Santoro, F.; Blancafort, L. Quantum Mechanical Studies on the Photophysics and the Photochemistry of Nucleic Acids and Nucleobases. *Chem. Rev.* **2016**, *116*, 3540–3593. [CrossRef]
4. Middleton, C.T.; de La Harpe, K.; Su, C.; Law, Y.K.; Crespo-Hernández, C.E.; Kohler, B. DNA Excited-State Dynamics: From Single Bases to the Double Helix. *Annu. Rev. Phys. Chem.* **2009**, *60*, 217–239. [CrossRef] [PubMed]
5. Crespo-Hernández, C.E.; Cohen, B.; Hare, P.M.; Kohler, B. Ultrafast Excited-State Dynamics in Nucleic Acids. *Chem. Rev.* **2004**, *104*, 1977–2020. [CrossRef] [PubMed]
6. Spata, V.A.; Lee, W.; Matsika, S. Excimers and Exciplexes in Photoinitiated Processes of Oligonucleotides. *J. Phys. Chem. Lett.* **2016**, *7*, 976–984. [CrossRef] [PubMed]
7. Markovitsi, D.; Gustavsson, T.; Vayá, I. Fluorescence of DNA Duplexes: From Model Helices to Natural DNA. *J. Phys. Chem. Lett.* **2010**, *1*, 3271–3276.
8. Chen, J.; Zhang, Y.; Kohler, B. Excited States in DNA Strands Investigated by Ultrafast Laser Spectroscopy. *Top. Curr. Chem.* **2015**, *356*, 39–87.
9. Martínez-Fernández, L.; Esposito, L.; Improta, R. Studying the excited electronic states of guanine rich DNA quadruplexes by quantum mechanical methods: Main achievements and perspectives. *Photochem. Photobiol. Sci.* **2020**, *19*, 436–444. [CrossRef] [PubMed]
10. Balanikas, E.; Banyasz, A.; Douki, T.; Baldacchino, G.; Markovitsi, D. Guanine Radicals Induced in DNA by Low-Energy Photoionization. *Acc. Chem. Res.* **2020**, *53*, 1511–1519. [CrossRef]
11. Gustavsson, T.; Markovitsi, D. Fundamentals of the Intrinsic DNA Fluorescence. *Acc. Chem. Res.* **2021**, *54*, 1226–1235. [CrossRef]
12. Kleinermanns, K.; Nachtigallová, D.; de Vries, M.S. Excited state dynamics of DNA bases. *Int. Rev. Phys. Chem.* **2013**, *32*, 308–342.
13. Beckstead, A.A.; Zhang, Y.; de Vries, M.S.; Kohler, B. Life in the light: Nucleic acid photoproperties as a legacy of chemical evolution. *Phys. Chem. Chem. Phys.* **2016**, *18*, 24228–24238. [CrossRef] [PubMed]
14. Boldissar, S.; de Vries, M.S. How nature covers its bases. *Phys. Chem. Chem. Phys.* **2018**, *20*, 9701–9716. [CrossRef] [PubMed]
15. Pfeifer, G.P.; Besaratinia, A. UV wavelength-dependent DNA damage and human non-melanoma and melanoma skin cancer. *Photochem. Photobiol. Sci.* **2012**, *11*, 90–97. [CrossRef]
16. Cadet, J.; Mouret, S.; Ravanat, J.L.; Douki, T. Photoinduced Damage to Cellular DNA: Direct and Photosensitized Reactions†. *Photochem. Photobiol.* **2012**, *88*, 1048–1065.
17. Nisoli, M.; Decleva, P.; Calegari, F.; Palacios, A.; Martín, F. Attosecond Electron Dynamics in Molecules. *Chem. Rev.* **2017**, *117*, 10760–10825. [CrossRef]
18. Calegari, F.; Trabattoni, A.; Palacios, A.; Ayuso, D.; Castrovilli, M.C.; Greenwood, J.B.; Decleva, P.; Martín, F.; Nisoli, M. Charge migration induced by attosecond pulses in bio-relevant molecules. *J. Phys. B At. Mol. Opt. Phys.* **2016**, *49*, 142001. [CrossRef]
19. Bucher, D.B.; Pilles, B.M.; Carell, T.; Zinth, W. Charge separation and charge delocalization identified in long-living states of photoexcited DNA. *Proc. Natl. Acad. Sci. USA* **2014**, *111*, 4369–4374.
20. Takaya, T.; Su, C.; de La Harpe, K.; Crespo-Hernández, C.E.; Kohler, B. UV excitation of single DNA and RNA strands produces high yields of exciplex states between two stacked bases. *Proc. Natl. Acad. Sci. USA* **2008**, *105*, 10285–10290.
21. Vayá, I.; Gustavsson, T.; Douki, T.; Berlin, Y.; Markovitsi, D. Electronic Excitation Energy Transfer between Nucleobases of Natural DNA. *J. Am. Chem. Soc.* **2012**, *134*, 11366–11368. [CrossRef] [PubMed]
22. Bucher, D.B.; Kufner, C.L.; Schlueter, A.; Carell, T.; Zinth, W. UV-Induced Charge Transfer States in DNA Promote Sequence Selective Self-Repair. *J. Am. Chem. Soc.* **2016**, *138*, 186–190. [CrossRef]
23. Banyasz, A.; Ketola, T.; Martínez-Fernández, L.; Improta, R.; Markovitsi, D. Adenine radicals generated in alternating AT duplexes by direct absorption of low-energy UV radiation. *Faraday Discuss.* **2018**, *207*, 181–197. [CrossRef]
24. Roca-Sanjuán, D.; Rubio, M.; Merchán, M.; Serrano-Andrés, L. Ab initio determination of the ionization potentials of DNA and RNA nucleobases. *J. Chem. Phys.* **2006**, *125*, 084302.
25. Wolf, T.J.A.; Gühr, M. Photochemical pathways in nucleobases measured with an X-ray FEL. *Philos. Trans. R. Soc. A Math. Phys. Eng. Sci.* **2019**, *377*, 20170473.
26. Wolf, T.J.A.; Holzmeier, F.; Wagner, I.; Berrah, N.; Bostedt, C.; Bozek, J.; Buckbaum, P.; Coffee, R.; Cryan, J.; Farrell, J.; et al. Observing Femtosecond Fragmentation Using Ultrafast X-ray-Induced Auger Spectra. *Appl. Sci.* **2017**, *7*, 681. [CrossRef]
27. Assmann, M.; Köppel, H.; Matsika, S. Photoelectron Spectrum and Dynamics of the Uracil Cation. *J. Phys. Chem. A* **2015**, *119*, 866–875. [CrossRef] [PubMed]
28. Assmann, M.; Weinacht, T.; Matsika, S. Surface hopping investigation of the relaxation dynamics in radical cations. *J. Chem. Phys.* **2016**, *144*, 034301.
29. Segarra-Martí, J.; Tran, T.; Bearpark, M.J. Ultrafast and radiationless electronic excited state decay of uracil and thymine cations: Computing the effects of dynamic electron correlation. *Phys. Chem. Chem. Phys.* **2019**, *21*, 14322–14330. [CrossRef] [PubMed]

30. Segarra-Martí, J.; Tran, T.; Bearpark, M.J. Computing the Ultrafast and Radiationless Electronic Excited State Decay of Cytosine and 5-methyl-cytosine Cations: Uncovering the Role of Dynamic Electron Correlation. *ChemPhotoChem* **2019**, *3*, 856–865. [CrossRef]
31. Crespo-Hernández, C.E.; Martínez-Fernández, L.; Rauer, C.; Reichardt, C.; Mai, S.; Pollum, M.; Marquetand, P.; González, L.; Corral, I. Electronic and Structural Elements That Regulate the Excited-State Dynamics in Purine Nucleobase Derivatives. *J. Am. Chem. Soc.* **2015**, *137*, 4368–4381. [CrossRef] [PubMed]
32. De Vries, M.S. Tautomer-Selective Spectroscopy of Nucleobases, Isolated in the Gas Phase. In *Tautomerism*; John Wiley & Sons, Ltd.: Hoboken, NJ, USA, 2013; Chapter 7, pp. 177–196.
33. Serrano-Andrés, L.; Merchán, M. Are the five natural DNA/RNA base monomers a good choice from natural selection?: A photochemical perspective. *J. Photochem. Photobiol. C Photochem. Rev.* **2009**, *10*, 21–32. [CrossRef]
34. Huang, Y.; Kenttamaa, H. Theoretical Estimations of the 298 K Gas-Phase Acidities of the Purine-Based Nucleobases Adenine and Guanine. *J. Phys. Chem. A* **2004**, *108*, 4485–4490.
35. Bravaya, K.B.; Kostko, O.; Dolgikh, S.; Landau, A.; Ahmed, M.; Krylov, A.I. Electronic Structure and Spectroscopy of Nucleic Acid Bases: Ionization Energies, Ionization-Induced Structural Changes, and Photoelectron Spectra. *J. Phys. Chem. A* **2010**, *114*, 12305–12317. [CrossRef]
36. Marian, C.M. The Guanine Tautomer Puzzle: Quantum Chemical Investigation of Ground and Excited States. *J. Phys. Chem. A* **2007**, *111*, 1545–1553. [CrossRef]
37. Zaytseva, I.L.; Trofimov, A.B.; Schirmer, J.; Plekan, O.; Feyer, V.; Richter, R.; Coreno, M.; Prince, K.C. Theoretical and Experimental Study of Valence-Shell Ionization Spectra of Guanine. *J. Phys. Chem. A* **2009**, *113*, 15142–15149. [CrossRef]
38. Dolgounitcheva, O.; Zakrzewski, V.G.; Ortiz, J.V. Electron Propagator Theory of Guanine and Its Cations: Tautomerism and Photoelectron Spectra. *J. Am. Chem. Soc.* **2000**, *122*, 12304–12309.
39. Mons, M.; Piuzzi, F.; Dimicoli, I.; Gorb, L.; Leszczynski, J. Near-UV Resonant Two-Photon Ionization Spectroscopy of Gas Phase Guanine: Evidence for the Observation of Three Rare Tautomers. *J. Phys. Chem. A* **2006**, *110*, 10921–10924. [CrossRef]
40. Lin, J.; Yu, C.; Peng, S.; Akiyama, I.; Li, K.; Lee, L.K.; LeBreton, P.R. Ultraviolet photoelectron studies of the ground-state electronic structure and gas-phase tautomerism of hypoxanthine and guanine. *J. Phys. Chem.* **1980**, *84*, 1006–1012.
41. Trofimov, A.B.; Schirmer, J.; Kobychev, V.B.; Potts, A.W.; Holland, D.M.P.; Karlsson, L. Photoelectron spectra of the nucleobases cytosine, thymine and adenine. *J. Phys. B At. Mol. Opt. Phys.* **2005**, *39*, 305–329. [CrossRef]
42. Shiozaki, T.; Woywod, C.; Werner, H.J. Pyrazine excited states revisited using the extended multi-state complete active space second-order perturbation method. *Phys. Chem. Chem. Phys.* **2013**, *15*, 262–269. [CrossRef]
43. Kotur, M.; Zhou, C.; Matsika, S.; Patchkovskii, S.; Spanner, M.; Weinacht, T.C. Neutral-Ionic State Correlations in Strong-Field Molecular Ionization. *Phys. Rev. Lett.* **2012**, *109*, 203007. [CrossRef]
44. Kotur, M.; Weinacht, T.; Zhou, C.; Matsika, S. Following Ultrafast Radiationless Relaxation Dynamics With Strong Field Dissociative Ionization: A Comparison Between Adenine, Uracil and Cytosine. *IEEE J. Sel. Top. Quantum Electron.* **2012**, *18*, 187–194. [CrossRef]
45. Segarra-Martí, J.; Bearpark, M.J. Modelling photoionisation in isocytosine: Potential formation of longer-lived excited state cations in its keto form. *ChemPhysChem* **2021**.
46. Matsika, S. Two- and three-state conical intersections in the uracil cation. *Chem. Phys.* **2008**, *349*, 356–362. [CrossRef]
47. Brash, D.E.; Goncalves, L.C.; Bechara, E.J. Chemiexcitation and Its Implications for Disease. *Trends Mol. Med.* **2018**, *24*, 527–541. [CrossRef]
48. Fdez. Galván, I.; Delcey, M.G.; Pedersen, T.B.; Aquilante, F.; Lindh, R. Analytical State-Average Complete-Active-Space Self-Consistent Field Nonadiabatic Coupling Vectors: Implementation with Density-Fitted Two-Electron Integrals and Application to Conical Intersections. *J. Chem. Theory Comput.* **2016**, *12*, 3636–3653. [CrossRef]
49. Atchity, G.J.; Xantheas, S.S.; Ruedenberg, K. Potential energy surfaces near intersections. *J. Chem. Phys.* **1991**, *95*, 1862–1876.
50. Hall, K.F.; Boggio-Pasqua, M.; Bearpark, M.J.; Robb, M.A. Photostability Via Sloped Conical Intersections: A Computational Study of the Excited States of the Naphthalene Radical Cation. *J. Phys. Chem. A* **2006**, *110*, 13591–13599. [CrossRef]
51. Tokmachev, A.M.; Boggio-Pasqua, M.; Bearpark, M.J.; Robb, M.A. Photostability via Sloped Conical Intersections: A Computational Study of the Pyrene Radical Cation. *J. Phys. Chem. A* **2008**, *112*, 10881–10886. [CrossRef]
52. Segarra-Martí, J.; Garavelli, M.; Aquilante, F. Multiconfigurational second-order perturbation theory with frozen natural orbitals extended to the treatment of photochemical problems. *J. Chem. Theory Comput.* **2015**, *11*, 3772–3784. [CrossRef] [PubMed]
53. Segarra-Martí, J.; Francés-Monerris, A.; Roca-Sanjuán, D.; Merchán, M. Assessment of the Potential Energy Hypersurfaces in Thymine within Multiconfigurational Theory: CASSCF vs. CASPT2. *Molecules* **2016**, *21*, 1666. [CrossRef]
54. Segarra-Martí, J.; Mukamel, S.; Garavelli, M.; Nenov, A.; Rivalta, I. Towards Accurate Simulation of Two-Dimensional Electronic Spectroscopy. *Top. Curr. Chem.* **2018**, *376*, 24. [CrossRef]
55. Serrano-Andrés, L.; Merchán, M.; Lindh, R. Computation of conical intersections by using perturbation techniques. *J. Chem. Phys.* **2005**, *122*, 104107.
56. Díaz-Tinoco, M.; Dolgounitcheva, O.; Zakrzewski, V.G.; Ortiz, J.V. Composite electron propagator methods for calculating ionization energies. *J. Chem. Phys.* **2016**, *144*, 224110.
57. Dolgounitcheva, O.; Zakrzewski, V.G.; Ortiz, J.V. Vertical Ionization Energies of Adenine and 9-Methyl Adenine. *J. Phys. Chem. A* **2009**, *113*, 14630–14635. [CrossRef]

58. Bartlett, R.J.; Musiał, M. Coupled-cluster theory in quantum chemistry. *Rev. Mod. Phys.* **2007**, *79*, 291–352. [CrossRef]
59. Krylov, A.I. Equation-of-Motion Coupled-Cluster Methods for Open-Shell and Electronically Excited Species: The Hitchhiker's Guide to Fock Space. *Annu. Rev. Phys. Chem.* **2008**, *59*, 433–462. [CrossRef]
60. Roca-Sanjuán, D.; Aquilante, F.; Lindh, R. Multiconfiguration second-order perturbation theory approach to strong electron correlation in chemistry and photochemistry. *WIREs Comput. Mol. Sci.* **2012**, *2*, 585–603. [CrossRef]
61. Giussani, A.; Segarra-Martí, J.; Nenov, A.; Rivalta, I.; Tolomelli, A.; Mukamel, S.; Garavelli, M. Spectroscopic fingerprints of DNA/RNA pyrimidine nucleobases in third-order nonlinear electronic spectra. *Theor. Chem. Acc.* **2016**, *135*, 121. [CrossRef]
62. Segarra-Martí, J.; Zvereva, E.; Marazzi, M.; Brazard, J.; Dumont, E.; Assfeld, X.; Haacke, S.; Garavelli, M.; Monari, A.; Léonard, J.; et al. Resolving the Singlet Excited State Manifold of Benzophenone by First-Principles Simulations and Ultrafast Spectroscopy. *J. Chem. Theory Comput.* **2018**, *14*, 2570–2585. [CrossRef] [PubMed]
63. Close, D.M.; Crespo-Hernández, C.E.; Gorb, L.; Leszczynski, J. Ionization Energy Thresholds of Microhydrated Adenine and Its Tautomers. *J. Phys. Chem. A* **2008**, *112*, 12702–12706.
64. Slavíček, P.; Winter, B.; Faubel, M.; Bradforth, S.E.; Jungwirth, P. Ionization Energies of Aqueous Nucleic Acids: Photoelectron Spectroscopy of Pyrimidine Nucleosides and ab Initio Calculations. *J. Am. Chem. Soc.* **2009**, *131*, 6460–6467. [CrossRef]
65. Pluhařová, E.; Jungwirth, P.; Bradforth, S.E.; Slavíček, P. Ionization of Purine Tautomers in Nucleobases, Nucleosides, and Nucleotides: From the Gas Phase to the Aqueous Environment. *J. Phys. Chem. B* **2011**, *115*, 1294–1305. [CrossRef] [PubMed]
66. Kumar, A.; Sevilla, M.D. Proton-Coupled Electron Transfer in DNA on Formation of Radiation-Produced Ion Radicals. *Chem. Rev.* **2010**, *110*, 7002–7023. [CrossRef] [PubMed]
67. Altavilla, S.F.; Segarra-Martí, J.; Nenov, A.; Conti, I.; Rivalta, I.; Garavelli, M. Deciphering the photochemical mechanisms describing the UV-induced processes occurring in solvated guanine monophosphate. *Front. Chem.* **2015**, *3*, 29. [CrossRef]
68. Weingart, O.; Nenov, A.; Altoè, P.; Rivalta, I.; Segarra-Martí, J.; Dokukina, I.; Garavelli, M. COBRAMM 2.0—A software interface for tailoring molecular electronic structure calculations and running nanoscale (QM/MM) simulations. *J. Mol. Model.* **2018**, *24*, 271. [CrossRef]
69. Tomaník, L.; Muchová, E.; Slavíček, P. Solvation energies of ions with ensemble cluster-continuum approach. *Phys. Chem. Chem. Phys.* **2020**, *22*, 22357–22368. [CrossRef]
70. Zhao, H.Y.; Lau, K.C.; Garcia, G.A.; Nahon, L.; Carniato, S.; Poisson, L.; Schwell, M.; Al-Mogren, M.M.; Hochlaf, M. Unveiling the complex vibronic structure of the canonical adenine cation. *Phys. Chem. Chem. Phys.* **2018**, *20*, 20756–20765. [CrossRef]
71. Martinez-Fernandez, L.; Gavvala, K.; Sharma, R.; Didier, P.; Richert, L.; Segarra Martì, J.; Mori, M.; Mely, Y.; Improta, R. Excited-State Dynamics of Thienoguanosine, an Isomorphic Highly Fluorescent Analogue of Guanosine. *Chem. Eur. J.* **2019**, *25*, 7375–7386.
72. Serrano-Andrés, L.; Merchán, M.; Borin, A.C. Adenine and 2-aminopurine: Paradigms of modern theoretical photochemistry. *Proc. Natl. Acad. Sci. USA* **2006**, *103*, 8691–8696.
73. Barbatti, M. Photorelaxation Induced by Water–Chromophore Electron Transfer. *J. Am. Chem. Soc.* **2014**, *136*, 10246–10249. [CrossRef] [PubMed]
74. Serrano-Andrés, L.; Merchán, M.; Borin, A.C. A Three-State Model for the Photophysics of Guanine. *J. Am. Chem. Soc.* **2008**, *130*, 2473–2484. [CrossRef]
75. Cohen, B.; Hare, P.M.; Kohler, B. Ultrafast Excited-State Dynamics of Adenine and Monomethylated Adenines in Solution: Implications for the Nonradiative Decay Mechanism. *J. Am. Chem. Soc.* **2003**, *125*, 13594–13601. [CrossRef] [PubMed]
76. Barbatti, M.; Aquino, A.J.A.; Szymczak, J.J.; Nachtigallová, D.; Hobza, P.; Lischka, H. Relaxation mechanisms of UV-photoexcited DNA and RNA nucleobases. *Proc. Natl. Acad. Sci. USA* **2010**, *107*, 21453–21458. [CrossRef]
77. Yarkony, D.R. Nonadiabatic Quantum Chemistry—Past, Present, and Future. *Chem. Rev.* **2012**, *112*, 481–498. [CrossRef] [PubMed]
78. Domcke, W.; Yarkony, D.R. Role of Conical Intersections in Molecular Spectroscopy and Photoinduced Chemical Dynamics. *Annu. Rev. Phys. Chem.* **2012**, *63*, 325–352. [CrossRef]
79. Blancafort, L. Photochemistry and Photophysics at Extended Seams of Conical Intersection. *ChemPhysChem* **2014**, *15*, 3166–3181.
80. Pepino, A.J.; Segarra-Martí, J.; Nenov, A.; Improta, R.; Garavelli, M. Resolving Ultrafast Photoinduced Deactivations in Water-Solvated Pyrimidine Nucleosides. *J. Phys. Chem. Lett.* **2017**, *8*, 1777–1783. [CrossRef]
81. Pepino, A.J.; Segarra-Martí, J.; Nenov, A.; Rivalta, I.; Improta, R.; Garavelli, M. UV-induced long-lived decays in solvated pyrimidine nucleosides resolved at the MS-CASPT2/MM level. *Phys. Chem. Chem. Phys.* **2018**, *20*, 6877–6890. [CrossRef]
82. Schuurman, M.S.; Stolow, A. Dynamics at Conical Intersections. *Annu. Rev. Phys. Chem.* **2018**, *69*, 427–450. [CrossRef]
83. Schalk, O.; Boguslavskiy, A.E.; Stolow, A. Substituent Effects on Dynamics at Conical Intersections: Cyclopentadienes. *J. Phys. Chem. A* **2010**, *114*, 4058–4064. [CrossRef]
84. Fdez. Galván, I.; Vacher, M.; Alavi, A.; Angeli, C.; Aquilante, F.; Autschbach, J.; Bao, J.J.; Bokarev, S.I.; Bogdanov, N.A.; Carlson, R.K.; et al. OpenMolcas: From Source Code to Insight. *J. Chem. Theory Comput.* **2019**, *15*, 5925–5964. [CrossRef] [PubMed]
85. Aquilante, F.; Autschbach, J.; Baiardi, A.; Battaglia, S.; Borin, V.A.; Chibotaru, L.F.; Conti, I.; De Vico, L.; Delcey, M.; Fdez. Galván, I.; et al. Modern quantum chemistry with [Open]Molcas. *J. Chem. Phys.* **2020**, *152*, 214117.
86. Olsen, J. The CASSCF method: A perspective and commentary. *Int. J. Quantum Chem.* **2011**, *111*, 3267–3272.
87. Robb, M.A. *Theoretical Chemistry for Electronic Excited States*; Theoretical and Computational Chemistry Series; The Royal Society of Chemistry: London, UK, 2018; pp. P001–P225. [CrossRef]

88. Widmark, P.O.; Malmqvist, P.Å.; Roos, B.O. Density matrix averaged atomic natural orbital (ANO) basis sets for correlated molecular wave functions. I. First row atoms. *Theor. Chim. Acta* **1990**, *77*, 291. [CrossRef]
89. Widmark, P.O.; Persson, B.J.; Roos, B.O. Density matrix averaged atomic natural orbital (ANO) basis sets for correlated molecular wave functions. II. Second row atoms. *Theor. Chim. Acta* **1991**, *79*, 419. [CrossRef]
90. Andersson, K.; Malmqvist, P.A.; Roos, B.O.; Sadlej, A.J.; Wolinski, K. Second-order perturbation theory with a CASSCF reference function. *J. Phys. Chem.* **1990**, *94*, 5483–5488.
91. Andersson, K.; Malmqvist, P.Å.; Roos, B.O. Second-order perturbation theory with a complete active space self-consistent field reference function. *J. Chem. Phys.* **1992**, *96*, 1218–1226. [CrossRef]
92. Finley, J.; Malmqvist, P.Å.; Roos, B.O.; Serrano-Andrés, L. The multi-state CASPT2 method. *Chem. Phys. Lett.* **1998**, *288*, 299–306. [CrossRef]
93. Granovsky, A.A. Extended multi-configuration quasi-degenerate perturbation theory: The new approach to multi-state multi-reference perturbation theory. *J. Chem. Phys.* **2011**, *134*, 214113. [CrossRef]
94. Forsberg, N.; Malmqvist, P.Å. Multiconfiguration perturbation theory with imaginary level shift. *Chem. Phys. Lett.* **1997**, *274*, 196–204. [CrossRef]
95. Ghigo, G.; Roos, B.O.; Malmqvist, P.Å. A modified definition of the zeroth-order Hamiltonian in multiconfigurational perturbation theory (CASPT2). *Chem. Phys. Lett.* **2004**, *396*, 142–149. [CrossRef]
96. Shiozaki, T. BAGEL: Brilliantly Advanced General Electronic-structure Library. *Wiley Interdisc. Rev. Comput. Mol. Sci.* **2018**, *8*, e1331,
97. Park, J.W.; Al-Saadon, R.; MacLeod, M.K.; Shiozaki, T.; Vlaisavljevich, B. Multireference Electron Correlation Methods: Journeys along Potential Energy Surfaces. *Chem. Rev.* **2020**, *120*, 5909. [CrossRef]
98. Bearpark, M.J.; Robb, M.A.; Schlegel, H.B. A direct method for the location of the lowest energy point on a potential surface crossing. *Chem. Phys. Lett.* **1994**, *223*, 269–274. [CrossRef]
99. Shiozaki, T.; Győrffy, W.; Celani, P.; Werner, H.J. Communication: Extended multi-state complete active space second-order perturbation theory: Energy and nuclear gradients. *J. Chem. Phys.* **2011**, *135*, 081106.
100. MacLeod, M.K.; Shiozaki, T. Communication: Automatic code generation enables nuclear gradient computations for fully internally contracted multireference theory. *J. Chem. Phys.* **2015**, *142*, 051103.
101. Vlaisavljevich, B.; Shiozaki, T. Nuclear Energy Gradients for Internally Contracted Complete Active Space Second-Order Perturbation Theory: Multistate Extensions. *J. Chem. Theory Comput.* **2016**, *12*, 3781–3787. [CrossRef]
102. Park, J.W.; Shiozaki, T. Analytical Derivative Coupling for Multistate CASPT2 Theory. *J. Chem. Theory Comput.* **2017**, *13*, 2561–2570. [CrossRef]
103. Schaftenaar, G.; Noordik, J. Molden: A pre- and post-processing program for molecular and electronic structures*. *J. Comput. Aided Mol. Des.* **2000**, *14*, 123–134. [CrossRef]

Article

Photoprotective Effects of Selected Polyphenols and Antioxidants on Naproxen Photodegradability in the Solid-State

Kohei Kawabata *, Ayano Miyoshi and Hiroyuki Nishi

Faculty of Pharmacy, Yasuda Women's University, Yasuhigashi 6-13-1, Asaminami-ku, Hiroshima 731-0153, Japan
* Correspondence: kawabata-k@yasuda-u.ac.jp; Tel.: +81-82-878-9440; Fax: +81-82-878-9540

Abstract: Photostabilization is an important methodology to ensure both the quality and quantity of photodegradable pharmaceuticals. The purpose of our study is to develop a photostabilization strategy focused on the addition of photostabilizers. In this study, the protective effects of selected polyphenols and antioxidants on naproxen (NPX) photodegradation in the solid state were evaluated. Residual amounts of NPX were determined by high-performance liquid chromatography (HPLC), and the protective effects of tested additives on NPX photodegradation induced by ultraviolet light (UV) irradiation were evaluated. As a result, quercetin, curcumin, and resveratrol suppressed NPX photodegradation completely. When they were mixed with NPX, the residual amounts of NPX after UV irradiation were significantly higher compared to that without additives, and comparable to those of their control samples. In addition, to clarify the mechanisms of the highly protective effects of these additives on NPX photodegradation, their antioxidative potencies, and UV filtering potencies were determined. There was no correlation between photoprotective effects and antioxidative potencies among selected polyphenols and antioxidants although photoprotective additives showed more significant UV absorption compared to NPX. From these results, it is clarified that a higher UV filtering activity is necessary for a better photostabilizer to photodegradable pharmaceuticals in the solid state.

Keywords: naproxen; photodegradation; photoproduct; photoprotective effect; polyphenols; antioxidants; HPLC

1. Introduction

Naproxen (NPX, Figure 1A) is a non-steroidal anti-inflammatory drug (NSAID) and chemically (2S)-2-(6-Methoxynaphthalen-2-yl)propanoic acid. This pharmaceutical is categorized as 2-aryl propionic acid family including ibuprofen, ketoprofen, and loxoprofen, and utilized for the relief of pains and fevers by the inhibition of cyclooxygenase resulting in the suppression of the generation of prostaglandins. Naixan® tablets, which are the original NPX tablets, have been widely used in clinical situations. It is well known that some NSAID medicines are photosensitive due to their photodegradable active pharmaceutical ingredients (APIs). There are several reports showing the photodegradabilities of NSAIDs such as diclofenac, indometacin, and sulindac [1–5]. Our previous reports showed that sulindac was photodegraded by both ultraviolet-light (UV) irradiation and sunlight irradiation [5] and converted to trans-sulindac by photoisomerization reaction [6]. NPX is also one of the photosensitive pharmaceuticals and photodegraded resulting in the generation of some photo products including 2-acetyl-6-methoxy-naphthalene (Figure 1B), which is the main photoproduct of NPX [7–10]. Amounts of APIs of NPX tablets were much decreased when these tablets were crushed or suspended following UV irradiation [11].

Furthermore, some pharmaceuticals might exert toxicological potencies as a result of the generation of their photoproducts in addition to the loss of beneficial effects derived from the decrease of APIs. For example, 5-diazoimidazole-4-carboxamide is a photoproduct of dacarbazine, which is known as an anti-cancer drug, and induces vascular pain [12]. In addition, other reports showed the change in biological activities of several pharmaceuticals induced by their photodegradation [5,13–17]. The change of chemical structures, due to the elimination, addition, and rearrangement, might induce changes in biological properties. The photostability of pharmaceuticals is a crucial determinant of their quality and quantity when they are photo-irradiated.

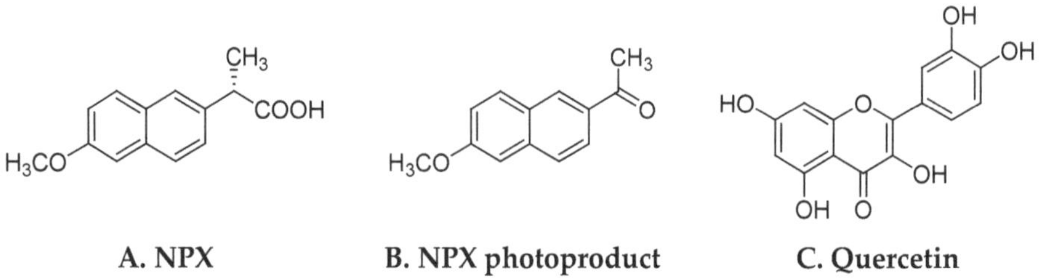

A. NPX   B. NPX photoproduct   C. Quercetin

Figure 1. Chemical structures of NPX (**A**), NPX photoproduct (**B**) and quercetin (**C**).

In recent years, photostabilization strategies of photodegradable pharmaceuticals have been developed for their safe use in clinical situations [18]. UV filtering is an efficient method to disrupt photo exposure for APIs. Solar filters and encapsulation have been utilized as a protective barrier to envelop photodegradable pharmaceuticals. Cyclodextrin and its modified forms are major photoprotective careers to perform photostabilization [19,20]. Also, the addition of some antioxidants such as ascorbic acid is one of the photostabilization strategies. Added antioxidants deactivate the excited state of pharmaceuticals, reactive oxygen species, and free radicals generated by UV irradiation, resulting in the exertion of photoprotective effects. Several studies indicated that ascorbic acid attenuates the photodegradation of several photodegradable pharmaceuticals [21–23]. Furthermore, the combination of UV filters and antioxidants showed significant protective effects on the photodegrdation of some pharmaceuticals [21,24]. However, in some cases, the stabilization effect of encapsulation was decreased [25]. These reports suggest that the use of a photostabilizer is limited and the further development of various photostabilization strategies is needed for the photoprotection of lots of photodegradable pharmaceuticals.

In this study, the protective effects of selected polyphenols and antioxidants on NPX photodegradation in the solid state were evaluated. To the best of our knowledge, there are no reports focused on the photostabilization of NPX in the solid state. First, the photoprotective effect of quercetin (Figure 1C) and its dose-dependency were evaluated. Our previous study showed that quercetin, which is one of the antioxidative polyphenols, significantly suppressed NPX photodegradation in an aqueous media [23]. The photoprotective effect of quercetin on NPX photodegradation in the solid state was evaluated. The residual amounts of NPX and the generation rates of NPX photoproducts were estimated utilizing high-performance liquid chromatography (HPLC). Secondly, the photoprotective effects of selected polyphenols and antioxidants (Figure 2) were determined. Finally, both UV absorptive potencies and antioxidative potencies of tested polyphenols were evaluated by means of UV spectral analysis and a test kit for the potential antioxidant (PAO test). The aim of this study is to clarify the efficient photostabilizer for the crushed tablets of a photodegradable pharmaceutical.

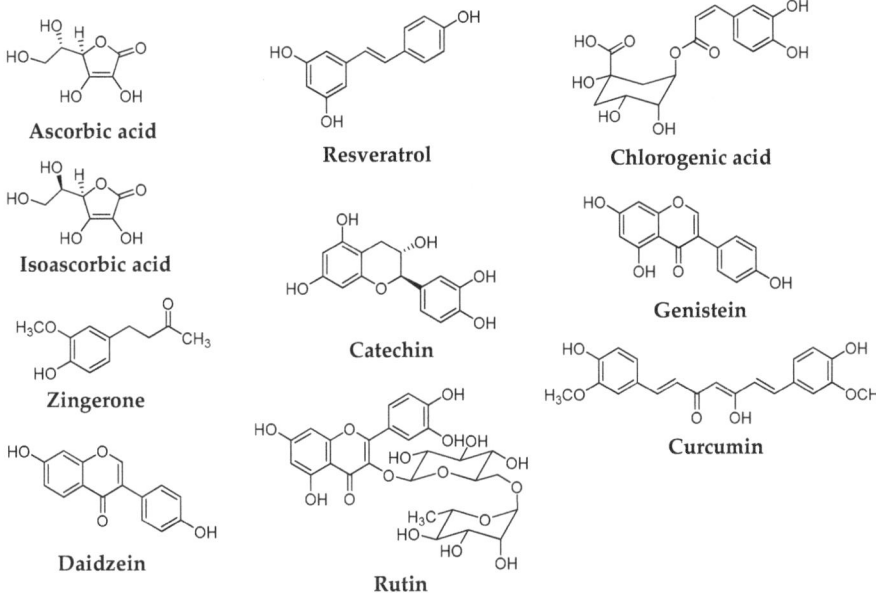

Figure 2. Chemical Structures of Selected Polyphenols and Antioxidants.

2. Materials and Methods

2.1. Materials

NPX, ascorbic acid, isoascorbic acid, zingerone, rutin, daidzein, chlorogenic acid, genistein, resveratrol, methanol, and formic acid were purchased from Fujifilm Wako Pure Chemical Corporation (Osaka, Japan). Catechin, curcumin, and quercetin were purchased from Tokyo Chemical Industry Corporation (Tokyo, Japan). All reagents and organic solvents were of special or HPLC grade. Milli-Q (18.2 Ω/cm) water was prepared by using a Milli-Q water purification system (Merck, Darmstadt, Germany).

2.2. Methods

The scheme of the experimental procedure of the evaluation of photostabilization potencies of selected additives for NPX photodegradation is shown in Figure S1 (see Supplementary Materials). Briefly, the difference in residual amounts of NPX in the presence or absence of selected additives after UV irradiation was evaluated.

2.2.1. Preparation of a Test Sample

To determine the dose dependency of quercetin, NPX (10 mmol) and quercetin (1.25 mmol, 2.5 mmol, 5 mmol and 10 mmol) were mixed to make molar ratios for NPX 0.125–1. Also, NPX (10 mmol) and each polyphenol or antioxidant (5 mmol) were mixed to make a molar ratio of NPX 0.5, and these mixtures were used as test samples. 10 mg of test samples were exposed to black light. Control samples were prepared using the same procedures but covered with aluminum foil to interrupt the photo exposure. UV-irradiated samples were dissolved in 100 mL of 50% (v/v) methanol and sonicated for 10 min for extraction. The extractions were analyzed by HPLC. All experiments were carried out in quadruplicate.

2.2.2. UV Irradiation Experiment

UV irradiation was carried out in a light cabinet with a black light lamp (20W FL20S BLB, Toshiba, Tokyo, Japan). The most abundant wavelength of this lamp is 365 nm, which is a component of sunlight. UV irradiation intensity at 365 nm was 300 µW/cm^2/sec as measured by a digital radiometer with a 365 nm sensor (UVX-36, UVP, Upland, CA, USA). UV irradiation was carried out at a temperature of 20 °C, and a distance from the lamp source of about 20 cm. Irradiation times were up to 24 h.

2.2.3. Evaluation of the Residual Amounts of NPX in UV-Irradiated Samples

The degradation of NPX and the generation of NPX photoproducts were monitored with an HPLC system, which was composed of an LC-20AB pump, a SIL-20AC autosampler, an SPD-M20A photodiode array (PDA) detector with LCsolution software, a CBM-20A system controller, a DGU-20A3 degasser, and a CTO-20A column oven (Shimadzu Corp., Kyoto, Japan). Shim-pack Arata C18 column (4.6 × 150 mm, particle size 5 µm, Shimadzu Corp., Kyoto, Japan) was used for HPLC analysis. The column was kept at 40 °C during analysis. The mobile phase was a mixture of methanol and 0.1% formic acid (5:5, v/v). Isocratic separations were achieved using this mobile phase. The flow rate was maintained at 1.0 mL/min, and the injection volume was 10 µL. The detection wavelength of NPX and NPX photoproducts was 260 nm. The retention time of NPX was ca. 21 min. Amounts of NPX are shown as the residual rate for amounts of NPX before UV irradiation, calculated by their peak areas.

2.2.4. Evaluation of Antioxidative Potencies

The antioxidative potencies of selected polyphenols and antioxidants were evaluated by the PAO test (Nikken SEIL Corp., Shizuoka, Japan). This assay evaluated the antioxidative potencies based on the reduction of Cu^{2+} to Cu^+ by tested compounds using the spectrophotometer. The antioxidative potencies of tested compounds were calculated as copper-reducing power (µmol/L). The tested compounds were dissolved in 50% (v/v) methanol to make a concentration of 10 mg/L (16.87–60.75 µmol/L) for the PAO test. All experiments were performed in triplicates.

2.2.5. UV Spectral Analysis

Tested polyphenols and antioxidants were dissolved in methanol at the final concentration of 10–100 µmol/L. UV absorption spectra were recorded with a V-670 UV/Vis spectrophotometer (JASCO, Tokyo, Japan), interfaced to a PC for data processing. The absorption-maximum wavelength (λ_{max}, nm) of each compound was obtained from these results. The molar absorption coefficients (ε, L mol^{-1} cm^{-1}) were calculated from the absorption of λ_{max}.

2.3. Statistical Analysis

Data are expressed as the mean ± standard deviation (S.D.). The statistical significance of a difference between two groups was estimated by Student's t-test, and between more than three groups was estimated by Tukey's test. The threshold for assessing the significance was $p < 0.05$, $p < 0.01$, or $p < 0.001$.

3. Results and Discussion

3.1. Dose Dependency of the Photoprotective Effect of Quercetin

In this study, the photoprotective effects of selected polyphenols and antioxidants on the photodegradation of NPX, which is well-known as a photodegradable pharmaceutical [7–10], were evaluated in the solid state. Our previous study indicated that selected polyphenols and antioxidants, such as quercetin, ascorbic acid, and isoascorbic acid, suppressed NPX photodegradation in an aqueous media [23]. Quercetin showed a significant photoprotective effect, so its photoprotective potency for NPX in the solid state was investigated first.

The dose dependency of the protective effect of quercetin on the photodegradability of NPX after UV irradiation was evaluated. HPLC chromatograms of NPX and UV-irradiated NPX with or without quercetin (molar ratio for NPX was 0.5) are shown in Figure 3. The effects of quercetin on residual amounts of NPX and generation rates of NPX photoproducts are shown in Figure 4. UV irradiation-induced the decrease of NPX peak concomitant with the generation of NPX photoproducts (retention time of the main photoproduct was ca. 18 min, Figure 3B). When a powder of NPX was UV-irradiated without quercetin for 24 h, its residual amount (84.1 ± 3.8%) was significantly less than that of the control sample (101.7 ± 1.1%) as shown in Figure 4A, and the generation rate of the main NPX photoproduct was 4.9 ± 0.7% (Figure 4B). As the same as in our previous study [11], NPX was excited by absorption of the energy derived from UV irradiation, resulting in the elimination of its carboxylic group followed by oxidation for the conversion to the NPX photoproduct in the solid state. Also, both NPX photodegradation and the generation of NPX photoproducts were not observed in the control sample (Figure 3A), showing that other factors such as hydrolysis and temperature did not contribute to the degradation of NPX.

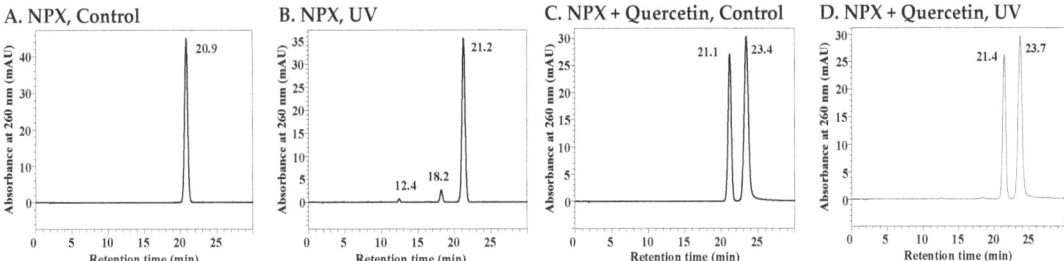

Figure 3. HPLC chromatograms of NPX powder and a mixture of NPX and quercetin (molar ratio for NPX is 0.5) with and without UV irradiation for 24 h. (**A**) an NPX powder, (**B**) a UV-irradiated NPX powder, (**C**) a mixture of NPX and quercetin, (**D**) a UV-irradiated mixture of NPX and quercetin. Detection wavelength: 260 nm.

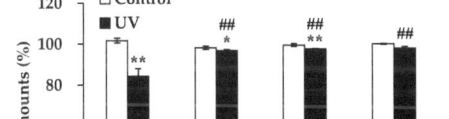

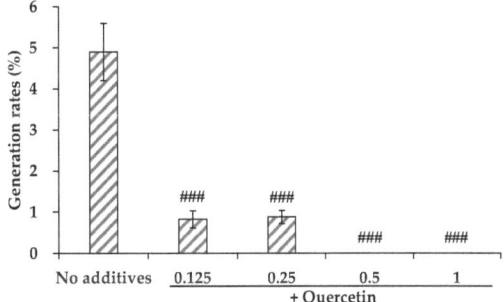

Figure 4. Dose dependency of the effect of quercetin (molar ratios for NPX were 0.125–1) on the residual amounts of NPX and the generation rates of the main NPX photoproduct in the UV-irradiated powder. (**A**) Residual amounts of NPX, (**B**) generation rates of the main NPX photoproduct. Values represent mean ± S.D. (n = 4). * Difference compared with control ($p < 0.05$), ** difference compared with control ($p < 0.01$) and ## difference compared with no additives ($p < 0.01$) (**A**). ### Difference compared with no additives ($p < 0.001$) (**B**).

On the other hand, both NPX photodegradation and the generation of NPX photoproducts induced by UV irradiation were suppressed completely in the presence of quercetin (molar ratio for NPX was 0.5, Figure 3D). Retention time of quercetin was ca. 23 min. The peaks of two NPX photoproducts were slightly detected but their peak area was less than the limit of quantification. The residual amount of NPX after UV irradiation (97.9 ± 0.7%) was significantly higher compared to that in the absence of quercetin

(84.1 ± 3.8%, Figure 4A). It is indicated that quercetin showed a photoprotective effect on NPX in the solid state. From the evaluation of the dose dependency of quercetin, NPX photodegradation and the generation of NPX photoproducts were suppressed partially even at a lower concentration (molar ratios for NPX was 0.125–0.25). Completely suppression was observed at a higher concentration (molar ratios for NPX was 0.5–1) as shown in Figure 4. From these results, it is suggested that quercetin might show the photoprotective effect on NPX by antioxidative potency, resulting in quenching the excitation of NPX and deactivating the hydroxyl radicals, or by UV filtering potency, resulting in disruption of UV irradiation for NPX, when powders of an NPX-quercetin mixture were UV-irradiated.

3.2. Comparison of the Photoprotective Effect of Selected Polyphenols and Antioxidants

Based on the results obtained from the evaluation of the photostabilization potency of quercetin, the comparative experiments of the photoprotective effects of selected polyphenols (catechin, chlorogenic acid, curcumin, daidzein, genistein, quercetin, resveratrol, rutin, and zingerone), and antioxidants (ascorbic acid and isoascorbic acid) were performed (Figure 2). Our previous study showed that ascorbic acid, isoascorbic acid, catechin, and curcumin showed photoprotective effects on NPX photodegradation in an aqueous media as the same as in quercetin [23]. So, we evaluated their photoprotective potencies of them and additional polyphenols for UV-irradiated NPX in the solid state. NPX and each polyphenol or antioxidant were mixed at a molar ratio of NPX 0.5 because this dose of quercetin completely suppressed NPX photodegradation. A comparison of the residual amounts of NPX after UV irradiation for 24 h in the absence or presence of selected additives is shown in Figure 5. Four additives, including ascorbic acid, zingerone, catechin, and isoascorbic acid, showed no protective effects on NPX photodegradation judged by the residual amounts of NPX in their presence. These values were almost the same as in that in the absence of additives. In the presence of these four additives, the residual amounts of NPX after UV irradiation for 24 h were significantly less than those of control samples. On the other hand, curcumin and resveratrol suppressed NPX photodegradation completely, the same as in quercetin. When quercetin, curcumin, and resveratrol were mixed with NPX, the residual amounts of NPX after UV irradiation (97.9 ± 0.7%, 98.9 ± 0.6% and 99.4 ± 1.1%, respectively) were significantly higher compared to that without additives (84.1 ± 3.8%), and comparable to those of their control samples (100.0 ± 0.3%, 99.2 ± 0.7% and 100.6 ± 1.2%, respectively). Other additives, including rutin, daidzein, chlorogenic acid, and genistein, showed photoprotective effects moderately on NPX photodegradation. The residual amounts of NPX after UV irradiation in the presence of rutin, daidzein, chlorogenic acid, and genistein were 91.5 ± 3.0%, 92.7 ± 1.1%, 94.1 ± 0.7%, and 97.3 ± 0.8%, respectively, which were significantly higher compared to that without additives, but these values were significantly less than those of their control samples (100.0 ± 0.4%, 99.0 ± 1.1%, 98.5 ± 0.7% and 100.7 ± 1.1%, respectively). These results indicated that some of the polyphenols tested in this study showed significant photoprotective effects on NPX photodegradation in the solid state. Quercetin, curcumin, and resveratrol showed remarkable protective potencies for NPX photodegradation.

In addition, quercetin and resveratrol showed high photostability because their residual amounts after UV irradiation were 98.0 ± 0.7% and 101.4 ± 1.3%, respectively (Figure S2), which were the same values as in those of control samples (98.5 ± 0.6% and 100.9 ± 1.1%, respectively). Daidzein, chlorogenic acid, and genistein were also photostable, but some polyphenols, including zingerone, catechin, rutin, and curcumin, were photodegradable. Several studies have been reported for the photodegdabilitiy of catechin and curcumin [26,27]. UV irradiation-induced the photodegradation of these polyphenols in addition to NPX. Furthermore, the residual amounts of curcumin, ascorbic acid, and isoascorbic acid in their control samples were 82.8 ± 1.2%, 91.8 ± 1.7%, and 93.2 ± 2.8%, respectively, indicating that these additives were degraded by the factors except for UV irradiation (probably oxidation, hygroscopicity and photodegradation by room-light irradiation during sample preparation, and so on). Based on results from the evaluations of both

NPX photostabilization and photostability of tested additives, it is tempting to speculate that quercetin and resveratrol might be good photostabilizers for powders and granules of photodegradable pharmaceuticals.

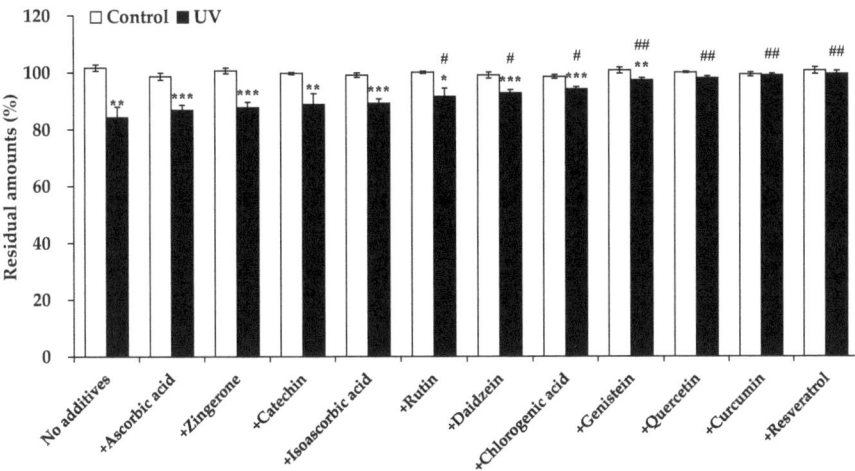

Figure 5. Photoprotective effects of selected additives on NPX photodegradation in a powder. Values represent mean ± S.D. (n = 4). * Difference compared with control ($p < 0.05$), **difference compared with control ($p < 0.01$), *** difference compared with control ($p < 0.001$), # difference compared with no additives ($p < 0.05$) and ## difference compared with no additives ($p < 0.01$).

3.3. Mechanism Elucidation of the Photoprotective Effects of Selected Additives

To clarify the mechanisms of the highly protective effects of quercetin, curcumin, and resveratrol on NPX photodegradation in the solid state, their antioxidative potencies were determined by the PAO test, as shown in Figure 6. From the results of the PAO test, there was no correlation between photoprotective effects and antioxidative potencies among selected polyphenols and antioxidants. Quercetin and resveratrol showed more significant antioxidative activities compared to the other tested compounds except for catechin and rutin. On the other hand, the antioxidative activity of curcumin was the same as in those of ascorbic acid and isoascorbic acid although curcumin showed a more photoprotective effect on NPX photodegradation compared to ascorbic acid and isoascorbic acid (Figure 5). In the case of catechin, its antioxidative activity was comparable to quercetin but their photoprotective effects on NPX were different (Figure 5). These results make it possible to confirm that an antioxidative potency has no contribution to the suppression of NPX photodegradation induced by UV irradiation in the solid state, and quercetin and resveratrol show highly protective effects by other natures. Interestingly, catechin showed a significant protective effect on NPX photodegradation in an aqueous media as shown in our previous study [23]. It is suggested that the status of photodegradable pharmaceuticals is an important factor to improve their photodegradability through the addition of photostabilizers.

Next, the UV absorption spectra of selected polyphenols and antioxidants were recorded to determine their absorption-maximum wavelengths (λ_{max}, nm) and molar absorption coefficients (ε, L mol^{-1} cm^{-1}) as an indicator of UV filtering potency. NPX showed characteristic absorption in the wavelength above 260 nm although ascorbic acid and isoascorbic acid had no absorption as shown in Table 1. Photoprotective polyphenols for NPX showed more significant UV absorption compared to NPX due to their higher ε values. Especially, quercetin, curcumin, and resveratrol had bigger ε values around 260 nm and 450 nm, compared to the other tested compounds and NPX. Their λ_{max} and ε were as follows; quercetin 370 nm (23,971 L mol^{-1} cm^{-1}), curcumin 424 nm and 263 nm

(72,442 L mol^{-1} cm^{-1} and 16,933 L mol^{-1} cm^{-1}, respectively) and resveratrol 308 nm (34,140 L mol^{-1} cm^{-1}). It is suggested that these three polyphenols might act as an efficient UV filter, which suppresses the excitation of NPX by disruption of UV irradiation, but poor photostabilizers such as ascorbic acid and isoascorbic acid might have no protective effects on account of their low UV filtering activities. It is proved that catechin had no photoprotective effect on NPX photodegradation due to showing insufficient UV absorption in the longer wavelength otherwise its higher antioxidative potencies.

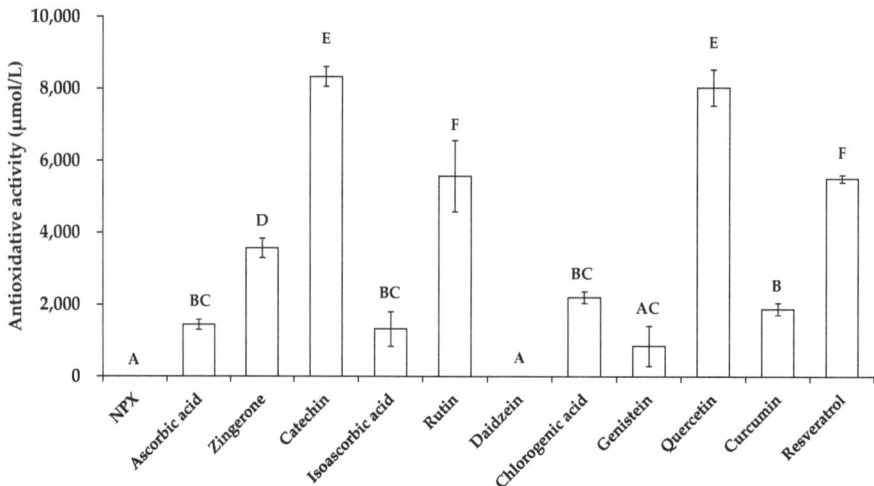

Figure 6. Antioxidative activities of NPX and selected additives. Values represent mean ± S.D. (n = 3). $^{A-F}$ Means without common superscripts are significantly different ($p < 0.05$).

Table 1. Absorption-maximum wavelengths (λ_{max}, nm) and molar absorbance coefficients (ε, L mol^{-1} cm^{-1}) of NPX and tested additives were analyzed using UV/Vis spectrophotometer.

	λ_{max} (nm) and ε (L mol^{-1} cm^{-1}) in the Wavelength above 260 nm
NPX	332 nm (1853), 318 nm (1481), 272 nm (5178), 263 nm (5144)
Ascorbic acid	-
Zingerone	282 nm (2984)
Catechin	280 nm (4114)
Isoascorbic acid	-
Rutin	360 nm (17,607)
Daidzein	303 nm (10,279)
Chlorogenic acid	329 nm (20,032), 301 nm (15,213)
Genistein	262 nm (44,251)
Quercetin	370 nm (23,971)
Curcumin	424 nm (72,442), 263 nm (16,933)
Resveratrol	308 nm (34,140)

From these results, it is clarified that a higher UV filtering activity is necessary for a better photostabilizer to photodegradable pharmaceuticals in the solid state. The summary of obtained results from photostabilization evaluation, determination of antioxidative potencies, and UV spectral analysis are shown in Table S1. Some of the selected additives suppressed NPX photodegradation due to their higher UV filtering activities, not antioxidative potencies. Quercetin, curcumin, and resveratrol completely suppressed NPX photodegradation on account of their remarkable UV absorption in the longer wavelength. Especially, quercetin and resveratrol disrupt the UV absorption of a compound, and they are photostable, suggesting that these polyphenols might suppress the photodegradation and the generation of photoproducts for various photodegradable pharmaceuticals in the solid state, in addition to NPX. The photoprotective effect of curcumin was remarkable, but it might be weakened depending on photo-exposure time due to its lower stability. Several reports show the expression of toxicological potencies of photosensitive pharmaceuticals derived from the generation of photoproducts by UV irradiation [28–31], and the NPX photoproduct is known as toxicological potent in ecotoxicological tests [8,32]. In the case of NPX, it is proved indirectly that most selected polyphenols suppress the generation of NPX photoproduct partially or completely resulting in the reduction of ecotoxicological potencies induced by UV irradiation.

4. Conclusions

This work revealed that it is important for additives for the purpose of the photostabilization of NPX in the solid state to have sufficient UV filtering activity. In the absence of a UV absorptive additive, NPX in the solid state is excited by UV irradiation following to proceeding the photodegradation (Figure 7A). In contrast, in the presence of some UV absorptive substances such as quercetin, UV irradiation for NPX is disrupted resulting in the suppression of its photodegradation (Figure 7B). The addition of UV absorptive additives including selected polyphenols might be an efficient tool for the improvement of the photostability of photodegradable pharmaceuticals in the solid state. It might be a useful method to make a photostabilization for crushed or decapsulated medicines, of which APIs are photosensitive, in clinical situations. Additional research focused on the photoprotective effects of other additives is required for the development of a photostabilization strategy.

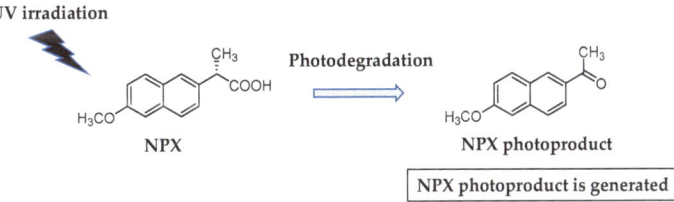

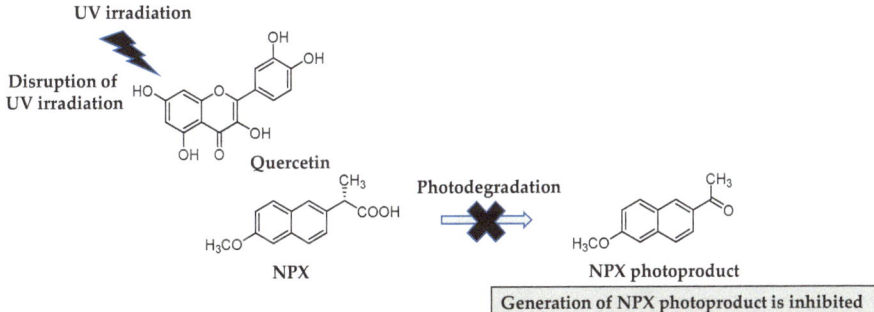

Figure 7. NPX is photodegraded by UV irradiation in the absence of UV absorptive substances (**A**) but they inhibit NPX photodegradation in a UV-irradiated powder by disruption of UV irradiation (**B**).

Supplementary Materials: The following supporting information can be downloaded at: https://www.mdpi.com/article/10.3390/photochem2040056/s1, Table S1: Summary of obtained results from photostabilization evaluation, determination of antioxidative potencies and UV spectral analysis; Figure S1: The scheme of experimental procedure of the evaluation of photostabilization potencies of selected additives for NPX photodegradation; Figure S2: Residual amounts of selected additives in the UV-irradiated powder in the presence of NPX. Values represent mean ± S.D. (n = 4). * Difference compared with control ($p < 0.05$) and *** difference compared with control ($p < 0.001$).

Author Contributions: Conceptualization, K.K. and H.N.; Methodology, K.K. and A.M.; Formal Analysis, K.K.; Investigation, K.K.; Resources, H.N.; Data Curation, K.K.; Writing—Original Draft Preparation, K.K.; Writing—Review & Editing, H.N.; Supervision, H.N.; Project Administration, K.K. and H.N.; Funding Acquisition, K.K. All authors have read and agreed to the published version of the manuscript.

Funding: This study was supported by JSPS KAKENHI Grant Number JP 20K15980.

Conflicts of Interest: The authors declare no conflict of interest.

References

1. Zhang, N.; Li, J.M.; Liu, G.G.; Chen, X.L.; Jiang, K. Photodegradation of diclofenac in aqueous solution by simulated sunlight irradiation: Kinetics, thermodynamics and pathways. *Water Sci. Technol.* **2017**, *75*, 2163–2170. [CrossRef] [PubMed]
2. Li, J.; Ma, L.Y.; Li, L.S.; Xu, L. Photodegradation kinetics, transformation, and toxicity prediction of ketoprofen, carprofen, and diclofenac acid in aqueous solutions. *Environ. Toxicol. Chem.* **2017**, *36*, 3232–3239. [CrossRef] [PubMed]
3. Temussi, F.; Cermola, F.; Dellagreca, M.; Iesce, M.R.; Passananti, M.; Previtera, L.; Zarrelli, A. Determination of photostability and photodegradation products of indomethacin in aqueous media. *J. Pharm. Biomed. Anal.* **2011**, *56*, 678–683. [CrossRef] [PubMed]
4. Iovino, P.; Chianese, S.; Canzano, S.; Prisciandaro, M.; Musmarra, D. Ibuprofen photodegradation in aqueous solutions. *Environ. Sci. Pollut. Res. Int.* **2016**, *23*, 22993–23004. [CrossRef]
5. Kawabata, K.; Sugihara, K.; Sanoh, S.; Kitamura, S.; Ohta, S. Photodegradation of pharmaceuticals in the aquatic environment by sunlight and UV-A, -B and -C irradiation. *J. Toxicol. Sci.* **2013**, *38*, 215–223. [CrossRef]
6. Kawabata, K.; Akimoto, S.; Nishi, H. Cis-trans isomerization reaction of sulindac induced by UV irradiation in the aqueous media. *Chromatography* **2018**, *39*, 139–146. [CrossRef]
7. Moore, D.E.; Chappuis, P.P. A comparative study of the photochemistry of the non-steroidal anti-inflammatory drugs, naproxen, benoxaprofen and indomethacin. *Photochem. Photobiol.* **1988**, *47*, 173–180. [CrossRef]
8. Ishidori, M.; Lavorgna, M.; Nardelli, A.; Parrella, A.; Previtera, L.; Rubino, M. Ecotoxicity of naproxen and its phototransformation products. *Sci. Total Environ.* **2005**, *348*, 93–101. [CrossRef]
9. Tu, N.; Liu, Y.; Li, R.; Lv, W.; Liu, G.; Ma, D. Experimental and theoretical investigation on photodegradation mechanisms of naproxen and its photoproducts. *Chemosphere* **2019**, *227*, 142–150. [CrossRef]
10. Liang, R.; Sun, S.S.; Huang, G.; Li, M.D. Unveiling the photophysical and photochemical reaction process of naproxen via ultrafast femtosecond to nanosecond laser flash photolysis. *Chem. Res. Toxicol.* **2019**, *32*, 613–620. [CrossRef]
11. Kawabata, K.; Mizuta, Y.; Ishihara, K.; Takato, O.; Oshima, S.; Akimoto, S.; Inagaki, M.; Nishi, H. Structure determination of naproxen photoproducts in the tablet generated by the UV irradiation. *Chromatography* **2019**, *40*, 157–162. [CrossRef]
12. Asahi, M.; Matsubara, R.; Kawahara, M.; Ishida, T.; Emoto, C.; Suzuki, N.; Kataoka, O.; Mukai, C.; Hanaoka, M.; Ishizaki, J.; et al. Causative agent of vascular pain among photodegradation products of dacarbazine. *J. Pharm. Pharmacol.* **2002**, *54*, 1117–1122. [CrossRef] [PubMed]
13. Klementová, Š.; Poncarová, M.; Kahoun, D.; Šorf, M.; Dokoupilová, E.; Fojtíková, P. Toxicity assessment of verapamil and its photodegradation products. *Environ. Sci. Pollut. Res. Int.* **2020**, *27*, 35650–35660. [CrossRef] [PubMed]
14. Jiang, F.; Wu, W.; Zhu, Z.; Zhu, S.; Wang, H.; Zhang, L.; Fan, Z.; Chen, Y. Structure identification and toxicity evaluation of one newly-discovered dechlorinated photoproducts of chlorpyrifos. *Chemosphere* **2022**, *301*, 134822. [CrossRef]
15. Klementová, Š.; Poncarová, M.; Langhansová, H.; Lieskovská, J.; Kahoun, D.; Fojtíková, P. Photodegradation of fluoroquinolones in aqueous solution under light conditions relevant to surface waters, toxicity assessment of photoproduct mixtures. *Environ. Sci. Pollut. Res. Int.* **2022**, *29*, 13941–13962. [CrossRef]
16. Lassalle, Y.; Nicol, É.; Genty, C.; Bourcier, S.; Bouchonnet, S. Structural elucidation and estimation of the acute toxicity of the major UV-visible photoproduct of fludioxonil–Detection in both skin and flesh samples of grape. *J. Mass Spectrom.* **2015**, *50*, 864–869. [CrossRef]
17. Nataraj, B.; Maharajan, K.; Hemalatha, D.; Rangasamy, B.; Arul, N.; Ramesh, M. Comparative toxicity of UV-filter Octyl methoxycinnamate and its photoproducts on zebrafish development. *Sci. Total Environ.* **2020**, *718*, 134546. [CrossRef]
18. Coelho, L.; Almeida, I.F.; Sousa Lobo, J.M.; Sousa E Silva, J.P. Photostabilization strategies of photosensitive drugs. *Int. J. Pharm.* **2018**, *41*, 19–25. [CrossRef]
19. Cagno, M.P.D. The potential of cyclodextrins as novel active pharmaceutical ingredients: A short overview. *Molecules* **2017**, *22*, 1. [CrossRef]

20. Iacovino, R.; Caso, J.V.; Di Donato, C.; Malgieri, G.; Palmieri, M.; Russo, L.; Isernia, C. Cyclodextrins as complexing agents: Preparation and applications. *Curr. Org. Chem.* **2017**, *21*, 162–176. [CrossRef]
21. Ioele, G.; Tavano, L.; De Luca, M. Photostability and ex-vivo permeation studies on diclofenac in topical niosomal formulations. *Int. J. Pharm.* **2015**, *494*, 490–497. [CrossRef]
22. León, C.; Henríquez, C.; López, N.; Sanchez, G.; Pastén, B.; Baeza, P.; Ojeda, J. Inhibitory effect of the Ascorbic Acid on photodegradation of pharmaceuticals compounds exposed to UV-B radiation. *J. Photochem. Photobiol.* **2021**, *7*, 100035. [CrossRef]
23. Kawabata, K.; Takato, A.; Oshima, S.; Akimoto, S.; Inagaki, M.; Nishi, H. Protective effect of selected antioxidants on naproxen photodegradation in aqueous media. *Antioxidants* **2019**, *8*, 424. [CrossRef] [PubMed]
24. Savian, A.L.; Rodrigues, D.; Weber, J. Dithranol-loaded lipid-core nanocapsules improve the photostability and reduce the in vitro irritation potential of this drug. *Mater. Sci. Eng. C Mater. Biol. Appl.* **2015**, *46*, 69–76. [CrossRef] [PubMed]
25. Detoni, C.B.; Souto, G.D.; da Silva, A.L.M. Photostability and skin penetration of different E-resveratrol-loaded supramolecular structures. *Photochem. Photobiol.* **2012**, *88*, 913–921. [CrossRef] [PubMed]
26. Shi, M.; Nie, Y.; Zheng, X.Q.; Lu, J.L.; Liang, Y.R.; Ye, J.H. Ultraviolet B (UVB) photosensitivities of tea catechins and relevant chemical conversions. *Molecules* **2016**, *21*, E1345. [CrossRef]
27. Sakamoto-Sasaki, S.; Sato, K.; Abe, M.; Sugimoto, N.; Maitani, T. Components of turmeric oleoresin preparations and photostability of curcumin. *Jpn. J. Food Chem.* **1998**, *5*, 57–63.
28. Dabić, D.; Hanževački, M.; Škorić, I.; Žegura, B.; Ivanković, K.; Biošić, M.; Tolić, K.; Babić, S. Photodegradation, toxicity and density functional theory study of pharmaceutical metoclopramide and its photoproducts. *Sci. Total Environ.* **2022**, *807*, 150694. [CrossRef]
29. Siciliano, A.; Guida, M.; Iesce, M.R.; Libralato, G.; Temussi, F.; Galdiero, E.; Carraturo, F.; Cermola, F.; DellaGreca, M. Ecotoxicity and photodegradation of Montelukast (a drug to treat asthma) in water. *Environ Res.* **2021**, *202*, 111680. [CrossRef]
30. DellaGreca, M.; Fiorentino, A.; Isidori, M.; Lavorgna, M.; Previtera, L.; Rubino, M.; Temussi, F. Toxicity of prednisolone, dexamethasone and their photochemical derivatives on aquatic organisms. *Chemosphere* **2004**, *54*, 629–637. [CrossRef]
31. Isidori, M.; Nardelli, A.; Pascarella, L.; Rubino, M.; Parrella, A. Toxic and genotoxic impact of fibrates and their photoproducts on non-target organisms. *Environ. Int.* **2007**, *33*, 635–641. [CrossRef] [PubMed]
32. DellaGreca, M.; Brigante, M.; Isidori, M.; Nardelli, A.; Previtera, L.; Rubino, M.; Temussi, F. Phototransformation and ecotoxicity of the drug Naproxen-Na. *Environ. Chem. Lett.* **2003**, *1*, 237–241. [CrossRef]

Article

Effect of Benzophenone Type UV Filters on Photodegradation of Co-existing Sulfamethoxazole in Water

Dilini Kodikara, Zhongyu Guo and Chihiro Yoshimura *

Department of Civil and Environmental Engineering, Tokyo Institute of Technology, Tokyo 152-8552, Japan
* Correspondence: yoshimura.c.aa@m.titech.ac.jp

Abstract: Benzophenones (BPs) frequently occur in water environments, and they are able to both screen UV light and to sensitize reactive intermediate (RI) production. However, BPs have largely been overlooked as a background water component when studying photodegradation of co-existing organic micropollutants (OMPs). Therefore, in this study, we investigated the influence of BP and its derivative oxybenzone (BP3) on the degradation of the co-existing model OMP sulfamethoxazole (SMX). A series of photodegradation experiments were conducted covering a range of BPs concentrations in µg/L levels, and the degradation of 1.00 µM of SMX was studied. The addition of BP at 0.10 µM, 0.25 µM, and 0.30 µM, and BP3 at 0.10 µM and 0.25 µM, significantly increased the first order degradation rate constant of 1.00 µM of SMX ($k_{obs(BP)}$) by 36.2%, 50.0%, 7.3%, 31.5%, and 36.2% respectively, compared to that in the absence of any BPs. The maximum indirect photodegradation induced by BP and BP3 reached 33.8% and 27.7%, respectively, as a percentage of the observed SMX degradation rate at the [BPs]/[SMX] ratio of 0.25. In general, triplet excited dissolved organic matter (^{3}SMX*, ^{3}BP*, and ^{3}BP3*) played the major role in the photosensitizing ability of BPs. The results further implied that the increase of SMX degradation at the molar ratio of 0.25 was possibly due to ^{3}BP* for the mixture of SMX and BP. Overall, this study revealed the sensitizing ability of BP and BP3 on the co-existing OMP, SMX, in water for the first time. Our findings can be applied to other BP type UV filters which are similar to BP and PB3 in molecular structure.

Keywords: indirect photolysis; reactive intermediates; photodegradation; benzophenone; sulfamethoxazole

Citation: Kodikara, D.; Guo, Z.; Yoshimura, C. Effect of Benzophenone Type UV Filters on Photodegradation of Co-existing Sulfamethoxazole in Water. *Photochem* **2023**, *3*, 288–300. https://doi.org/10.3390/photochem3020017

Academic Editors: Marcelo Guzman, Ioannis Konstantinou, Vincenzo Vaiano and Rui Fausto

Received: 12 January 2023
Revised: 28 April 2023
Accepted: 8 May 2023
Published: 1 June 2023

Copyright: © 2023 by the authors. Licensee MDPI, Basel, Switzerland. This article is an open access article distributed under the terms and conditions of the Creative Commons Attribution (CC BY) license (https://creativecommons.org/licenses/by/4.0/).

1. Introduction

Organic micropollutants (OMPs) are being increasingly detected in the environment, raising public as well as scientific concern [1]. They undergo physical (e.g., sorption, dilution), biological (e.g., biodegradation), and chemical (e.g., hydrolysis, photolysis) transformations in the environmental water [2,3]. Some OMPs are degraded through either biodegradation or photodegradation, while some degrade through both and some by neither [3]. For many OMPs, the photodegradation rate under sunlit conditions is known to be larger than their biodegradation rate [3]. In environmental water, OMP photodegradation rate is influenced by co-existing components either positively due to the generation of reactive intermediates (RIs) that promote indirect photolysis or negatively due to scavenging of RIs and the attenuation of light [4]. For instance, NO_3^- [5–11], Cl^- [10,11] and dissolved organic matter (DOM) [6–10] in the water column possibly contribute to generating RIs. In addition, some OMPs have also exhibited promoting effects on the photodegradation of other co-existing OMPs [12–14]. For instance, triplet excited aromatic ketones are able to react well with other aromatic compounds which contain electron-donating groups [15], such as phenolate ions [16] and phenols [17].

Among the OMPs detected in the water environment are UV filters/sunscreen products such as benzophenone derivatives, *p*-Aminobenzoic acid derivatives, camphor derivatives, benzotriazole derivatives, and salicylate derivatives [18]. Sunscreen products are widely used for the protection of skin from carcinogenic effects that may result from the exposure to UV radiation. Among them, benzophenone (BP) derivatives are actively used components of organic sunscreens [16–19]. Besides, BP is used during the production of insecticides, agrochemicals, and pharmaceuticals as it serves as photo-initiator, UV curing agent, fragrance enhancer, and a flavoring agent. It is also used as an additive in plastics, coating products, and adhesives [19]. As a result of such extensive use, BP and its derivatives have been widely detected in surface water and wastewater treatment plants [20–24] among other organic micropollutants (OMPs).

Upon reception of photon energy in aqueous media, BP produces $^1O_2$, $O_2^{\bullet-}$ and $^{\bullet}OH$ through type I and II photodynamic reactions in a concentration dependent manner [25]. Furthermore, BP can abstract H from coexisting molecules through its excited triplet state [12]. In addition, fenofibrate, amiodarone [12], and fenofibric acid [13,14], which contain the carbonyl moiety as in BP and its derivatives, have shown photosensitization capabilities, sometimes exceeding the sensitizing effect of natural organic matter [13,14]. For instance, fenofibric acid, a chlorobenzophenone, accelerates the photodegradation process of the gemfibrozil fibrate drug under UV radiation via fenofibric acid's triplet excited state and $^1O_2$ [13]. Fenofibric acid also increases the degradation rate of coexisting bezafibrate via (1) triplet excited fenofibric acid and $^1O_2$, (2) the energy transfer of singlet excited fenofibric acid or degradation byproducts with dissolved oxygen, (3) e_{aq}^-, and (4) $O_2^{\bullet-}$ to a smaller extent, under solar irradiation [14].

However, despite their occurrence in environmental waters in a range of concentrations, the influence of BP itself and its sun-screening derivatives such as oxybenzone (BP3) has been largely overlooked as possible photosensitizers. For instance, BP has been acknowledged as one of the aromatic ketones, whose excited triplet state can oxidize other compounds such as phenylurea herbicides [15], phenols [17,26], phenolate ions [16], methoxybenzenes [27], amino acids and aminopolycarboxylic acids [28], and amines [29,30]. However, those studies have particularly focused on only BP's triplet excited state often through laser flash photolysis and have not investigated its influence in wide ranges of their concentrations. To be specific, considering BP's simultaneous ability to screen UV irradiation and sensitize RI production, their concentration dependent influence on the photodegradation of co-existing OMPs is yet to be understood in environmentally relevant concentrations. BPs typically have been detected in the environmental waters in ng/L to µg/L levels [31–40]. For instance, BP and BP3 have been detected in surface water up to 82 ng/L [40,41] and 44,000 ng/L [31–43], respectively. Nevertheless, studies focusing on the sensitizing ability of BP or BP-like OMPs have employed degradation-target pollutants in mg/L level (i.e., 1–10 mg/L [13,14] and sensitizing OMPs also in mg/L level (i.e., 0.5–20.0 mg/L [13,14,25]). Considering that RI production by BP is concentration dependent [25], research covering a wide range of concentrations in ng/L or µg/L is required to understand the sensitizing effect of BPs on co-existing OMPs in the environmental water.

Therefore, the objective of this research was to investigate the promotional/inhibitory influence of BP and its derivative BP3 on the degradation of co-existing OMPs at a range of concentrations in µg/L level. It should be noted that this study did not include elucidating the reaction pathway of SMX degradation in the presence of BPs. To this end, we selected sulfamethoxazole (SMX) as a model target micropollutant and investigated processes responsible for the influence of BP and BP3 on photodegradation of SMX in a range of BP and BP3 concentrations. The selection of SMX as the model OMP was based on its frequent detection in surface water worldwide [44–49], its high dependence on photodegradation compared to other processes such as biodegradation [3], and its capability to undergo indirect photodegradation via $^{\bullet}OH$, $^1O_2$ and $O_2^{\bullet-}$ [50–52]. Series of photodegradation experiments under simulated sunlight were conducted in combination with RI quenching experiments to understand dominant physical and chemical processes for the co-exposure degradation of SMX with BPs.

2. Materials and Methods

2.1. Material

SMX (≥98%), BP (analytical grade), BP3 (≥98%), methanol (MeOH) (≥99.9%, high-performance liquid chromatography (HPLC) grade), sorbic alcohol (SA) (97%) were purchased from Sigma-Aldrich (Tokyo, Japan). HPLC grade acetonitrile (ACN) (≥99.9%), p-nitroanisole (PNA, 97%), 2-propanol (IPA, ≥99.7% for gas chromatography (GC)), pyridine (PYR, ≥99.5% for GC), acetic acid (≥99.7%) and superoxide dismutase (SOD, 5550 U/mg) were purchased form Kanto Chemicals (Tokyo, Japan). Phosphate buffer powder (1/15 mol/L, pH 7.0) and NaN$_3$ (99%) were purchased from Fujifilm Wako Pure Chemical corporation (Osaka, Japan) and Arcos Organics (Japan) respectively.

2.2. Chemical Actinometry Experiment

Photodegradation experiments were conducted using a photoreactor (HELIOS.Xe, Koike Precision Instruments, Hyogo, Japan), which is equipped with a 300 W Xe short arc lamp (Ushio, Tokyo, Japan), an optical filter between the lamp and samples to cutoff wavelengths less than 280 nm, and a temperature regulator which maintained the experimental system at 20 °C. In this reactor 10 quartz tubes placed around the lamp were irradiated simultaneously. The distance between the lamp and the inner-most surface of each quartz tube was 6 cm.

To determine photon irradiance ($E^0_{p,tot}$) and spectral photon irradiance ($E^0_{p,\lambda}$) for the wavelength range 280–400 nm at each tube position of the photoreactor, the chemical actinometry experiment was carried out using the chemical actinometer system containing PNA and PYR at initial concentrations of 10.00 μM ([PNA]$_0$) and 10.00 mM ([PYR]$_0$), respectively [53]. In this experiment PYR was added to accelerate the photoreaction of PNA. This experiment was conducted in triplicate with 10 mL of the actinometer in quartz tubes. After initiating radiation, samples were collected at 10-min intervals, up to 50 min, and the PNA concentrations were determined using an HPLC system (Prominence UFLC, Shimadzu, Kyoto, Japan) equipped with a UV-Vis absorbance detector (SPD-20 UFLC, Shimadzu, Kyoto, Japan) and a C18 column (Kinetex, Phenomenex Co., Torrance, CA, USA; 5 μm, 4.6 × 250 mm) using a mobile phase of 60/40 MeOH/Milli-Q(MQ) (v/v %). After the measurement, the initial degradation rate constant of PNA for each tube position was determined and applied for calculating the irradiance at each of specific locations of the quartz tubes using the equations in Table S1.

2.3. Photodegradation Experiments

Photodegradation experiments were carried out at 20 °C using 10.00 mL solutions of SMX alone (1.00 μM) and with BP or BP3 (hereafter BPs) mixtures in phosphate buffer at pH 7 using the above-mentioned photoreactor. Samples were continuously stirred during the 10 h of each experiment. The molar ratio [BP or BP3]/[SMX] was varied in the range 0.00–1.00 by fixing [SMX] at 1.00 μM to investigate the influence of the proportion of BPs on SMX degradation. In addition, quenching experiments of RIs were conducted for SMX/BPs mixtures in the presence of IPA (2000 μM), NaN$_3$ (500 μM) SA (2500 μM) and SOD (3000 U/L) for quenching •OH, ^{1}O$_2$, ^{3}DOM* (i.e., ^{3}SMX*, ^{3}BP*, and ^{3}BP3* in this study) and O$_2^{•-}$ respectively.

Samples of 0.6 mL were collected at 1, 2, 4, 6, 8, and 10 h during photodegradation experiments with and without quenchers. SMX concentration in the collected samples were determined using an HPLC system (Prominence UFLC, Shimadzu, Kyoto, Japan) equipped with a UV-Vis absorbance detector (SPD-20 UFLC, Shimadzu, Kyoto, Japan) and a C18 column (Kinetex, Phenomenex Co., Torrance, CA, USA; 5 μm, 4.6 × 250 mm). A 50/50 ACN/MQ (v/v %) mobile phase containing 0.1% acetic acid was used for determining SMX concentration. The flow rate and column temperature were set at 0.60 mL/min and 40 °C, respectively. The average relative error in the HPLC measurement was 4%.

2.4. Data Analysis

The experimental results were analyzed to quantify the inner filter effect and the sensitizing effect of BPs on the photodegradation of SMX. Wavelength dependent light screening factor caused by BPs (s_λ) is given by Equation (1) [54,55] where α_λ is the light attenuation coefficient of BP (/cm), ε'_λ is the molar absorption coefficient of SMX (L/cm. mol), l is the light path length (cm) calculated according to Equation (2) using the internal radius of the cylindrical quartz tube r (r = 0.8 cm) and [SMX$_0$] is the initial SMX concentration (1.00 × 10^{-6} M). Equation (1) can be modified into Equation (3) by substituting α_λ with 2.303 ε_λ[BP$_0$] where ε_λ is the molar absorption coefficients of BPs (L/cm. mol) and [BP$_0$] is the initial BPs concentration (0.10 × 10^{-6} M–1.00 × 10^{-6} M) [56]. Then, s_λ and the total light screening coefficient (S) induced by each concentration of BPs were calculated using Equations (3) and (4), respectively [54–56].

$$s_\lambda = \frac{1 - (10^{-(\alpha_\lambda + \varepsilon'_\lambda[\text{SMX}_0])l})}{2.303 \, (\alpha_\lambda + \varepsilon'_\lambda[\text{SMX}_0])l} \tag{1}$$

$$l = \frac{\pi r^2}{2r} \tag{2}$$

$$s_\lambda = \frac{1 - (10^{-(2.303\,\varepsilon_\lambda[\text{BP}_0] + \varepsilon'_\lambda[\text{SMX}_0])l})}{2.303 \, (2.303\,\varepsilon_\lambda[\text{BP}_0] + \varepsilon'_\lambda[\text{SMX}_0])l} \tag{3}$$

$$S = \frac{\sum I_\lambda s_\lambda \varepsilon'_\lambda}{\sum I_\lambda \varepsilon'_\lambda} \tag{4}$$

where I_λ is the lamp irradiance at a specific wavelength λ (280–400 nm).

Since the absorbance of tested OMP mixtures was minor in the visible light region (400–800 nm), it was excluded from the analysis. UV-visible absorption spectra of BPs and SMX individually at each concentration used for experiments were obtained using a UV-vis spectrophotometer (UV1800, Shimadzu, Kyoto, Japan). Subsequently, the obtained absorbance spectra were used in calculating ε'_λ using the Beer-Lamber law (Equation (5)),

$$A_\lambda = \varepsilon'_\lambda \, L \, [SMX_0] \tag{5}$$

where A_λ is the measured absorbance at a given wavelength λ and L is the optical path length (1 cm). Then, the calculated S value was used to obtain the direct and self-sensitized photolysis rate constant of SMX (k_{d+s}, Equation (6)) and indirect photolysis rate constant initiated by BPs (k_{ind}) was estimated using Equation (7).

$$k_{d+s} = S \, k_{obs(non\,BPs)} \tag{6}$$

$$k_{obs(BPs)} = k_{d+s} + k_{ind} \tag{7}$$

where $k_{obs(non\,BPs)}$ is the observed photolysis rate constants of SMX in the absence of BP (h^{-1}) and $k_{obs(BPs)}$ is the observed photolysis rate constants of SMX in the presence of BP (h^{-1}). In the presence of BPs, $k_{obs(non\,BPs)}$ is reduced to k_{d+s} by the inner filter effect/light screening of BPs and this reduction is accounted for by S (Equations (3) and (6)). Hence, the enhanced SMX degradation owing to the sensitization by BPs was estimated by k_{ind} in Equation (7).

3. Results and Discussion

3.1. Effect of BP and BP3 on the Photodegradation of SMX

In the absence and presence of BPs, SMX degradation followed 1st order photodegradation kinetics with $R^2 \geq 0.97$. Figure 1 shows the photodegradation kinetics for some selected [BPs]/[SMX] molar ratios 0.00–0.30.

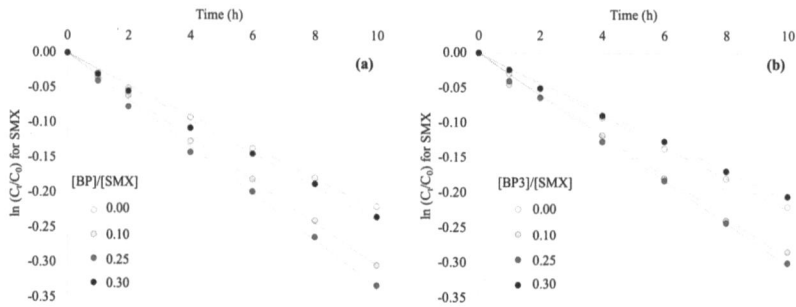

Figure 1. First order photodegradation kinetics for SMX in the absence and presence of (**a**) BP and (**b**) BP3 (for selected [BPs]/[SMX] ratios). Experimental conditions: $[SMX]_0 = 1.00$ μM, pH = 7.0, temperature = 20 °C, degradation time = 10 h.

Consequently, the first order degradation rate constant of SMX was influenced by the coexisting BPs depending on the molar ratio (Figure 2) (hereafter [BP]/[SMX] and [BP3]/[SMX] will be referred to as BP ratio and BP3 ratio respectively) (Figure 2a). At BP ratios of 0.10 and 0.25, an obvious promotion of the SMX degradation by BP was evident. For BP ratios of 0.10 and 0.25, $k_{obs\,(BP)}$ increased by 36.2% and 50.0%, respectively, relative to the negative control (i.e., in the absence of BP) (Figure 2b). At the BP ratio 0.25, the degradation rate of SMX was the maximum at 3.34×10^{-2} h^{-1} for the tested range of [BP]/[SMX].

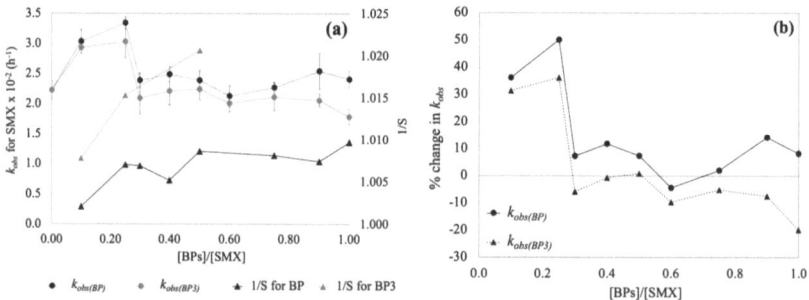

Figure 2. The relation between [BPs]/[SMX] and observed rate constant (k_{obs}). (**a**) k_{obs} for SMX (circles) and the inverse of total light screening coefficient (S) (triangles) at different molar ratios. The inverse of S corresponds to the light absorption. Error bars indicate a standard deviation of triplicated experiments. (**b**) Percentage change in $k_{obs(BPs)}$ for SMX compared to $k_{obs(non\,BPs)}$.

Similarly, for SMX irradiated with BP3, a significant increase in the first order degradation rate constant for SMX was observed for BP3 ratios of 0.10 and 0.25 (Figure 2b). The increase was 31.5% and 36.2% for the ratios of 0.10 and 0.25, respectively. The maximum SMX degradation rate was observed to be 3.03×10^{-2} h^{-1} at the BP3 ratio 0.25. All other BP3 ratios exhibited a decrease of the SMX degradation rate compared to the negative control except for the BP3 ratio 0.50 where the rate remained almost similar to that of SMX irradiated alone. The largest inhibition of SMX degradation was observed at the BP3 ratio of 1.00. Overall, at the lower ratios of 0.10 and 0.25, BP promoted SMX degradation slightly more than BP3 (Figure 2b).

At the BP ratios of 0.10, 0.25, and 0.30, k_{ind} accounted for 26.7%, 33.8%, and 7.4% of the corresponding $k_{obs\,(BP)}$ values, respectively (Figure 3, Table S2). At BP3 ratios of 0.10 and 0.25, k_{ind} accounted for 24.5% and 27.7% of the corresponding $k_{ob\,(BP3)}$ values, respectively. The contribution of BP to enhancing SMX degradation at the lower molar ratios (i.e., 0.10 and 0.25) was slightly higher than those of BP3 (Table S2). It should be noted that the calculation was not carried out for BPs molar ratios that inhibited SMX degradation compared to $k_{obs\ (non\ BP)}$. Furthermore, k_{ind} as a percentage of $k_{obs\ (BPs)}$ in the presence of BP and BP3 increased as BPs ratios increased in the low range (i.e., 0.00–0.25) (Figure 3). This indicates that, within such low range, the sensitizing effect of BP dominantly determined the overall degradation rate of SMX, making the inner filter effect minor. Beyond the BP ratio of 0.25, k_{ind} as a percentage of $k_{obs\ (BP)}$ remained more or less stable approximately around 7.5%, indicating that the sensitizing effect no longer significantly increased with the increase of BP. As for BP3, the ratio of 0.25 showed the maximum $k_{ob\ (BP3)}$ value and all other ratios beyond 0.25 inhibited the SMX degradation. As BP3 shows a higher absorbance than BP in the wavelength range of 280–400 nm (Figure S1), it likely showed a more significant inner filter effect on SMX than BP.

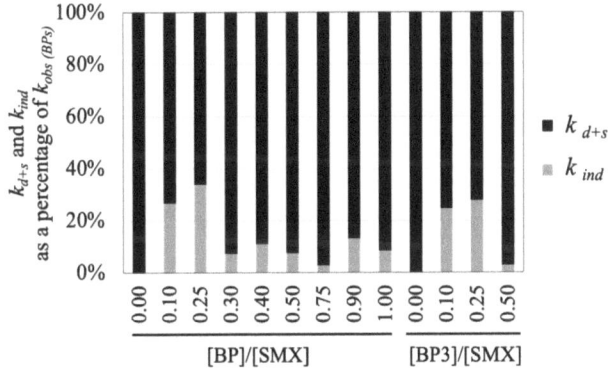

Figure 3. k_{d+s} and k_{ind} as a percentage of $k_{obs\ (BPs)}$ for BP and BP3 molar ratios, indicating the enhanced SMX degradation.

In a relevant study, k_{ind} initiated by 20 mg/L of fulvic acid in the Suwannee River accounted for 11% of k_{obs} of 3.65 µM of 2-phenylbenzimidazole-5-sulfonic acid (between 290–350 nm) [54]. Another study reported that k_{ind} initiated by 5 mg-C/L of fulvic acid and humic acid in the Suwannee River, Pony Lake fulvic acid, and Nordic aquatic fulvic acid, respectively, accounted for 33.0%, 38.6%, 35.7%, and 45.4% of the k_{obs} value of 2,3-dibromopropyl-2,4,6-tribromophenylether (between 290–320 nm) [55]. These experimental results indicate that BP and BP3 can be of similar significance as natural organic matter which are considered an important source of photosensitization for photodegrading co-existing OMPs in the water environment.

Overall, the shift of the SMX photodegradation rate in the presence of BPs was owing to the inner filter effect and the sensitizing effect caused by the BPs. According to the UV visible spectra, the BPs showed higher absorbance than SMX within the wavelength range 280–400 nm (Figure S1). Thus, light screening by BPs significantly affected the degradation rate constant of SMX when the concentration of BPs was relatively high (Figure 2b). According to the UV-visible spectra of BP and BP3 (Figure S1), BP3 has a larger absorbance than BP. This must have resulted in the inhibition of SMX degradation at higher BP3 ratios, in comparison to BP (Figure 2a).

Furthermore, fenofibric acid, which is a pharmaceutical containing the BP moiety, has also been reported to accelerate the degradation of 20.0 µM of gemfibrozil [13] and 13.8 µM of bezafibrate [14]. The acceleration of the degradation of those OMPs by fenofibric acid was in the order of 3.14 µM > 1.57 µM > 0.31 µM of fenofibric acid. This means that the sensitization ability continues to increase when the molar ratios [fenofibric acid]/[gemfibrozil] increases from 0.02 to 0.16 and [fenofibric acid]/[bezafibrate] increases from 0.02 to 0.23. These enhanced effects are identical to our experimental results for BPs and interestingly the molar ratios of OMPs are similar between the present and those studies. At the same time, the concentration of SMX (1.00 µM) in our study was much lower than those in the reported investigations [13,14]. Nevertheless, in the current study, BPs exhibited the ability to screen light, which caused inhibition of SMX degradation at most of the higher BPs molar ratios alongside acting as a sensitizer. Thus, among the limited studies focusing on co-exposure photodegradation of OMPs, our study added new evidence enhancing our understanding on their interactive reactions under sunlit environment.

3.2. Effect of RIs in SMX Photodegradation with BP and BP3

In the absence of any BPs, quenchers SA, SOD, NaN$_3$, and IPA reduced the degradation coefficient of SMX (k_{d+s}) by 45.0%, 15.6%, 9.3%, and 6.4%, respectively. Thus, the contribution of RIs for SMX photodegradation followed the order ^{3}SMX* > $O_2^{\bullet-}$ > $^1O_2$ > $^{\bullet}$OH (Figure 4). This result agreed with the observation by Zhou et al., (2015) [51], where ^{3}SMX* exhibited the most contribution to SMX degradation and $^1O_2$ and $^{\bullet}$OH showed smaller contributions at pH 8, when SMX (the initial concentration, 3.95 µM) was irradiated in UV-visible light including UV-C. According to Zhou et al., (2015) and Lin et al., (2023) [52], SMX is capable of undergoing self-sensitized indirect photodegradation through $^1O_2$, $^{\bullet}$OH, and $O_2^{\bullet-}$ produced in the presence of O_2 and H_2O via ^{3}SMX*. In addition, SMX undergoes direct photolysis through ^{3}SMX* when irradiated [51,52]. Considering that SA quenching caused the largest inhibition of SMX degradation rate, direct photolysis is the more prominent process for the degradation of irradiated SMX than its self-sensitized photodegradation, which was also supported by the largest reduction in $k_{obs\ (non\ BP)}$ in the SA quenching (Figure 4).

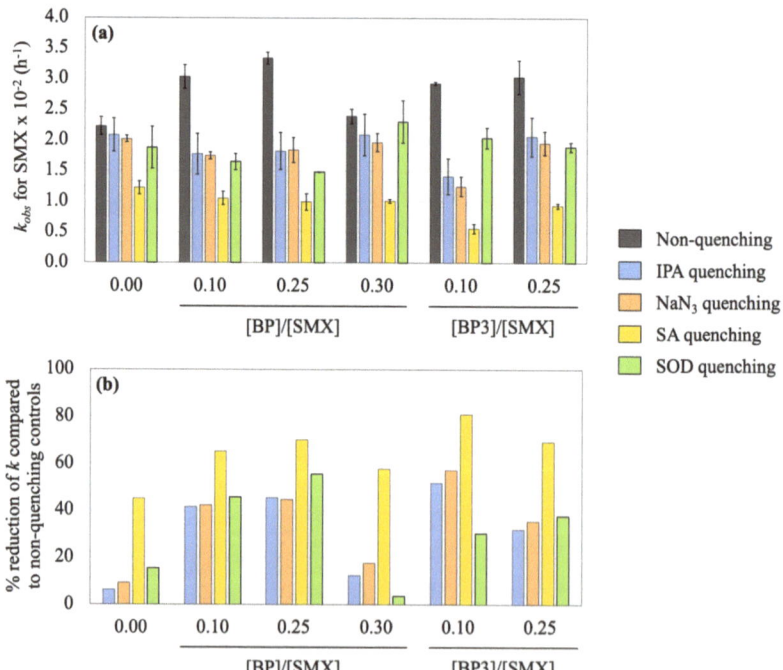

Figure 4. The change of k_{obs} in the quenching experiment. (**a**) First order degradation rate constant of SMX in the absence ($k_{obs(non\ BPs)}$) and presence ($k_{obs(BPs)}$) of BPs with and without quenchers and (**b**) the reduction of $k_{obs(non\ BPs)}$ and $k_{obs(BPs)}$ in the presence of quenchers, as a percentage of those in the absence of quenching.

In case of the presence of BPs, ^{3}SMX*, ^{3}BP*, and ^{3}BP3* quenching by SA caused the largest reduction of degradation rate of SMX for all tested BP and BP3 ratios (Figures 4b and 5). Hence, triplet excited states are the most significant for the degradation of SMX in the presence of BPs. The addition of BP at ratios 0.10 and 0.25 significantly contributed to SMX degradation via $^{\bullet}OH$, $^1O_2$ and $O_2^{\bullet -}$ radicals as shown in IPA, NaN_3, and SOD quenching, respectively. Compared to their respective non-quenching controls, each of $^{\bullet}OH$, $^1O_2$, and $O_2^{\bullet -}$ respectively contributed to 41.5%, 42.3%, and 45.6% of SMX degradation at the BP ratio of 0.10 and to 45.4%, 44.8%, and 55.7% at the BP ratio of 0.25. The largest reduction of k_{obs} was observed at the BP ratio 0.25 for all quenchers. In fact, BP produces $^3BP^*$ upon irradiation, and then $^3BP^*$ can produce $^1O_2$ by energy transfer to O_2, abstract H from C-H bonds of alcohols, and produce $^{\bullet}OH$ in the presence of H_2O [12,57]. Therefore, the experiment of quenching $^3SMX^*$ and $^3BP^*$ reasonably displayed the highest inhibition of SMX degradation. Furthermore, the percentage reduction of $k_{obs(BP)}$ (Figure 4b) has increased when BP ratio increased from 0.00 to 0.25. Hence, the increase of k_{ind} from the BP ratio 0.10 to BP ratio 0.25 (Figure 3) is likely attributed mainly to the production of $^3BP^*$.

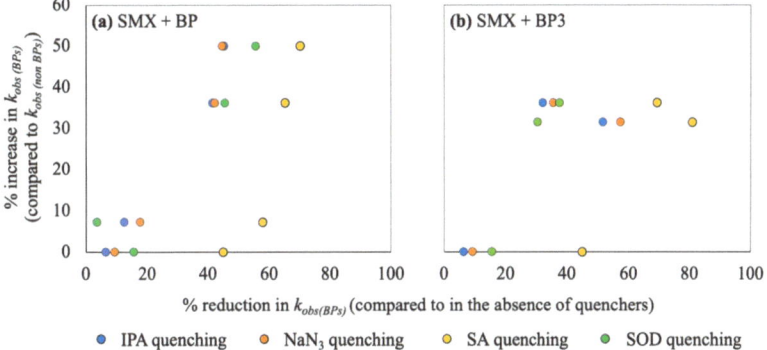

Figure 5. Relations between the percentage reduction in $k_{obs(BPs)}$ compared to in the absence of quenchers and the percentage increase in $k_{obs(BPs)}$ compared to $k_{obs(non\ BPs)}$ and for (**a**) BP ratios 0.00, 0.10, 0.25 and 0.30 and (**b**) BP3 ratios 0.00, 0.10 and 0.25.

Similar to BP, SA quenching experiments in the presence of BP3 in ratios 0.10 and 0.25 caused the largest inhibition of SMX degradation out of the four quenchers. This indicates that BP3 also promotes SMX degradation through pathways involving triplet excited states (Figures 4b and 5). Compared to the non-quenching controls, $^{\bullet}OH$ quenching caused the reduction in SMX degradation rate by 51.7% and 32.1% at BP3 ratios 0.10 and 0.25, respectively. $^1O_2$ quenching at those ratios reduced the SMX degradation by 57.3% and 35.5% and $O_2^{\bullet -}$ quenching by 30.4% and 37.6%. At the same time, the highest contribution by $^{\bullet}OH$, $^1O_2$ and $^3SMX^*/^3BP3^*$ was observed at BP3 ratio 0.10 while that for $O_2^{\bullet -}$ was highest at BP3 ratio 0.25. The contribution of each RI to SMX degradation decreased when the BP3 ratio was increased from 0.10 to 0.25 except for the $O_2^{\bullet -}$. The contribution of $O_2^{\bullet -}$ increased by about 7% when the BP3 ratio increased from 0.10 to 0.25. It should be also noted that the sum of the reductions of $k_{obs(BPs)}$ in the presence of quenchers for some BP or BP3 ratios exceeded 100%, possibly due to simultaneous quenching of non-targeted RIs [58]. Nevertheless, BP3 exhibited a better sensitizing ability for SMX degradation at the tested lower BP3 ratios.

In the degradation process of SMX, $^{\bullet}OH$, $O_2^{\bullet -}$ and $^1O_2$ can attack the isoxazole ring, the benzene ring, or the S-N bond of SMX molecules to form several intermediate products [50,59]. Those reactions go through three main pathways, which are hydrogen abstraction, hydroxylation, and electron transfer. For $^{\bullet}OH$ specifically, hydrogen abstraction and hydroxylation play an important role to initiate SMX degradation. The intermediates produced through reaction of $^{\bullet}OH$ with SMX contain free radicals and thus they can undergo further reactions [50]. In addition, triplet aromatic carbonyls such as $^3BP^*$ are known to react with amines in aqueous media [30]. SMX contains an amine group and thus it is likely to react directly with the triplet states of BP and BP3. Hence, the direct reactions of ^{3}BP* and ^{3}BP3* with SMX in addition to the production of RIs justify the overall sensitizing effect of BPs as well as those triplet states being the most important species for promoting SMX degradation.

Overall, the quenching experiments revealed that the most important RI for sensitizing SMX degradation by BP and BP3 is their triplet excited states. This result is reasonable as SMX produces $^1O_2$, $^{\bullet}OH$, and $O_2^{\bullet-}$ in the presence of O_2 and H_2O via ^{3}SMX* [51,52]. Furthermore, BP produces $^1O_2$ in the presence of O_2 and $^{\bullet}OH$ [12,57] and $^{\bullet}OH$ is an effective non-selective RI for the photodegradation of most OMPs [4,60]. As explained earlier in the discussion, these BPs have been proven to be as significant as natural organic matters in sensitizing indirect photodegradation of OMPs. Therefore, our study could be a steppingstone for acknowledging these commonly found BPs and other similar organic pollutants as important background constituents in assessing OMP degradation.

Our experiments were conducted under environmentally relevant μg/L level concentrations of SMX and BPs at pH 7 and 20 °C. The outcome indicated that the influence of BPs on SMX degradation was a combination of the inner filter effect and sensitization of RIs initiated by BPs and SMX, which appeared to be concentration dependent. Both BP and BP3 sensitized SMX degradation significantly at concentrations 0.10 and 0.25 μM. In addition, SMX degradation was enhanced at all tested BP ratios except 0.60 and BP3 did not show any significant enhancement in SMX degradation at other tested ratios. The improvement of the photodegradation of SMX in these lower ranges of BPs ratios indicates that the sensitizing effect by BPs overpowers their light screening effect. In previous studies [13,14], the target OMP's concentration was always higher than the concentration of sensitizing OMPs, for instance 5.0 mg/L of gemfibrozil in the presence of 0.1, 0.5, and 1.0 mg/L of fenofibric acid, where gemfibrozil degradation has been promoted by fenofibric acid at all tested concentrations [13]. In our study, for the first time, inhibitory or promotional behavior of BPs was revealed by covering molar concentrations equal to and greater than the target OMP (i.e., SMX).

Overall, this experimental study demonstrated the ability of BP and BP3 to sensitize the production of RIs and thus promote degradation of co-existing OMPs in water. In the current study, the experiments were conducted in pH 7.0 buffer containing only the BPs and SMX, which is a relatively simple water matrix compared to environmental water. Our outcomes might not always be valid in the context of actual waters matrices. Thus, the concentration-dependent mechanism of BPs to sensitize RIs production is yet to be investigated to fully understand the competitive actions of RIs production and light screening, which influences the degradation of coexisting OMPs. To this end, for instance, complex water matrices including natural organic matter should be employed in further experimental study. Furthermore, the presented study did not elucidate the reaction pathway of SMX with the produced intermediates in the presence of BP or BP3. Therefore, the elucidation of the reaction pathway of SMX in the presence of BPs using mass spectrophotometry would be one of the next challenges in further study.

Nevertheless, major outcomes from this study are possibly applicable to other BP type UV filters (e.g., sulisobenzone (BP4) and dioxybenzone (BP8)) as they have similar molecular structures to BP and BP3. It would be worthwhile further exploring those similar compounds in the context of photodegradation of OMPs (i.e., SMX and others) considering their widespread use and frequent detection in water environments. In this regard, this study should be extended to other types of OMPs that are resistant to direct photodegradation in the natural environment in further investigation.

4. Conclusions

The overall objective of this study was to investigate the influence of co-existing benzophenone UV filters (BP and BP3) on photodegradation of SMX. The experiments using the photoreactor clearly showed that photodegradation of SMX was influenced by the co-existing BP and BP3 and this effect was concentration dependent. Both BP and BP3 enhanced SMX degradation in the low ranges of concentration including 0.10 and

0.25 µM. In addition, the quenching experiments further revealed that the photolysis of SMX is promoted by radicals •OH, $^1O_2$, $O_2^{•-}$ and $^3SMX^*/^3BP^*/^3BP3^*$, which also depends on the concentration of BPs. Among them, $^3SMX^*/^3BP^*$ played the dominant role in the sensitizing process of SMX/BP mixtures.

This is the first investigation showing the sensitizing ability of BP and BP3 in the context of coexisting OMP photodegradation, employing a wide range of concentrations in µg/L levels. The outcomes of this study could be applied to other BP type UV filters considering their molecular similarity to the studied BPs. Further investigation is expected regarding the sensitizing behavior of the BPs in the presence of natural organic matters and other complex water matrices to gain sufficient understanding on the fate of OMPs in the water environment.

Supplementary Materials: The following supporting information can be downloaded at: https://www.mdpi.com/article/10.3390/photochem3020017/s1, Table S1: Equations and calculations used in chemical actinometry experiment; Table S2: Results of the data analysis of indirect photodegradation induced by the BPs in the experimental system; Figure S1: UV absorbance spectra from 280 nm to 400 nm for 1.00 µM of SMX, BP, and BP3 measured using UV-vis spectrophotometer (UV2600, Shimadzu, Kyoto, Japan).

Author Contributions: Conceptualization, D.K.; methodology, D.K. and Z.G.; formal analysis, D.K.; investigation, D.K.; resources, D.K., Z.G. and C.Y.; data curation, D.K.; writing—original draft preparation, D.K.; writing—review and editing, D.K. and C.Y.; visualization, D.K. and C.Y.; supervision, C.Y.; funding acquisition, C.Y. All authors have read and agreed to the published version of the manuscript.

Funding: This research was funded by Japan Society for the Promotion of Science, JSPS KAKENHI, grant numbers 21H01462 and 22J12590, and Asia-Pacific Network for Global Change Research, CRRP2019-08MY-Khanal.

Data Availability Statement: The data presented in this study are available either in this article itself and Supplementary Materials.

Acknowledgments: Authors would like to acknowledge the financial supports of the Japan Society for the Promotion of Science and the Water Resources Environmental Center, Japan and the technical support provided by Manna Wang and Manami Miyamoto for this study. In addition, Kodikara Dilini acknowledges the financial support from the International EMECS center (Kobe, Japan).

Conflicts of Interest: The authors declare no conflict of interest.

References

1. Nakata, H.; Shinohara, R.I.; Nakazawa, Y.; Isobe, T.; Sudaryanto, A.; Subramanian, A.; Tanabe, S.; Zakaria, M.P.; Zheng, G.J.; Lam, P.K.S.; et al. Asia-Pacific mussel watch for emerging pollutants: Distribution of synthetic musks and benzotriazole UV stabilizers in Asian and US coastal waters. *Mar. Pollut. Bull.* **2012**, *64*, 2211–2218. [CrossRef] [PubMed]
2. Garcia-Rodríguez, A.; Matamoros, V.; Fontàs, C.; Salvadó, V. The ability of biologically based wastewater treatment systems to remove emerging organic contaminants—A review. *Environ. Sci. Pollut. Res.* **2014**, *21*, 11708–11728. [CrossRef] [PubMed]
3. Baena-Nogueras, R.M.; González-Mazo, E.; Lara-Martín, P.A. Degradation kinetics of pharmaceuticals and personal care products in surface waters: Photolysis vs biodegradation. *Sci. Total Environ.* **2017**, *590–591*, 643–654. [CrossRef] [PubMed]
4. Ribeiro, A.R.L.; Moreira, N.F.F.; Puma, G.L.; Silva, A.M.T. Impact of water matrix on the removal of micropollutants by advanced oxidation technologies. *Chem. Eng. J.* **2019**, *363*, 155–173. [CrossRef]
5. Castro, G.; Rodríguez, I.; Ramil, M.; Cela, R. Evaluation of nitrate effects in the aqueous photodegradability of selected phenolic pollutants. *Chemosphere* **2017**, *185*, 127–136. [CrossRef]
6. Wang, Y.; Roddick, F.A.; Fan, L. Direct and indirect photolysis of seven micropollutants in secondary effluent from a wastewater lagoon. *Chemosphere* **2017**, *185*, 297–308. [CrossRef]
7. Zhang, Y.; Zhang, J.; Xiao, Y.; Chang, V.W.C.; Lim, T.-T. Direct and indirect photodegradation pathways of cytostatic drugs under UV germicidal irradiation: Process kinetics and influences of water matrix species and oxidant dosing. *J. Hazard. Mater.* **2017**, *324*, 481–488. [CrossRef]
8. Cédat, B.; de Brauer, C.; Métivier, H.; Dumont, N.; Tutundjan, R. Are UV photolysis and UV/H2O2 process efficient to treat estrogens in waters? Chemical and biological assessment at pilot scale. *Water Res.* **2016**, *100*, 357–366. [CrossRef]
9. Sun, P.; Pavlostathis, S.G.; Huang, C.H. Photodegradation of veterinary ionophore antibiotics under UV and solar irradiation. *Environ. Sci. Technol.* **2014**, *48*, 13188–13196. [CrossRef]

10. Yao, H.; Sun, P.; Minakata, D.; Crittenden, J.C.; Huang, C.H. Kinetics and modeling of degradation of ionophore antibiotics by UV and UV/H_2O_2. *Environ. Sci. Technol.* **2013**, *47*, 4581–4589. [CrossRef]
11. Mouamfon, M.V.N.; Li, W.; Lu, S.; Qiu, Z.; Chen, N.; Lin, K. Photodegradation of sulphamethoxazole under UV-light irradiation at 254 nm. *Environ. Technol.* **2010**, *31*, 489–494. [CrossRef]
12. Boscá, F.; Miranda, M.A. New Trends in Photobiology (Invited Review) Photosensitizing drugs containing the benzophenone chromophore. *J. Photochem. Photobiol. B* **1998**, *43*, 1–26. [CrossRef]
13. Zhang, Y.N.; Zhou, Y.; Qu, J.; Chen, J.; Zhao, J.; Lu, Y.; Li, C.; Xie, Q.; Peijnenburg, W.J.G.M. Unveiling the important roles of coexisting contaminants on photochemical transformations of pharmaceuticals: Fibrate drugs as a case study. *J. Hazard. Mater.* **2018**, *358*, 216–221. [CrossRef]
14. Zhou, Y.; Zhao, J.; Zhang, Y.N.; Qu, J.; Li, C.; Qin, W.; Zhao, Y.; Chen, J.; Peijnenburg, W.J.G.M. Trace amounts of fenofibrate acid sensitize the photodegradation of bezafibrate in effluents: Mechanisms, degradation pathways, and toxicity evaluation. *Chemosphere* **2019**, *235*, 900–907. [CrossRef]
15. Canonica, S.; Hellrung, B.; Müller, P.; Wirz, J. Aqueous Oxidation of Phenylurea Herbicides by Triplet Aromatic Ketones. *Environ. Sci. Technol.* **2006**, *40*, 6636–6641. [CrossRef]
16. Das, P.K.; Bhattacharyya, S.N. Laser flash photolysis study of electron transfer reactions of phenolate ions with aromatic carbonyl triplets. *J. Phys. Chem.* **1981**, *85*, 1391–1395. [CrossRef]
17. Canonica, S.; Hellrung, B.; Wirz, J. Oxidation of Phenols by Triplet Aromatic Ketones in Aqueous Solution. *J. Phys. Chem. A* **2000**, *104*, 1226–1232. [CrossRef]
18. Ramos, S.; Homem, V.; Alves, A.; Santos, L. Advances in analytical methods and occurrence of organic UV-filters in the environment—A review. *Sci. Total Environ.* **2015**, *526*, 278–311. [CrossRef] [PubMed]
19. Chhabra, R.S. *National Toxicology Program Technical Report on the Toxicity Studies of Benzophenone (CAS No. 119-61-9). Administered in Feed to F344/N Rats and B6C3F Mice*; National Toxicology Program: Research Triangle Park, NC, USA, 2000.
20. Mao, F.; You, L.; Reinhard, M.; He, Y.; Gin, K.Y.H. Occurrence and Fate of Benzophenone-Type UV Filters in a Tropical Urban Watershed. *Environ. Sci. Technol.* **2018**, *52*, 3960–3967. [CrossRef] [PubMed]
21. Wu, M.H.; Li, J.; Xu, G.; Ma, L.D.; Li, J.J.; Li, J.S.; Tang, L. Pollution patterns and underlying relationships of benzophenone-type UV-filters in wastewater treatment plants and their receiving surface water. *Ecotoxicol. Environ. Saf.* **2018**, *152*, 98–103. [CrossRef]
22. Guo, Q.; Wei, D.; Zhao, H.; Du, Y. Predicted no-effect concentrations determination and ecological risk assessment for benzophenone-type UV filters in aquatic environment. *Environ. Pollut.* **2020**, *256*, 113460. [CrossRef]
23. Langford, K.H.; Reid, M.J.; Fjeld, E.; Øxnevad, S.; Thomas, K.V. Environmental occurrence and risk of organic UV filters and stabilizers in multiple matrices in Norway. *Environ. Int.* **2015**, *80*, 1–7. [CrossRef]
24. Negreira, N.; Rodríguez, I.; Rubí, E.; Cela, R. Dispersive liquid–liquid microextraction followed by gas chromatography–mass spectrometry for the rapid and sensitive determination of UV filters in environmental water samples. *Anal. Bioanal. Chem.* **2010**, *398*, 995–1004. [CrossRef]
25. Amar, S.K.; Goyal, S.; Mujtaba, S.F.; Dwivedi, A.; Kushwaha, H.N.; Verma, A.; Chopra, D.; Chaturvedi, R.K.; Ray, R.S. Role of type I & type II reactions in DNA damage and activation of Caspase 3 via mitochondrial pathway induced by photosensitized benzophenone. *Toxicol. Lett.* **2015**, *235*, 84–95. [CrossRef]
26. Das, P.K.; Encinas, M.V.; Scaiano, J.C. Laser Flash Photolysis Study of the Reactions of Carbonyl Triplets with Phenols and Photochemistry of p-Hydroxypropiophenone. *J. Am. Chem. Soc.* **1981**, *103*, 4154–4162. [CrossRef]
27. Das, P.K.; Bobrowski, K. Charge-transfer reactions of methoxybenzenes with aromatic carbonyl triplets. A laser flash photolytic study. *J. Chem. Soc. Faraday Trans. 2 Mol. Chem. Phys.* **1981**, *77*, 1009–1027. [CrossRef]
28. Battacharyya, S.N.; Das, P.K. Photoreduction of benzophenone by amino acids, aminopolycarboxylic acids and their metal complexes. A laser-flash-photolysis study. *J. Chem. Soc. Faraday Trans. 2 Mol. Chem. Phys.* **1984**, *80*, 1107–1116. [CrossRef]
29. Bhattacharyya, K.; Das, P.K. Nanosecond Transient Processes in the Triethylamlne Quenching of Benzophenone Triplets in Aqueous Alkaline Media. Substituent Effect, Ketyl Radical Deprotonation, and Secondary Photoreduction Kinetics. *J. Phys. Chem.* **1986**, *90*, 3987–3993. [CrossRef]
30. Cohen, S.G.; Parola, A.; Parsons, G.H. Photoreduction by Amines. *Chem. Rev.* **1973**, *73*, 141–161. [CrossRef]
31. Liu, H.; Liu, L.; Xiong, Y.; Yang, X.; Luan, T. Simultaneous determination of UV filters and polycyclic musks in aqueous samples by solid-phase microextraction and gas chromatography–mass spectrometry. *J. Chromatogr. A* **2010**, *1217*, 6747–6753. [CrossRef]
32. Fent, K.; Zenker, A.; Rapp, M. Widespread occurrence of estrogenic UV-filters in aquatic ecosystems in Switzerland. *Environ. Pollut.* **2010**, *158*, 1817–1824. [CrossRef] [PubMed]
33. Pedrouzo, M.; Borrull, F.; Marcé, R.M.; Pocurull, E. Stir-bar-sorptive extraction and ultra-high-performance liquid chromatography-tandem mass spectrometry for simultaneous analysis of UV filters and antimicrobial agents in water samples. *Anal. Bioanal. Chem.* **2010**, *397*, 2833–2839. [CrossRef] [PubMed]
34. Román, I.P.; Alberto, A.C.; Canals, A. Dispersive solid-phase extraction based on oleic acid-coated magnetic nanoparticles followed by gas chromatography–mass spectrometry for UV-filter determination in water samples. *J. Chromatogr. A* **2011**, *1218*, 2467–2475. [CrossRef] [PubMed]
35. Ho, Y.C.; Ding, W.H. Solid-phase extraction coupled simple on-line derivatization gas chromatography-tandem mass spectrometry for the determination of benzophenone-type UV filters in aqueous samples. *J. Chin. Chem. Soc.* **2012**, *59*, 107–113. [CrossRef]

36. Wu, J.W.; Chen, H.C.; Ding, W.H. Ultrasound-assisted dispersive liquid–liquid microextraction plus simultaneous silylation for rapid determination of salicylate and benzophenone-type ultraviolet filters in aqueous samples. *J. Chromatogr. A* **2013**, *1302*, 20–27. [CrossRef] [PubMed]
37. Grabicova, K.; Fedorova, G.; Burkina, V.; Steinbach, C.; Schmidt-Posthaus, H.; Zlabek, V.; Kroupova, H.K.; Grabic, R.; Randak, T. Presence of UV filters in surface water and the effects of phenylbenzimidazole sulfonic acid on rainbow trout (Oncorhynchus mykiss) following a chronic toxicity test. *Ecotoxicol. Environ. Saf.* **2013**, *96*, 41–47. [CrossRef]
38. Gago-Ferrero, P.; Mastroianni, N.; Díaz-Cruz, M.S.; Barceló, D. Fully automated determination of nine ultraviolet filters and transformation products in natural waters and wastewaters by on-line solid phase extraction–liquid chromatography–tandem mass spectrometry. *J. Chromatogr. A* **2013**, *1294*, 106–116. [CrossRef]
39. Tsui, M.M.P.; Leung, H.W.; Wai, T.C.; Yamashita, N.; Taniyasu, S.; Liu, W.; Lam, P.K.S.; Murphy, M.B. Occurrence, distribution and ecological risk assessment of multiple classes of UV filters in surface waters from different countries. *Water Res.* **2014**, *67*, 55–65. [CrossRef]
40. Kameda, Y.; Kimura, K.; Miyazaki, M. Occurrence and profiles of organic sun-blocking agents in surface waters and sediments in Japanese rivers and lakes. *Environ. Pollut.* **2011**, *159*, 1570–1576. [CrossRef]
41. Kawaguchi, M.; Ito, R.; Honda, H.; Endo, N.; Okanouchi, N.; Saito, K.; Seto, Y.; Nakazawa, H. Simultaneous analysis of benzophenone sunscreen compounds in water sample by stir bar sorptive extraction with in situ derivatization and thermal desorption–gas chromatography–mass spectrometry. *J. Chromatogr. A* **2008**, *1200*, 260–263. [CrossRef]
42. Wick, A.; Fink, G.; Ternes, T.A. Comparison of electrospray ionization and atmospheric pressure chemical ionization for multi-residue analysis of biocides, UV-filters and benzothiazoles in aqueous matrices and activated sludge by liquid chromatography–tandem mass spectrometry. *J. Chromatogr. A* **2010**, *1217*, 2088–2103. [CrossRef]
43. Kasprzyk-Hordern, B.; Dinsdale, R.M.; Guwy, A.J. The removal of pharmaceuticals, personal care products, endocrine disruptors and illicit drugs during wastewater treatment and its impact on the quality of receiving waters. *Water Res.* **2009**, *43*, 363–380. [CrossRef]
44. Lindim, C.; van Gils, J.; Georgieva, D.; Mekenyan, O.; Cousins, I.T. Evaluation of human pharmaceutical emissions and concentrations in Swedish river basins. *Sci. Total Environ.* **2016**, *572*, 508–519. [CrossRef]
45. Paíga, P.; Santos, L.H.M.L.M.; Ramos, S.; Jorge, S.; Silva, J.G.; Delerue-Matos, C. Presence of pharmaceuticals in the Lis river (Portugal): Sources, fate and seasonal variation. *Sci. Total Environ.* **2016**, *573*, 164–177. [CrossRef]
46. Yan, Q.; Zhang, Y.X.; Kang, J.; Gan, X.M.; Xu-Y, P.; Guo, J.S.; Gao, X. A Preliminary Study on the Occurrence of Pharmaceutically Active Compounds in the River Basins and Their Removal in Two Conventional Drinking Water Treatment Plants in Chongqing, China. *Clean* **2015**, *43*, 794–803. [CrossRef]
47. Wang, Z.; Zhang, X.H.; Huang, Y.; Wang, H. Comprehensive evaluation of pharmaceuticals and personal care products (PPCPs) in typical highly urbanized regions across China. *Environ. Pollut.* **2015**, *204*, 223–232. [CrossRef]
48. Subedi, B.; Codru, N.; Dziewulski, D.M.; Wilson, L.R.; Xue, J.; Yun, S.; Braun-Howland, E.; Minihane, C.; Kannan, K. A pilot study on the assessment of trace organic contaminants including pharmaceuticals and personal care products from on-site wastewater treatment systems along Skaneateles Lake in New York State, USA. *Water Res.* **2015**, *72*, 28–39. [CrossRef]
49. Patel, M.; Kumar, R.; Kishor, K.; Mlsna, T.; Pittman, C.U.; Mohan, D. Pharmaceuticals of emerging concern in aquatic systems: Chemistry, occurrence, effects, and removal methods. *Chem. Rev.* **2019**, *119*, 3510–3673. [CrossRef]
50. Yang, J.; Lv, G.; Zhang, C.; Wang, Z.; Sun, X. Indirect photodegradation of sulfamethoxazole and trimethoprim by hydroxyl radicals in aquatic environment: Mechanisms, transformation products and eco-toxicity evaluation. *Int. J. Mol. Sci.* **2020**, *21*, 6276. [CrossRef]
51. Zhou, L.; Deng, H.; Zhang, W.; Gao, Y. Photodegradation of sulfamethoxazole and photolysis active species in water under Uv-Vis light irradiation. *Fresenius Environ. Bull.* **2015**, *24*, 1685–1691. Available online: https://www.researchgate.net/publication/292391936 (accessed on 13 March 2023).
52. Lin, X.; Zhou, W.; Li, S.; Fang, H.; Fu, S.; Xu, J.; Huang, J. Photodegradation of Sulfamethoxazole and Enrofloxacin under UV and Simulated Solar Light Irradiation. *Water* **2023**, *15*, 517. [CrossRef]
53. Laszakovits, J.R.; Berg, S.M.; Anderson, B.G.; O'Brien, J.E.; Wammer, K.H.; Sharpless, C.M. P-Nitroanisole/pyridine and p-Nitroacetophenone/pyridine actinometers revisited: Quantum yield in comparison to ferrioxalate. *Environ. Sci. Technol. Lett.* **2017**, *4*, 11–14. [CrossRef]
54. Zhang, S.; Chen, J.; Qiao, X.; Ge, L.; Cai, X.; Na, G. Quantum chemical investigation and experimental verification on the aquatic photochemistry of the sunscreen 2-phenylbenzimidazole-5-sulfonic acid. *Environ. Sci. Technol.* **2010**, *44*, 7484–7490. [CrossRef]
55. Zhang, Y.N.; Wang, J.; Chen, J.; Zhou, C.; Xie, Q. Phototransformation of 2,3-Dibromopropyl-2,4,6-tribromophenyl ether (DPTE) in Natural Waters: Important Roles of Dissolved Organic Matter and Chloride Ion. *Environ. Sci. Technol.* **2018**, *52*, 10490–10499. [CrossRef] [PubMed]
56. Liu, S.; Cui, Z.; Bai, Y.; Ding, D.; Yin, J.; Su, R.; Qu, K. Indirect photodegradation of ofloxacin in simulated seawater: Important roles of DOM and environmental factors. *Front. Mar. Sci.* **2023**, *10*, 351. [CrossRef]
57. Delatour, T.; Douki, T.; D'Ham, C.; Cadet, J. Photosensitization of thymine nucleobase by benzophenone through energy transfer, hydrogen abstraction and one-electron oxidation. *J. Photochem. Photobiol. B* **1998**, *44*, 191–198. [CrossRef]
58. Guo, Y.; Long, J.; Huang, J.; Yu, G.; Wang, Y. Can the commonly used quenching method really evaluate the role of reactive oxygen species in pollutant abatement during catalytic ozonation? *Water Res.* **2022**, *215*, 118275. [CrossRef]

59. Musial, J.; Mlynarczyk, D.T.; Stanisz, B.J. Photocatalytic degradation of sulfamethoxazole using TiO$_2$-based materials—Perspectives for the development of a sustainable water treatment technology. *Sci. Total Environ.* **2023**, *856*, 159122. [CrossRef]
60. Lian, L.; Yao, B.; Hou, S.; Fang, J.; Yan, S.; Song, W. Kinetic Study of Hydroxyl and Sulfate Radical-Mediated Oxidation of Pharmaceuticals in Wastewater Effluents. *Environ. Sci. Technol.* **2017**, *51*, 2954–2962. [CrossRef]

Disclaimer/Publisher's Note: The statements, opinions and data contained in all publications are solely those of the individual author(s) and contributor(s) and not of MDPI and/or the editor(s). MDPI and/or the editor(s) disclaim responsibility for any injury to people or property resulting from any ideas, methods, instructions or products referred to in the content.

Article

Evaluation of MAA Analogues as Potential Candidates to Increase Photostability in Sunscreen Formulations

Jacobo Soilán [1], Leonardo López-Cóndor [1], Beatriz Peñín [1], José Aguilera [2,3], María Victoria de Gálvez [2,3], Diego Sampedro [1,*] and Raúl Losantos [1,*]

[1] Departamento de Química, Instituto de Investigación en Química (IQUR), Complejo Científico Tecnológico, Universidad de La Rioja, Madre de Dios 53, 26006 Logroño, Spain

[2] Department of Dermatology and Medicine, Faculty of Medicine, University of Malaga, Campus Universitario de Teatinos s/n, 29071 Malaga, Spain; jaguilera@uma.es (J.A.); mga@uma.es (M.V.d.G.)

[3] Photodermatology Laboratory, Medical Research Center, University of Malaga, Campus Universitario de Teatinos s/n, 29071 Malaga, Spain

* Correspondence: diego.sampedro@unirioja.es (D.S.); raul.losantosc@unirioja.es (R.L.)

Abstract: Avobenzone is one of the most widely used sunscreens in skin care formulations, but suffers from some drawbacks, including photo instability. To mitigate this critical issue, the use of octocrylene as a stabilizer is a common approach in these products. However, octocrylene has been recently demonstrated to show potential phototoxicity. The aim of this work is to analyze the performance of a series of mycosporine-like amino acid (MAA)-inspired compounds to act as avobenzone stabilizers as an alternative to octocrylene. Different avobenzone/MAA analogue combinations included in galenic formulations were followed under increasing doses of solar-simulated UV radiation. Some of the synthetic MAA analogues analyzed were able to increase by up to two times the UV dose required for 50% of avobenzone photobleaching. We propose some of these MAA analogues as new candidates to act as avobenzone-stabilizing compounds in addition to their UV absorbance and antioxidant properties, together with a facile synthesis.

Keywords: sunscreens; photoprotection; UV radiation; MAAs; photochemistry

1. Introduction

Sunlight exposition has been increasing for the last decades. This is induced by our outdoors habits and, in combination with the decrease of the ozone layer, it causes increased UV exposure in humans. Due to this, the impact of melanoma and other types of skin cancer has been increasing through the last decades [1]. To avoid UV radiation damaging effects, scientists have develop a series of sunscreen formulations to protect humans from potential UV damage [2], in addition to natural protection [3]. Nowadays, none of the commercially available sunscreen molecules present ideal behavior as a photoprotector, as these molecules should fulfill a long list of requirements [4,5]. These compounds should present, among other features, very large photostability together with broad and intense absorption in the UV region. These properties should also be combined with negligible toxicity or phototoxicity. Potential side effects, like photosensitization or photoreactivity, that could induce toxic effects in living organisms, should be considered as a major drawback and should be avoided. Complementarily, the environmental impact and bioaccumulation effect on the ocean's ecosystem should be taken into consideration. Due to these side problems, the development of safer and environmentally friendly systems has gained a lot of interest, mainly from the industrial sector. The previously mentioned problems are only related to human care, but equally important is the environmental impact that can induce the massive use of sunscreen lotions.

Even if there are many ingredients that can be used in sunscreen formulations [2], the actual legislation in Europe allows 48 compounds to be used as photoprotectors (Annex

VI, Regulation 1223/2009/EC on Cosmetic Products, as amended by Regulation (EU) 2022/2195, OJ L 292, 11 November 2022). Despite this large number of UV blockers, the most used ones are still the avobenzone/octocrylene pair. These compounds are quite far from behaving as ideal sunscreens, but their combinations provide adequate practical results. In particular, avobenzone is known to undergo fast photobleaching due to the population of the keto-keto form, which leads to undesired photo instability [6,7]. Due to this, the combined use of avobenzone with octocrylene is almost mandatory due to the strong stabilization that can be achieved in combination. However, octocrylene presents its own drawbacks, as it is known that it can cause phototoxicity [8,9] and act as an endocrine disruptor [10]. Therefore, it is urgently necessary to find new alternatives for the design of sunscreen formulations which could provide a photostable mixture using safer substances.

Using photoprotective natural compounds as inspiration, we developed a versatile and efficient route to prepare synthetic analogues of the well-known natural compounds called mycosporine-like amino acids (MAA). These low molecular weight compounds are widely available on the planet and present really efficient sunscreen capabilities [11]. These compounds are thermally and photochemically stable, and present a considerable absorption coefficient and lack of fluorescence, photoreactivity, and toxicity [12]. These properties make these compounds quite close to the ideal features of sunscreens [12–14]. In this context, we aimed to study these compounds as avobenzone stabilizers. In the literature, there have been a considerable number of attempts to stabilize avobenzone [15]. The most common approach is the use of octocrylene, with the previously mentioned drawbacks. Other employed strategies are the combination with (2-hydroxy)propyl-β-cyclodextrin, which also provides a significant stabilization due to encapsulation [16], as well as micellar encapsulation [17]. In addition, the use of antioxidant molecules [18,19], other sunscreens, like bisethylhexyloxyphenol methoxyphenyltriazine (BEMT) [20], or complex structures like zeolites have been explored in this topic. Additionally, an innovative approach was the use of light to induce and control avobenzone production through a photochemical transformation [21]; this is a potent strategy but could have serious flaws due to the photo reactivity in a complex and more rigid environment, i.e., a cosmetic formulation.

In this paper, we aim to explore the capabilities of MAA analogues to stabilize cosmetic formulations containing avobenzone by chemical combination. For this, we have prepared a series of cosmetic formulations and studied their behavior against different doses of light. For our experiments, we have prepared a series of compounds with large photostability (negligible degradation with an equivalent dose of irradiation higher than 6 h, according to previously carried out experiments in solution detecting through ^{1}H NMR spectrometry) aiming to stabilize avobenzone in cosmetic formulations under realistic conditions in galenic formulations and to study the possible effects of concentration for the photostability of avobenzone.

2. Results and Discussion

A series of four compounds (Scheme 1) was prepared, aiming to provide high absorbance in the UV region. For this, we used an amine condensation with the 1,3-diketo compounds under Dean–Stark conditions as described in more detail in Section 4 [12]. In a preliminary step, the UV-Vis spectra of all the compounds were measured in solution. All the compounds featured an absorbance maximum at the UVB region (Figure 1). Noteworthily, all of them also featured impressively high photostability in solution, as previously measured by ^{1}H-NMR [12]. This extremely stable behavior suggests great potential to be used as a sunscreen ingredient. Also, the facile synthesis, with only one step, easy purification, the use of precipitation, and high yields offers a great alternative to approach the excellent properties of natural MAAs in a more sustainable way with respect to extraction from algae. This work presents a preliminary screening of those candidates in formulation, aiming at the stabilization of other ingredients present in the formula.

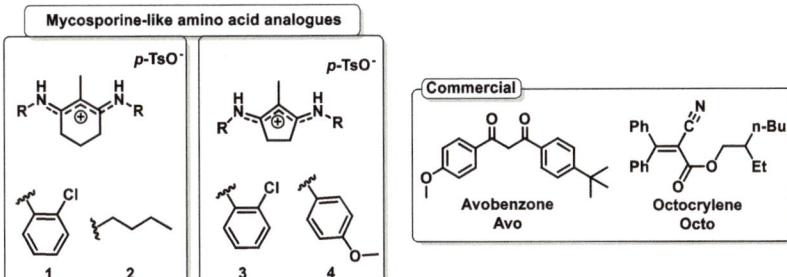

Scheme 1. Structures of the commercial and the prepared MAA analogues. Note the difference between the core between **1–2** and **3–4**.

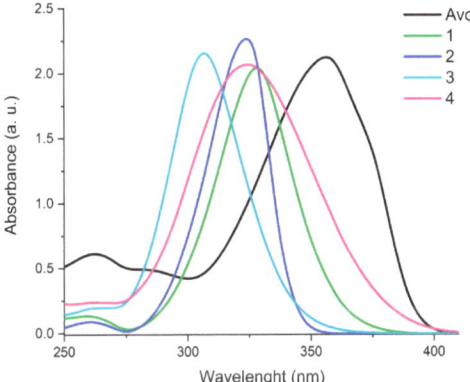

Figure 1. UV-Vis spectra of the studied compounds in 1×10^{-4} M solution in methanol.

Aiming to prove the ability of our compounds to be formulated in stable cosmetic formulations, we have produced a series of galenic formulations using different mixtures of avobenzone (Avo), octocrylene (Octo), and **1** to **4**. We intended to test these combinations in the same interval of concentrations as the standard filters approved by international legislation regarding cosmetic products (for instance, the 76/768/CEE directive of the European Union), using between 5 and 10% w/w, similar to the current practical use for avobenzone and octocrylene. Due to this, the compounds were studied at 5% in a prototype 10 g of NeoPCL formulation prepared as described in Section 4. All the compounds presented an appropriate absorption for use as UV blockers, as shown in Figure 1. But in contrast to the high absorbance found in solution, the absorbance or transmittance measured in formulation decreased considerably; see Figure 2. It is also worth mentioning that there is a minimal absorbance overlap with avobenzone in formulation, allowing a proper monitorization of the photobleaching of Avo against UV doses, while the range of photon absorption is increased in the mixtures compared with the isolated compounds. With these initial promising features, we aimed for the incorporation of these combinations into more complex galenic formulations. To further evaluate their photochemical properties, we irradiated the samples using the standardized methacrylate (PMMA) plates using a solar-like irradiation source with a combination of UVB–UVA lamps (see more details in Section 4). The temporal evolution of the samples was monitored by absolute irradiance measurements at increasing UV doses of radiation. To provide a standardized analysis, the irradiance data were converted to transmittance, dividing the irradiance of the lamp through the sample by the irradiance of the plate impregnated with an equivalent amount of Vaseline. At that point, the different compounds could be compared independently and could be converted to absorbance directly. Looking at the results shown in Figure 2, we found a clear behavior, as expected for commercial sunscreen Avo, which presented low photostability. In the case of avobenzone, it almost fully decomposes under the experiment dose, depicting a non-efficient deactivation mechanism (i.e., photoreactivity). In contrast, the four prepared compounds **1–4** exhibited higher photostability, remaining unaltered after comparable doses. The temporal or dose evolution can be seen in Figure 2. The irradiation of those samples yielded to a critical bleaching of avobenzone, decreasing to 50% after only 200 kJ/m^2 and being fully decomposed after 450 kJ/m^2 of total UV dose. Compounds **1** to **4** did not present any noticeable decomposition measured at their corresponding absorption maxima after the maximum dose. The initial increase in Avo is attributable to insufficient dryness before the first measurement, possibly due to the more viscous character of the formulation respect to the others. This can yield a higher evaporation during the irradiation, which is only notable at short times (5 min) and no longer noted afterwards.

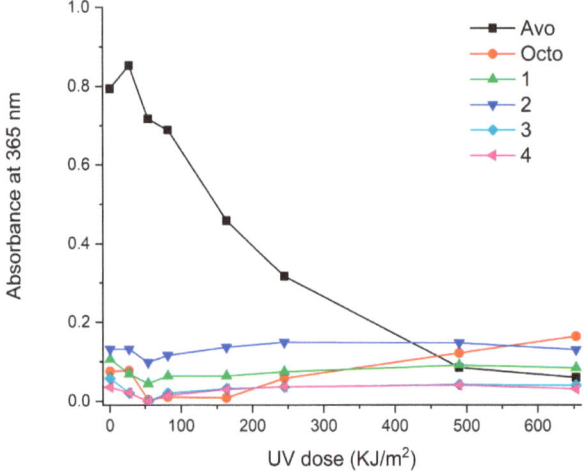

Figure 2. Photostability of the studied samples at 5%. Monitored by absorbance measurements at Avo maximum.

Even with this poor photostability, avobenzone is one of the most widely used sunscreens in cosmetics. In commercial formulations, this is possibly due to the inclusion of other ingredients as stabilizers. The combination of at least two UV absorbents is the most used strategy to provide photostability to cosmetic formulations. In this scenario, it is possible to decrease the absolute quantity of each compound in the formulation. As already mentioned, octocrylene is broadly used as an avobenzone stabilizer, so we moved to assay our compounds with this role, aiming to propose some alternatives to the use of the problematic octocrylene. In the literature, it is widely reported that octocrylene reacts with amines [22] which, in humans, are mainly found in the form of lysine residues. Then, it can undergo an initial Michael-type addition followed by a reaction sequence corresponding to a retro-aldol reaction and the formation of immunogenic hapten–protein complexes. This follows a mechanism closely related to the Schiff base mechanism and is often considered when aldehydes form those immunogenic complexes [23].

Notwithstanding, the presented compounds have not been tested in biotoxicity assays, but we hypothesized that they cannot undergo this reaction because they do not have an α,β-unsaturated carbonyl structure nor a strong nucleophilic or electrophilic center. Even if octocrylene has strong absorption in the UVB, its main application is as a stabilizer of other photoprotectors such as avobenzone, with the previously mentioned disadvantages [8].

Therefore, we irradiated a series of formulations containing the different ingredients in combination with Avo at 5% and 10% to study their photochemical behavior in cosmetic formulations. In Figure 3, we can see that Octo is in fact the best performer for avobenzone stabilization at high UV doses. However, it is clear that compound **4** performs quite well in comparison with free Avo. In fact, the absorption provided by mixtures of **4** + Avo is even better at low UV doses, especially for low concentrations (5%), where it increases the absorption of Avo with respect to the result achieved in the sample with Octo at equivalent concentration.

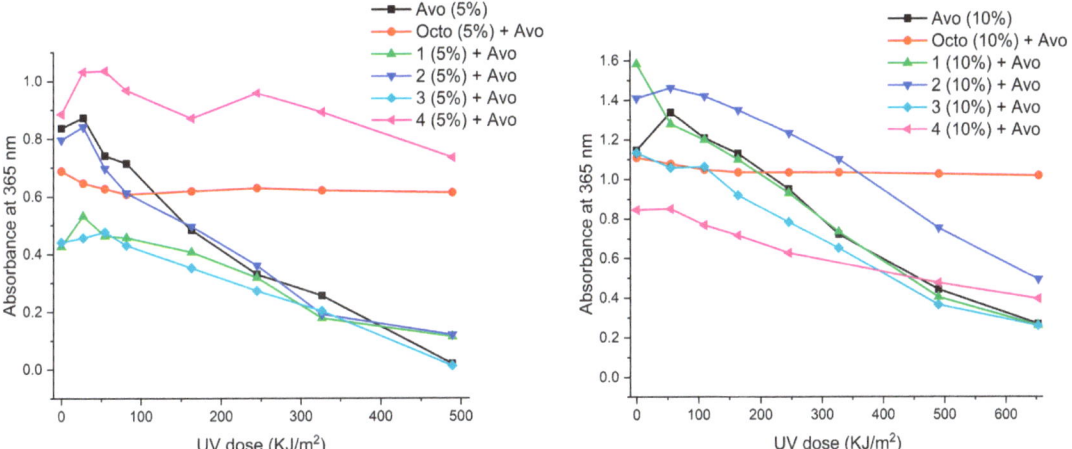

Figure 3. Time evolution of mixtures of Avo with **1–4** and Octo, at 5% on the left and 10% (w/w) on the right for all the ingredients. All the values were obtained from the transmittance spectra and converted to absorbance for easier interpretation. In case of sample **4** (10%) + Avo (right graph in purple), we discarded the spectrum at 60 min (326 kJ/m^2) due to an unknown drastic change in the transmission spectrum.

As can be seen in Figure 3, left, at 5% we can observe that compound **4** offers an extra boost to the absorption (ca. 16%), which will yield an increase in the eventual solar protective factor of the formula. This appears to be in combination with a noticeable stabilization in Avo photobleaching, increasing its stability by at least a factor of two. In this line, a small effect can also be noted for compounds **1** and **3**, which looking the slope of degradation offers slightly slower bleaching. This is in contrast with the behavior found at 10%, which resulted in lower stabilization from almost all the studied MAA analogues. This was quite surprising, but we were not able to provide a rationalization according to the principles ruling the conventional absorption processes occurring in diluted solutions. The only compound that exhibited an observable change in the slope of the decomposition trend was compound **4** but the observed boost in the absorbance at 5% was completely quenched here, even yielding into a decrease in the Avo intensity. In contrast to this, a small potentiation was induced by compound **2**, but at the end it did not result in effective stabilization. The broad conclusion that we can extract from these experiments is that we are quite far off a model to predict how the cosmetic ingredients could behave in cosmetic formulations. This can be justified from different points; firstly, the matrix effect, with all the additives and excipients, could play a substantial role and will vary completely between different lotions. On the other hand, the photophysical properties and photochemical mechanisms could be drastically affected by having the sunscreen molecules in a less flexible environment. This could drastically hinder the molecular movement, resulting in a potential diminution of the sunscreen efficiency, which would be a great inconvenience in the molecules, which would dissipate the exceeding energy by vibrational relaxation, as happens in MAA analogues.

As a final summary, we have calculated the required doses to reach 50% decomposition of Avo; this offers a proper overview of the activity of each ingredient and will highlight the most interesting results. Those values are shown in Table 1.

Table 1. Estimated 50% decomposition times of Avo and the required UV doses to induce it. * The value is approximated from the measured values represented in Figures 2 and 3.

Formulation	50% Decomposition Time *	UV Dose (kJ/m^2) *
Avobenzone (5%)	35 min	200
Avobenzone (10%)	75 min	400
Octocrylene (5%)	>120 min	>652
1 (5%)	>120 min	>652
2 (5%)	>120 min	>652
3 (5%)	>120 min	>652
4 (5%)	>120 min	>652
Avo (5%) + Octo (5%)	>120 min	>652
Avo (5%) + **1** (5%)	65 min	350
Avo (5%) + **2** (5%)	35 min	200
Avo (5%) + **3** (5%)	65 min	350
Avo (5%) + **4** (5%)	>120 min	>652
Avo (10%) + Octo (10%)	>120 min	>652
Avo (10%) + **1** (10%)	75 min	400
Avo (10%) + **2** (10%)	100 min	550
Avo (10%) + **3** (10%)	75 min	400
Avo (10%) + **4** (10%)	>120 min	650

The presented results suggest that the proposed compounds could be used to prepare formulations providing different degrees of stabilization of avobenzone under high UV doses. In the case of compound **2**, it shows poor stabilizing ability, not affecting the photodecomposition of Avo. This suggests that this specific compound does not interact with Avo, at least in an efficient manner. In contrast, a larger effect is observed for **4**. In fact, this compound performs quite similarly to Octo for high UV doses or could be even better at lower concentrations due to the obtained boost to the absorption. This turns this compound into a promising candidate to replace octocrylene as an avobenzone photostabilizer for potentially safer commercial formulations.

We have shown that the prepared compounds are compatible with avobenzone in real formulations. In addition, avobenzone is stabilized by our new sunscreens and some of them could be used to replace octocrylene. Further studies to understand the complex interactions between Avo and the MAAs analogues are underway. This should allow us to design new and improved versions of our compounds with increased performance.

3. Conclusions

In this contribution, we have studied the use of mycosporine-like amino acid analogues as potential replacements for octocrylene in cosmetic formulations as avobenzone stabilizers. Our results have shown that the overall stabilization provided by the new compounds is lower than the one found for octocrylene. However, the new compounds also have a significant effect on the avobenzone photodecomposition, enhancing the stability by a factor of two in some cases and approaching octocrylene in terms of supported UV dose. The best candidate is capable of maintaining the same stabilization of avobenzone in typical erythematic doses required during the real application of cosmetic formulations. Therefore, we have proved a series of MAA analogues to act as stabilizers for avobenzone comparable to octocrylene while offering an expected lower induced toxicity.

4. Materials and Methods

The studied compounds were synthetized according to the previously reported protocol by refluxing a mixture of the corresponding amine with *p*-toluene sulfonic acid and 1,3-cyclopentadione or 1,3-cyclohexadione in dry toluene using Dean–Stark apparatus to remove the water from the condensation for 24–48 h. Afterwards, the toluene was removed under rotatory evaporation and the crude residue was crystallized from CH_2Cl_2 by n-hexane addition and collected by filtration as a solid. In case of need, some excess CH_2Cl_2 can be used for washing to fully remove the exceeding amine, obtaining the desired compounds as powder solids with high yields between 75 and 95% [12]. The present counterion (*p*-toluene sulfonate) was observed using NMR. All reagents and solvents were used as received from commercial sources without further purification steps. ^{1}H and ^{13}C NMR spectra were recorded on a Bruker ARX-300 spectrometer (Billerica, MA, USA). Methanol-*d4* has been used as the usual deuterated solvent, using its signal as standard. Chemical shifts are given in ppm and coupling constants in Hertz. High Resolution Mass Spectrometry was performed using a Microtof-Q electrospray source in positive-ion mode. Absorption molecular spectra were recorded on an Ocean Optics USB4000 UV-Vis diode array spectrophotometer (200–850 nm). All the experiments were carried out in quartz cuvettes (1 cm path length) using methanol 1×10^{-4} M solutions. Those compounds are very soluble in alcohols, moderately in halogenated ones and mainly insoluble in water and non-polar non-protic solvents.

All the assays in galenic formulation were made using a solid phase cream-like substance based on a self-emulsifiable O/W "NeoPCL" base at 20% (*w/w*) + propylene glycol (PEG) (0.05 mL) (Acofarma, Barcelona, Spain) and distilled water to reach the total weight of cream (*ca.* 1 g). A detailed list of components is shown in ESI. We have chosen the simplest and easiest preparation base for the formulation, aiming to minimize the possible matrix effect and maximize the observation of our compounds of interest. It is also a nonionic formula broadly approved by cosmetic regulations worldwide.

We have prepared the formulas using a two-component mixture. The water phase (water + PEG+ **1–4**) and the hydrophobic phase (NeoPCL) were both heated to 60 °C.

Once they were liquified and under intense mechanical agitation, the hydrophilic phase was slowly added to the hydrophobic one until we obtained a homogenous formula through continuous agitation. The color of the galenic formulas depends on the color of the MAA analogue used. Also, the texture of the formula slightly changed between the different samples. Galenic formulations were always freshly prepared prior to analysis and base formulation (vehicle) was used as control. But even with these cautions, the preparation of these simple formulations is always challenging for achieving robust replicas due to the huge impact of the emulsion on the physical properties, which can motivate some uncontrollable effects, like different scattering patterns.

All the measurements were made according to the international standard ISO-24443:2021 [24], using standardized 5×5 cm PMMA plates with 1.3 mg/cm^2 of cream. The spectral distribution of the light source as well as its transmitted spectral distribution by control and samples probes in PMMA plates was measured by means of a double monochromator attached to a Ulbrich sphere (MACAM SR9910-v7, Irradian, Scotland, UK). The use of an irradiation sphere offers a partial solution to minimize the effects of inhomogeneity from the samples, minimizing the influence of scattering in the obtained data.

The irradiation was implemented by a solar simulating combination of fluorescent lamps in a Daavlin irradiator model consisting of 2 Qpanel-340 36 W lamps and a Philips TL 10/36 W black light lamp [25,26]. The spectral irradiance distribution of the lamp system compared to that of the Sun at the Earth's surface in southern Europe at midday in a typical summer day is shown in ESI, Figure S3. The total UV irradiance (290–400 nm) emitted by the lamp was 90.64 W/m^2 and the transmitted spectral distribution of the samples and control was measured at intervals of approximately 15 min in a total exposure time of 2 h, corresponding to a total UV dose of 652 kJ/m^2. In terms of erythemal dose, total UV exposure corresponded to 8 minimal erythemal doses for phototype II (MED was defined as 250 J/m^2). The irradiation time was not extended due to the appearance of some small cracks on the top of the PMMA plates, which promoted the appearance of higher absorbance spots, making the analysis of the formulation extremely dependent on the chosen spot. Due to this, the longer acquired times at 3 h were discarded and not included in the results.

The transmittance was calculated using as reference a PMMA plate impregnated with 1.3 mg/cm^2 of glycerol (Vaseline). The total transmittance was obtained using the relation between the sample and blank irradiances, according to the following equation:

$$\text{Total transmittance}_{(\lambda)} = \text{measured Irradiance}_{(\lambda)} / \text{glycerol Irradiance}_{(\lambda)}$$

Also, the conversion to absorbance was undertaken by A = $-\log$(T) to present the data in a more intuitive way.

Analytical Data

The prepared compounds present the following characterization data.

Compound **1**:

^{1}H-NMR (300 MHz, MeOD) δ ppm 7.70 (d, J = 8.2 Hz, 2H), 7.65 (m, J = 5.4, 3.4, 1.2 Hz, 2H), 7.55–7.41 (m, 6H), 7.27–7.19 (m, 2H), 2.42 (t, J = 6.3 Hz, 4H), 2.37 (s, 3H), 2.15 (s, 3H), 1.82 (q, J = 6.3 Hz, 2H).
^{13}C-NMR (75 MHz, MeOD) δ = 173.5, 143.7, 141.5, 135.9, 133.0, 131.5, 131.4, 130.8, 129.7, 129.5, 126.9, 101.6, 28.3, 21.3, 21.1, 9.7.
UV-Vis (CH$_3$CN): λ (nm) = 330 (ε = 26,400 M^{-1}cm^{-1}).
ES-MS (+) (C$_{19}$H$_{18}$Cl$_2$N$_2$ + H): calc. 345.0920, found 345.0927

Compound **2**:

^{1}H-NMR (300 MHz, MeOD) δ ppm 7.70 (bs, 2H), 7.24 (bs, 2H), 2.89 (bs, 4H), 2.60 (bs, 4H), 2.37 (bs, 6H), 1.94–1.62 (m, 2H), 1.61 (bs, 4H), 1.39 (bs, 4H), 0.96 (bs, 6H).
^{13}C-NMR (75 MHz, MeOD) δ ppm 170.1, 143.5, 141.7, 129.8, 126.9, 79.4, 40.5, 33.1, 30.6, 26.2, 21.3, 20.6, 13.8, 8.9.
UV-Vis (CH$_3$CN): λ (nm) = 324 (ε = 4200 M^{-1}cm^{-1}).
ES-MS (+) (C$_{15}$H$_{28}$N$_2$ + H): calc. 237.2325, found 237.2327

Compound **3**:

^{1}H-NMR (300 MHz, MeOD) δ 7.68 (d, J = 8.2 Hz, 2H), 7.64–7.57 (m, 2H), 7.55–7.39 (m, 6H), 7.24–7.17 (m, 2H), 2.80–2.51 (m, 4H), 2.35 (s, 3H), 2.09 (s, 3H).
^{13}C-NMR (75 MHz, MeOD) δ = 142.3, 140.2, 134.8, 131.1, 130.2, 130.1, 128.8, 128.4, 128.2, 125.6, 27.7, 20.0, 5.8.
UV-Vis (CH$_3$CN): λ (nm) = 306 (ε = 32,400 M^{-1}cm^{-1}).
ES-EM (+) (C$_{18}$H$_{16}$Cl$_2$N$_2$ + H): calc. 331.0763, found 331.0776.

Compound **4**:

^{1}H-NMR (400 MHz, MeOD) δ ppm 7.70 (d, J = 7.9 Hz, 2H), 7.26 (d, J = 8.6 Hz, 4H), 7.22 (d, J = 7.8 Hz, 2H), 7.01 (d, J = 8.6 Hz, 4H), 3.82 (s, 6H), 2.79 (s, 4H), 2.36 (s, 3H), 1.97 (s, 3H).
^{13}C-NMR (100 MHz, MeOD) δ ppm 180.5, 160.5, 143.7, 141.6, 131.9, 129.8, 127.5, 126.9, 115.7, 106.9, 56.0, 29.1, 21.3, 7.3.
UV-Vis (CH$_3$CN): λ (nm) = 328 (ε = 34,560 M^{-1}cm^{-1}).
ES-EM (+) (C$_{20}$H$_{22}$N$_2$O$_2$ + H): calc. 323.1754, found 323.1760.

5. Patents

The synthesis of these compounds was included in a Spanish patent application: ES2550374A1.

Supplementary Materials: The following supporting information can be downloaded at: https://www.mdpi.com/article/10.3390/photochem4010007/s1, Figure S1: NMR spectra of **1**. Figure S2: NMR spectra of **3**. Figure S3: Spectral irradiance used for the irradiation experiments. Figure S4: PMMA plates with the prepared formulations of **1–4** at 5%.

Author Contributions: Conceptualization, R.L., J.A., M.V.d.G. and D.S.; methodology, J.S., J.A. and R.L.; synthesis, L.L.-C. and B.P.; writing—original draft preparation, J.S. and R.L.; writing—review and editing, all authors; supervision, J.A., M.V.d.G., D.S. and R.L.; project administration, D.S.; funding acquisition, D.S. All authors have read and agreed to the published version of the manuscript.

Funding: This research was funded by the Spanish Ministerio de Ciencia, Innovación y Universidades (MICIN), grant number PDC2021-121410-I00.

Data Availability Statement: The data presented in this study are available upon request from the corresponding author.

Acknowledgments: R.L. thanks Universidad de La Rioja for his Margarita Salas grant. B.P. thanks Universidad de La Rioja for her fellowship.

Conflicts of Interest: The authors declare no conflicts of interest.

References

1. Yang, J.-W.; Fan, G.-B.; Tan, F.; Kong, H.-M.; Liu, Q.; Zou, Y.; Tan, Y.-M. The Role and Safety of UVA and UVB in UV-Induced Skin Erythema. *Front. Med.* **2023**, *10*, 1163697. [CrossRef]
2. Shaath, N.A. Ultraviolet Filters. *Photochem. Photobiol. Sci.* **2010**, *9*, 464–469. [CrossRef]
3. Brenner, M.; Hearing, V.J. The Protective Role of Melanin Against UV Damage in Human Skin [†]. *Photochem. Photobiol.* **2008**, *84*, 539–549. [CrossRef]
4. Aguilera, J.; Gracia-Cazaña, T.; Gilaberte, Y. New Developments in Sunscreens. *Photochem. Photobiol. Sci.* **2023**, *22*, 2473–2482. [CrossRef] [PubMed]
5. Osterwalder, U.; Hareng, L. Global UV Filters: Current Technologies and Future Innovations. In *Principles and Practice of Photoprotection*; Wang, S.Q., Lim, H.W., Eds.; Springer International Publishing: Cham, Switzerland, 2016; pp. 179–197. ISBN 978-3-319-29381-3.
6. Vallejo, J.J.; Mesa, M.; Gallardo, C. Evaluation of the Avobenzone Photostability in Solvents Used in Cosmetic Formulations. *Vitae* **2011**, *18*, 63–71. [CrossRef]
7. Kockler, J.; Oelgemöller, M.; Robertson, S.; Glass, B.D. Photostability of Sunscreens. *J. Photochem. Photobiol. C Photochem. Rev.* **2012**, *13*, 91–110. [CrossRef]
8. De Groot, A.C.; Roberts, D.W. Contact and Photocontact Allergy to Octocrylene: A Review. *Contact Dermat.* **2014**, *70*, 193–204. [CrossRef] [PubMed]
9. Berardesca, E.; Zuberbier, T.; Sanchez Viera, M.; Marinovich, M. Review of the Safety of Octocrylene Used as an Ultraviolet Filter in Cosmetics. *J. Eur. Acad. Dermatol. Venereol.* **2019**, *33*, 25–33. [CrossRef] [PubMed]
10. Carrotte-Lefebvre, I.; Bonnevalle, A.; Segard, M.; Delaporte, E.; Thomas, P. Contact Allergy to Octocrylene: First 2 Cases. *Contact Dermat.* **2003**, *48*, 45–55. [CrossRef] [PubMed]
11. De La Coba, F.; Aguilera, J.; De Gálvez, M.V.; Álvarez, M.; Gallego, E.; Figueroa, F.L.; Herrera, E. Prevention of the Ultraviolet Effects on Clinical and Histopathological Changes, as Well as the Heat Shock Protein-70 Expression in Mouse Skin by Topical Application of Algal UV-Absorbing Compounds. *J. Dermatol. Sci.* **2009**, *55*, 161–169. [CrossRef] [PubMed]
12. Losantos, R.; Funes-Ardoiz, I.; Aguilera, J.; Herrera-Ceballos, E.; García-Iriepa, C.; Campos, P.J.; Sampedro, D. Rational Design and Synthesis of Efficient Sunscreens To Boost the Solar Protection Factor. *Angew. Chem. Int. Ed.* **2017**, *56*, 2632–2635. [CrossRef]
13. Cowden, A.M.; Losantos, R.; Whittock, A.L.; Peñín, B.; Sampedro, D.; Stavros, V.G. Ring Buckling and C=N Isomerization Pathways for Efficient Photoprotection in Two Nature-inspired UVA Sunscreens Revealed through Ultrafast Dynamics and High-level Calculations. *Photochem. Photobiol.* **2023**. [CrossRef]
14. Woolley, J.M.; Losantos, R.; Sampedro, D.; Stavros, V.G. Computational and Experimental Characterization of Novel Ultraviolet Filters. *Phys. Chem. Chem. Phys.* **2020**, *22*, 25390–25395. [CrossRef]
15. Piccinino, D.; Capecchi, E.; Trifero, V.; Tomaino, E.; Marconi, C.; Del Giudice, A.; Galantini, L.; Poponi, S.; Ruggieri, A.; Saladino, R. Lignin Nanoparticles as Sustainable Photoprotective Carriers for Sunscreen Filters. *ACS Omega* **2022**, *7*, 37070–37077. [CrossRef]
16. Yuan, L.; Li, S.; Huo, D.; Zhou, W.; Wang, X.; Bai, D.; Hu, J. Studies on the Preparation and Photostability of Avobenzone and (2-Hydroxy)Propyl-β-Cyclodextrin Inclusion Complex. *J. Photochem. Photobiol. Chem.* **2019**, *369*, 174–180. [CrossRef]

17. Hanson, K.M.; Cutuli, M.; Rivas, T.; Antuna, M.; Saoub, J.; Tierce, N.T.; Bardeen, C.J. Effects of Solvent and Micellar Encapsulation on the Photostability of Avobenzone. *Photochem. Photobiol. Sci.* **2020**, *19*, 390–398. [CrossRef] [PubMed]
18. Afonso, S.; Horita, K.; Sousa e Silva, J.P.; Almeida, I.F.; Amaral, M.H.; Lobão, P.A.; Costa, P.C.; Miranda, M.S.; Esteves da Silva, J.C.G.; Sousa Lobo, J.M. Photodegradation of Avobenzone: Stabilization Effect of Antioxidants. *J. Photochem. Photobiol. B* **2014**, *140*, 36–40. [CrossRef] [PubMed]
19. Govindu, P.C.V.; Hosamani, B.; Moi, S.; Venkatachalam, D.; Asha, S.; John, V.N.; Sandeep, V.; Hanumae Gowd, K. Glutathione as a Photo-Stabilizer of Avobenzone: An Evaluation under Glass-Filtered Sunlight Using UV-Spectroscopy. *Photochem. Photobiol. Sci.* **2019**, *18*, 198–207. [CrossRef] [PubMed]
20. Herzog, B.; Giesinger, J.; Settels, V. Insights into the Stabilization of Photolabile UV-Absorbers in Sunscreens. *Photochem. Photobiol. Sci.* **2020**, *19*, 1636–1649. [CrossRef] [PubMed]
21. Termer, M.; Carola, C.; Salazar, A.; Keck, C.M.; Von Hagen, J. Methoxy-Monobenzoylmethane Protects Human Skin against UV-Induced Damage by Conversion to Avobenzone and Radical Scavenging. *Molecules* **2021**, *26*, 6141. [CrossRef] [PubMed]
22. Karlsson, I.; Vanden Broecke, K.; Mårtensson, J.; Goossens, A.; Börje, A. Clinical and Experimental Studies of Octocrylene's Allergenic Potency. *Contact Dermat.* **2011**, *64*, 343–352. [CrossRef]
23. Karlsson, I.; Persson, E.; Mårtensson, J.; Börje, A. Investigation of the Sunscreen Octocrylene's Interaction with Amino Acid Analogs in the Presence of UV Radiation. *Photochem. Photobiol.* **2012**, *88*, 904–912. [CrossRef] [PubMed]
24. *ISO 24443:2021*; Cosmetics—Determination of Sunscreen UVA Photoprotection In Vitro. ISO: Geneva, Switzerland, 2021. Available online: https://www.iso.org/standard/75059.html (accessed on 7 December 2023).
25. Tripp, C.S.; Blomme, E.A.G.; Chinn, K.S.; Hardy, M.M.; LaCelle, P.; Pentland, A.P. Epidermal COX-2 Induction Following Ultraviolet Irradiation: Suggested Mechanism for the Role of COX-2 Inhibition in Photoprotection. *J. Investig. Dermatol.* **2003**, *121*, 853–861. [CrossRef] [PubMed]
26. Pentland, A.P.; Scott, G.; VanBuskirk, J.; Tanck, C.; LaRossa, G.; Brouxhon, S. Cyclooxygenase-1 Deletion Enhances Apoptosis but Does Not Protect Against Ultraviolet Light-Induced Tumors. *Cancer Res.* **2004**, *64*, 5587–5591. [CrossRef] [PubMed]

Disclaimer/Publisher's Note: The statements, opinions and data contained in all publications are solely those of the individual author(s) and contributor(s) and not of MDPI and/or the editor(s). MDPI and/or the editor(s) disclaim responsibility for any injury to people or property resulting from any ideas, methods, instructions or products referred to in the content.

MDPI AG
Grosspeteranlage 5
4052 Basel
Switzerland
Tel.: +41 61 683 77 34

MDPI Books Editorial Office
E-mail: books@mdpi.com
www.mdpi.com/books

Disclaimer/Publisher's Note: The title and front matter of this reprint are at the discretion of Marcelo Guzman, the compiler. The publisher is not responsible for their content or any associated concerns. The statements, opinions and data contained in all individual articles are solely those of the individual Editor and contributors and not of MDPI. MDPI disclaims responsibility for any injury to people or property resulting from any ideas, methods, instructions or products referred to in the content.

www.ingramcontent.com/pod-product-compliance
Lightning Source LLC
LaVergne TN
LVHW072314090526
838202LV00019B/2283